Springer Geography

The Springer Geography series seeks to publish a broad portfolio of scientific books, aiming at researchers, students, and everyone interested in geographical research.

The series includes peer-reviewed monographs, edited volumes, textbooks, and conference proceedings. It covers the major topics in geography and geographical sciences including, but not limited to; Economic Geography, Landscape and Urban Planning, Urban Geography, Physical Geography and Environmental Geography.

Springer Geography—now indexed in Scopus

Magdalena Vicuña · Emanuele Giorgi
Editors

Intersections

Interdisciplinary Research on Architecture, Design, City and Territory

Springer

Editors
Magdalena Vicuña
Faculty of Architecture, Design and Urban Studies
Pontificia Universidad Católica de Chile
Santiago, Chile

Emanuele Giorgi
School of Architecture, Art and Design
Tecnológico de Monterrey
Chihuahua, Mexico

ISSN 2194-315X ISSN 2194-3168 (electronic)
Springer Geography
ISBN 978-3-031-76401-1 ISBN 978-3-031-76402-8 (eBook)
https://doi.org/10.1007/978-3-031-76402-8

The original submitted manuscript has been translated into English. The translation was done using artificial intelligence. A subsequent revision was performed by the author(s) to further refine the work and to ensure that the translation is appropriate concerning content and scientific correctness. It may, however, read stylistically different from a conventional translation.

Translation from the Spanish language edition: "IV CONGRESO INTERDISCIPLINARIO DE INVESTIGACIÓN EN ARQUITECTURA, DISEÑO, CIUDAD Y TERRITORIO INTERSECCIONES 2023 Entre escalas, agencias y territorios en Latinoamérica" by Magdalena Vicuña and Emanuele Giorgi,

This book has been edited and published thanks to the great desire of the Pontificia Universidad Católica de Chile and the Tecnológico de Monterrey.

This Springer imprint is published by the registered company Springer Nature Switzerland AG
The registered company address is: Gewerbestrasse 11, 6330 Cham, Switzerland

For all those people who believe in the intersections among individuals, institutions, and dreams.

Foreword by Mario Ubilla S.

Intersections has evolved over time and space as a platform promoting discussion and academic collaboration across various levels and formats. Initially, we showcased research conducted at the Faculty of Architecture, Design, and Urban Studies of the Pontificia Universidad Católica de Chile, aiming to understand the intellectual pursuits and research trajectories of our academics. Subsequently, we invited participation from faculties of architecture, design, and urbanism within Chile to foster networks among academics and postgraduate students researching similar topics. This initiative aimed to broaden, sophisticate, and strengthen our collective exploration.

Seeking to extend our reach beyond national borders, we invited the Faculty of Architecture and Design of the Universidad de Los Andes in Colombia and the Faculty of Architecture, Design, and Arts of the Tecnológico de Monterrey in Mexico, with whom we had previously collaborated successfully. We also aimed to continue our collaboration with the Universidad de Chile's Faculty of Architecture and Urbanism, sharing a commitment to education and knowledge creation centered on people, with the goal of influencing society, culture, the nation, the region, and the world. The involvement of these four universities as organizers, alongside various other Latin American universities, has elevated the Intersections Congress to a new dimension, providing a Latin American regional perspective that enriches the selected works presented here.

This edition of the Congress, notable for the post-pandemic opportunity to showcase the research advancements of our academic staff, researchers, and postgraduate students at a continental level, has generated significant participation and interest. It has provided a platform to explore the diverse topics each of us is investigating and the future opportunities these proposals present. A particularly intriguing aspect has been the variety of research forms exhibited, ranging from traditional research to industry and private sector applications, public sector and civil society initiatives, and research-creation. This book celebrates the diversity and disciplinary breadth of architecture, design, and urban studies across ten thematic axes.

I extend my gratitude to the organizers of Intersections for their complex task of coordinating activities among four university faculties located in distant regions, where fraternity and a shared objective prevailed. The research directions of the participating faculties were crucial to the event's success.

We celebrate this dissemination material as a product-platform that fosters critical reflection on our work and documents the state of the art of our research. We hope that many of these ideas and reflections will be implemented in our societies, cities, and territories, where the contributions of our disciplines are essential to promoting well-being, quality of life, and the sustainability of our development as a society.

Mario Ubilla S.
Dean, Faculty of Architecture, Design and Urban Studies
Pontificia Universidad Católica de Chile
Santiago, Chile

Foreword by Roberto Iñiguez Flores

Prologue Roberto Iñiguez

The organization of the "IV Congreso Interdisciplinario de Investigación en Arquitectura, Diseño, Ciudad y Territorio" has been an important moment in the relations between the School of Architecture, Art and Design of the Tecnológico de Monterrey and the Faculty of Architecture, Design and Urban Studies of the Pontificia Universidad Católica de Chile. The invitation we received to be part of this interesting effort, led by the Pontificia Universidad Católica de Chile was received with much gratitude and enthusiasm, because it was not only an invitation to the organization of a scientific event, but also an invitation to strengthen an important collaboration, already active for a long time, and an invitation to establish a deep reflection on the very meaning of conducting research in a society that evolves so rapidly and that, more than ever, needs spaces for *intersections* between contexts, experiences and visions.

This exercise of sharing experiences and visions among researchers, led by four Latin American universities, has undoubtedly been a great opportunity for institutional and personal growth: more than 400 researchers from 11 different countries contributed to shedding light on complex phenomena that characterize a region as rich and complex as Latin America, which, due to its peculiarities, deserves very particular attention from our design, architecture and urban studies disciplines. It is precisely the peculiar complexity that characterizes the territories of this part of the world that imposes the responsibility of developing new ways of researching more consistent with the geographical contexts and with the current emergencies.

For these reasons we have to acknowledge that it was a great achievement to weave—and find—new *intersections*: *intersections* between the strategies and efforts that each institution is developing to promote the culture of impactful research, capable of transforming the society in which we live; *intersections* between the themes and challenges of the projects that were presented at the congress that glimpse a pattern of similarity between our countries; *intersections* between the methodologies and approaches increasingly beyond the disciplinary limits and embrace multidisciplinary experiences, to achieve a more frank understanding of the phenomena studied. "Intersections 2023" was a great achievement, as well as a great basis for the generation of reflections and shared projects that, a few months later, are already taking shape.

The essence of this interdisciplinary experience is compiled here, in this book that aims to be a collection of some of the most appreciated presentations of the congress and a space for reflection on future opportunities for collaboration between researchers and institutions. I thank the editors of this book who did not want to miss the opportunity to take this interesting experience to a high level of dissemination.

Roberto Iñiguez Flores
National Dean, School of Architecture,
Art and Design
Tecnológico de Monterrey
Chihuahua, Mexico

Acknowledgments

This book has been edited and published thanks to the strong desire of the Pontificia Universidad Católica de Chile and the Instituto Tecnologico de Monterrey, to materialize in an international publication the ideas and experiences that were generated during the “IV Intersections Congress 2023”, also organized thanks to the University of Chile and University of the Andes (Colombia).

The editors would like to express their sincere thanks to these four institutions and their directors or deans, who have made this possible. In particular, sincere thanks go to Mario Ubilla (Dean, Faculty of Architecture, Design and Urban Studies, Pontificia Universidad Católica de Chile) and Roberto Iñiguez Flores (National Dean, School of Architecture, Art and Design, Instituto Tecnologico de Monterrey).

Moreover, we would like to also thank all the members of the Organizing Committee: Alfredo Hidalgo (Instituto Tecnologico de Monterrey), Pablo Ortiz (Pontificia Universidad Católica de Chile), César Peña (Universidad de los Andes), José Allard (Pontificia Universidad Católica de Chile), Luciana Pastor (Universidad de Chile), David Sánchez (Instituto Tecnologico de Monterrey), María Luisa Vela (University of the Andes), Felipe Ladrón de Guevara (Pontificia Universidad Católica de Chile), Cristian Gómez-Moya (Universidad de Chile), Natalia García (Instituto Tecnologico de Monterrey), Eduardo Mazuera (Universidad de los Andes), Ximena Arizaga (Pontificia Universidad Católica de Chile), and Eliana Sánchez (Universidad de los Andes). A special thanks is also addressed to the 76 members of the Scientific Committee, who participated in the review process of the conference and of this book. And to the Faculty of Architecture and Urbanism of Universidad de Chile for also hosting the Congress.

In addition to the members of the committees, thanks also go to all those researchers (more than 400) who presented their work during the IV Intersections Congress 2023, some of whom were invited to expand their contribution in this book.

The publishers also want to thank, and in a special way, Berenice De Dios Sandoval who helped impeccably in the realization of this work, playing a key role in the coordination between the publishers and the 109 authors.

Finally, we are especially grateful also to Dr. Robert K. Doe (Editorial Director) and Arumugam Deivasigamani from the Springer publishing team, who constantly helped us in the publication process.

Contents

Contributors

Izaskun Álvarez-Aguado Universidad de las Américas, Mexico City, Mexico

Oscar Álvarez Solano Department of Chemical and Food Engineering, Universidad de los Andes, Bogotá, Colombia

Camilo Hernandez Acevedo Department of Mechanical Engineering, Escuela Colombiana de Ingeniería Julio Garavito, Bogotá, Colombia

Leonardo Parra Agudelo Department of Design, Universidad de los Andes, Bogotá, Colombia

Osvaldo Almarza Collective of popular educators, Santiago, Chile

Piroska Angel Institute of Urban and Territorial Studies, Pontificia Universidad Católica de Chile, Santiago, Chile

Diego Apablaza Díaz Escuela de Arquitectura y Diseño, Pontificia Universidad Católica de Valparaíso, Valparaíso, Chile

Javier León Aravena Department of Social Sciences, Universidad del Bío Bío, Concepción, Chile

Ximena Arizaga School of Architecture, Researcher at the Green Landscapes and Infrastructures Laboratory (LAB_PAI), Pontificia Universidad Católica de Chile, Santiago, Chile

María Cristina Avilés Collective of popular educators, Santiago, Chile

Camilo Ayala-Garcia Department of Design, Universidad de los Andes, Bogotá, Colombia

Francisca Balbontín Design School, Pontificia Universidad Católica de Chile, Santiago, Chile

Christian Sebastián Balmaceda School of Psychology, Pontificia Universidad Católica de Chile, Santiago, Chile

Paulina F. Gonzalez-Valenzuela Universidad de Chile, Santiago, Chile

Pilar Grant Design School, Pontificia Universidad Católica de Chile, Santiago, Chile

Margarita Greene Center for Sustainable Urban Development (CEDEUS), Pontificia Universidad Católica de Chile, Santiago, Chile

Edgar Alan Egurrola Hernández School of Architecture, Art and Design, Tecnologico de Monterrey, Lopez Mateos, Mexico;
School of Architecture, Art and Design, Tecnologico de Monterrey, Guadalajara, Mexico

Alfredo Hidalgo Rasmussen School of Architecture, Art and Design, Tecnologico de Monterrey, Monterrey, México;
IAE Paris Sorbonne, Panthéon Sorbonne University, Paris, France

Karen Hinojosa Hinojosa School of Architecture, Art and Design, Tecnologico de Monterrey, Monterrey, México

Alicia Porras Holguín Department of Chemical and Food Engineering, Universidad de los Andes, Bogotá, Colombia

Magdalena Manriquez Hörmann Design School, Pontificia Universidad Católica de Chile, Santiago, Chile

Felipe Igualt Escuela de Arquitectura y Diseño, Pontificia Universidad Católica de Valparaíso, Valparaíso, Chile

Jorge Inzulza Contardo Departamento de Urbanismo, Facultad de Arquitectura y Urbanismo, Universidad de Chile, Santiago, Chile

María Camila Jiménez Frias Universidad de los Andes, Bogotá, Colombia

Aleksandra Krstikj School of Architecture, Art and Design, Tecnologico de Monterrey, Lopez Mateos, Mexico

Carlos Lange Facultad de Arquitectura y Urbanismo, Instituto de La Vivienda, Universidad de Chile, Santiago, Chile

Jorge Sánchez Lara School of Architecture, Art and Design, Tecnologico de Monterrey, Guadalajara, México

Katia Soto Liebe Faculty of Economics and Business, Universidad de Chile, Santiago, Chile

Lucía Martín López Universidad Rey Juan Carlos (URJC), Madrid, Spain;
School of Architecture, Art and Design, Tecnologico de Monterrey, Monterrey, México

Rodrigo Durán López Universidad Anáhuac México, Huixquilucan, México

Sandro Maino Departamento de Arquitectura, Universidad Técnica Federico Santa María, Valparaíso, Chile

Alejandro Marañón Department of Mechanical Engineering, Universidad de los Andes, Bogotá, Colombia

Alejandro Marañón León Department of Mechanical Engineering, Universidad de los Andes, Bogotá, Colombia

Francisco Álvarez Martín Collective of popular educators, Santiago, Chile

José Ignacio Medina Ministry of Energy, Providencia, Chile

Laura Mena Design School, Pontificia Universidad Católica de Chile, Santiago, Chile

Katherine Mollenhauer School of Design, Research Center on Governance and Territorial Planning (NuGOT), Pontificia Universidad Católica de Chile, Santiago, Chile

Daniel Moreno Research Center on Governance and Territorial Planning (NuGOT), Pontificia Universidad Católica de Chile, Santiago, Chile

Osvaldo Moreno School of Architecture, Researcher at the Green Landscapes and Infrastructures Laboratory (LAB_PAI), Pontificia Universidad Católica de Chile, Santiago, Chile

Francisca Molinos Moyano Pontificia Universidad Católica de Chile, Santiago, Chile

Luis Natera School of Architecture, Art and Design, Tecnologico de Monterrey, Monterrey, México

Barbara Ojeda Departament of Urban Planning and Design, Universidad del Bío Bío, Concepción, Chile;
Department of Social Sciences, Universidad del Bío Bío, Concepción, Chile

Arturo Orellana Institute of Urban and Territorial Studies, Research Center on Governance and Territorial Planning (NuGOT), Pontificia Universidad Católica de Chile, Santiago, Chile

Aaron Napadensky Pastene Departament of Urban Planning and Design, Universidad del Bío Bío, Concepción, Chile

Erick Abdel Figueroa Pereira Universidad del Atlántico, Barranquilla, Colombia

Andrea Wechsler Pizarro Facultad de Arquitectura y Urbanismo, Universidad de Chile, Santiago, Chile

Juan Sebastian Porras Cardenas Department of Mechanical Engineering, Universidad de los Andes, Bogotá, Colombia

Alicia Porras Holguín Department of Chemical and Food Engineering, Universidad de los Andes, Bogotá, Colombia

Quentin Ramond Centre for Economics and Social Policy (CEAS), Universidad Mayor, Santiago, Chile

Renée Rodo Iunnissi Pontificia Universidad Católica de Valparaíso, Valparaíso, Chile

Carolina Pérez Rodríguez Department of Design, Universidad de los Andes, Bogotá, Colombia

Javiera Rodríguez School of Design, Pontificia Universidad Católica de Chile, Santiago, Chile

Nadia Rodríguez Faculty of Engineering, Universidad Autónoma de Querétaro, Santiago de Querétaro, Mexico

Jessica Rodríguez Briceño Department of Design, Universidad de los Andes, Bogotá, Colombia

Gonzalo Rodríguez Paillape Faculty of Social Sciences and Arts, Universidad Mayor, Temuco, Chile

Daniel Ronderos-Lopez Universidad de los Andes, Bogotá, Colombia

Vanessa Rugiero Institute of Urban and Territorial Studies, Pontificia Universidad Católica de Chile, Santiago, Chile

Patricia Sánchez-Hidalgo School of Design, Pontificia Universidad Católica de Chile, Santiago, Chile

Pilar Saavedra Design School, Pontificia Universidad Católica de Chile, Santiago, Chile

Elke Schlack Faculty of Architecture, Design and Urban Studies, Pontificia Universidad Catolica de Chile, Santiago, Chile

Herbert Spencer González Pontificia Universidad Católica de Valparaíso, Valparaíso, Chile

Alfo Puppo Stuardo Facultad de Arquitectura y Urbanismo, Universidad de Chile, Santiago, Chile

Josefina Tapia School of Architecture, Researcher at the Green Landscapes and Infrastructures Laboratory (LAB_PAI), Pontificia Universidad Católica de Chile, Santiago, Chile

Claudia Torres Departamento de Arquitectura, Facultad de Arquitectura y Urbanismo, Universidad de Chile, Santiago, Chile

Ricardo Trufello Center for Sustainable Urban Development (CEDEUS), Pontificia Universidad Católica de Chile, Santiago, Chile

Juan Ugalde Center for Bioinformatics and Integrative Biology, Facultad de Ciencias de la Vida, Universidad Andrés Bello, Santiago, Chile

Pablo Hermansen Ulibarri School of Design, Pontificia Universidad Católica de Chile, Santiago, Chile

Carlos Lange Valdes Facultad de Arquitectura y Urbanismo, Instituto de la Vivienda, Universidad de Chile, Santiago, Chile

Rossana Valdivia School of Architecture, Art and Design, Tecnologico de Monterrey, Monterrey, México

Abraham Vallejos Arias Escuela de Arquitectura y Diseño, Pontificia Universidad Católica de Valparaíso, Valparaíso, Chile

Giovanni Vecchio Institute of Urban and Territorial Studies, Pontificia Universidad Católica de Chile, Santiago, Chile

Vanessa Vega Córdova Millennium Institute for Care Research (MICARE), Santiago, Chile

Rodrigo Vera School of Design, Pontificia Universidad Católica de Chile, Santiago, Chile

Daniela Villouta Department of Civil Engineering, Universidad Católica de la Santísima Concepción, Concepción, Chile

Chapter 1
Research in Architecture, Design, City, and Territory: Among Scales, Agencies, and Territories in Latin America

Magdalena Vicuña del Río and Emanuele Giorgi

Abstract This chapter introduces the structure of the book, organized with ten thematic axes that collect the best presentations of the "IV Congreso Intersecciones—2023", organized in Santiago de Chile by Pontificia Universidad Católica de Chile, Universidad de Chile, Tecnológico de Monterrey (Mexico) and Universidad de los Andes (Colombia). The chapter expresses the deep conviction of the editors about the importance of moments of intersection: spaces of meeting and common reflection between different institutions, disciplines and contexts, which contribute to generating new visions on the role of research in architecture, design, cities and territories. This sharing is more important than ever for a region like Latin America which finds itself facing many common challenges and which experiences just as many opportunities to present itself as an international reference on how to deal with the necessary changes in research activities and sustainable development.

Keywords Research practice · Latin America · Conference · Urban studies · Architecture · Design · Mexico · Chile · Colombia

1.1 The Meaning of Intersection

As a geometric point where one or more elements meet, the term "intersection" refers to the encounter, the confluence, the crossing. At the "IV Congreso Intersecciones—2023", four faculties from the areas of architecture, art, design, and urban studies belonging to prominent universities in the Latin American context met: Pontificia Universidad Católica de Chile, Universidad de Chile, Tecnológico de Monterrey

M. V. del Río (✉)
Faculty of Architecture, Design and Urban Studies, Pontificia Universidad Católica de Chile, Santiago, Chile
e-mail: mvicunad@uc.cl

E. Giorgi
School of Architecture, Art and Design, Tecnologico de Monterrey, Monterrey, Mexico
e-mail: egiorgi@tec.mx

M. Vicuña and E. Giorgi (eds.), *Intersections*, Springer Geography,
https://doi.org/10.1007/978-3-031-76402-8_1

(Mexico), and Universidad de los Andes (Colombia). This meeting took place at a time of significant transformations in the economic and development model, which directly impact how we understand society, the cultural manifestations that represent it, and, consequently, the way we produce and think about design, architecture, cities, and territories. The "IV Congreso Intersecciones—2023" was developed around ten thematic axes that called the academic world to critically address the tensions that emerge in the encounter between different scales, agencies, and territories, between the local, regional, and global, and how architecture, design, city, and territory can constitute relevant intersections for these encounters in the Latin American context.

1.2 Ten Axes to Interpret Scales, Agencies, and Territories in Latin America

This book is organized around ten thematic axes, presenting a selection of contributions that dialogue around the editorial proposal of the congress.

1.2.1 Heritage Turns, Cultures, and Micro-histories in Disciplinary Discussion and Project Practices

The first thematic axis corresponds to Heritage Turns, Cultures, and Micro-Histories in Disciplinary Discussion and Project Practices. The works presented here are inserted in various dimensions of Chilean cultural identity. In "Landscape and Artifice: the natural context as a discursive strategy of contemporary Chilean architecture", Felipe Corvalán discusses how the imaginary associated with the landscape is recovered in the process of disseminating architecture recently produced in Chile. In "Memory as a Territory in dispute: Tension between Heritage and Neoliberal Urban Development from the Memory Site Venda Sexy", Francisca Molinos addresses the materialization of collective memory through the detention centers that operated during the dictatorship that are part of the urban heritage. With the aim of recognizing their contribution to the recent discussion on this term in areas such as visual arts, design, and crafts, Paulina González and Eduardo Castillo analyze the curatorial policies of Chilean university museums in "Folk art and University Extension: Some Categories in Tension during the 1960s in Chile". Relying on anthropology for sociocultural analysis and with an interdisciplinary and inter-scalar approach, Claudia Torres, Laura Gallardo, Carlos Lange, and Sandro Maino, address the Chilean-Antarctic culture emerged through the colonization with military camps (Bases) in "Antarctic Architecture a contemporary heritage: Analysis of architecture and life at the Eduardo Frei Montalva Antarctic Air Base, 1969–2022".

1.2.2 Sustainable Landscapes, Resilience, and Regenerative Cultures: Emerging Approaches to the Climate and Ecological Crisis

The second thematic axis that structures this book corresponds to "Sustainable landscapes, resilience, and regenerative cultures: emerging approaches to the climate and ecological crisis". From various perspectives and geographical contexts, innovative proposals are presented to advance territorial sustainability. It's the cases of the chapter "Regenerative and understanding the place for urban–rural contexts in Mexico. Methodological proposal", by Carlos Cobreros and Nadia Rodríguez and the chapter "The environmental observatory for mining projects: a system for the analysis of public environmental management information", by Kay Bergamini, Piroska Angel, Katherine Mollenhauer, José Ignacio Medina and Vanessa Rugiero, and "Huasco Wetland Observatory. The chapter "Huasco wetland observatory. Promoting the environmental and cultural values of urban coastal wetlands in the Atacama Desert", by Osvaldo Moreno and Ximena Arizaga, address critical issues in environmental management in Chile, from access to public information and generation of indicators, to landscape architecture strategies for the visibility and conservation of environmental and cultural values. In the same country, the contribution "Evaluation criteria for resilient civic design for response capacity in residential communities affected by the San Ramón Fault, in Santiago de Chile", by Jorge Inzulza Contardo explores specific criteria for evaluating resilient civic design in two residential communities, in order to determine their response capacity to seismic risk. Finally, "Adaptability of architectural forms for irregular buildings in the flood-prone coastal zone of Laguna Verde, Valparaíso", Felipe Igualt Jara, Abraham Vallejos and Diego Apablaza propose a prototype of residential module, whose purpose is to guide an initial approach towards the design of progressive homes adapted for areas of irregular occupation on the way to settlement.

1.2.3 Right to Housing, to the City and Services: Alternatives for Territorial Inclusion and Social Integration

In the third thematic axis, four works address the challenges for architecture and urbanism implicit in the Right to housing, to the city and services: alternatives for territorial inclusion and social integration. With the intention of a base for planning decisions that target the improvement of access to a healthy diet in informal settings, Aleksandra Krstikj and Edgar Alan Egurrola Hernández, in "Nutritious cities: an exploration of the food environments in five Mexican cities", conclude that informal markets reduce nutritional vulnerability, especially in the urban periphery. In "PICTOS: A service focused on cognitive accessibility for navigation and evaluation of services in Chile", Katherine Exss Cid, Herbert Spencer González, Vanessa

Vega Córdova, Izaskun Álvarez-Aguado and Renée Rodo Lunnissi present the inclusive research process through the co-design of a web service that provides supports to ensure cognitive accessibility in the use of services. In "Walking in deprived neighborhoods: understanding the role of the built environment", Cristhian Figueroa Martinez presents evidence on how the built environment influences the walkability of vulnerable neighborhoods in Santiago. Finally, Quentin Ramond, in "Separated and unequal childhoods: Residential segregation and neighborhood inequalities of children in Chile", shows how age and socioeconomic factors interact in the spatial sorting process and specify its relationship with contextual inequalities that crucially affect the distribution of life opportunities, especially among children.

1.2.4 Gender Approach and Ethics of Care: New Disciplinary and Project Perspectives in Design, Architecture and Urbanism

The fourth axis proposes a discussion around the Gender approach and ethics of care: new disciplinary and project perspectives in design, architecture and urbanism. In "Public Health as Public Space: Feminist design for spaces of care, encounter and collaboration between State and Citizenship around well-being", Patricia Sánchez addresses a Family Health Center in La Pintana, Santiago (Chile). Through qualitative field research, the author proposes feminist-focused design recommendations and reflects on the capacity of design to postulate political approaches that promote change. With a collaborative approach and based on co-design to reduce communication and movement gaps, in "Sensing the City. Co-designing Urban Space with Deaf Children and their Caregivers", Carolina Carrasco Pizarro presents an innovative initiative that addresses the accessibility and inclusion of deaf children and their caregivers in public space. In "Openly Caring: Integration of Sustainability and Inclusion in Urban Education", Karen Hinojosa Hinojosa and Nélida Astrid Escobedo Ruíz demonstrate that, despite consensus on the importance of education for sustainable development and inclusion, spatial expressions of gender issues are still perceived as strange by design and urban planning students. Alessandra Cireddu, in "Urban proximity: a methodological approach from a gender perspective" proposes the construction of a methodology for qualitative analysis at the neighborhood level to measure proximity from a gender perspective. This section concludes with the work "House on Kings Road: the house of Pauline Gibling Schindler". Here, Andrea Castro-Marcucci reviews the definition of the authorship of the architectural project during the modern movement, and the particular omission in the historiography of the intervention of important actors.

1.2.5 Processes, Technologies and New Materialities: Creation of Possible Futures, Speculative Design, New Aesthetic Universes, New Ecosystemic Relationships

The fifth axis presents five works on the theme Processes, technologies and new materialities: creation of possible futures, speculative design, new aesthetic universes, new ecosystemic relationships. In "Metaversal Space. Methodological for the definition of space in the Metaverse", Daniel Ronderos López and María Camila Jiménez Frías reflect on the need to learn from the different ways in which the spatialities of the real world are understood, including the diversity offered by different types of users. Rodrigo Vera, in "New landscape and the emerging dynamics of interaction between users and the urban space produced by digital delivery platforms", exposes the potential of research through design for urban studies, in contexts of collaborative economy and digitalization. Also, with a methodological approach from design, Katherine Mollenhauer, Daniel Moreno and Arturo Orellana, in "Interdisciplinary Implementation of the National Rural Development Policy in local and regional strategic planning instruments in Chile", discuss how the components, methods, techniques and instruments that make up toolkits, become relevant tools for public officials and private consultants.

From another point of view, Andrea Wechsler Pizarro, Yesenia Briones Castro, Pablo Domínguez González and Roberto Osses Flores, in "New biobased materials: Exploratory analysis of the actors developing new biobased materials in Chile" analyze the state of the art of this topic in Chile, to then propose profiles of actors who value solid and organic waste generated by small and medium national industries and that generate research and development of new materialities. On the other hand, the contribution "Multidisciplinary approach for the development of materials from the byproduct of the opening of the cashew nut in Vichada", by Camilo Ayala García, Clara Forero Lesmes, Santiago De Francisco Vela, Leonardo Parra Agudelo, Alejandro Marañon León, Camilo Hernández Acevedo, Oscar Álvarez Solano and Alicia Porras Holguín, addresses a multidisciplinary approach to the development of materials from the agro-industrial byproducts of the cashew nut.

1.2.6 Territories, Intermediate Cities and People in Motion

The second half of the book begins with the works of the sixth axis on Territories, intermediate cities and people in motion. With the chapter "Characterization for the management of urban parks and green areas of mid-sized cities", Arizaga, Moreno and Tapia study the territorial, socioeconomic and financial conditions of the communes of La Araucanía to propose recommendations so that public investment in parks and green areas may be related to the peculiar characteristics of the territory. In the chapter "The Scales of Vulnerability: Mobility and Accessibility of the Vulnerable Active Population in Santiago de Chile" is a study presented by an inter-institutional team, composed of researchers from the Universidad de las Americas, Universidad Andres Bello, and the Pontificia Universidad Católica de Chile. The study explores the relationship between urban-scale accessibility and vulnerable contexts. Correa, Ladrón, and Vecchio provide evidence that this relationship is key to prioritizing interventions in the public transport network and invite consideration of different scales to address the problem of accessibility. The following chapter, "The Construction of a Segregated Metropolis: Urban Segregation and Natural Pre-existences in the Metropolitan Area of Concepción", developed between the Pontifical Catholic University of Chile and the University of Concepción, focuses on the issue of urban segregation that is greatly influencing the Latin American debate in recent years. Using quantitative methodologies and spatial statistics, the evolution of socio-spatial segregation in the Metropolitan Area of Concepción is observed, comparing it with its representation in space. This study by Villouta, Trufello, Greene, and Fernandez aims to improve samples for survey generation and case study selection in the Metropolitan Area of Concepción. The last chapter of this axis, "Community knowledge in action: participatory design for the transformation of the cashew value chain in the Vichada region, Colombia" presents the results of the work developed in the Vichada region, by the Universidad de los Andes and the Escuela Colombiana de Ingeniería. A participatory approach allowed the development of inclusive technological transfer processes, attentive to the peculiarities of the productive context, co-creating models for the use of by-products and the development of a prototype opening machine.

1.2.7 Critical Methodology, Project, and Didactics of Research in Architecture, Design, City, and Territory

The seventh thematic axis collects works on Critical Methodology, Project, and Didactics of Research in Architecture, Design, City, and Territory. This thematic axis begins with the chapter "Narratives as a research technique to understand urban nature production" by Alfonsina Puppo Stuardo, which addresses a very critical issue for several Latin American countries: deregulated urbanization, simultaneously a product and cause of social and ecological transformations of contemporary ecosystems. Through the research technique of Narrative Productions, it seeks to understand the perception of urbanization in Chile, identifying socio-cultural valuations towards the urban nature of Chiguayante (Chile). In the presented case, it is evident how the population expresses the concern to rediscover vernacular wisdom on the subject of food provision in the face of the threat of loss of native forest due to wild urbanization. The following chapter "Photoethnography as a Research Methodology for Manifestations in Public Space in Times of Revolt" reflects on the research methodology of photo-ethnography, as a tool for approaching the study of public spaces subject to interventions and transformations by current social revolts. The authors Hermansen and Fernández present both the technical and problematic aspects of this methodology, also evidencing the ethical–political consequences that the authorial and contextual conditions of photographic processes can bring. Starting from the recent Chilean experience of attempts (failed) to draft a new constitution, the chapter "Playability, Democracy, and Design: Experience of Designing Popular Educational Materials on Constitution Proposal in Chile" shows how design can facilitate the civic involvement of the most marginalized sectors. This chapter, prepared by authors from the *Pontificia Universidad Católica de Chile* and the *Colectivo de educadores y educadoras populares*, investigates the "role of design in creating spaces that enable shared power and democratic reflection." In 2022, designers, educators, and psychologists developed four games to promote reflection, participation, and dialogue. This article describes the methodology used in these processes, evidencing the principles, needs, and scopes.

1.2.8 Perceptions, Emotions, and Imaginaries in Architecture, Design, City and Territory

The eighth axis of this book reflects on Perceptions, emotions and imaginaries in architecture, design, the city and the territory. Studies from history allow the development of critical thoughts to face challenges of contemporaneity. Chapter 33 "Narratives and public discourses of the Right to the City. A review from the press to the Bicentennial Bridge conflict. Concepción, Chile 2010–2020" by Napadensky, Leon and Ojeda (from the University of Bío-Bío) argues that the narrative of the "Right to the City" fails to significantly influence the processes of urbanization and city development. A combination of framing theory and content analysis identifies the communicative frames of the different groups of actors involved in the conflict of the Bicentennial Bridge in Concepción, Chile, which faced informal settlers and the public-state sector. The hegemony of the capitalist narrative limits the scope and influence of discourses based on the "Right to the City", which affects the material development and the future of the disputed place.

The following chapter "The projected city. Urban analysis from historical cartography. Temuco 1888–1919" confronts the role that urban planning has in the development of a "foundation city" like Temuco. The chapter explores the first urban cartographies made between 1888 and 1892 to understand the construction of the city and determines the preponderance of the 1888 foundational plan in the subsequent development of the urban grid, anticipating the occupation of space and the programmatic use of the lots.

Also, the following chapter "Affective frameworks, or the construction of a Place in the housing complexes in Providencia carried out by Luciano Kulczewski." takes a historical study as an opportunity to reflect on very relevant issues in the current architectural and urban debate: character and place. The chapter addresses the architectural aesthetics in Luciano Kulczewski through the study of two residential complexes—still little studied—built between 1929 and 1935 in Providencia, Santiago de Chile, inspired by the concept of "garden-neighborhood". The authors Harris and Veraguas present the architectural strategies adopted by Kulczewski to make a "space" become a "place": the configuration of the autonomy of space and the search for recognition by the inhabitants as belonging to the built place.

In "Stereotypes and gender roles in Unidad Independencia through Nacho López's photographic archive" Lucía Martín and Rodrigo Durán study, through the photographs of Nacho López, the case of the Independencia Unit (1951) in the Mexican Federal District. This Unit was part of the construction plan for the Mexican Social Security Institute (IMSS) with the aim of providing suitable buildings for social security services throughout the country. Currently, the National Housing Program (2019–2024) of Mexico proposes the Independencia Unit as a reference for affordable and quality housing. The chapter seeks to analyze gender roles in relation to the space of the Independencia Unit through the photographs of Nacho López (hired by the IMSS to document everyday life in the Independencia Unit and use his work as propaganda), to determine if it can be considered a reference in this aspect.

The chapter "BRAT'YA. The architecture of Dostoyevsky" by Jorge Sánchez moves to the analysis of the texts of Fyodor Dostoyevsky, promoting a reflection on how the spaces described by the Russian writer are means to elaborate a social reflection, which can teach a lot to contemporary designers. Dostoyevsky's works reflect universal social concerns that still persist today, such as the struggle for dignity, the beauty of youth, the vulnerability of children and social oppression. His reflections can be interpreted from an architectural perspective, as they address issues such as the need to preserve nature, design shelters and prevent social and family breakdown. Each incident in Dostoyevsky's works occurs in a specific place, which invites the reader to reflect on how we live and could live our spaces. In particular, Sánchez shows how Dostoyevsky intertwines the material creations and the actions and thoughts of the characters in his novels, highlighting how society creates its own environment, and vice versa.

1.2.9 Cultures and Aesthetics of the Public: Discussions and Alternative Paradigms in Latin America

The ninth axis collects two works on Cultures and Aesthetics of the Public: Discussions and Alternative Paradigms in Latin America that investigate the recurring phenomena in the contemporary Latin American panorama of self-construction and standardization.

The chapter "The public and the community in tension in Self-Built urban spaces: the case of the "Operation Site" settlements in Santiago de Chile" by Schlack and Fariña presents a research that seeks to contribute to the discussion about the influence of urban morphology in defining the character of spaces, particularly those that emerge in contexts of self-construction. Starting from the case of "Operation Site" in Santiago de Chile (1960–70), Schlack and Fariña focus on understanding the complexity of phenomena of cohesion, attachment, appropriation, and domination of space.

With a focus on the republican state in La Araucanía, Fuentes and Rodríguez analyze the practices of standardization acted within the framework of the building processes during the occupation of the region of La Araucanía between the end of the 19th century and the beginning of the 20th. In particular, focus is made on modern school establishments, considered by the State as civilizing triggers. This work is presented in the chapter "Beginnings of standardization in educational architecture: Forms of expansion of the republican state in La Araucanía".

1.2.10 Pandemic and Post-pandemic: Challenges for Architecture, Design, the City, and the Territory

The book closes with the works of the tenth axis that collects research on the challenges that our disciplines found to discuss as a result of the covid pandemic. This last session contains works on Pandemic and post-pandemic: challenges for architecture, design, the city, and the territory.

The chapter "(Non) Proximate Cities, lessons from the pandemic" presents the operation of the algorithm and the tool that were generated by the Observatory of Cities of the Tecnológico de Monterrey to measure the proximity to different facilities and urban services, taking into account the travel time in different areas of the cities. This chapter by Hidalgo, Natera, Egurrola, and Valdivia explains how the need for this instrument lies in the experience of the Covid-19 pandemic that has generated a renewed interest in urban health, highlighting the importance of aspects such as walkability and the diversity of uses and services in the neighborhoods.

Carlos Lange, in the chapter "Exploring Architecture as an Exit Device in Times of Pandemic" highlights the relationship between the crisis, the revaluation of common spaces, and the role of architecture as a facilitator of solutions and adaptations in situations of uncertainty and precariousness in the urban context. This demonstrates the ability of architecture to give life to new common spaces, proving to be a key force in facing crises like the pandemic. According to Lange, the everyday, conflicts, negotiations, and agreements are the promoters of a new understanding of common spaces, configured from shared resources.

The last chapter "Urban Microbiome: Architecture and Microbiology for a Healthy Built Environment" by Encinas, Soto, Freed, and Ugalde presents a multidisciplinary work between the Faculty of Architecture, Design and Urban Studies of the Pontificia Universidad Católica de Chile, the Faculty of Economics and Business of the Universidad de Chile and the Faculty of Life Sciences of the Unviersidad Andrés Bello. This multidisciplinarity allows a very interesting approach to the health of living spaces. Three sampling campaigns of microorganisms in the homes of Santiago allowed to show how the microbiome is very variable and influenced by the built environment. This research, started with concerns about the quality of microbial colonization due to the indoor confinement of built spaces, highlighting how design and architectural project are key in the healthiness of spaces and in the life of residents.

1.3 The Hope of This Book

With the works collected in this book, the editors want to give a sample of the experience of discussion and reflection lived during the "IV Congreso Intersecciones—2023". In this book, we want to present high-impact research projects that (although driven by disciplines that have their own peculiar characteristics) highlight how the boundaries of the discipline are increasingly blurred and that the complexity of the contemporary world requires an evolution of research practice. It is no longer remotely possible to think that the key to a detailed interpretation of a phenomenon belongs to a single discipline and that the solutions that can give hope to the contemporary world are rooted in a single field. Discussion, reflection, and collaboration emerge, today more than ever, from the intersection between different areas of knowledge. This collection of works comes from the conviction of the editors that it is the intersection (this meeting, confluence, and crossing) of experiences, methodologies, and visions that allows formulating proposals for this new world, in which changes are stronger and faster.

With this vision, we hope that these contributions inspire new debates and actions that enrich our understanding and practice in the field of design, architecture, and urban studies in Latin America and beyond.

Part I
Heritage Turns, Cultures and Micro-histories in the Disciplinary Discussion and Project Practices

Chapter 2
Landscape and Artifice. The Natural Context as a Discursive Strategy of Contemporary Chilean Architecture

Felipe Corvalán Tapia

Abstract Due to its geographical location, the difficulty of access to an extreme territory by travelers and conquerors, the presence of the landscape has become a 'mythical' quality that defines Chile and its national identity. Throughout this research we will try to discuss how this imaginary associated with the landscape is recovered in the process of disseminating the architecture recently produced in Chile. In this regard, we argue that in such a process and from an essentially photographic perspective, the landscape becomes an artifice and is at the service of the construction of an appreciative discourse that attempts to offer a coherent body of works that would define the contemporary Chilean architecture. It is this discourse, the role of the landscape and its representations, what we will try to critically think about in this work. Regarding the results of the research, at this opportunity an advance of the interpretive framework of the study in dialogue with attention to specific cases that exemplify the proposed field of inquiry is presented.

Keywords Contemporary Chilean architecture · Architectural field · Landscape · Dissemination strategies · Architectural discourse

2.1 Introduction. Field Readings

Along with the usual importance given to the object-work, architecture, as a field of knowledge, has increasingly recognized the importance of other types of materials in the constitution of its limits of meaning. Texts, images and other representations,[1] inform us about architecture, about its debates and practices, allowing us to read in them the state of the discipline: its disputes, changes and continuities.[2]

As has been intensely discussed by Beatriz Colomina (1994), from the consolidation of modernity we can hardly think of architecture without attending to those

F. Corvalán Tapia (✉)
Departamento de Arquitectura, Facultad de Arquitectura y Urbanismo, Universidad de Chile, Santiago, Chile
e-mail: fecorva@u.uchile.cl

M. Vicuña and E. Giorgi (eds.), *Intersections*, Springer Geography,
https://doi.org/10.1007/978-3-031-76402-8_2

means, own or foreign, that allow its visibility. We speak of a graphic, photographic, textual or expository mediation, which according to what is indicated by Colomina herself, directly affects the valuation that architectural works receive. Attending to this matter and its repercussions, this work proposes a critical approach to the role given to the landscape in the construction of an appreciative discourse around contemporary Chilean architecture and its attributes.

Although throughout this text we will use the expression 'Chilean architecture', such use does not imply its uncritical acceptance. On the contrary, the research presented here discusses the use and scope of explanatory categories around the architectural works recently produced in Chile, especially when they have an identity claim. In this line, it is suggested that the representation of the landscape has become a key material to decipher the scope of the architectural works recently produced in Chile and the interpretations made around them. To carry out such reflection, following the work of the French sociologist Pierre Bourdieu (2013), we propose to carry out here a 'field reading'. That is, a reading that seeks to discuss how and through what mechanisms the landscape has been used and represented, by what strategies and to say what things about the architecture made in Chile.

From a methodological point of view, the research is built through a qualitative approach, which with an interpretative aim seeks to analyze the studied materials: photographs, exhibitions and editorial proposals that address the dialogue between architectures and landscapes. Considering what is indicated by Jane Rendell (2004), the work proposes a theoretical-critical exercise that aims to identify and debate the observation parameters of the studied phenomenon: the role assigned to the landscape in the construction of an explanatory discourse, also appreciative, about the architectural production that has taken place in Chile over the last few decades.

The argumentative thread of this text is woven from the development of three sections: (i) firstly, an approach to the idea of landscape as an identity trait of the formation of Chile as nation and territory; (ii) secondly, a contextualization of the diffusion process experienced by Chilean architecture in recent decades; and, finally, (iii) a more detailed approach to the use of landscape as an interpretive argument regarding the works selected to be exhibited and published. An approach that, while it addresses specific cases, does not, for the purposes of this text, have a claim to exhaustiveness. Rather, on this occasion the work seeks to trace the margins of a field of action, fundamentally representational and discursive, in which the link between architecture and landscape has been tightened.

In this sense, although the research focuses its attention on a specific context, which has a body of works and representations—the production of architecture in Chile—the proposed view also suggests a reflection with a somewhat larger scope. That is, a reflection that allows us to think and discuss how discourses are constructed in architecture, those appreciative frameworks that guide our interpretations regarding the works and their value.

As will be discussed throughout this text, what we usually name or receive under the title 'Chilean architecture' is a partiality; a type of sample articulated under certain editorial or curatorial lines, which we can hardly read as the expression of a totality.

In the development of this research, the group of works referred is associated, precisely, with those filters that guide the emission and reception of the architecture made in Chile. That is, to talk about 'Chilean architecture' is equivalent, in the field that interests us to analyze, to a partial sample specially chosen to be sent and received, especially, by the international architectural field.

2.2 Landscapes of Identity

Probably, due to its remote and extreme geographical location, difficult access for conquerors and travelers, the natural landscape in Chile has acquired a sort of mythical character. A character that, while it is not entirely 'objectifiable', ends up establishing a common sense that guides our reading of that observed reality (Barthes, 2010). We are certainly talking about the establishment of a common sense that is not produced by spontaneous generation, that is fed, extends and reproduces in each new representation that refers us to that initial mythical voice. In the case of Chile, it does not seem to be coincidence that the origin of the nation and the beginning of its life as an independent country coincide with a significant development of landscape painting. A pictorial genre that settles in Chile in the nineteenth century, from the work of Mauricio Rugendas, Alejandro Ciccarelli or Antonio Smith (Ahumada, 2012; Valdés, 2012a).

Moreover, in the first decades of the 20th century, the representation of the landscape, its geopolitical description, will be fundamental to establish the scope and understanding—in territorial and symbolic terms—of Chile as a nation (Pérez Oyarzun, 2017).

If we move forward in time, without intending to be exhaustive, and we move towards an even closer scene, we can recognize the persistence of this attention to the landscape by contemporary art. Examples of this are the intervention *Gran Sur* by the artist Fernando Prats or the photographic work of Gaspar Abrilot which, with specific accents and nuances, address the role of the landscape and its representations in the construction of a country's imaginary.[3] The case of the investigation *Mundo Mágico. Formas de representación de un paisaje alterado* [*Magic World: Forms of representation of an altered landscape*] by Gaspar Abrilot (2020) is especially significant, as we explicitly recognize in it an exercise of reflection on the forms of representation of the landscape. Abrilot portrays the ruins of some models that are part of the Mundo Mágico attraction center and that reproduce scaling the Andes Mountain Range (Fig. 2.1), a fundamental territory in paying attention to the natural context as a sign of identity of the Chilean nation.

Beyond the strictly artistic space, the presence of landscape representations has extended to much more every day or popular areas, a matter that allows us to insist on its recurrent use as an image carrying national identity. The presence of an image of the Andes Mountain Range on a box of matches (Fig. 2.2) commonly and massively used in the homes of families in Chile or the literal sending of a piece of Chilean iceberg to the Universal Exhibition of Seville in 1992, are added to the list of examples

Fig. 2.1 Mountain Range XXII, from the series MundoMágico [MagicWorld], 2019. Analog photography. Author: Gaspar Abrilot. *Source* Courtesy of the author

that attend to the landscape as a distinctive sign of Chilean culture. On the other hand, from an audiovisual and also political perspective, the relatively recent documentary by Patricio Guzmán titled *La cordillera de los sueños* [*The Cordillera of Dreams*] (2019) is added as another link in this chain of attention to the mountain range, its visuality and symbolic dimension. As Catalina Valdés (2012b, p. 139) indicates, the image of the mountain range:

> (...) is inscribed in the collective retina: there are the typical matchboxes brand Los Andes and there is no child from the central and southern area of Chile who does not include the high white mountains of snow as the background of their drawings

The geographical peculiarities of the country, its coordinates of location become something that exceeds the 'simple' material description, as such variables are read and exposed as signs of differentiation with respect to other contexts. Beyond its physical or territorial extension, the landscape also reveals itself as a cultural fact

Fig. 2.2 Label box Compañía chilena de fósforos [Chilean match company]. *Source* Digital reproduction from original object

(Maderuelo, 2020), associated as it has been mentioned to the constitution of Chile as a country. That is, the natural landscape is read as a sign of identity.

Throughout this research we are interested in reflecting on how this imaginary linked to the natural landscape is recovered in the process of dissemination and reception of contemporary Chilean architecture. In this regard, we argue that in such a process an appreciative discourse is articulated around the works of architecture and their value. A discourse that builds and sustains, to a large extent, from the dissemination of a set of photographic images that enhance the relationship between the works of architecture and the natural context in which they are located, insisting on the landscape as a sign of national identity. These are images, but also, a series of textual statements that, when naming publications or exhibitions, refer to the landscape as that particularity that distinguishes the local architectural production. Thus, the natural context becomes a sort of ideal backdrop to expose the presence of the new architectural pieces that colonize extreme territories, as can be seen in the photographic record of Cristóbal Palma, Pablo Casals or Guy Wenborne.[4]

2.3 Export Architectures

In light of the above, it is appropriate to ask ourselves what is this interest in recording and disseminating the link between architecture and landscape? How can we read this recurrent attention to the natural context when exposing the works of architecture? And, beyond the obvious, what do these images tell us, what can we see in them?

Although the portrayed works are indeed located in those natural landscapes captured by the photographic lens, their choice as a representative and significant sample of the architecture that is being carried out in Chile does not seem to be a coincidence and offers a space for reflection. In this direction, we can say, as a hypothesis, that in this attention and use of images that expose the relationship works-landscapes is the fundamental intention to internationalize contemporary Chilean architecture and thus expand its reception space. We refer to a will to internationalize that, at times, exceeds the strictly architectural, in whose impulse and activation have participated different institutions—among them university and editorials—and that has even been governmentally sponsored through the agency *ProChile*, linked to the Ministry of Foreign Affairs.[5] Following again Pierre Bourdieu (2013), this participation of agents and institutions can also be read as a 'field problem'. This, insofar as this participation can influence, directly or indirectly, what is said or not about the selected works and put into circulation.

The architecture-landscape relationship acts as one more input of the so-called 'country image', which precisely seeks to stage the benefits of Chile especially in the eyes of foreigners. That the landscape is exposed as an essential part of the country image is not new, as demonstrated by the usual advertising campaigns of tourism that try to increase the number of visits to the Chilean territory. Now, that architecture is also considered in this narrative of exposing the qualities of the country does deserve, in our opinion, more attention. It is no longer about considering strictly disciplinary issues in the exhibition of architectural works, but also their possible coincidence with the country narrative put into circulation. Thus, while the architectural works made in Chile are exposed, the country and one of its most internationally recognized attractions are also promoted: its landscapes and natural resources.

Beyond the interpretations that we can make or suggest, the truth is that, in the growing diffusion experienced by the contemporary Chilean architecture, references to the landscape are not a minor or marginal issue. We are talking about a process of diffusion and consolidation of a 'local architectural field' that, as has been exposed by different authors (Liernur, 2009; Torrent, 2010), is activated especially from the nineties of the last century.

The return to democracy, the gradual unlocking of the political-cultural distance that isolated the dictatorship from international spheres, supposes a fertile space to start this diffusion of the architecture made in Chile. First at a local level, for example, through the magazine *CA* of the College of Architects or *Ediciones ARQ* of the Pontifical Catholic University of Chile.[6] In such publications and their contents, a space for diffusion 'among architects' is woven, but that will also manage to advance beyond strictly disciplinary margins. This is the case, among others, of the continuous publication of architectural works in the supplement *Vivienda y Decoración* [*Housing and Decoration*], in a high impact and circulation newspaper such as *El Mercurio*.[7]

As we approach the year 2000, the presence of works of architecture made in Chile will become increasingly frequent. And not only at a local level, but also in the international context. A context in which the attention to the link between architecture and landscape that we have been commenting on will become evident. "Último Chile. Paisajes próximos de una tierra remotaÚltimo" ["Last Chile. Near landscapes of a remote land"], edition number 85 of the Spanish publication *Arquitectura Viva* published in 2002 (Fig. 2.3) or "Chile: Deep South", number 430 of the Japanese magazine *a + u Architecture and Urbanism Magazine* of the year 2006, are two significant examples of this period. In both cases, as can be observed on their covers and interior pages, the reference to the landscape is double. On the one hand, we observe its textual and explicit enunciation through titles that speak to us of 'remote' and 'deep' landscapes located in the south of the world. On the other hand, the evidence from the photographic record visually places us in those distant territories where the portrayed works are located.

Complementing and also expanding the circulation space defined by the publications, from the first decade of the new 21st century, it is also possible to notice an increasingly intense presence of Chilean architecture at different international events. Participation in the *Biennale di Venezia* since 2002 onwards and the awarding of its own pavilion in this space, constitute a turning point on the path towards the internationalization of architecture made in Chile. Along with this significant presence in Venice, other exhibitions, also spread beyond local borders, have contributed to constitute this space for dissemination and reception. This is the case, for example, of the exhibition "Chile, territorio para la arquitectura" ["Chile, territory for architecture"] held in 2009 at the Puro Chile Gallery in New York, under the curatorship

Fig. 2.3 Cover edition number 85 magazine Arquitectura Viva. *Source* Courtesy of Arquitectura Viva

of Alberto Sato, or "White Mountain. Recent Architecture in Chile", whose edition was in charge of Miquel Adrià, initially exhibited at the Gabriela Mistral Center in Santiago in 2012, to then circulate around different parts of the world. Two exhibitions, two ways of showing and staging the architecture made in Chile that, once again, contain in their titles the appeal to the landscape or territory as that central space in which Chilean architecture seems to operate repeatedly.

Beyond the exhibition strategies used, the cases referred to ultimately define a type of work that 'deserves' to be shown. In the field we are analyzing, it is a series of works marked by the natural conditions of their location, by the exuberance of those landscapes in which the new buildings are inserted. Thus, what is named under the title 'contemporary Chilean architecture' or 'recent architecture in Chile', is a specific type of work, which coincides with that story that exalts the link between buildings and landscapes. In this way, the process of disseminating the 'Chilean architecture' that we are observing acquires a certain degree of recursivity. This, insofar as, once the argumentative thread based on the architecture-landscape relationship is defined, the selection of works to be exhibited or published feeds the resonance of that previously installed story. These are 'export' architectures, exposed as the spearhead of local work.

2.4 Architecture, Landscape and Artifice

As has already been stated, a particularly significant work for the analysis that we propose here is *White Mountain. Recent Architecture in Chile*. It is a publication, also an exhibition, made from Chile, but in whose content, we can recognize the intention to disseminate internationally a set of specially selected architectural works.

There are several signs that account for the intention of international reception that guides this editorial project. The book was edited by *Ediciones Puro* Chile, an entity interested in promoting the cultural and artistic production made in Chile abroad. In addition, already in its first edition of 2010 the book was published in a completely bilingual version, Spanish–English, while its second edition, from 2013, had the participation of the German publisher Hatje Cantz.

The appearance of *White Mountain* aims to be a sort of illustrative summary of the architecture made in Chile from the nineties. Although it is true that not all the works selected by *White Mountain* are located or are registered in exuberant landscapes, the inclusion of the natural context as an argumentative thread is relevant.

Already in the title of the work we can recognize, as we have seen, an explicit appeal to the natural landscape. In this case, the 'white mountain' refers us to the Andes Mountain Range, the same geographical accident that appears in pictorial paintings, documentaries and that transits without problems towards the *pop* imaginary of the nation. On this occasion, in addition, the appeal to the mountain range rescues some verses that are present in the fifth stanza of the national anthem of Chile: "(…) majestic is the white mountain/that the Lord gave you as a bulwark", written by Eusebio Lillo in 1847. Thus, the identity role that the landscape plays in the narrative that explains the scope of the architecture made in Chile seems even more evident and explicit.

On the cover of the first edition of the book *White Mountain*, in which a photograph of the trail *Punta Pite* by Teresa Möller (Fig. 2.4), or on the interior pages that contain material, for example, from the *Casa Poli* by the office Pezo von Ellrichshausen, or from the *Casa en Bahía Azul* by the architect Cecilia Puga, the reference to the landscape and the presence in it of the works of architecture emerge with total clarity. In the case of the mentioned works and their photographs, we can initially observe that the link between architecture and landscape is produced from the staging of a contrast. A contrast between the natural landscape, its own and irregular forms, and the abstraction of the architectural volumes that are located in them. Thus, one of the possible contradictions that are part of the process we are analyzing becomes evident: the coexistence between 'identity landscapes' and decidedly delocalized architectural forms. That is, architectural forms that, in their desire for internationalization, are usually based on a formal or aesthetic repertoire of global scope.

At this point, it is worth considering the scope of the word 'artifice', present in the title that names this research. A word that, based on what has recently been commented, can help us understand the role that the representation of the landscape plays in the process of disseminating the architecture made in Chile.

If we consider the third meaning of the term established by the RAE, an artifice is "an object constructed for a specific purpose".[8] That is, an object that, from its condition of being 'something constructed', differs from that which is 'natural'. It is a difference that as pointed out by Iñaki Ábalos (2009), has allowed to separate waters between buildings, read as the manifestation of a constructed will—that is, as 'artifices'—and the natural context. We are talking about a distinction that is seen in the photographs, exhibitions and titles of publications that we have discussed here, but at the same time can be complemented with another layer of analysis and interpretation. This, insofar as, in the case of Chilean architecture, the representation of the landscape also becomes an artifice, in a set of images constructed with a specific purpose: to articulate a graphic narrative capable of staging a group of works, guiding their reception and interpretation. We refer to a representational construction that proposes meaning and that aims to weave a coherent narrative around the architecture made in Chile during the last decades.

As indicated by the editor of the project *White Mountain*, Miquel Adrià, the images of contemporary Chilean architecture, the chosen works, become true postcards.[9] In presentation letters sent to the possible spaces of reception of the local production, articulating a panorama that allows to expose "the richness of the contemporary Chilean architectural culture" (Adrià, 2010, p. 25). However, like every panoramic view that shows and stages a general view, its extension always runs the risk of neglecting the different, that which may be out of focus, but not for that reason be less relevant or, even, representative.

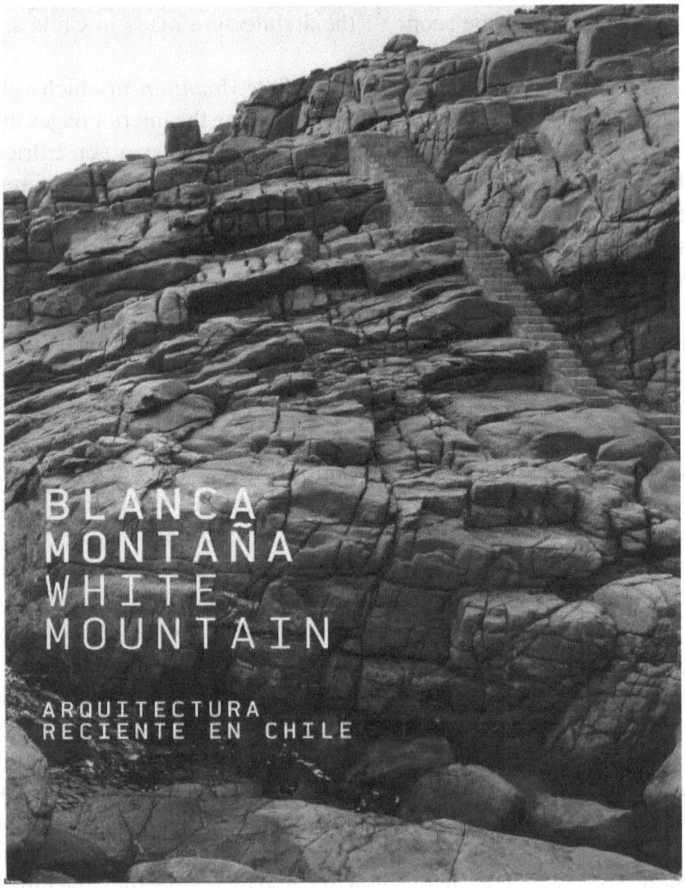

Fig. 2.4 Invitation to the launch of the book White Mountain. Recent Architecture in Chile *Source* Digital reproduction from original document. Personal archive

2.5 Final Considerations

In March of the year 2009, the *Design Issue* of the supplement *T: The New York Times Style Magazine* published on its cover a photograph of the *Casa Bahía Azul* by Cecilia Puga.[10] Under the title "On the edge", the cover is defined by a photographic record made by Jason Schmidt that shows the aforementioned work of Puga peering over the edge of the sea. In addition, the report includes references to other works by Alejandro Aravena, Smiljan Radic or Pezo Von Ellrichshausen, and insists on that inescapable relationship between architecture and landscape that would define the Chilean architectural scene.

Observing this publication, it seems appropriate to ask ourselves how to have these images reached that place? Not only the specific photographs of the referred works, but rather all that imaginary that distinguishes in the landscape a fundamental attribute to understand the architecture made in Chile and its diffusion. From an even broader perspective in terms of its audience, the case of this supplement of *The New York Times* adds to those of other international magazines previously discussed, which allows us to recognize the existence of a specific reception space that reads the architecture recently produced in Chile under the prism of its relationship with the landscape. A relationship that indeed 'exists', 'occurs', we do not intend to say otherwise here, but that in our opinion deserves to be thought critically, demystified. We refer to the need to promote critical attention capable of focusing on the content, but also on the specific way in which appreciations or positions are presented. Following Manfredo Tafuri (Liernur, 2019), it is about activating a critical exercise associated with the dismantling of those representations that install ideas or wills, in order to understand their true scope, what they include and, also, what they leave out.

As was pointed out at the beginning of this discussion, every myth is a way of constructing and installing 'truth' about things and reality itself. Beyond the degree of 'objectivity' that a story may have, what matters when anchoring a certain mythical voice is the veracity that it manages to achieve, its ability to be effectively received.

In the case of the architecture made in Chile during the last decades, the reading and reception of its presence in the landscape has ended up installing a concrete way of understanding its qualities. Moreover, the images that show the link between architecture and landscape have also been understood as a feature of identity, a factor that differentiates the architecture made in Chile from other architectures. We speak of a look that although, as was said, cannot be cataloged as a mere 'invention', it is, at least, a partial observation. Especially, if we consider that the stories put into circulation have an identity vocation in relation to the chosen works.

In this regard, it is worth considering what was pointed out by Cristóbal Palma, one of the most relevant photographers of the works of architecture recently produced in Chile. When commenting on the content of the mentioned edition number 430 of the Japanese magazine $a + u$, Palma (2006, p. 32) points out:

> Anyone who has leafed through the latest issue of the prestigious Japanese publication a+u (...) probably ended up with the impression that Chile is basically a region of incredible landscapes (many of them barely explored), populated by architectural gems inhabited by a middle-upper class public. Inevitably any selection has a bias and establishes a particular angle, which is not necessarily representative, but without a doubt this is not the first time this angle has been used to talk about Chile

Following Palma's story, it is about an installed reality that has managed to define a reception space. However, as the photographer himself indicates, the recognition of this established reality allows us it should also allow us to intuit that, beyond it and its repercussions, there are other possible realities, other ways of thinking about the practice of architecture carried out in Chile. In this direction, it seems necessary to insist on the need, also the opportunity, to activate a deep reflection that, without ignoring the value that the works frequently published or exhibited have, allows us to discuss in depth about the field of action of what we usually call 'contemporary Chilean architecture' and its appreciative discourses.

End Notes

1. Although in the case of architecture we usually link the term representation to the development of graphic documents or to the expressive capacity of the works themselves, in this text we adopt the use proposed by Stuart Hall (2013). For Hall, a representation is everything capable of producing and sharing meaning around reality; a definition that certainly includes a wide spectrum of possible materials: images, texts, oral sources, etc.
2. In this regard, the list of authors who in recent decades have dealt with these other products of architecture have defined a relevant area of disciplinary discussion. In this space and without ignoring the differences or specificities of each case, we can inscribe, for example, the work carried out by Beatriz Colomina (1994), Felicity D. Scott (2016), Mario Carpo (2001) or Pérez Gómez and Pelletier (1997).
3. Since it is a somewhat elusive term, it is important to specify what we mean by using the word 'imaginary'. Following Néstor García Canclini (1997), we can think of imaginaries as a shared field of appreciation around reality, in which 'objective' facts coexist with appreciations, emotions, ideas and inventions produced around them, which are expressed, precisely, through a wide range of representations.
4. The importance achieved by photography in the process of disseminating the works carried out in the country can also be seen in the work Winter: images of contemporary Chilean architecture (2020). Although in this case the landscape coexists with other focuses of attention, the gathered photographs account for the imaginary that has been constituted around the architecture recently produced in Chile.
5. As stated on its website, https://www.prochile.gob.cl/, ProChile is an organization that aims to encourage the supply of services and goods made in Chile abroad. Within this offer, architecture has been recognized as an 'exportable' good or service.
6. Although in the case of the magazine CA it is a publication that began its journey in 1968, we mention it because in the nineties it is absolutely current and devotes significant efforts to publish contemporary works of architecture made in Chile. For its part, in the case of Ediciones ARQ, founded in the early eighties, to the leading role assumed by the magazine ARQ, we must add a series of monographic publications that contribute to weave the field of meaning of local architecture. In this line we can mention, for example, the publications Arquitectura reciente en Chile: las lógicas del proyecto [Recent Architecture in Chile. The logics of the Project] (2000) or Compendium ARQ. Generación del'90 [Compendium ARQ. Generation of the '90] (2001).
7. In this work we are only accounting for the publication of architectural works in a medium that does not come from academia or trade union activity. However, the case of Housing and Decoration would merit a deep analysis regarding the type of work published in this medium and the interpretations exposed,

especially considering the conservative profile of the newspaper El Mercurio and its target audience.

8. Definition obtained from the Dictionary of the Spanish language of the Royal Spanish Academy, in its online version: https://dle.rae.es/artificio?m=form.
9. When thinking in these terms and using the expression 'postcards', Miquel Adrià is considering another particularly significant exhibition work: I WAS THERE—Chilean Souvenirs. This refers to the installation carried out by Mauricio Pezo and Sofía Von Ellrichshausen for the presentation of Chile at the 11th Venice Architecture Biennale, held in 2008. On that occasion, the exhibition consisted of the display of 100 small pieces or models of architectural works carried out in Chile that, literally, acted as the sending of souvenirs.
10. The cover and content of the report can be consulted at the following link: https://www.nytimes.com/2019/03/19/t-magazine/chile-architecture.html.

References

Ábalos, I. (2009). Introducción. In I. Ábalos (Ed.), Naturaleza y artificio. El ideal pintoresco en la arquitectura y el paisajismo contemporáneo [Nature and artifice. The picturesque ideal in contemporary architecture and landscape] (pp. 7–18). Editorial Gustavo Gili.

Abrilot, G. (2020). Mundo Mágico. Formas de representación de un paisaje alterado [Magic World. Forms of representation of an altered landscape]. Bex Autores, Bariloche

Adrià, M. (2010). Chilean postcards. In M. Adrià (Ed.), *White mountain. Recent architecture in Chile* (pp. 12–25). Ediciones Puro Chile.

Ahumada, P. (2012). Paisaje y nación: La majestuosa montaña en el imaginario del siglo XIX [Landscape and nation: The majestic mountain in the imaginary of the 19th century]. *Artelogie, 3*, 1–27. https://doi.org/10.4000/artelogie.6841

Arquitectura Viva. (2002). Último Chile: Paisajes próximos de una tierra remota [Last Chile: Nearby landscapes of a remote land]. Arquitectura Viva (85). Arquitectura Viva.

Arquitectura Viva. (2006). Chile: Deep South. a+u Architecture and Urbanism Magazine (403). A+U Publishing.

Barthes, R. (2010). Mitologías [Mythologies]. Siglo Veintiuno.

Bourdieu, P. (2013). *Cuestiones de sociología [Questions of sociology]*. Akal.

Carpo, M. (2001). *Architecture in the age of printing: Orality, writing, typography, and printed images in the history of architectural theory*. MIT Press.

Colomina, B. (1994). *Privacy and publicity: Modern architecture as mass media*. MIT Press.

García Canclini, N. (1997). Imaginarios urbanos [Urban imaginaries]. Eudeba.

Guzmán, P. (Director) (2019). La cordillera de los sueños [The cordillera of dreams]. Atacama Productions.

Hall, S. (2013). The work of representation. In S. Hall, J. Evans, & S. Nixon (Eds.), *Representation* (pp. 1–47). The Open University.

Liernur, J. F. (Curator) (2009). Portales del laberinto. Arquitectura y ciudad en Chile 1977–2009 [Portals of the labyrinth. Architecture and city in Chile 1977–2009]. UNAB Editions.

Liernur, J. F. (2019). Conversación con Manfredo Tafuri. Buenos Aires, Agosto 1981. In M. Tafuri (Ed.), Tafuri en Argentina [M. Tafuri in Argentina] (pp. 24–53). Ediciones ARQ.

Maderuelo, J. (2020). *El espectáculo del mundo: Una historia cultural del paisaje [The spectacle of the world: A cultural history of the landscape]*. Abada Editores.

Molina, C. (Ed.) (2020). Winter: Images of Contemporary Chilean Architecture. Arquine.

Palma, C. (2006). Chile: Truly, madly, deeply. *ARQ, 64*, 32–35. https://doi.org/10.4067/S0717-69962006000300007

Palmer, M. (Introd). (2001). Compendium ARQ. Generación del '90 [Compendium ARQ. Generation of the '90s]. Ediciones ARQ.

Pérez Gómez, A., & Pelletier, L. (1997). *Architectural representation and the perspective hinge*. MIT Press.

Pérez Oyarzun, F. (2017). El territorio chileno y su representación. In F. Pérez Oyarzun (Ed.), Arquitectura en el Chile del siglo XX [Architecture in Chile of the 20th century] (pp. 16–21). Ediciones ARQ.

Rendell, J. (2004). Architectural research and disciplinarity. *ARQ: Architectural Research Quarterly, 8*(2), 141–147. https://doi.org/10.1017/S135913550400017X

Scott, F. D. E. (2016). ¿Qué salió mal? [What went wrong?]; Fuera de lugar [Out of place]. Ediciones ARQ.

Torrent, H. (Introd). (2000). Arquitectura reciente en Chile: las lógicas del proyecto [Recent Architecture in Chile: the logics of the project]. Ediciones ARQ.

Torrent, H. (2010). The nineties: Articulations of Chilean architectural culture. In M. Adrià (Ed.), *White mountain*. Recent architecture in Chile (pp. 26–49). Ediciones Puro Chile.

Valdés, C. (2012a). Comienzo y deriva de un paisaje. Alessandro Ciccarelli, Antonio Smith y los historiadores del arte chileno [Beginning and drift of a landscape. Alessandro Ciccarelli, Antonio Smith and the historians of Chilean art]. *Artelogie, 3*, 1–24. https://doi.org/10.4000/artelogie.6872

Valdés, C. (2012b). La medida de lo sublime. La Cordillera de los Andes vista desde Chile durante el siglo XIX [The measure of the sublime. The Andes Mountain Range seen from Chile during the 19th century]. *Concinnitas, 2*(21), 139–168. https://www.e-publicacoes.uerj.br/concinnitas/article/view/12425

Chapter 3
Memory as a Territory in Dispute: Tension Between the Urban Heritage and Neoliberal Urban Development from the Memory Site Venda Sexy

Francisca Molinos Moyano

Abstract Since the nineties, a process of materializing memory through detention centers that operated during the dictatorship has begun, becoming part of the urban heritage. The National Institute of Human Rights (INDH) proposes four phases of recovery of the buildings: identification (1), declaration (2), recovery (3), and enablement (4). However, memory policies have not yet managed to promote a legislative framework that allows everyone to reach the final stage, being violated by the neoliberal logic that structures the urban development of Chilean cities. This study focuses on the Venda Sexy memory site, where the building, despite its declaration and the State's attempts to buy it, was sold to a real estate company in 2019, becoming a disputed territory and leading communities to develop various strategies of participation and resistance for the conservation of memory in cities.

Keywords Urban heritage · Neoliberal urban development · Memory sites · Patrimonialization

3.1 Introduction

After the coup led by Augusto Pinochet on September 11, 1973 in Chile, economic and social reforms were carried out towards a neoliberal economic model, whose principles are established in the 1980 constitution that perpetuates the imposed political-social order (Cornejo et al., 2013). The neoliberal logic promotes an institutional framework that encourages private property rights, free markets, and freedom of trade (Harvey, 2005). In this way, the neoliberal logic is expressed in cities through a neoliberal urbanization, which for De Mattos (2002) focuses mainly on policies of liberalization and deregulation of land, focused on private capital being the protagonist of urban development.

F. M. Moyano (✉)
Pontificia Universidad Católica de Chile, Santiago, Chile
e-mail: fjmolinos@uc.cl

M. Vicuña and E. Giorgi (eds.), *Intersections*, Springer Geography,
https://doi.org/10.1007/978-3-031-76402-8_3

The logic of a neoliberal urban development and the great transformations of cities result in the conflict between what is built and what is wanted to be built, with access to land being a factor of dispute, where the tension between urban heritage and the logics of neoliberal urban development results in one of the most important conflicts regarding access to land (Carrasco, 2017). In this line, the tension between urban heritage, understood as the "set of meanings and interpretations that arise from the media relationship between the object-support and individuals" (Dormaels, 2011, p. 8) and real estate development has been one of the most important conflicts regarding access to land in recent years. For Carrasco (2017), the neoliberal model favors the real estate sector and has not managed to systematically protect buildings and heritage areas, but follows a logic of demolition and construction, where new buildings are prioritized over the conservation and restoration of existing buildings.

Regarding Chilean heritage, since the return to democracy in the 90s and as a measure of reparative justice, it is established to declare the main torture centers as Historical Monuments, turning them into memory sites (Fernández et al., 2016). As Rebolledo (2022) points out, "the collective memory materialized through these sites constitutes an urban heritage, which sought to challenge civil society as a whole". However, as Escariz (2022) mentions, there has not been a push from the State to declare the spaces, and it has happened mainly because there is no investment, nor incentives for inter-institutional articulation in matters of memory sites.

Regarding the reports and programs that promote the conservation of memory sites, contributing to the promotion of a culture of respect and promotion of human rights in our society (Red Sitios de Memoria, 2014), the annual human rights situation reports prepared by the National Institute of Human Rights (INDH) stand out, precisely in its annual report of 2018 it proposes a process of recovery of memory sites, contemplating four total stages that make up the process of recovering properties: identification (1), application and declaration (2), recovery (3) and enablement (4). However, as the INDH (2018) suggests, most spaces get stuck between the second stage of heritage declaration and the third of site recovery, which are in tension due to the limitations of the current regulatory framework, private property, and the expansion and densification of cities under a neoliberal logic. Specifically, sites with (peri)central locations and highly attractive for real estate development that could be linked with the neoliberal transformations of cities in a conflictive scenario, generating a disputed territory.

Former Venda Sexy Detention Center

The Former Venda Sexy Detention Center is located in the commune of Macul, specifically in the northeastern sector of the commune, at the intersection of Iran with Av. Los Plátanos. The Communal Regulatory Plan (PRC) of Macul (2021) defines this sector as a low density residential zone 8 (ZR-8) where a maximum height of two floors is allowed. Regarding the enclosure, as the National Monuments Council (2016) points out, it was characterized by carrying out various types of sexual tortures, some with animals in charge of Ingrid Olderöck. For this reason, the names "Venda Sexy" or "Discoteque" derive due to the high volume of music that

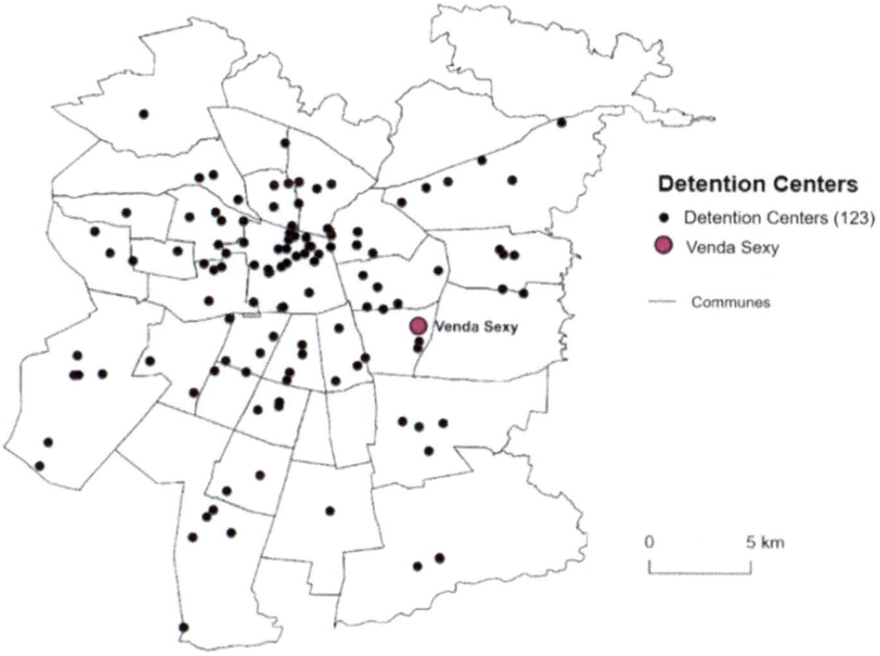

Fig. 3.1 Venda sexy location in the metropolitan area of Santiago (AMS). (Own elaboration, 2022)

sounded during the day so that the noises of the tortured were not heard (Consejo de Monumentos Nacionales, 2016) (Fig. 3.1).

The conflict that occurs in the recovery of the memory site lies in the fact that the property owner is private, and that the enclosure has a residential use. Legally, there are no powers to give another use to space. Added to this, the property, despite being declared and the State's attempts to buy, was sold to a real estate company in 2019 (Reyes, 2019). In this line, there are several press media that evidenced what happened in Venda Sexy and that started a series of positions and statements by different actors who have a voice in the conflict. On the one hand, National Assets stated that "conversations were held with the owner for the acquisition of the house, from April of this year until the end of August. It reports that no agreement was reached, ending the negotiations on the part of the State, which in practice means not being able to access what was resolved by the National Monuments Council. It also states that given the Constitution we have, ["which sanctifies private property, above any other good (our opinion)"], it is impossible to expropriate" (Sánchez, 2019).

Therefore, the following article proposes the hypothesis that the absence of a legislative framework on memory and the logics of the neoliberal urbanization model could tension the process of materializing memory in the city, thus generating disputed territories and leading communities to develop various strategies of participation and resistance for the conservation of memory in cities. In this context, the question that guides the research relates to the manifestations of the tensions between the logics of neoliberal urban development and urban heritage through the memory site Venda Sexy, focused on the heritage strategy proposed by the INDH (2018).

Logics of Neoliberal Urban Development

As for what is understood as neoliberal urban development, large cities in the world, and particularly in Chile, have been transformed through this neoliberal logic. This model can be described as "a theory of political-economic practices that asserts that the best way to promote human welfare is by not restricting the free development of individual capabilities and entrepreneurial freedoms" (Harvey, 2005, p. 6). In this line, for Brenner et al (2009) "the neoliberal ideology rests on a starkly utopian vision of market supremacy, a vision that has "its roots in an idealized conception of competitive individualism and in a deep antipathy towards various forms of social and institutional solidarity" (Brenner et al., 2009, p. 3). In this sense, neoliberal logic has had a strong impact on urban development, with access to land and private property being the main logics that produce tensions in the processes of declaration and management of urban heritage.

Memory, Postmemory and Gender

Heritage is considered as a social construction, where subjects assign symbolic values to objects produced in the past (Delgadillo, 2015), so heritage linked to memory falls within the appropriation of buildings. In this sense, authors such as Halbwachs (1925), Nora and Erll (1997), Jelin (2003) argue that memory is both individual and collective, being a reconstruction of the past, through collective experiences and narratives. Now, memory from the Chilean perspective is loaded with history. In this line, Del Valle (2018) with a more current vision defines memory in Chile as "A construction that changes by virtue of its agonistic forms: the struggles between memories and forgetfulness forge memory, but also memory is defined through its confrontation with other memories" (pp. 303–304).

In addition, it is necessary to understand memory from two perspectives. First, considering that memory can be transmitted to new generations, where authors such as Beatriz Sarló (2007) Marianne Hirsch (2012) and Laia Quílez (2014), define it as "*postmemory*", understood as the relationship of a later generation to the one that receives a trauma, where the relationship with these traumatic experiences is through stories and images, where the transmitted experiences become their own experiences and their own memories (Hirsch, 2012). In this line, postmemory is proposed as a structure of intergenerational and transgenerational transmission of knowledge and traumatic experiences (Hirsch, 2012). On the other hand, it is also necessary to understand the relationship between gender and memory, where in the words of Maceira (2017) memory is marked by power relations, which generates that the "unequal position of women as a collective in contemporary patriarchal societies implies a situation of disadvantage with respect to social memory, both as a remembered group and as a group that remembers" (Maceira, 2017, p. 110).

Sites of Memory: Recovery Processes in Cities of Neoliberal Logic

As a first approach, it is necessary to understand the meaning of places of memory. For Pierre Nora "places of memory are, first of all, remains. The extreme form where a commemorative consciousness survives in a history that summons it because it ignores it" (1997, p. 24). Now, in the Latin American case, as pointed out by López González (2010), Seguel (2019) and Mora (2011), places used as detention and torture centers, materialize physically leaving a material trace in the urban fabric, where discourses and heritage practices are produced around the past, establishing commemoration strategies. Therefore, sites of memory are not left out of heritage conflicts in the city, where territorial disputes are generated between what should be preserved and what should be destroyed. For Lee (2016) urban heritage as it loses its meaning accelerates the process of destruction increasingly, which Dormaels (2012) reaffirms by pointing out that people are the protagonists of heritage recovery. However, sites of memory materialize traumatic events in space, where although memory requires a physical place to leave a trace, the repair of these events also has a spatial dimension and the city becomes one of the main agents to preserve memory by maintaining historical and cultural events (Montealegre and Rozas-Krause, 2018).

One of the main tools for the protection of heritage in cities corresponds to the process of declaring heritage by a community, called "heritagization", understood as "a voluntary process of incorporating socially constructed values, contained in the space-time of a particular society and is part of the processes of territorialization that are at the base "of the relationship between territory and culture" (Bustos, 2004, p. 11). In this line, these processes of patrimonialization for Dormaels (2012) respond to a form of "democratizing" heritage, where the different processes and social uses of this have allowed a takeover by communities, where they have become protagonists. In the words of Durán et al. (2014) "These processes, like the political practices associated with them, aim to recover the uniqueness of the events and the actors remembered and to inscribe themselves in the urban fabric" (p. 6).

3.2 Methodological Design

In methodological terms, and regarding data collection, the article collects information from various organizations. On the one hand, the identification and location of detention centers through the databases of the Network of Memory Sites and the Museum of Memory and Human Rights, it should be noted that the research focuses only on detention centers within the AMS corresponding to properties and that it was possible to assign them an exact location. The states of patrimonial declaration through the National Monuments Council and the National Institute of Human Rights and regarding the type of property through the review of files of the Conservator of Real Estate of Santiago. Finally, it is necessary to mention as part of the methodology of this research the link with the Museum of Memory and Human Rights as part of the "Thesis Contest 2022" that the institution imparts each year, which has allowed to have a direct link with the different activities that the museum has carried out and that have been an important part for the development of the research.

On the other hand, following an inductive logic, interviews were conducted with various actors that make up the dispute of Venda Sexy. On the one hand, institutional actors (National Monuments Council, Ministry of National Assets and Municipality of Macul) and regarding social actors, representatives of the collectives of the case study: Women Survivors Always Resistant and Memories of Feminist Rebellions who led the process of memorialization of Venda Sexy. Also, the review of secondary sources, especially academic literature, press material and social networks was carried out. In addition, data and spatial representation techniques are used to deepen the analysis regarding the state of patrimonialization formulated by the INDH. However, it should be stated that the research from the beginning sought the use of a mixed methodology, seeking to contrast information about the expansion of urban development and the locations of the memory sites, where no major results were obtained, before that the focus that this research has was oriented, through the qualitative approach.

3.3 Results

Geography and Institutionality of Memory Disputes in the Metropolitan Area of Santiago (AMS)

In the Metropolitan Area of Santiago (AMS), 123 detention centers were identified that were operational during the military dictatorship. Most of these facilities (90) correspond to police stations or military facilities. The distribution of the centers is presented in Fig. 3.2. Most of the detention centers were concentrated in the central area of the region, with the commune of Santiago standing out, which has the largest number of centers, mainly finding the police stations in the most peripheral areas.

In the figure, the levels at which each building is located regarding the phases proposed by the INDH can be seen. Most of the detention centers are in phase 1, that is, they are identified, but do not have a patrimonial declaration, where a large part no longer exist. Regarding phase 2, where Venda Sexy is located, there are only six buildings that have obtained their patrimonial declaration, but do not have the property or administration of the site. Finally, in the final stages (three and four) there are eight memory sites, some being among the most nationally recognized as Villa Grimaldi, Londres 38 or Nido 20, among others.

The main factor influencing whether detention centers manage to reach phases 3 and 4 proposed by the INDH, corresponds to the type of property ownership (public or private). In the case of Venda Sexy, it is stuck between phase 2 of heritage declaration and phases 3 and 4 of recovery and enabling of the property, where the tension is directly produced with a logic of neoliberal urban development, since being sold in 2019, there is an "illegal sale" where the State was not part, being that as mentioned in Law 17.288 it is the one that has the first priority in case of sale, and despite

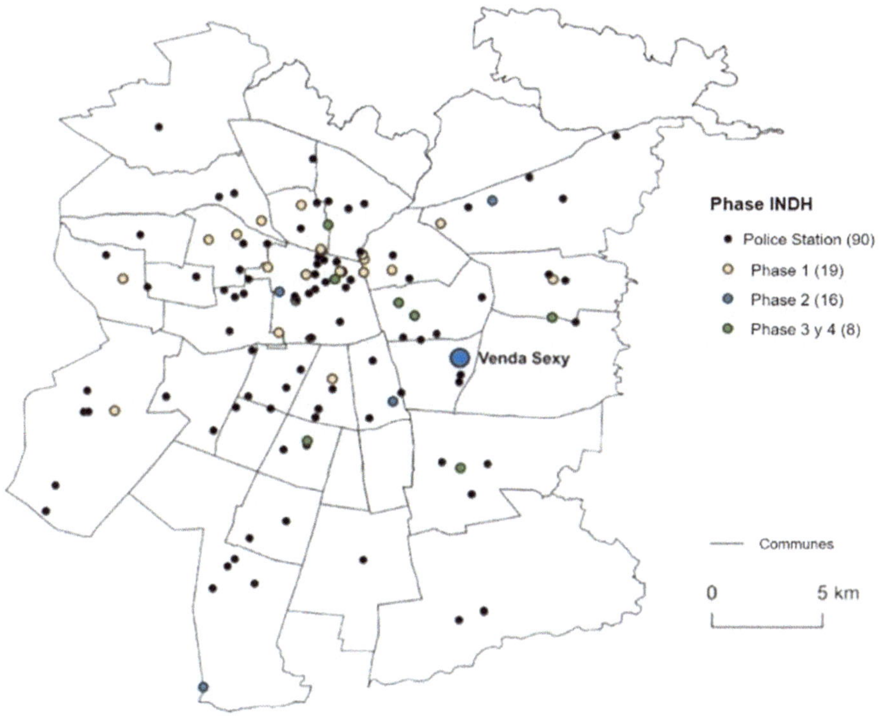

Fig. 3.2 Detention centers in the AMS and their classification (Own elaboration, 2022)

the fact that the real estate company corresponds to the same owner of the property located at Iran 3037, has indicated to the media that it is a strategy of "real estate business" (24 hours, 2019), being this business that is carried out with a memory site, a real estate speculation with urban heritage. Where by putting value on the material recovery of memory, it caused the collectives of Women Survivors and Memories of Feminist Rebellions do not want to recover the property, generating a territory in dispute, which goes beyond the property and its physical characteristics, but it is sought to have a physical space to materialize the events occurred in Venda Sexy.

Chile does not have a specific law for memory sites, which prevents detention centers from being categorized as such. According to Escariz (2022) memory sites have been cataloged in an existing category, losing their characteristic attributes, which should be treated and managed in a specific way, since they have different needs, due to their commemorative character or recovery by the State, since the properties can be in the hands of private individuals or the Armed Forces. For the same reason, the detention centers that maintain a territorial dispute, as well as the properties that have been demolished to generate new constructions, respond to a weak heritage management regarding memory sites and the absence of a specific law for these, which reaffirms the National Monuments Council (CMN) "From the legal point of view, there are problems that there is no category of memory sites in brings a problem between the institution and the collectives, since the women belonging to these organizations today do not want to recover the property, because they argue that the premises is no longer the place that represents them, where real estate speculation has been generated, where a monetary value has been given with the installation of a swimming pool, jacuzzi, etc. But that in the words of the collectives has lost the true sense of memory. For this reason, they propose the installation of a memorial in Plaza Arabia (located meters from the property). The memorial is the new construction of the heritage discourse by the Venda Sexy collectives, being the way to leave a material mark in the urban space in a dispute with the property, situating a memory from the public, a construction of memory with the neighborhood and with the different generations of society.

Therefore, for the collectives "heritage is not only buildings and monuments, it is also the memory of the struggle of the past of the survivors of the civic-military dictatorship that is intertwined with the struggles of the present and future" (Facebook CMSSR[2] May 27, 2016), they generate heritage beyond materiality, but from practices and commemorations, in that sense public space becomes vitally important, where it has been a meeting place for the different commemorations that have been held at the premises, where they have been able to make themselves known and establish ties with passers-by.

Through the recovery and conservation of detention centers an intergenerational memory has been developed: a postmemory, which Hirsch (2012) defines as "the relationship with the experiences that they remember through the stories, images, and behaviors in the midst of which they grew up. But these experiences were transmitted to them so deeply and affectively that it seems to construct their own memories" (p. 19). In this sense, the youth have appropriated the struggle to recover detention centers as sites of memory, as in the case of Venda Sexy where the collectives encompass women of all ages and not only survivors.

Following this line, it cannot be overlooked that one of the main struggles of the collectives corresponds to the visibility of political sexual violence as a practice of torture that occurred in Venda Sexy and that is today part of the history and memory of women, of their history of fighting women and survivors of a dictatorship that systematically incorporated gender violence in the precinct located at Iran 3037, which is why they have promoted this dialogue from feminism, forming alliances with human rights and feminist organizations, one of the most iconic corresponds to the link with the 8M coordinator for the recovery of the property.

The work then is collective, it is of all women and responds to the need of a city that today allocates memory spaces to forgetfulness and not to their recovery, causing the women of Venda Sexy to build their own memory among them, among various groups, of different ages, forming a memory of the future, enhancing a postmemory through intergenerational dialogue.

3.4 Conclusions

In the case of Venda Sexy, it can be concluded that the main dispute is generated in wanting to recover the precincts, where it is left to the will of individual owners, so the heritage declaration emerges as an emergency response to the dispute. On the other hand, all centers that do not have a heritage declaration and are private today do not exist, which reaffirms the hypothesis that detention centers are vulnerable under neoliberal logics that promote private property and privatize the land market.

In addition, the absence of a Law of Memory Sites means that memory sites do not have a specific category, leaving aside the specificities of memory sites. On the other hand, the non-existence of a Memory Policy at the national level, means that the different institutions (National Assets, Council of Monuments and Municipalities) regulate through their own powers, where there is a fragmentation in the role that each institution plays in the conservation of the sites. In addition, it accounts for the importance that the property is acquired by the State to move from a phase of declaration to recovery and enabling of the precincts.

Finally, the collectives belonging to Venda Sexy in the face of the dispute have generated ways of construct memory from the territory, from the individual to the collective and from the private to the public, reflected in the different commemorations they hold outside the property or in the proposal for a memorial located in Plaza Arabia. On the other hand, these constructions are focused on creating a memory of the future, for the next generations, for women, so they promote an intergenerational and feminist dialogue in all their commemorations, transmitting their memories and generating a postmemory in the new generations.

The case of Venda Sexy has shown that the materiality of the property is not the important thing to preserve, but to leave a mark in the public space, where it seeks to generate new ways of materializing memory, beyond the precinct. Finally, heritage is not only the property, but the practices carried out by the communities, where they seek to materialize a space to build their own heritage discourses.

End Notes

1. Collective of Surviving Women Always Resistant (CMSSR) and Memory of Feminist Rebellions (MRF).
2. Facebook Collective of Surviving Women Always Resistant.

References

24 horas cl. (2019, September 12). 24 Reports: The controversial sale of the torture house "Venda sexy" | 24 Hours TVN Chile [Video]. YouTube. https://www.youtube.com/watch?v=Ifq9Jb_NyMI

Annual Report on the Human Rights Situation in Chile. (2018). National Institute of Human Rights. https://www.indh.cl/destacados/informe-anual-2018/

Bustos, R. (2004). Patrimonialization of territorial values. Tourism, productive systems and local development. *Contributions and Transfers, 8*(2), 11–24. ISSN: 0329-2045

Brenner, N., Theodore, N., & Peck, J. (2009). Neoliberal urbanism: The city and the empire of markets. *Social Issues*, 66 (Santiago).

Carrasco, G. (2017). Urban development and heritage: Between renewal and conservation. In W. Imilan, J. Larenas, G. Carrasco, & S. Rivera (Eds.), Where is housing going in Chile?: New challenges in residential habitat (pp. 81–93). INVI, MINVU, Adrede.

Cornejo, M., Reyes, M., Cruz, M., Villaroel, N., Vivanco, A., Cáceres, E., & Rocha, C. (2013). Stories of the Chilean military dictatorship from generational voices. *Psykhe (Santiago), 22*(2), 49–65. https://doi.org/10.7764/psykhe.22.2.603

Decree 277. (2016). Consejo de Monumentos Nacionales.

Decree 347. (2016). Consejo de Monumentos Nacionales.

Del Valle, N. (2018). Memories of the (post)dictatorship: Practices, dates and sites of memory in recent Chile. *Mexican Journal of Political and Social Sciences, 63*(232), 301–322. https://doi.org/10.22201/fcpys.2448492xe.2018.232.5811

De Mattos, C. (2002). Transformation of Latin American cities: Impacts of globalization? *EURE (Santiago), 28*(85), 5–10. https://doi.org/10.4067/S0250-71612002008500001

Delgadillo, V. (2015). Urban heritage, tourism and gentrification. *Perspectives on the Study of Gentrification in Mexico and Latin America, 113*.

Dormaels, M. (2011). Heritage, patrimonialization and identity. Towards a hermeneutics of heritage. *Heritage Journal, 24*(1–2).

Dormaels, M. (2012). Identity, communities and local heritage: A new social legitimacy. *Alterities, 22*(43), 9–19.

Durán, V., Messina, L., & Salvi, V. (2014). Dossier 'Spaces of memory: Controversies around the uses and strategies of representation'. Clepsidra. *Interdisciplinary Journal of Studies on Memory, 2*, 5–11.

Elgueta, G. (2018). Institutionalization and patrimonialization of memory sites in Chile. A reading from the experience of London 38. *Aletheia, 16*, 1–17.

Escariz, M. (2022). Regulation of memory sites in the Southern Cone: Critical analysis and proposals for Chile. *Latin American Legal Studies, 10*(1), 1–56.

Fernández, G., Ricci, S., Valenzuela, S., & Ramos, A. (2016). Reappropriation and resignification of territory and heritage: Application of the reception evaluation to the two-hole zone Argentina. UNICEN.

Harvey, D. (2005). Brief history of neoliberalism. Akal Editions.

Hirsch, M. (2012). *The generation of postmemory. Writing and visual culture after the holocaust*. Columbia University Press.

Jelin, E. (2003). Human rights and the memory of political violence and repression: The construction of a new field in the social sciences. *IDES Notebooks, 2*, 1–27.

Lee, M. (2016). On the concept of urban heritage. Studies on conservation, restoration and museology, 3. https://revistas.inah.gob.mx/index.php/estudiosconservacion/article/view/7741

López González, L. (2010). Places of memory: Narratives of repression.

Maceira, L. (2017). Gender and memory in cities. In *Policies of memory, gender and city* (pp. 105–157).

Memory Sites Network. (2014). Memory Sites | Memory Sites Network. https://sitiosdememoria.cl/

Mora, M. (2011). Museum of memory and human rights: An aesthetic-political bet of legibility of the dictatorial experience. *Chair of Arts, 11*, 63–76.

Montealegre, P., & Rozas-Krause, V. (Eds.). (2018). *Dispute the city: Submission, resistance, memorialization, repair*. Bifurcations Publishing.

Nash, C. (2014). The debts of the Chilean transition in terms of human rights. An uncomfortable debate. *Annals, 15*(141), 162.

Nora, P., & Erll, A. (1997). *Les lieux de mémoire* (Vol. 3, pp. 1984–1986). Gallimard.

Quílez, L. (2014). Towards a theory of postmemory. Reflections on the representations of generational memory. *Historiographies*, 57–75.

Rebolledo, D. (2022). Memories in the city: The integration of sites of conscience into the territory as urban heritage. In X. Abarca, D. Rebolledo, C. Sánchez, & O. Sagredo (Eds.), *Places of memory and sites of conscience: Building heritage and memory for action in human rights* (pp. 41–58)

Reyes, R. (2019, September 11). The sale of Venda sexy: The fight to build a feminist memory. The Clinic - Reports, columns, interviews and humor. https://www.theclinic.cl/2019/09/10/la-venta-de-venda-sexy-la-pelea-por-construir-una-memoria-feminista/

Sánchez, N. (2019, August 18). Memory site and national monument "Venda Sexy" for sale. In *The left daily - international network.* https://www.laizquierdadiario.cl/Sitio-de-memoria-y-monumento-nacional-Venda-Sexy-en-venta

Sarlo, B. (2007). Past time. Culture of memory and subjective turn. A discussion. 21st Century.

Seguel, P. (2019). Human rights and heritage: Stories/memories of (para) state repression in Chile. Editions of the National Cultural Heritage Service.

Stern, S. (2012). Memories under construction: The challenges of the present past in Chile, 1989–2011.

Chapter 4
Folk Art and University Extension: Some Categories in Tension During the 1960s in Chile

Paulina F. González-Valenzuela and Eduardo E. Castillo-Espinoza

Abstract In 1964, under the direction of the artist Nemesio Antúnez, the Museum of Contemporary Art held the exhibition Chilean Folk Art which included craft pieces from different genres and territories of the country. To revitalize the discussion in this area and establish a connection between "cultured art" and "people's art" (MAC, 1964), the organization proposed a vision of these objects distant from the folkloric view that governed them until then, instead highlighting the social value of the artist-craftsman as a producer of plastic expressions. This perspective, in line with the ideal of a "living museum" promoted by Antúnez (Lihn, 1963; Novoa & Avaria, 2022), generated tensions within the Institute of Extension of Plastic Arts of the University of Chile (IEAP), both for the questioning of the work of the Museum of Popular American Art (MAPA), which had more than two decades of activity, as well as for the administration of the IEAP of the museums under its charge. The implications of this dispute exemplify the problematization of the concept of Folk Art within the University of Chile in the years prior to the University Reform, confronting the anthropological and traditional vision led by MAPA, with a more dynamic and inclusive one, akin to contemporary artistic institutionalism. This chapter analyzes the curatorial policies of these university museums in terms of Folk Art, with the aim of recognizing their contribution to the recent discussion on this term in areas such as visual arts, design and crafts. For this, a historiographic and qualitative approach will be taken from perspectives from cultural studies, art history and design.

Keywords Contemporary art · Folk art · Museum · University extension · Plastic arts

P. F. González-Valenzuela (✉)
Universidad de Chile, Santiago, Chile
e-mail: pauligonzalez@uchile.cl

E. E. Castillo-Espinoza
Departamento de Diseño, Facultad de Arquitectura y Urbanismo, Universidad de Chile, Santiago, Chile
e-mail: eduardo.castillo@uchilefau.cl

M. Vicuña and E. Giorgi (eds.), *Intersections*, Springer Geography,
https://doi.org/10.1007/978-3-031-76402-8_4

4.1 Introduction

The recognition of university extension as one of the pillars in the formative mission of the University of Chile, was materialized during the rectorship of Juvenal Hernández (1933–1953), following the axes imposed in the Organic Statute of 1931.[1] This was carried out through the creation of multiple institutes and centers for research and cultural dissemination, which promoted, among other subjects, the study of Folk Art in the country.

Chile's adherence to the International Commission for Intellectual Cooperation, established in 1922 as part of the League of Nations, guided efforts in this area. Through a national committee, led by the academic Amanda Labarca,[2] the teaching of folklore in the Summer Schools of the University of Chile was promoted, and a series of exhibitions between 1936 and 1943 were sponsored. The last of these, held in the premises of the National Museum of Fine Arts under the title *American Exhibition of Popular Arts*, was organized with the occasion of the centenary celebrations of the house of studies, when the participating pieces, coming from Central and South American countries, were donated to the institution. This fact motivated the foundation the following year of the Museum of Popular American Art (MAPA) in the Hidalgo Castle of Cerro Santa Lucía, following international recommendations on the preservation of folklore and the material culture of the peoples, their artistic value and the underlying social benefits to the wage-earning classes (Lago, 1945).

For Tomás Lago, who was director of the museum between 1944 and 1968, the rescue of these objects linked popular life—and mainly that of the rural environment—with the surviving tradition in their crafts and manual trades, enabling a reconsideration of the role of art in society (Acuña, 2011), along with highlighting the importance of the popular subject as a producer of culture (Berríos et al., 2009). In this sense, Lago considered that such pieces functioned as "living documents" due to the economic and racial reality of the continent, so the MAPA focused its activity on the exhibition, research and dissemination of this practice (Quijada, 2018).

This anthropological character that marked the first decades of the museum reflects the romantic ideal of "the popular" that Ticio Escobar links to the principles of primitivism (preservation of traditions), communitarianism (collective and anonymous practice), and purism (uncontaminated by foreign habits) (Escobar, 2008). A nationalist and identity-based perspective that also had its correlate in the Latin American interpretation of Marxist thought by authors such as José Carlos Mariátegui[3] and Víctor Raúl Haya de la Torre,[4] who vindicate a common indigenous and colonial past that in Chile was replicated by left-wing intellectuals linked to Popular Art. This is the case of Oreste Plath, director of the MAPA between 1968 and 1973, who edited in Valparaíso the magazine *Gong* (1929–1931). Both the content and the form of this medium closely dialogued with *Amauta*, the periodic publication led by Mariátegui that had started its circulation three years earlier.[5]

Following the emergence of the Institute of Extension of Plastic Arts (IEAP) as the body in charge, among other tasks, of coordinating the management of university museums, the Museum of Contemporary Art (MAC) was created in 1947, located in the Parthenon of Quinta Normal. Unlike the MAPA, the specific task of this body directed by Marco Bontá (1949–1962) was to show the most representative of the work of living artists, intending to capture the complete panorama of current Chilean art, without discrimination of styles or major differences between Fine Arts and Applied Arts [6] (Arqueros, 2012).

Both museums were the first of their kind at the continental level, and they established their respective museographic itinerary by defining their specificity in the practices and agencies that originated their collection. However, the arrival of Nemesio Antúnez to the direction of the MAC (1962–1964), with the slogan of installing a "living museum" capable of bringing art closer to the population in a didactic way (Lihn, 1963; Novoa & Avaria, 2022), implied a change of orientation in the programmatic structure of the institution, carrying out exhibitions and activities that gave room to other artistic expressions, among which Popular Art was also found.

Both this ideology and the participation of the private world in the management of the MAC, particularly through the creation of a Society of Friends in charge of financing the continuous development of exhibitions, itineraries and activities, were determining aspects in the approach to these manifestations from a renewed sense. Thus, the definition of the term Popular Art, formerly linked to a conservationist view, was now built in virtue of the modernizing discourses of the current decade, which were promoted by an institutionality permeated by the Cold War and the dispute for cultural hegemony on the continent (Allende, 2021).

This link between Popular Art and Plastic Arts has a clear precedent in the Educational Reform of 1928 and the flank of the artistic education, led by the director of the School of Fine Arts, Carlos Isamitt, who advocated for the rescue of the popular-local, encouraged by the nationalist scenario of the Interwar period (Castillo et al., 2010). The proposal, which perceived in these objects a suggestive character intended not only for contemplation, but also for their assimilation, would be subsequently taken up by other artists, who diversified this practice.

In this sense, the role of Antúnez is particularly noteworthy. Firstly, in his role as a creator, he included representations of Quinchamalí ceramics in his lithographs, murals, and mosaics. In addition, he shared his taste for these expressions by having a craft shop with his first wife, Inés Figueroa, in the premises of his house on Guardia Vieja Street, Providencia, where Taller 99 also operated.[7]

On the other hand, as director of the MAC, he subscribed to cooperation initiatives emanating from Pan-Americanism and public–private alliances, which promoted the development of activities associated with Popular Art, such as the realization of an Easter Island festival of music, dance, and crafts (1963); the management of the Chilean shipment to the exhibition *Popular Art of America and the Philippines*, which was to be held at the Hispanic Culture Institute in Madrid; and the canceled exhibitions *Peru, gold, ceramics and pre-Columbian textiles from the Mujica Gallo collection*, promoted by the Art Center of Lima, and *Arauco*, defined as a synthesis of Mapuche art.[8] Finally, on September 10, 1964, the exhibition *Chilean Popular Art* (Fig. 4.1), which marked a turning point in the relationship between both university museums, was inaugurated.

Fig. 4.1 Front and back cover of the exhibition catalog Chilean Folk Art, held from September 10 to 30, 1964. Collection National Library of Chile

4.2 Chilean Popular Art, a Controversial Exhibition

The exhibition *Chilean Popular Art* was managed by the Society of Friends of the MAC, in collaboration with public and private entities such as the Araucano Museum of Temuco, the Archaeological Museum of La Serena, the Penitentiary, and the MAPA. In addition, the Amancai store and collectors Alicia Oportot, Lina Levine, Maruja Burchard, Jorge Landea, and archaeologist Ruperto Vargas, who facilitated part of their collection, participated.

In order to highlight the aesthetic sense of the pieces and promote their production, the Society of Friends established as a priority to generate a more permanent coexistence between the "cultured art" and the "art of the people".[9] (MAC, 1964), replicating the strategy of the Plastic Arts Fairs of Parque Forestal (Fig. 4.2) in which the exchange, dissemination, and sale of these works in the public space gave artisans and popular artists a new social value (Cáceres & Reyes, 2008; Soto, 2021). This unprecedented instance, promoted initially by lawyer Germán Gasman and channeled in 1959 by the Institute of Modern Art,[10] included the participation of academy artists, artisans, and students of the School of Fine Arts, which marked a milestone for the disciplinary opening by questioning the hegemonic view of Fine Arts regarding Popular Art (Cancino, 2019).

This recognition of subaltern practices by an erudite culture that had previously downplayed them as activities of creation, implied that the reading of these objects was carried out under the codes of the artistic medium, where the use of the word "author" to define an anonymous practice, or "unique piece" to allude to a collective tradition, went against the approach defended by specialized institutions that sought to respect the origin of these cultural manifestations, and even distance themselves from the dialogue with Western art in aesthetic and productive terms (Acuña & Arqueros, 2015).

Fig. 4.2 Cover of the catalog of the 1st Plastic Arts Fair, held in Forestal Park in 1959

In this regard, one of the main detractors of the exhibition was Tomás Lago, who, in his capacity as director of MAPA, sent two letters to Jorge Elliott, director of IEAP, accounting for his disagreements.

The first of these was written in Paris on June 23, 1964, after receiving a report on the situation from the painter and researcher Ida González. In the letter, Lago expressed his reservations about the incorporation of the term "Popular Art" in the title of the show, since the curatorial proposal that had been communicated to him proposed a combination of elements that was contrary to studies on the subject, which could cause confusion in the viewer. Among these aspects, was the participation of the Jesuit priest Gustavo Le Paige, a pioneer in archaeological research in northern Chile, as well as the presentation of original ceramic works by the artist Ricardo Irarrázaval.[11]

To support his arguments, Lago used as an example an exhibition he had recently visited at the *Musée National des Arts et Traditions Populaires* in Paris, whose level of quality, the result of four years of research, made it impossible to imagine among its pieces any "hybrid product of the *faubourg* or tourist industries".[12] The achievement of a result of these characteristics in Chile then depended on the level of relevance that the IEAP would give to the university work of guidance and study in the field of Popular Art, a specialized work that the MAPA had been leading for the last two decades.

Although the museum's trajectory was not questioned, and the MAC had requested its participation in this instance, Lago considered appropriate to establish some conditions beforehand: to modify the title of the exhibition for one that accurately reflected what was there exposed; to give the MAPA collection the most prominent room of the show; and to clarify that it was not about commercial business or unfair policies between both institutions. The truth is that none of these approaches had a response, so the reasons that prompted the museum to collaborate regardless are unknown.

On September 14, 1964, just four days after the inauguration, Lago sent a second letter, where he was emphatic in denouncing the lack of relevance of the MAC in the treatment of Popular Art. One of the reasons was the use of strange concepts that distorted and made hybrid the genuine expressions of traditional groups, ignoring the recommendations disseminated by the own house of studies through its extension bodies.[13] This is especially relevant given that at that time a large part of the publications on the subject originated from the MAPA. In addition, it was in this institution where to date the only course dedicated to the subject, titled "Theory of American Popular Art", was taught. This course was approved in 1951 as an elective subject of the program of the Faculty of Sciences and Plastic Arts,[14] and later, in 1961, it had become a mandatory requirement for fourth-year students of the degree in Art Education.

Such conceptual divergence had already been widely discussed during the Round Table on Chilean Folk Art in 1959, which was organized by MAPA in collaboration with UNESCO.[15] Precisely one of the focuses of this meeting was centered on the elaboration of national definitions on this term, making explicit its differences with crafts, applied art, primitive art, and popular industries[16] (Alarcón et al., 1959). Although the fundamental features of these objects were emphasized as the "anonymity" of their productions, the "traditional" nature of their craft, and the "popular" nature of their class condition, on this occasion the need was raised to make known the current state of these expressions and identify the variables that had caused their deformation, such as industrialization, tourism, social mobility, and cultural extension (Plath, 1959).

In addition to challenging the content of the exhibition, Lago held IEAP responsible for the administrative failures caused by its lack of action, as they had praised the MAC project to the detriment of the recognition of the University of Chile as a pioneering agent in the dissemination of these practices. The media impact of the exhibition *Chilean Folk Art*, and its incursion into the institutionality of cultured art, positioned it as an unprecedented event in the recent history of the country, even being recognized by the critics as "the most extensive set of national popular crafts ever organized".[17]

4.3 Curatorial Aspects: Selection and Assembly

The curatorial narrative of the exhibition, prepared on the occasion of the national holidays in Chile, aimed to highlight the diversity of artisan traditions developed throughout the territory. The selection of the pieces, which ranged from Calama to Chiloé and included Rapa Nui, was carried out in collaboration with collectors and regional institutions that identified the techniques, formats, and representatives of each locality. Notable were the glazed ceramic works from Concepción, the horsehair textiles from Panimávida, the pottery from Quinchamalí and Llolleo, the textiles from La Ligua, the clay pieces from Pomaire, the chamantos from Doñihue, the silverware from Temuco, the leather carvings from Linares, among others, which dialogued with archaeological pieces from the northern part of the country, and whose characteristics and institutional origin are unknown.

In the province of Coquimbo, it was the Archaeological Museum of La Serena who performed this task. Through its intermediary, four popular artists from Vicuña, Tongoy, and the sector of Compañía Alta, in the provincial capital, were invited. Marta Cáceres and María F. Pinto represented Vicuña, the former with a basket of flowers and a small basket with vegetables made through painted and modeled sugar alcorzas, while the latter presented a bouquet of flowers and a pair of child's slippers crocheted. C. Estela Marín, a resident of the Tongoy resort, sent paper flowers, and Clara Contreras, from La Serena, presented figurative pieces in ceramics.[18]

In Santiago, part of this selection was made through the Penitentiary, which facilitated objects of prison art. This term was coined by Oreste Plath during the presentation of the Catalog of the *II Exhibition of Art and Prison Crafts Works*, which took place at the Main House of the University of Chile in 1951, and which allowed the inscription of this practice in the category of Popular Art.[19] It is important to note that Antúnez's management was already familiar with these objects, due to their participation as a qualifying jury in the Prison Craft Shows that were developed on the occasion of the IV Penitentiary Week.[20]

The few photographs of the exhibition (Fig. 4.3) allow us to observe how the assembly of some of these objects sought a direct interaction with the public, in contrast to the traditional museographic vision that usually adopted a more conservative approach. This could explain the letter sent from the MAC to Yolanda Monteyano, a decoration professor at the University of Chile, requesting her collaboration with the event,[21] probably with the intention of addressing these objects from a more aesthetic perspective and less focused on their functionality.

Fig. 4.3 Photograph of the inauguration of the exhibition Chilean Folk Art, where director Nemesio Antúnez is seen with members of the Friends of the MAC Society. El Mercurio, Santiago, September 11, 1964. Nemesio Antúnez Foundation Archive

The above, along with the invitation to renowned artisans such as Guillermo Prado, an aeromodelling professor from the Chilean Air Force who innovated in the design of kites, and Luis Manzano, "Manzanito", a wicker weaver who integrated the creation of ornamental figures such as fish and birds into a utilitarian tradition, reflects a curation that relates without hesitation different aspects of popular practices, resting on the permeability of their boundaries in the art space.

Unlike the MAC where financial resources and the scale of the building allowed the development of an exhibition in association with external agents, the MAPA lacked any possibility of exhibiting in its facilities anything different from the small rotating exhibitions of its collection. The lack of adequate infrastructure to exhibit its collection, led to the design of functional showcases (Fig. 4.4). As a rule, these showed a single collection of objects distributed in the three rooms that made up its headquarters, or presented a single artifact for several countries simultaneously, highlighting their similarities and differences. After two months, it was replaced by another (Lago, 1964). This caused the main exhibitions carried out by the museum during the sixties, such as the *Instinctive Painting Exhibition* (1963) and the exhibition *Popular Art of Brazil* (1964), to be held in places other than Castillo Hidalgo, like the University Hall or the Brazilian Culture Center.

This institutional dispersion, along with the communication strategies implemented by Antúnez, facilitated the positioning of the MAC as an active agent in the promotion of these expressions, promoting the dissemination not only of the works but also of the artisans. In turn, the displacement of these objects into the artistic field opened the way to new perspectives of cultural exchange and creative emphasis, which could be considered as a prelude to the changes that will take place

Fig. 4.4 One of the rooms of the Museum of American Folk Art at its headquarters in Castillo Hidalgo, ca. 1970. Courtesy of José Moreno

later with the University Reform (Castillo et al., 2010; Soto, 2021) and then with the government of the Popular Unity, led by President Salvador Allende.

4.4 Conclusions

The new perspectives on Popular Art, during the 1960s, motivated the agents of the MAC to organize an exhibition from a contrasting view to the one defended by the MAPA, transferring to the institution a cultural exchange that was being established directly in the public space with the Plastic Arts Fairs since 1959 (Cáceres & Reyes, 2008). This apparent overlap of functions, which caused a discrepancy within the IEAP, belonging to the University of Chile, was based on the personal will of Antúnez to consolidate his museum project, and to give meaning to a view of the concept of Popular Art from the artistic world. The exhibition alluded to the identity and aesthetic perspective of these expressions, but highlighting the role of the artisans and how their productions can become visual and creative references for new artists.

On the other hand, Lago's latent concern for the disappearance of this practice under modernizing currents, which promoted the abandonment of old techniques and encouraged exchange with disciplines foreign to the artisanal world, was a constant theme in his research and preservation work of these objects, even after his departure from MAPA in 1968. In this sense, the MAC exhibition, its curatorial innovations, and the diversity of participating agents, both public and private, contributed to the extension of a discourse in line with the social, economic, and cultural impulses of the current era, whose hegemony and media power even managed to make the educational work of a university museum with an extensive track record on the subject invisible.

This conflict around the categorization of the concept was also linked to the guardianship of this traditional imaginary and its consequent link to certain ideological orientations: one in agreement with the spirit of the Latin American left, and another in line with the changes proposed by the Alliance for Progress, which sought the depoliticization of these cultural practices. Not in vain, the ideas proposed by Isamitt in 1928 to incorporate the study of folklore into artistic education and to promote the recognition of its different manifestations as a source of inspiration for the creative field, reappeared decades later as a result of the global interest in promoting the "heritage of the people", in whose authenticity the characteristics of the historical, cultural, and physical environment of the human being were expressed (Isamitt, 1962).

At the end of the same decade, an economic perspective was developed that saw in these practices an opportunity for resource generation and the promotion of local production, whose promotion in Chile, led by state entities such as the Corporation for the Promotion of Production (CORFO) and the Technical Cooperation Service (SERCOTEC), was resisted by academia and the artistic field (Castillo et al., 2010). Despite these different approaches, the development of both alternatives was established over the vision that guarded the purity of these expressions, thus enhancing transdisciplinary crossings and with it hybridization as a place of reading, production, and consumption of cultural practices (Zubieta, 2000) in the transition to the 1970s.

After the period of the Popular Unity, this reached another tone during the civic-military dictatorship, as most of the approaches that emerged in the sixties around Popular Art were obliterated in order to impose an official, commercial, and excessively conservative vision from 1974 onwards (Moreno Aliste, 1984; Silva & Moreno Aliste, 1986).

In this way, this article is inscribed as one more precedent of a discussion that reaches its peak at the end of the decade, and which at this point in history expresses its first arguments in the curatorial policies of the museums analyzed.

End Notes

1. The Organic Statute of 1931 sought to transform the University of Chile into a nucleus of knowledge and culture at a national level. For this, extension was a fundamental factor as through this it was possible to articulate an educational model capable of making the population participate both in professional training, as of research (Berríos et al., 2009).
2. Amanda Labarca was the first female academic at the University of Chile. She was appointed extraordinary professor of Psychology at the Pedagogical Institute in 1922, later becoming part of the Faculty of Philosophy and Humanities. Among other functions, she was in charge of organizing the Summer Schools of the study house, and was director of the Department of Cultural Extension, created during the rectorship of Juvenal Hernández. On behalf of the University, she chaired the Executive Committee of the Chilean Commission for Intellectual Cooperation. For more information, see: Stuven (2019).
3. José Carlos Mariátegui (1894–1930) was a Peruvian writer and intellectual, recognized for being the founder of indigenous socialism in Latin America. His work was embodied in numerous books, articles, and essays, as well as his own publishing house Amauta. The book, Seven Interpretive Essays on Peruvian Reality, published in 1928, is his most influential work at the regional level.
4. Víctor Raúl Haya de la Torre (1895–1979) was a Peruvian politician and thinker, founder of the political party American Popular Revolutionary Alliance (APRA) in 1930. This organization, of great influence in Latin America, advocated for a nationalist and anti-imperialist approach, reinforcing the idea of the "Indo-American" as the identity that united the oppressed in the region in the postcolonial scenario.
5. Contributions from the classes of the artist and engraver Hugo Rivera-Scott (1943).
6. Indeed, Bontá had previously served as the head teacher of the Graphic Arts Workshop at the School of Applied Arts of the University of Chile between 1931 and 1937, before leaving on a cultural mission to Venezuela to promote similar education between 1938 and 1943.

7. Engraving workshop created by Antúnez in 1956, after participating in Stanley William Hayter's Atelier 17 in New York.
8. It is likely that the main impediment to the realization of this exhibition at the MAC was the exhibition Araucanian Silverware, that MAPA presented in the University Room that same year.
9. A precedent for this objective can be found in the first American Biennial of Engraving, which took place at the end of 1963 in the MAC. The revaluation of this graphic production, as an art of mass circulation, also sought to break with the cleavage between high arts and popular arts (Allende, 2021).
10. The relationship between both events is not coincidental, as many of the members of the board of this institute, also known as the Museum of Modern Art, later became part of the Society of Friends. Among these, the architect Sergio Larraín García-Moreno, the dean of the Faculty of Fine Arts Luis Oyarzún, and Antúnez himself.
11. Despite this mention, no documentation has been found that allows linking them to the exhibition.
12. Letter from Tomás Lago, director of the Museum of Popular Art American, addressed to Jorge Elliott, director of the Institute of Extension of Plastic Arts. In: Subfund Correspondence, COR 1964.10, folio FH. a2315. Historical Archive Museum of Contemporary Art, Faculty of Arts, University of Chile.
13. Letter from Tomás Lago, director of the Museum of Popular American Art, addressed to Jorge Elliott, director of the Institute of Extension of Plastic Arts. In: Subfund Correspondence, COR 1964.14, folio FH. a2479. Historical Archive Museum of Contemporary Art, Faculty of Arts, University of Chile.
14. Despite its approval in 1951, the chair proposed by Lago was only taught in 1953, after the First Week of American Folklore was celebrated in Santiago, organized by the Department of Cultural Extension of the University of Chile (Quijada, 2018).
15. The program of the Round Table revolved around four topics. The first of them, presented by the Society of Friends of Popular Art, elaborated national definitions on concepts related to Popular Art. The second, led by Oreste Plath, addressed the conditions of Popular Art in Chile to date. The third, presented by María Bichon, accounted for the preservation and teaching of crafts and Popular Art. Finally, the fourth topic, presented by Roberto Montandón, alluded to the economic and cultural improvement in the communities dedicated to these practices and the maintenance of traditional forms. See: Round Table on Chilean Popular Art (1959). Chilean Popular Art. Definitions, problems, current reality. University of Chile, Santiago.
16. These distinctions included aspects such as the organization of work in collective workshops (craftsmanship), formal education and authorship of the works (applied arts), indigenous origin (primitive art), and the casual production of low-quality material objects (popular industries).
17. Bindis, R. (1964, September 27). "Exhibition of Popular Art". La Nación, p. 4.

18. "Four popular artists from the province participate in Folk Art Exhibition". (September 24, 1964). El Día. In: Press clippings collection, COL 1964. Historical Archive Museum of Contemporary Art, Faculty of Arts, University of Chile.
19. Plath's study of these objects was also presented to the academic and institutional world during the Round Table of Chilean Popular Art in 1959, and thanks to this the MAPA acquired for its collection a set of works created in prison (Bustos, 2019).
20. Rafael Silva Lastra, Director General of the Prison Service, thanks Nemesio Antúnez for his cooperation during the events of the IV Penitentiary Week. In: Subfund correspondence, COR 1962.5. Historical Archive Museum of Contemporary Art, Faculty of Arts, University of Chile.
21. Nemesio Antúnez requests Yolanda Monteyano's help for the assembly of the exhibition Chilean Popular Art. In: Subfund correspondence, COR 1964.13. Historical Archive Museum of Contemporary Art, Faculty of Arts, University of Chile.

Acknowledgements This project has been funded by the National Fund for Cultural and Arts Development, 2022 call, folio N° 628246, "Popular Art in Chile. A state of the matter".

References

Acuña, C. (2011). Origin and evolution of the Tomás Lago American Popular Art Museum. In *A way of looking: Treasures from the MAPA collection.* American Popular Art Museum, University of Chile.

Acuña, C., & Arqueros, G. (2015). *Tomás Lago.* Eight Books, Santiago.

Alarcón, N., Domínguez, M. J., & González, I. (1959). Popular Art and crafts. Manual arts in general. Applied art and primitive art. National definitions of these concepts. In *Chilean Popular Art. Definitions, problems, current reality* (pp. 25–33). University of Chile.

Allende, M. (2021). The Museum of contemporary art and the cold war: Nemesio Antúnez director. *Aisthesis, 71*, 137–152. https://doi.org/10.7764/Aisth.71.9

Arqueros, G. (2012). The museum of contemporary art of the University of Chile. Time, history and contemporaneity. In *MAC. 2005. Refoundation.* Museum of Contemporary Art, University of Chile.

Berríos, P., Cancino, E., & Santibáñez, K. (2009). The construction of the contemporary. The modern institution of art in Chile (1910–1947). Art Studies.

Bustos, V. (2019). Prison art: Another expression of popular art. Valuation of the MAAL collection. Under the Magnifying Glass, Subdirectorate of Research, National Service of Cultural Heritage. https://www.museodelinares.gob.cl/publicaciones/arte-carcelario-una-expresion-mas-del-arte-popular-puesta-en-valor-de-la-coleccion

Cáceres, A., & Reyes, J. (2008). History made with hands. In *We the Craftsmen and the Craft Fairs of the 20th Century.* CNCA.

Cancino, E. (2019). Nemesio, manager: Towards a living museum. *Cultural Observatory, 37*, 19–31. https://www.cultura.gob.cl/wp-content/uploads/2019/04/observatorio-cultural-nemesio-antunez.pdf

Castillo, E., Muñoz, M., Álvarez, P., Rivera-Scott, H., & Vico, M. (2010). Craftsmen, artists, artisans. In *The school of applied arts of the university of Chile (1928–1968)*. Ocho Libros Publishers.
Escobar, T. (2008). *The myth of art and the myth of the people*. Heavy Metals Editions.
Isamitt, C. (1962). Folklore as an element in teaching. *Chilean Musical Magazine, XVI, 27*, 75–94.
Lago, T. (1945). Mission of the museum of popular art. *Antarctica, 9*, 89–90.
Lago, T. (1964). The museum of American Folk Art, Santiago de Chile. *Museum, XVII, 2*, 105–112.
Lihn, E. (1963). MoyΣeion. Art Magazine of the Museum of Contemporary Art, 18: [s. n.].
MAC. (1964). *Catalogue of popular art exhibition*. University of Chile.
Moreno Aliste, A. (1984). *Urban marginal crafts*. CENECA.
Novoa, S., & Avaria, P. (2022). Nemesio Antúnez: Ideals for a living museum. In *MAC Memory*. 1962–1964 (pp. 30–85). Museum of Contemporary Art, University of Chile.
Plath, O. (1959). Current conditions of folk art in Chile. In *Chilean Folk Art. Definitions, problems, current reality* (pp. 35–45). University of Chile.
Quijada, F. (2018). An art for everyone. Applied and popular arts in Chilean artistic education. In *Yearbook of the MAC archive research group* (pp. 68–91). Museum of Contemporary Art, University of Chile.
Silva, J. A., & Moreno Aliste, A. (1986). *Craftsmanship in Chile (exploratory diagnosis)*. CENECA.
Soto, G. (2021). The convergence between applied arts, popular arts and fine arts. Master's Thesis, University of Chile.
Stuven, A. M. (2019). *Amanda Labarca*. Hueders Publishers.
Zubieta, A. M. (2000). Popular culture and mass culture. In *Concepts, journeys and controversies*. Paidós, Buenos Aires.

Chapter 5
Antarctic Architecture a Contemporary Heritage: Analysis of the Architecture and Life at the Eduardo Frei Montalva Air Base, 1969–2022

Claudia Torres, Laura Gallardo, Carlos Lange, and Sandro Maino

Abstract Antarctica is a young continent in its exploration and development. The current climate emergency accounts for a strategic future for the countries involved in the Antarctic territory, considering the responsibilities in its environmental care. In this context, Chilean Antarctic settlements respond to the particularities of climatic conditions, connectivity, and an isolated territory from the rest of the country, with limitations in displacements and restrictions of daily functions. After 75 years of Chileans' arrival on the Antarctic continent (1947), a 'Chilean-Antarctic culture' is outlined, emerged through colonization with military camps (Bases), expanding to territorial domain and sovereignty facilities with civilians and families, reaching a support system for research, with projections to a controlled Antarctic tourism. Through documentation and interviews with inhabitants, the architecture and life experiences at the Antarctic Air Base President Eduardo Frei Montalva (1969) are historically analyzed, considering the existing set and the family homes of Villa las Estrellas (1984–2018). With the aim of understanding the relevance of everyday life in the production of inhabited space, we rely on anthropology for sociocultural analysis. With an interdisciplinary approach, the analysis was addressed on three scales: geopolitical context and its relationship with the continent; territorial planning and interculturality and codependency relations with other Bases; and the

C. Torres (✉) · L. Gallardo
Departamento de Arquitectura, Facultad de Arquitectura y Urbanismo, Universidad de Chile, Santiago, Chile
e-mail: claudiatorres@uchilefau.cl

L. Gallardo
e-mail: lauragallardofrias@uchilefau.cl

C. Lange
Facultad de Arquitectura y Urbanismo, Instituto de La Vivienda, Universidad de Chile, Santiago, Chile
e-mail: clange@uchilefau.cl

S. Maino
Departamento de Arquitectura, Universidad Técnica Federico Santa María, Valparaíso, Chile
e-mail: sandro.maino@usm.cl

M. Vicuña and E. Giorgi (eds.), *Intersections*, Springer Geography,
https://doi.org/10.1007/978-3-031-76402-8_5

architectural-social space and its forms of habitability in everyday, collective, and intimate places.

Keywords Antarctic culture · Habitability · Interculturality · Ways of living · Place

5.1 Introduction

The settlement of the Antarctic Air Base Eduardo Frei Montalva (EFM) has been located since 1969 in Fildes Bay, King George Island of the South Shetland Archipelago. Emerging as a Meteorological and Telecommunications Center in charge of the FACH, it has become the largest Chilean settlement on the Antarctic continent and the one with the greatest socio-functional complexity.

The set of the EFM Air Base is mainly composed of facilities of the Chilean Air Force (FACH), the General Directorate of Civil Aeronautics (DGAC), the Professor Julio Escudero Base of the Chilean Antarctic Institute (INACH), the Navy Port Captaincy and the Lieutenant Rodolfo Marsh Aerodrome. Along with this institutional equipment is *Villa las Estrellas*, the housing complex where Chilean families have resided since 1984, giving the Base the character of a 'village'.

Currently, the Base is at a turning point regarding its future, with significant changes occurring in its organization. New buildings have been projected for the *Escudero Base* and for the Air Force equipment. Also, in 2018 the resident families were removed from *Villa las Estrellas* due to: deterioration problems of the houses, the need to comply with new environmental standards, and the high maintenance costs for the Air Force.

The focus of this research is based on a multidimensional and multiscale study centered on architecture as a link between the environment and the forms of habitability that condition the development of people's life and customs. For this, the conditions of geopolitical context and its relationship are analyzed logistics with the continent; the territorial organization of the EFM Base, with immediate references; the architectural space and its habitability, materialized in workspaces, community, service, and residential areas.

On the other hand, it should be mentioned that there are currently sites heritage protected in the Antarctic territory, either by 'National Monuments Law' (No. 17,288), such as the Bernardo O'Higgins Base or the Prat Base; or as 'Historic Monument Sites' (HMS) according to the Consultative Meeting of the Antarctic Treaty (1972) in whose list there are works of architecture, monoliths, and significant landmarks, such as the Shackleton's Hut. However, the *Eduardo Frei Base* does not have heritage protection of any kind, despite being, along with the *Esperanza Base* of Argentina, the only ones that have housed families in the Antarctica.

This research proposes that, the existing architecture at the EFM Air Base, although precarious and improvised in some aspects, is an expression of a 'Chilean-Antarctic culture', presenting ways of living that have allowed adaptation to the extreme conditions of isolation and confinement.

The analysis of this set of variables is considered a necessary and fundamental contribution, as it is a field of study that has been little addressed in architectural and anthropological research, constructing a historical-analytical narrative of the ways of living in said territory, to understand its scope and heritage significance.

5.2 Methodological Approach

The 'Chilean-Antarctic culture' refers to a set of experiences, ways of life and forms of living that allow the adaptation of people in an environment of extreme climatic conditions, through the appropriation of predefined architectural spaces for this purpose.

Articulating the hermeneutic approach of Martin Heidegger (1975), the Marxist-humanist of Henri Lefebvre (1973) and the theoretical-methodological of Angela Giglia (2012), the experience of living is understood from three axes:

- The response to the socio-functional requirements of subsistence and habitability, associated with the generation of a place.
- The production and materialization of the architectural space.
- Everyday social practices.

Thus, the architectural work is capable of materializing the spaces, allowing its inhabitants to grant them the quality of a place to live (Azara, 2005; Gallardo & Toledo, 2020; Morales, 1999). On the other hand, Norberg-Schulz (2009) mentions that living is not only having a shelter, but also a sense of belonging and significance, as evidenced by the experiences shared by inhabitants of different generations at the EFM Base. In this, identity aspects are outlined that account for a local cultural heritage by delving into life experiences related to rituals, the production and valuation of architectural space.

The multi-scalar and relational approach of the study focuses on the architectural production as a significant and linking element between the geographical and environmental context; individual and collective experiences; in addition to the ways of living that condition the development of life.

Under these perspectives, the research comprises three phases:

- The review of official documents to contextualize the historical process of production and settlement of the EFM Base, especially archives of the Antarctic Division of the FACH, INACH, bibliography and photographs provided by the interviewees.

- The analysis of semi-structured interviews with 21 inhabitants: 'permanent' and 'transitory' residents, among them, officials from the FACH, the Navy, women (wife of the crew), professor, doctor, banker, students and researchers.
- Architectural analysis (based on the morphological and typological characteristics of the built work) linked to the anthropological study, in accordance with the categories of experiences that emerged in the interview narratives.

This intersection of relationships implies that in these circumstances, spaces are constructed that are suitable for different ways of living and the development of cultural expressions (Giglia, 2012). Thus, considering Norberg-Shulz's approach (2009), the analysis is organized through three orders: collective places, everyday places, and intimate places.

5.3 Historical Background

Before Chile signed the Antarctic Treaty (1959), as early as 1947, it had inaugurated on Greenwich Island, 'Sovereignty', the first permanent Chilean military base (Orrego, 1948), concretizing, even with its name, the political will of national dominion over the Antarctic territory. This Base is currently known as Naval Base *Arturo Prat* (INACH, 1965).

However, the main Chilean settlement in the Antarctic Territory in terms of dimensions and complexity has been the EFM Air Base since 1969. This Base emerged as an Antarctic Meteorological and Telecommunications Center when its location was reconsidered, after the *Pedro Aguirre Cerda Base*, located on Deception Island, was evacuated and abandoned following the volcanic eruption of 1967. After that, the move to Fildes Bay on King George Island, next to the already existing Russian Bellinghausen Base, was chosen for its geomorphological, topographical, shelter and availability of fresh water for supply (Iturriaga, 1999).

On March 7, 1969, with the presence of President *Eduardo Frei Montalva*, the President Frei Meteorological Center was inaugurated (Romero, 1985). From there, and in response to the need to develop a permanent territorial connectivity inherent to the institutional vision and national policy, various strategies were implemented and capabilities developed that culminated with the construction in 1980, by the FACH, of a landing strip that gave rise to the Lieutenant Rodolfo Marsh aerodrome. Since that year, permanent flights have been coordinated, both for connection with the continent and within Antarctica, providing support equipment and personnel for its operation and maintenance.

Although Article I of the Antarctic Treaty (1959, 37) prohibits the establishment of military bases and fortifications, according to the role assumed by Chilean bases as Antarctic operators that carry out logistics for the transfer, movements and permanence of scientific and exploration teams, it complies with the second title of the same article of the Treaty which indicates: "This Treaty does not prevent the use of military personnel or equipment for scientific research or for any other peaceful purpose".

With the Aerodrome operational, Chile dynamized accessibility to King George Island in a logistics complementary to the arrival of Navy ships in summer periods, which, to this day, allow the transfer of heavy cargo. This action strengthened Chilean sovereignty in the Antarctic territory, as with better connectivity, a geopolitical strategy of territorial dominance was initiated by moving families to live at the EFM Base and creating a more complex settlement.

The incorporation of families and homes equipped for long-term permanence gave rise in 1984 to the southernmost town of Chile, *Villa las Estrellas.* Initially inaugurated with six families, it expanded in the following years to a housing complex of 13 homes to accommodate relatives of FACH officers, DGAC personnel, doctors and teachers (Romero, 1985). "Parallel to this, social infrastructure was implemented, with a Nursing, Basic School, Chapel, Post Office, Gymnasium and others, to provide servitude to families […] and INACH Scientists" (FACH, 2019, p. 4).

Currently, and as reported in various official documents and from the FACH (Chile, 2017), the EFM Air Base presents serious technical conservation problems. Therefore, in December 2018, after 34 years of uninterrupted use, the families who lived permanently in *Villa las Estrellas* were removed and did not return.

The problems mentioned have modified the country's geopolitical strategies regarding the Antarctic Territory. On the one hand, it has been prioritized to direct existing resources to scientific work with the logistical operation that supports it, and on the other, the potential and impact of tourist development is observed, defining a 'National Antarctic Tourism Policy' (Chile, 2020).

The background and stories of the experiences of Chileans who have lived in Antarctica, and especially those who lived with families, allows us to construct a sociocultural and architectural history of ways of life in this distant territory.

5.4 Multi-scale Analysis

The integration of context, architectural and sociocultural variables is achieved through the approach of three scales of analysis, which arise from historical background, documentary, interviews and stories. These scales are: geopolitical, territorial and architectural (Fig. 5.1).

5.4.1 Geopolitical Scale

The EFM Base located on the peninsula, outside the Polar Circle, is considered the "gateway" to Antarctica, allowing permanence for those who carry out exploration and research tasks within the continent and at the same time is the logistical center of Chile in the area.

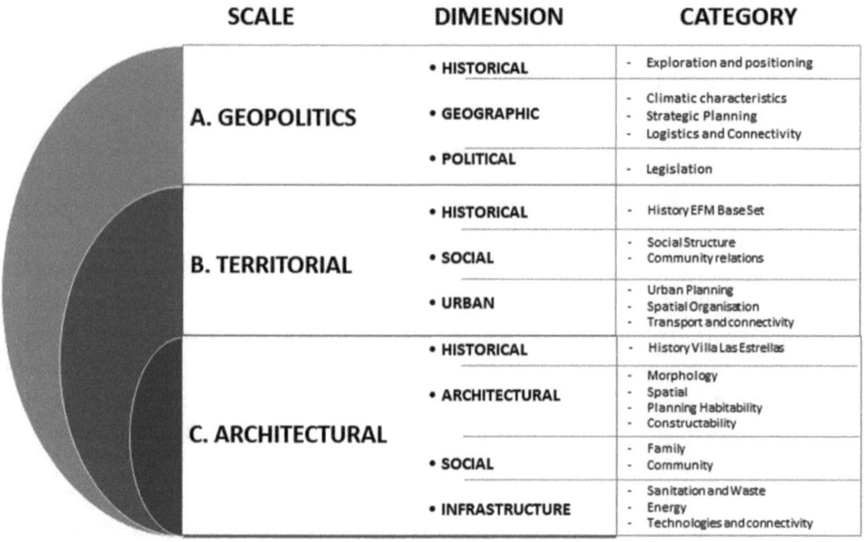

Fig. 5.1 Scheme of scales, dimensions and categories of analysis

5.4.1.1 Insularity, Isolation and Social Organization

The experience of 'insularity' refers to the geopolitical context of the Antarctic territory, which is characterized by a geographical isolation from the rest of the country and continents, but mainly it is given by the extreme climatic conditions of an inhospitable territory, basic access to equipment and services, limitations in movements and connectivities that restrict the development of daily activities, among others.

The perception of 'isolation' has been changing according to the implementation of new communication technologies, therefore, from television, landline, internet, cell phones, to video calls, have facilitated long stays with less longing and without weakening the bond with relatives.

Historically for Chile, the implementation of Antarctic Bases (in a strategy of dispersion rather than concentration of resources) has been a strategic challenge as it is associated with the establishment and reaffirmation of territorial, maritime and air sovereignty in the area. Despite this, in the Antarctic settlements there is no freedom or autonomy of permanence and there are no stable inhabitants, as subsistence depends logistically on the 'Continent' and the 'Antarctic operators', in this case, the Navy or the Air Force. Therefore, the settlements with the most development are those that are located in bays with easy access to boats and with aerodromes, such as the EFM Base. In accordance with this, the civic-social order of the town does not respond to the political-administrative logic of other insular territories of the country, but has a social order determined by a hierarchical system typical of the Armed Forces. The Air Force Commander is responsible for everything that happens at the Base, as they, in addition to being logistical operators, are the administrators of basic resources and services, such as water, energy and telecommunications. On the other hand, the people who live there depend economically on each institution to which they belong.

Particularly, in the Chilean case, to live in the EFM Air Base or in *Villa las Estrellas*, one had to be part of the crew or family of the FACH officers, DGAC, doctors or teachers. In this way, residents are classified into: 'permanent' inhabitants of long stays with biannual relief cycles, who fulfill roles and fundamental functions for the operation of the Base; 'transitory' inhabitants of short stays (generally summer months) who have roles and functions limited in time, whose character is associated with the fulfillment of specific objectives, generally they are researchers, maintenance personnel; political visits and tourists of daily and sporadic permanence.

The residents with the longest stay are selected and prepared to resist, on the one hand, the psychological pressure of isolation and group relations. At the same time, they must adapt to the modifications of the day-night light cycles and adjust to the rationing of services and supply. They are also trained for subsistence in the face of low temperatures, snow and blizzards, in addition to the actions to face fires and rescue in emergencies. Emergencies have been of different scope, the main ones in recent years being the fires that affected the gym and the Sea Captaincy.

In recent years, the Base has received an increase in tourists under the modality of aero-cruises carried out by private operators. They use as a boarding and disembarking platform the Lieutenant Marsh aerodrome and Fildes Bay, a matter of concern for the State expressed through the Antarctic Tourism Policy.

On tourism, the interviewees agreed that, so far, this does not take place in Antarctica itself, and could be considered minimally invasive. However, despite acceptance, they agree that tourism must be regulated due to the environmental impact that would imply a massive arrival of people without specific preparation.

5.4.2 Territorial and Urban Scale

5.4.2.1 Collaboration and Codependency

According to the Antarctic Treaty (1959), there are no national territorial limits in the area, sharing the resources and potentials of each settlement. The scope of the territorial relations of the inhabitants at Base EFM is associated with logistical, social and official links with the other Bases located in Fildes Bay (Russia, Uruguay, China), which can be accessed by land (Fig. 5.2).

The dependence of these bases on the Chilean aerodrome Lieutenant Marsh, which allows for supply and relatively continuous access throughout the year for everyone, is considered a shared resource among nations.

Among these Bases, exchange relationships, collaboration, mutual aid and cultural coexistence are established, given by the visits, official activities, celebrations of

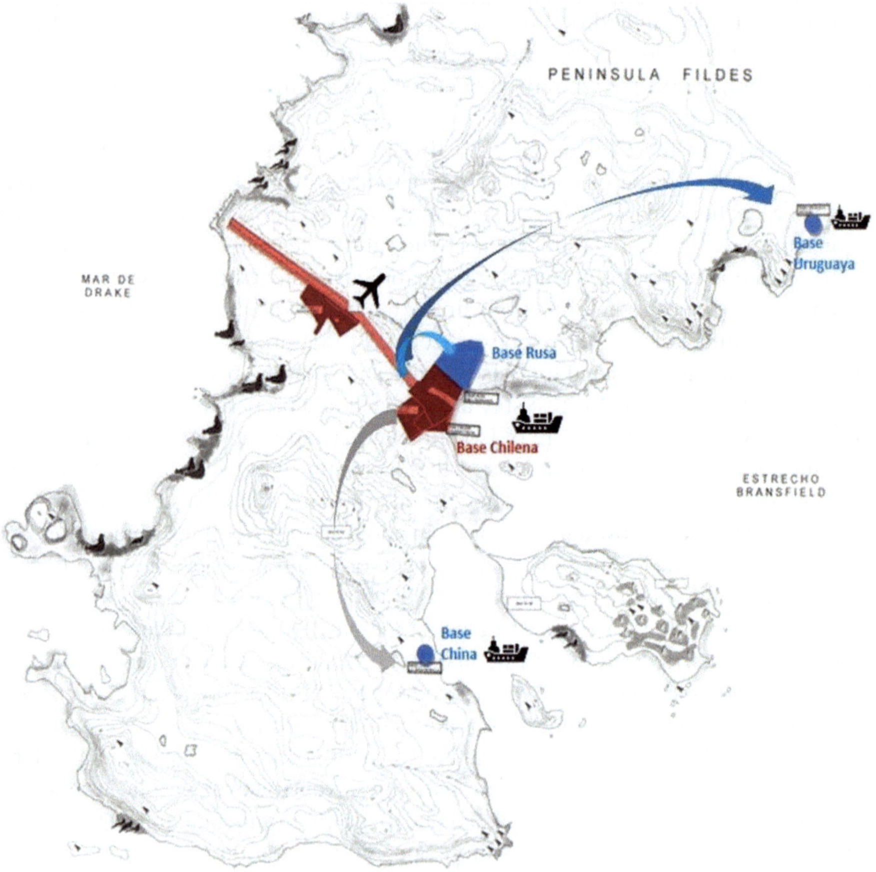

Fig. 5.2 Territorial plan in Fildes Bay. Chilean Bases (in red), Russian Base Bellingshausen, Uruguayan Base Artigas, Great Wall China Base. *Source* Own elaboration based on plans provided by Antarctic Division, FACH

the festivities of each country and by shared events typical of Antarctica, such as celebration of 'Midwinter' or the 'Antarctic Olympics'.

The particular presence of families with children at the Chilean base, made it an attraction pole, managing events and actions of coexistence especially to share with minors, as the crews of the other bases spent long seasons without seeing their families and therefore, sharing with children was a way of keeping the 'spirit high' to break the routines and sustain the general isolation.

5.4.2.2 Equipment and Everyday Places

The Chilean settlement borders the Russian Base Bellingshausen, separated by the stream that descends from the lagoon that provides water to both bases. The Chilean Base is composed of a series of low-height modular rooms located in two areas, the first, around Fildes Bay and the second in the high area where the aerodrome is located.

The group of buildings next to the bay does not present a structured arrangement, but rather inorganic. They are scattered volumes that have been incorporated over time. The exterior spaces are constituted as traces on the ground, taking advantage of the network systems for displacements.

Another peculiarity of this sector is the natural condition of the sea-border. There is no type of formal appropriation of the border, as there are no buildings or operational-logistical constructions, no docks, just a few traces that demarcate the areas used for the disembarkation of boats coming from the navy ships or tourist cruises.

Due to the blizzards and snowfalls, social life is concentrated in interior spaces, considering constant preparation rituals for going outside from the conditioned interiors, as the connection spaces between modules are exterior and are not enabled. Thus, displacements are subject to changes in weather conditions.

In addition to the functional and work spaces, since 1984, when families were integrated into *Villa las Estrellas*, complementary equipment was incorporated. These are the 'everyday places' where daily life's work and functional activities are carried out, made up of the modules of the Basic School, the Nursing office, the Post Office, the Bank office (BCI Bank), a radio station and the supermarket that includes a 'cafeteria' corner (Fig. 5.3). These services give the character of 'village' that distinguishes the EFM Air Base.

Another relevant aspect has to do with the synchronous operation of the inhabitants for the supply of basic services, such as the 'water ritual', to fill the tanks of each house and functional places, leaving the pipes empty and avoiding the freezing of water in the process; waste and garbage management systems; the electrical generation which is the primary energy source generated based on fossil fuels, especially for heating interior spaces; or the unloading processes of the boats that bring supplies from the ships, etc.

5.4.3 Architectural Scale

5.4.3.1 Events and Collective Places

According to the stories, it can be observed that during the stay in Antarctica, customs typical of Chilean culture are preserved, such as family celebrations and collective events (birthdays, Christmas, National Holidays, Armed Forces days), social structure, family and spatial organization (Fig. 5.4).

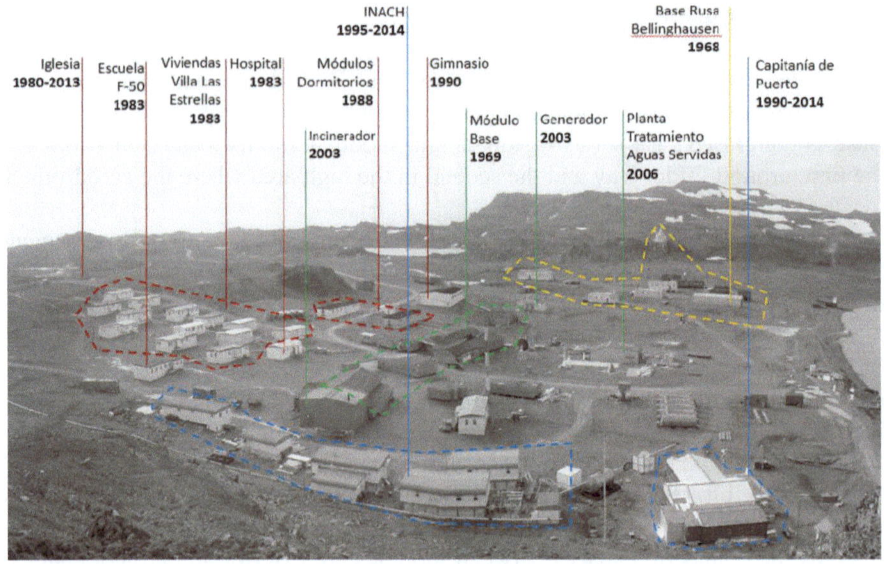

Fig. 5.3 Buildings of the places of the institutionality of the village. *Source* Elaboration based on FACH documents

The spaces for the development of community activities as 'collective places' that are significant of social and cultural life are organized in three areas:

- Exteriors or natural, are places of recreation (including ski slope) that can be shared with inhabitants of other bases, the appropriation of these spaces is only feasible when weather conditions allow it.
- Collective interiors, for the development of official activities, sports or celebrations of festivities, where the gymnasium, due to its dimensions, is the most used and multi-purpose, also the Base Module in which various social activities are carried out.
- Social interiors, give place to social life among families and close friends, these are smaller spaces that, however, allow interaction around shared moments of relaxation and food. The most characteristic are the casino, the inn, the supermarket and the same single-family homes.

Faced with precariousness, multifunctionality is a concept associated with spaces and people, that's why many spaces are used for different activities, just as residents perform several roles simultaneously.

Fig. 5.4 Traditional celebrations midwinter and national holidays. Images provided by MLP

5.4.3.2 Housing, Domesticity and Intimate Places

The experience of 'domesticity' refers to an adaptive response to the context of insularity, which allows the organization of daily activities based on the characteristics of the environment. People develop 'procedures and daily rituals' to function operationally, and domesticity allows the construction of a sense of identity and belonging.

The stories and documentation of living conditions account for an intense social life, where homes are the nerve center of coexistence. Living rooms and open kitchens (American model) allow interaction with guests in a certain condition of 'familiarity'. These possibilities and the series of scheduled activities contributed to an *esprit de corps*, and similar experiences among inhabitants of different periods. However, it is observed from the interviews, that the 'spaces of individuality' are scarce, there is practically nowhere to 'escape' from the collective, understanding that the exterior space is little habitable in certain seasons.

From the architectural analysis of domesticity spaces, three types are characterized:

- Single-family homes, which were built as isolated modules for the families of Air Force officers, doctors, DGAC professionals and teachers. They are equipped with three bedrooms considering the family character, meeting the basic conditions of functionality, comfort and habitability (Figs. 5.5 and 5.6). The houses are institutional, they are equipped and the residents adjust to the preexistences, incorporating some elements 'of their own' of family identity.
- The room pavilions are modules for the staff without family who are on Commission of Service and stay between one and two years. They consider a camp format, with single rooms and common bathrooms aligned in double bay with interior circulation. These pavilions are autonomous and have living spaces and all the services for survival in case of isolation.

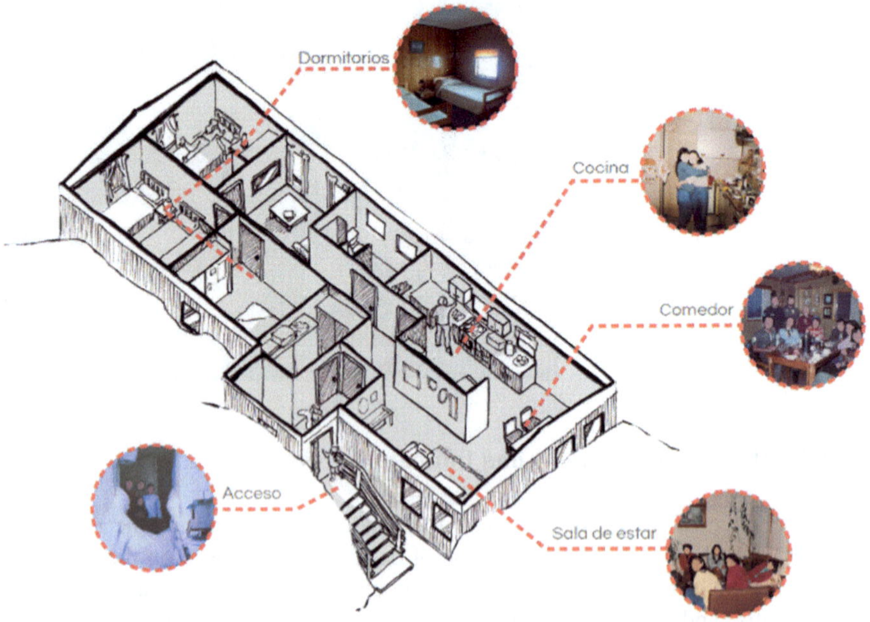

Fig. 5.5 Model and functionality of a single-family home. *Source* Isis Zuñiga based on FACH plans

Fig. 5.6 Images and characterization of domestic architecture. Homes of Villa las Estrellas. *Source* Carla Antognini

- The Hostel or Refuge, located in the area of the aerodrome, has conditions similar to the room modules, however, the rooms are shared with two beds and its use is for passengers, researchers, visitors or short-stay staff. It also operates on a double bay pavilion system, with interior circulation and rooms on both sides, concentrating areas of bathrooms and laundry.

The buildings erected over all these years present formal and spatial characteristics more or less homogeneous. They are introverted modules, with small openings that deny the relationship with the context, as greater hermeticity and efficiency is sought against low temperatures, on the other hand, small fenestrations facilitate the control of light in extreme periods of light or prolonged darkness.

The architectural expression of the buildings at the EFM Air Base responds to an industrial production and aesthetic, either of assemblable modules or prefabricated container-type systems, as in most of the Bases erected in the 1980s' (Gernandt & Meyer 2021; Pellegrino, 2021). Understanding the difficulties in the execution processes of the constructions, the transportation of materials by ship (without dock), the lack of specific machinery (such as cranes), the limitation of labor that must be specially transported and that the buildings must be erected quickly in summer seasons; the designs respond to this with standard volumes of low height, prefabricated, serialized, quick assembly and easy transport by ship, which can be expandable and modifiable, allowing flexibility in the modification of interiors.

The structures of the modules currently in existence are made of wood or steel, the envelopes of panels with insulation, internally lined with wooden plates and externally with metal sheeting. The roofs of little slope have been modified to improve their resistance to winds. The same goes for ground contact, leaving air chambers to improve the operation (avoiding freezing) of the facilities. Color is used as a way to stand out in the white or black (when there is thawing) and to express the institutional corporate colors.

From architecture, Peter Wilkinss (1993), Niall Patrick (2020) and Georgina Davis (2017) among others, argue that in the face of the obsolescence of the first scientific bases, the new urban-architectural proposals in Antarctica have to consider more controlled ways of life in terms of environmental impact, incorporating aspects of sustainable and architectural designs focused on new formal, aesthetic and technological proposals, leaving behind the modular and improvised buildings existing in most of the Bases.

5.5 Conclusions

By way of conclusions, it is worth noting that, although the EFM Base is created as a Meteorological Center ex novo, the strategic positioning for its growth and consolidation is based on the feasibility of the landing strip, the rest is linked to overcoming extreme weather conditions, access to maritime supply, provision of fresh water and the presence of other nearby communities.

The buildings, although elementary, have met the basic requirements to adapt to the climatic conditions. Thus, technological advances have reduced the condition of isolation and to a lesser extent, technologies have been implemented for better living conditions, using more sustainable resources.

In this sense, the architectural work in its simplicity constitutes a trace of human culture, considering from its planning and execution to its subsequent habitability and adaptations. This architectural expression, in addition to being a place of shelter, has enabled the appropriate interrelationships of coexistence, development of individual and collective capacities, as well as the cultural manifestations of the population, overcoming the, until now, conditions of isolation and confinement.

Interculturality develops predominantly in the Antarctic Continent as a form of exchange and codependence between inhabitants of different nations. Being there, it is understood that there are no territorial limits and spaces, and cultural manifestations are shared, creating new instances of multinational meetings and recreation.

It is shown that the families and staff of the FFAA preserve customs typical of Chilean culture, such as celebrations, social structure, family and spatial organization, so we can talk about a 'Chilean-Antarctic culture'.

Thus, this "Chilean-Antarctic culture" is expressed in a particular production of "places", which are nourished by everyday ways of living in contexts of insularity, codependence and extreme habitability.

The houses represent a field of multiple experiences, as they make up the intimate family space, granting some moments of individuality (very appreciated and required by the inhabitants) and at the same time it is a place of social gatherings among friends and colleagues, therefore, it plays a role in the domesticity of intimate places and at the same time can be considered, sporadically, as a collective space for celebrations.

Family life is a singularity that contributed to give socio-urban everydayness in the category of 'village' to the settlement. The Chilean Base stood out, beyond the aerodrome, as a center of attraction for the presence of children, compared to the other bases.

On the other hand, the successive relays of inhabitants limit the rooting, the tracing of a common history and the development of greater diversity of creative or cultural manifestations of the place.

However, in terms of its heritage relevance, according to the 'Guidelines for the designation and protection of historical sites and monuments' (2009), we can say that the characteristics for possible protection of the Base are met, as: "*it is representative or part of a wide-ranging activity that has been important in the development and knowledge of Antarctica*", this considering the permanence of families with children carrying out their daily activities in the Antarctic territory; "*its materials, design or construction method have a particular technical, historical, cultural or architectural value*", so we can say that the buildings are typical of the period of the 70s and 80s with the industrialized technologies that are adapted for the conditions of transport, assembly and habitability in that context; "*it offers the possibility of revealing information through study or educating people about important human activities in Antarctica*", the experiences of community life, in the format of a village, allows for anthropological, historical, political and cultural studies of various kinds, being a permanent place of attraction for people from different parts of the world for such experience.

This research opens a field of architectural study that intersects the logics of conservation of a cultural heritage, with the new approaches of territorial sovereignty and the requirements of sustainability in the face of a context of climate change.

References

Antarctic Treaty Secretariat. (2009). Guidelines for the designation and protection of historical sites and monuments. *Appendix of Resolution, 3.*

Azara, P. (2005). *Castillos en el aire: mito y arquitectura en Occidente.* Barcelona: Gustavo Gili.

Chile. (2017). Presidential Directive Report. To enhance the functionality of the defense bases in operational and logistical aspects, as support to the national presence in Antarctica. Government of Chile, Michelle Bachelet Jeria.

Chile. (2020). National Antarctic Tourism Policy. Ministry of Foreign Affairs.

Contracting Parties. (1959). *Antarctic Treaty.* United Nations.

Contracting Parties. (1972). Seventh Consultative Meeting of the Antarctic Treaty. Wellington

Davis, G. (2017). A history of McMurdo station through its architecture. *Polar Record, 53*(2), 167–185. https://doi.org/10.1017/S0032247416000747

FACH. (2019). Report "Background related to the state of the infrastructure of the Antarctic Air Base "President Frei". Chilean Air Force.

Gallardo, L., & Toledo, M. I. (2020). Ethnography for architectural projects: Inclusion of the inhabitant's perspective. *Arquitetura Revista, 16*(2), 197–216. https://doi.org/10.4013/arq.2020.162.02

Gernandt, H., & Meyer, H.-J. (2021). Containersation in polar architecture. In G. Foscari et al. (Eds.), *Antarctic resolution.* Lars Müller Publishers, Giulia Foscari, Wildmann Rezzonico, and UNLESS.

Giglia, A. (2012). *El habitar y la cultura: Perspectivas teóricas y de investigación.* Barcelona: Anthropos.

Heidegger, M. (1975). Building, dwelling, thinking. *Theory Journal*, (5–6), 150–162. https://revistas.uchile.cl/index.php/TRA/article/view/41564

INACH. (1965). Bulletin of the Chilean Antarctic Institute. May 1965.

Iturriaga, J. (1999). *Infierno en la Isla Decepción.* Santiago de Chile: Arturo Merino Benitez Foundation.

Lefebvre, H. (1973). *From the rural to the urban.* Ed. Peninsula.

Morales, J. (1999). *Arquitectónica. Sobre la idea y el sentido de la arquitectura.* Madrid : Biblioteca Nueva

Norberg-Schulz, C. (2009). The principles of modern architecture. In *On the New Tradition of the 20th Century.* Reverte Publishing.

Orrego, E. (1948). *Terra Australis. Diario de la primera expedición Antártica Chilena.* Santiago de Chile: Editorial Zig-Zag, S.A.

Patrick, N. (2020). In Antarctica, architecture is heating up. Architecture Platform 05/04/2020. In: https://www.plataformaarquitectura.cl/cl/935644/en-la-antartida-la-arquitectura-se-esta-calentando

Pellegrino, F. (2021). Recombining architecture. The role of the container at Mario Zucchelli Station. In G. Foscari et al. (Eds.), *Antarctic resolution.* Lars Müller Publishers, Giulia Foscari, Wildmann Rezzonico, and UNLESS.

Romero, P. (1985). *Síntesis de la Historia Antártica de Chile.* University of Santiago.

Wilkniss, P. (1993). The paradox of Mc Murdo. *Antarctic Journal of the United States, 28*(2), 3–11.

Part II
Sustainable Landscapes, Resilience and Regenerative Cultures: Emerging Approaches to the Climate and Ecological Crisis

Chapter 6
Regenerative Understanding of Place for Urban–Rural Contexts in Mexico. Methodological Proposal

Carlos Cobreros and Nadia Rodríguez

Abstract The emergence of new territorial expressions in Mexico, as a result of the urbanizing processes of mid-20th century modernity, dispersed the traditional dichotomy between urban and rural in a transformed environment. Likewise, the theorization, recovered from other contexts unrelated to the reality of the country, has scarcely addressed the opportunities within strategic planning for territorial development and its impact on the natural environment. The study of the urban–rural demands to be approached from a systemic vision that helps to understand and adequately conceptualize its complexity, therefore, the "Understanding of Place" from a regenerative approach, becomes an appropriate methodological tool for the study of urban–rural phenomena according to the essence and uniqueness of the place, contributing to decision-making for design and development within these contexts, which actively involves the inhabitants in understanding the vocation and potential of the place. This article proposes the reinterpretation of the regenerative methodological tool "Understanding of Place", for the study of the Mexican urban–rural phenomenon, makes a methodological proposal for the study of the urban–rural and presents results from a case of application.

Keywords Regenerative design · Regenerative development · Urban–rural · Regenerative understanding of place

N. Rodríguez
Faculty of Engineering, Universidad Autónoma de Querétaro, Santiago de Querétaro, Mexico

C. Cobreros (✉)
School of Architecture, Art and Design, Tecnologico de Monterrey, Santiago de Querétaro, Mexico
e-mail: ccobreros@tec.mx

M. Vicuña and E. Giorgi (eds.), *Intersections*, Springer Geography,
https://doi.org/10.1007/978-3-031-76402-8_6

6.1 Introduction

6.1.1 The Urban–Rural

The study of relationships and interaction between the urban and rural emerges from the theories of modernization (Dueñaz Checa, 2017). The most significant advance is mostly attributed to studies of the urban periphery as an urban–rural fringe that is constituted mainly by the "interface between two apparently oppositional and well differentiated geographical types, the countryside and the city" (Barsky, 2000, p. 28). Other perspectives also point to the presence of the phenomenon outside the periphery in the "municipalities that are not exactly rural, but not completely urban either" (Herrera, 2011, p. 115), these spaces should be treated as a bidirectional interaction according to which "both spaces receive and contribute building a unique dynamic" (Fernández & de la Vega, 2017).

An increasingly globalized economic market, directs the actions of urban–rural spaces both in urbanizing processes, as well as socio-spatial ones, which constitute their intermediation dynamics (Salazar et al., 2018), thus taking for granted that the countryside will eventually become city and leaving out all experience and perspective of its inhabitants (Bellet et al., 2015; Bolay, 2016).

With the identification of the phenomenon comes the need to develop theories for the understanding of its system of relationships, as the notion of the urban–rural is far from fitting into the dynamics and regional realities. In the theory around the urban–rural it is necessary to overcome the dichotomy of the social-spatial and attend to a relational, historical and dialectical view according to the particularities of each territory and place (Dalla-Torre et al., 2017).

6.1.2 The Urban–Rural Interface and the Processes of Urban Transformation in Latin America and Mexico

Between the 1930s and 1970s, territorial expansion, as a result of demographic growth, occurred in the most important urban centers in Latin America at an unprecedented speed (Cotrim & Lara, 2020). The promise of progress in the new modernity of the mid-20th century triggered the emergence of new territorial phenomena that manifested in new activities (economic, urban and transport infrastructure, services, etc.) favored by urban dispersion (Ávila Sánchez, 2009), inevitably reaching rural spaces and deconcentrating the functions of cities (Aguilar, 2006; Ornelas Delgado, 2000), unleashing multiple environmental impacts, territorial disintegrations, and social inequalities.

The transformations that occurred established links between urbanization, industrial development, and socio-spatial segregation, with a trend towards suburbanization that modeled the processes of North American and European urban expansion (Parussini, 2012), which meant the transfer of population to peripheral areas and an exaggerated expansion of cities over their rural environment (González-Gómez et al., 2012), absorbing them in their logic and operation (Ávila-Sánchez, 2005).

Although urban–rural relations have aroused great interest among academics in Latin America (Dalla Torre et al., 2017), the discourse recovered from these forums is generally rescued from foreign theory and adopting knowledge from other realities (Ávila Sánchez, 2005), so the understanding of the phenomenon in the Latin American scenario becomes almost imperceptible. In the case of Mexico, the body in charge of administering the classification of territories in the Mexican Republic, the National Institute of Statistics and Geography (INEGI), defines the limits of the territory in three large groups under the criterion of population density: (1) Rural localities, which cover a population range of 1 to 2499 inhabitants; (2) Urban localities or cities that house more than 15,000 inhabitants; and (3) those localities that are distinguished by being in between the first two and that have a population range of 2500 to 14,999 inhabitants (INEGI, 2020a). If we take only these values as a reason for characterization, the dynamics of interaction, the perspective and experience of the inhabitants, the particularities of the physical environment and its relationship with the inhabitants, culture, history, the legacy of its collective memory, and other fundamental elements that shape and define a territory are excluded. Avoiding all these elements in the definitions of urban–rural spaces encourages the naivety of their administrations within regional transformations.

In Mexico, spatial definitions are far from fitting into global conceptions, while global urban studies are only focusing mainly on metropolitan areas (Salazar et al., 2018), which worryingly makes more comprehensive studies of the urban–rural phenomenon invisible. Urban and rural should not be conceived as isolated elements of a system, but as a dialogic transition zone (Delgado-Castillo, 2017) that incorporates interwoven relationships of its own personality through a series of processes in the structuring of the territories in which it manifests, hence its study requires comprehensive approaches and transdisciplinary methods (Ávila Sánchez, 2009).

Urban–rural transformations are, for Latin America, and in Mexico specifically, with the region's insertion into international trade through primary activities (within the logic of neoliberal policy), where, at a structural level, there has been a greater flow of capital from rural areas. This has caused the accelerated demographic growth of intermediate urban settlements (or urban–rural), as mentioned by Salazar et al. (2018), since it houses a significant part of the labor force associated with primary activities. Salazar and co-authors also highlight the lack of reference to the experience and perspective of the inhabitants in territorial transformations, as the authors mention: "Usually, transformations have overlooked the dynamics associated with the experiences that city dwellers have with them and their perspectives" (Salazar et al., 2018), proposing participation models for the analysis of territories as part of the study of "intermediate cities".

6.1.3 Regenerative Design and Development

The socio-ecological, climatic and humanitarian crisis on a planetary scale accentuates the socio-environmental and territorial problems caused by new urban–rural realities. A fundamental rethinking of the way we develop territories (Roggema, 2022), landscapes, cities and their communities is necessary. A transition towards futures in which ecosocial systems are designed to balance anthropogenic activity and the integrity of natural ecosystems (Wals, 2019; Tafuni & Hess, 2019) is also necessary.

Although sustainability, based on a mechanistic, reductionist, deterministic, dualistic and anthropocentric worldview (Benne & Mang, 2015), promises not to affect future humanity, it does not contemplate the possibility of contributing and co-evolving with nature and is no longer sufficient (Wahl, 2016). The need for resilience provides a new definition of sustainability, which is not based on achievements but on a stable state that can regroup and advance when the balance has been disrupted (Mang & Haggard, 2016). The term regenerative sustainability emerges, which includes and transcends these objectives, aspiring to thriving living systems in which the health and well-being of the entire system continuously increase (Gibbons, 2020).

We need to move towards regenerative sustainability (Lyle, 1996; Reed, 2007), whose goal is to heal our damaged world, allowing it to evolve and thrive (Mang & Haggard, 2016). This requires a worldview shift from "mechanistic" to "ecological" (Mang & Reed, 2012). It is necessary to create a positive future through a worldview based on living systems (Du Plessis & Brandon, 2015), to another based on understanding systemic change, complexity and self-organization (Roggema, 2022).

With the 21st century comes a new paradigm and new processes of regenerative development and design (Mang & Reed, 2020), from the understanding of the functioning of living systems, from the understanding of place as something unique, singular, to be developed from its essence, potential and vocation. Regenerative Design is based on a systemic thinking, based on the place and oriented to contribute to positive outcomes (Svec et al., 2012); It allows increasing social and natural capital (Clegg, 2012); It seeks the co-evolutionary relationship between sociocultural and ecological systems (Cole et al., 2013). From the understanding of the essence, the unique and the singularity of the place, Regenerative Design, instead of presenting sets of strategies and specific design guidelines that could inhibit creative solutions, is populated with provocative questions that change and can adapt to different situations, from community commitment to detailed design explorations (Cole et al., 2012).

6.1.4 Understanding of Place for a Regenerative Design

Based on the idea of Regenerative Design and Development, Regenesis Group develops the basis of the methodological tool of understanding of the place based, in an original regenerative methodological framework (Mang & Reed, 2012). Their proposal explores the process through premises and basic concepts. Within the premises are contemplated (Mang & Reed, 2012): (a) the role of humans and human activities in co-evolution with living systems; (b) A change of perspectives and focus; (c) A new role of the designer; (d) The development of processes, from the capacity of regeneration. The concepts they handle are: (1) Regeneration, as a catalyst for human impacts on natural systems; (2) Establishment of broader and deeper design practices; (3) The Place, as a unique network of multiple layers of living systems; (4) Literacy of patterns, as the language of relationships; (5) The History of the Place, as relationships and connections that create a unique narrative; (6) The potential, as the inherent capacity of a place's unique essence. The importance of these premises and concepts lies in that they prevent a conflict of conceptions when addressing a design problem and develop a system of processes that characterizes three consecutive phases (Mang & Reed, 2012): (1) the understanding/conceptualization of the appropriate relationship with the place that allows identifying the potential for each place; (2) Technologies that consider theoretical systemic frameworks; (3) An ecological-based design system.

This approach requires a living systems thinking that allows thinking about organizations from how they organize, order, structure, and evolve (Mang & Haggard, 2016).

The study of the urban–rural phenomenon fits into this dimension (Ávila-Sánchez, 2009), as it is necessary to adopt a relational view of the particular dynamics of each territory, as urban–rural transformations are generally not associated with processes that address the experiences and perspectives of local inhabitants (Salazar et al., 2018).

6.1.5 Objective and Scope of the Article

The objective of this article is to conduct an analysis of the approach to the complexity of urban–rural relationships through a semi-urban locality in Mexico, using the methodological tool of regenerative understanding of Place, to understand and conceptualize the manifestation of the phenomenon from a local context.

6.2 Regenerative Understanding of Place for Urban–Rural Contexts in Mexico

Based on the methodological framework of Regenesis Group (Mang & Haggard, 2016), it is necessary to follow a flow of processes, for the understanding/conceptualization of the place, which aims at the understanding of phenomena with the recognition that each place is a unique dynamic entity. According to this, in the application to the urban–rural study, the flow is developed through the recognition of territories as the dialectical core between the community and its place of life.

The basis of nesting scales is proposed, which provide a view of interrelationships (Mang & Haggard, 2016). However, given the magnitude of the scales, it becomes difficult to discern the more specific networks, it is considered necessary to also direct the gaze to the most important sectors of urban cores (commerce, religion, politics, education, health, etc.), adding a new scale of interpretation called "Scale of Living Points". The use of this concept provides more effective management in the collection of information and human experiences of a specific urban–rural space.

The flow of understanding continues with the comprehensive evaluation, through a systemic analysis, and the identification of the emerging central patterns, from a comprehensive evaluation of the site's complete systems (Ecological Systems, Socio-Economic Systems and Cultural Systems). Due to the complex nature of the system of relationships, organizations, and dynamics, it requires hard work to distinguish the central patterns at all scales, making necessary the use of tools that allow a systematization of data, such as the Matrix of Influences proposed by Cole (Delgado-Castillo, 2017). This systematization is necessary, since the reality of each specific place is different from another. Once a study achieves the understanding of the place, its essence, and its relationships, fragmented problems can be addressed as part of a complete system in the global context of the vocation and potential.

The flow culminates once a transition is made from understanding to the concept of systems, grounding the regenerative role of a place and, from working from the potential of the place, the appropriate proposal can be made, on activation nodes and activation principles, to generate vital support and the beginning of a process of co-evolutionary and regenerative mutualism (Mang & Haggard, 2016).

6.2.1 Complementary Aspects for an Urban–Rural Context

With the flow of the Understanding of Place process from a regenerative approach, there are still some aspects to address if a complete study for the urban–rural phenomenon is sought. To begin with, the definition of a place with urban–rural dynamics not only depends on nesting scales, but on the criteria given to characterize specific places. In Mexico, spaces of both urban and rural interaction are identified as semi-urban localities or regional population centers classified as localities of 2,499 and 14,999 inhabitants, but addressing the classification of urban–rural spaces from numerical discrimination is not enough to characterize it (George et al., 1967). Therefore, it is necessary to establish criteria both by population and by functions and dynamics to characterize a place.

6.2.2 Methodological Proposal

The methodological proposal consists of 3 phases, based on the original methodological idea of Regenesis Group (Mang & Haggard, 2016). It proposes a first phase where the "Definition of the Place" is developed, which initially develops the selection of the empirical object of the research, then defines the scales of the nesting system. Phase 2 corresponds to an iterative process that recognizes the narrative of the place through a documentary and ethnographic analysis called in this research Critical Bioregionalism and, Ethnography and participation; which will carry out a systemic analysis in parallel where it is intended to discern the Central Patterns that will reveal the essence, potential, and vocation of the selected place. The last phase seeks to identify the activation nodes in coherence with the findings and discoveries of the second phase to find leverage opportunities from a regenerative approach, limiting itself to finding them without making change proposals, as the research does not intend to reach the Design phase (Fig. 6.1).

6.2.2.1 Case Studies and Place of Application

Querétaro has a population of 2,368,467 inhabitants distributed in 2,192 localities corresponding to: 16 urban localities with a total of 1,391,541 inhabitants, 93 semi-urban (or urban–rural) localities with a total of 478,319 inhabitants, and 2,083 rural localities with a total of 498,607 inhabitants (INEGI, 2020c). Among these semi-urban localities, this study has focused on the South Region of the State of Querétaro for its socio-territorial qualities and the process of its urban–rural transformations that are based on conservation and continuity, differing from the trend to peri-urbanization already widely studied. From the South Region that includes Amealco and Huimilpan, the municipality of Amealco de Bonfil and its head as a semi-urban locality have been chosen.

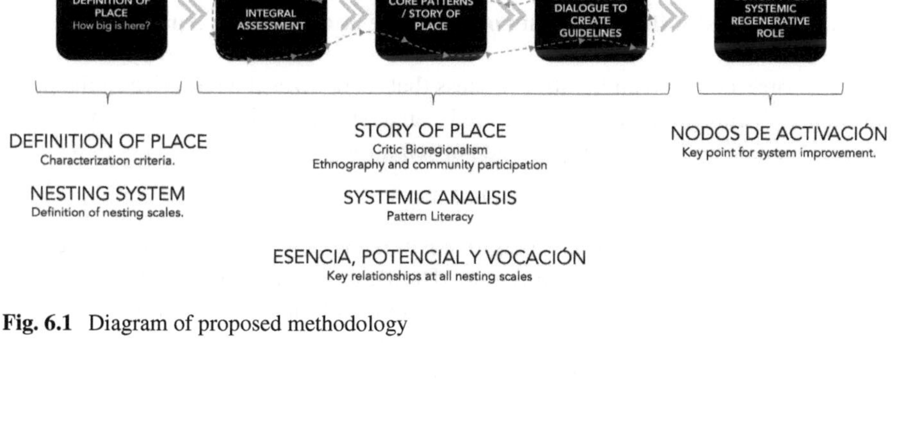

Fig. 6.1 Diagram of proposed methodology

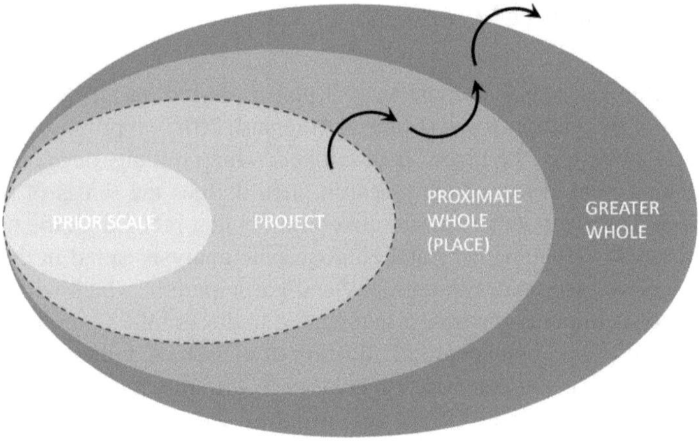

Fig. 6.2 Physical representation of three nesting scales. Own elaboration

Under a preliminary and partial judgment, the semi-urban locality Amealco de Bonfil is subjected to a validation with the objective of determining if it is an ideal locality to exercise the regenerative tool Reading and Understanding the Place adapted to the urban–rural study, from the criteria of population density; economic, productive and commercial activity; socio-territorial scope; environmental conservation and spatial typology.

6.2.2.2 Selection of the Project or Empirical Object of this Research and Definition of the Scales of the Nesting System

In phase 1 of the methodology, the following Nesting scales are proposed for the locality of Amealco de Bonfil. Own elaboration based on the nesting scales model of Regenesis Group (Fig. 6.2).

6.2.2.3 Understanding, Reading of the Place and Systemic Analysis

Phase 2 corresponds to an iterative process that recognizes the narrative of the place through a documentary and ethnographic analysis called in this research Critical Bioregionalism and, Ethnography and participation.

Phase 2a. Critical Bioregionalism—Desk Work

According to Martino (2005), Bioregionalism is the new scientific look at the relationship with the place, with the region. The Bioregional perspective is conceptualized from geography and consciousness, and such approach highlights that the natural conditions within a bioregion are similar and that these influence the forms of human occupation (Martino, 2005). This study seeks a bioregional look from a critical stance, where there is individual consciousness and a thought of scales that integrate not only political boundaries, but also the regional ones, through a regenerative approach under the systems and their respective 9 subsystems:

- Ecological System: Geological, Hydrographic and Biological.
- Social System: Settlements, Economy, Development.
- Cultural System: Psychological, Values and Beliefs.

Based on the above, the proposal organizes the systems, and their 9 subsystems as follows (Table 6.1).

In this sense, critical bioregionalism is proposed in the technical-documentary aspect to compile the necessary information proposed in Table 4 through the following tools: Documentary information from official and academically quality websites; National Geo-statistical Framework (INEGI, 2022b); Mexican Elevation Continuum Version 3.0 (CEM 3.0) (INEGI, 2013); SCINCE 2020 (INEGI, 2020b); National Statistical Directory of Economic Units 2020 (DENUE 2020) (INEGI, 2020a); Agricultural Census 2022 (CA 2022) (INEGI, 2022a); Geoinformation Portal of the National Commission for the Knowledge and Use of Biodiversity 2023 (CONABIO 2023); Open Data Portal of the Government of the Republic: These are the digital public data that are accessible online on the official website of the Government of Mexico.

Digital instruments for the management of geo-statistical information QGIS are also used.

Phase 2b. Ethnography, Participation, and Social Mapping

Two important elements of the stakeholders are integrated by the community and the researcher, and to ensure their participation, tools are necessary that allow them to actively work with the project, and to provide transactional information with the aim of generating a narrative in the qualitative character. The proposal of this research for the integration of the community and the researcher with the place is based on the practices of Ethnography and its various tools for the collection of information in the discovery of the narrative of the place. For this, it is necessary to briefly review the intentions of Ethnography and similar approaches for the urban–rural study.

For Eduardo Restrepo (2018), Ethnography allows qualitative studies and can be defined as "the description of what a people do from the perspective of the same people" (Restrepo, 2018). The author affirms that such a phrase means that both the practices (what people do) and the meanings that the practices acquire for those who perform them (the perspective of the people) are of important interest in the ethnographic study. These principles are extremely important in the urban–rural study, since, as Salazar et al. (2018) well mention, by ignoring the socio-spatial dynamics of intermediate (or urban–rural) Latin American cities, the experience and perspective of the inhabitants in the process of urban transformations are excluded, making the inhabitants alien to their places.

For this research, two techniques are used to obtain qualitative data around the systems and 9 subsystems shown in Table 4: Community mapping and Social cartographies, social cartography understood as "the methodological and conceptual

Table 6.1 Organization of the systems and the 9 subsystems in subsequent categories. Own elaboration

Ecological system			Social system			Cultural system		
Subsystem geological	Subsystem hydrographic	Subsystem biological	Settlement subsystem	Subsystem economic	Subsystem of development	Subsystem psychological	Subsystem of values	Subsystem of beliefs
– Geology – Topography – Climatology	– Hydrography	– Bioregion – Ecosystems – Biology	– Urban aspects – Human settlements	– Trade – Work – Resource management	– Society – Education – Sustainability	– Mindset – Culture	– Ancestral values – History – Ancestral knowledge	– Religion – Spirituality

Note The subsequent categories are intended to complement the subsystems that allows communities to know and build a comprehensive understanding of their territory so that they can choose a better way to live it" (Jiménez Ramos, 2019). For the purposes of this section in the research, six participatory instruments are highlighted: (a) Mapping our body-territory; (b) Oraliteratures, word, Memory and collective creation: (c) Maps of the territory; (d) Biocultural or integrated calendar; (e) Power maps and relationships between actors; (f) Collective mapping.

Phase 2c. Systemic Analysis Based on Pattern Reading

According to Mang and Haggard (2016) environmental problems must be addressed within their unique bio-socio-cultural contexts. For the authors "Each particular place" is a laboratory of complex forces in continuous evolution integrated into an organic whole, this means that, in their words, places are "dynamic and comprehensible" (Bénard Calva, 2019). However, these meanings also mean that there are special challenges, so learning to recognize and read patterns (Pattern Literacy), can facilitate the discernment of patterns. There are three types of patterns that allow identification and according to the identification of each pattern it will become easier to understand the uniqueness of places: patterns of nested systems across scales or holons, the patterns take up this concept to describe that smaller systems nest within larger systems, where systemic health, as Daniel Wahl points out, depends entirely on the linking of scales of interactions and relationships within "complex dynamic systems" (Wahl, 2016); Interaction patterns: Living systems interact with each other and these interactions can be traced back and forth, there is an exchange of information and they have a mutual influence; Essence patterns: The essence, as has already been mentioned in previous sections, is the nature or distinctive character of something.

For Mang (2020) the discernment of patterns is possible with graphic tools, such as mind maps, therefore, the methodological framework that Regenesis Group developed for the graphic mapping of patterns is effective for systematizing relationships as the findings of subphases 2a and 2b are discovered, with the intention of globally visualizing the central patterns that clarify the essence of a place.

For the discernment of the central patterns of this research, the model of Fig. 6.3 is adapted into a Gigamap of relationships between nested patterns, as an adaptation of the methodological framework of Regenesis Group for pattern mapping.

The framework of Fig. 6.3 shows the nested scales represented by 4 circles, starting from the smallest: the Living Points, and continuing with the project circle: the Urban Core, which nests in the Next Whole that corresponds to the municipality and which, in turn, nests in the circle of the Greater Whole, the Watershed. With this base, the Ecological, Social and Cultural systems are located. For the identification of the Living Points, a Collective Mapping exercise was designed based on the techniques of the Collective Mapping Manual (CMM) promoted by the ICONOCLASISTAS group. The exercise uses a combination of techniques: (1) Senses, perceptions and territories and, (2) Tables in public space. For the development of the exercise, two main questions were taken as a basis: "According to your activities, which places do you frequent the most?" And, "What is your favorite place?", the same ones that

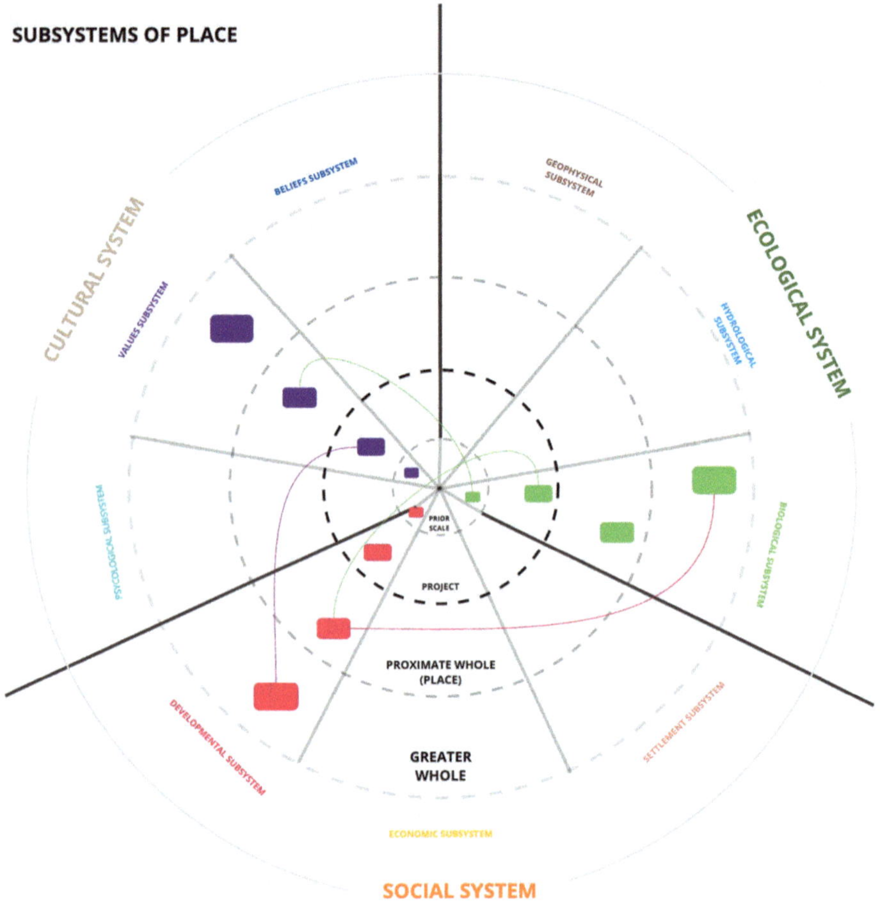

Fig. 6.3 Gigamap of relationships between nested patterns. *Note* This model is a figurative example, the final visualization will be found in the results of subphase 2c

when answered by the participants, were free to choose the specific points for their most important interactions according to the activities of their daily life.

Essence, Potential and Vocation of the Place

Phase 2 ends with the identification of the essence (uniqueness), potential, and vocation of the place.

Essence: The essence is the unique character and nature of a place. Architect Christopher Alexander refers to it as "The Quality Without a Name" (Alexander et al., 1977). In a place, the essence becomes key to identity and purposes, with the essence, a design project with a regenerative focus begins to take shape (Mang & Haggard, 2016). The discovery of the essence of a place, as Mang (2020) mentions, uses the power of storytelling. This process then necessarily requires the participation of the community.

Potential: To develop a regenerative project, one must understand the evolutionary dynamics of a place with the aim of identifying the potential to achieve systemic health and its viability, this with human presence. Mang and Reed (2020) define potential as "the capacity to leverage human interventions to achieve systemic health over time, by the place they occupy and on which they depend" (Mang & Reed, 2020).

Vocation: For Mang and Haggard (2016), vocation for humans is a source of meaning, so people regularly respond to their call in search of individual and collective satisfaction.

6.2.2.4 Identification of Activation Nodes

Under the systemic view that seeks to transform the understanding and appreciation of the living world, regenerative work validates its effectiveness through small, but powerful, conscious interventions that have a beneficial influence on the system and its nesting scales.

For this, it is necessary to start not only by understanding that everything is connected, but it is also necessary to understand how it is connected. For Mang and Haggard (2016), nodes are this point of connection, they are the points where material energy and information flows cross, or the key points located where information is exchanged and where transformations occur.

6.3 Results and Discussion of Results

Next, we show some of the results obtained within the proposed phases, showing, in phase 2, as an example, the work on the ecological system, showing some results of the work carried out in terms of ethnography and social cartography and then focusing on the discoveries for the study of the urban–rural phenomenon from a regenerative approach.

6.3.1 Selection of the Empirical Object of this Research and Definition of the Nesting System Scales

The following Nesting scales are proposed for the locality of Amealco de Bonfil. Own elaboration based on the nesting scales model of Regenesis Group (Fig. 6.4).

For the identification of the Living Points, as a result of the exercise, the following map was obtained (Figs. 6.5 and 6.6).

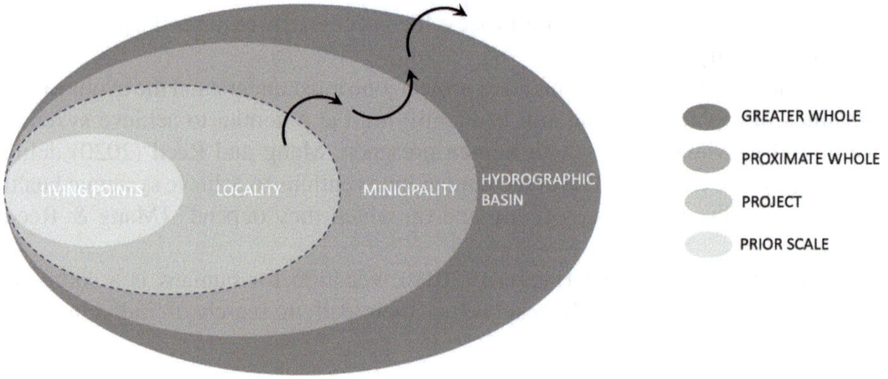

Fig. 6.4 Physical representation of three nesting scales. Own elaboration

Fig. 6.5 Collective mapping of Living Points of inhabitants of Amealco de Bonfil. Own cartographic elaboration based on the adaptation of the techniques proposed by the MMC 4 living points are defined on which to focus the exercise of understanding and reading the place, leaving the following physical relationship of the nesting scales

Regarding the Living Points that were identified within the municipal head, they can be seen in Fig. 6.7.

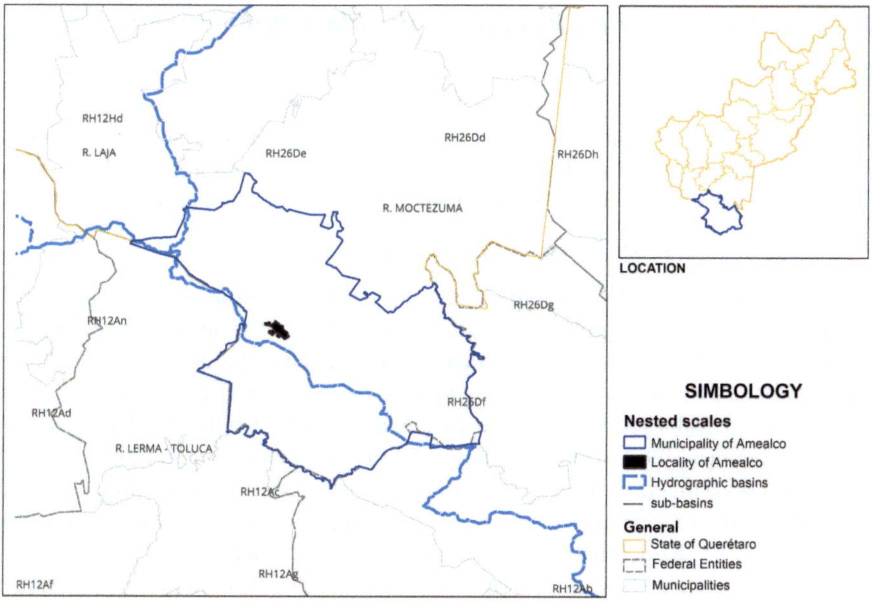

Fig. 6.6 Physical representation of three nesting scales. Own elaboration

Fig. 6.7 Physical representation of the Living Points in cartography of Amealco de Bonfil locality. Own elaboration based on the results of the exercise for the selection of Living Points

6.3.2 Understanding, Reading of the Place and Systemic Analysis

Phase 2 corresponds to an iterative and recursive process that recognizes the narrative of the place through a documentary, participatory and ethnographic analysis called in this research Critical Bioregionalism and, Ethnography and participation.

Phase 2a. Critical Bioregionalism—Desk Work

Some of the developed cartographies are shown below, to have a vision, even if partial, of the results of application of the methodology of understanding of place.

In the desk work phase for a critical bioregionalism, the defining elements of the 3 systems, Ecological System, Socio-economic System and Cultural System: Psychological, Values and Beliefs, are documented based on bibliographic, statistical and documentary study, with a vision of relationship with the territorial and spatial.

Work is done on the different scales of nested systems extracting defining elements that inform about the analyzed system. The information is read, and the relationships are established. The defining elements of this phase are poured into the gygamap to make a systemic analysis and discover patterns that lead us to identify the essence, the vocation and the potential of the place.

As an exemplifying element for this article, we use the study on the geology and edaphology of the place, where the different relationships of the characteristics of this subsystem, its defining elements, as an information layer, and its relationship with the different scales of the project, as well as with the urban fabric and the existing overlap with the different living points analyzed, characterizing each one of them in this way, contributing to the understanding of these living points from their marked singularity, in this case, by this subsystem, within the characterized ecological system. Within the Edaphology of Amealco (Fig. 6.8), we find that the largest surface is composed of Luvisol Feozem, in a lesser percentage Lithosol, Chromic Cambisol and Chromic Luvisol, this last one covers almost half of the locality of Amealco, and is where Cerro de los Gallos and the Sports Unit, two of the determined living points, are located (Fig. 6.9).

Within the ecological system, the hydrological subsystem is also exemplified, which marks a direct relationship with the urban fabric, the human settlements and the relationship between the different living points determined. As a main characterization, it is obtained as the localities of El Bothé and Mesillas within the municipality, which are located in the aquifer of the Valley of Amealco, rainwater harvesting is carried out in high functional areas (Aguilar Sánchez & Aguazul, 2021), therefore, the area is characterized by functioning as a recharge area for aquifers, in addition to supplying the community, a fundamental discovery for understanding the place (Fig. 6.10).

In the analysis of the socio-economic system, as the last system exemplified in its analysis, it is understood as Amealco de Bonfil, having a population of 66,841 people, where 34,606 are women and 32,235 are men, characterized by its scattered and diffused settlements throughout the territory, with a high percentage of indigenous

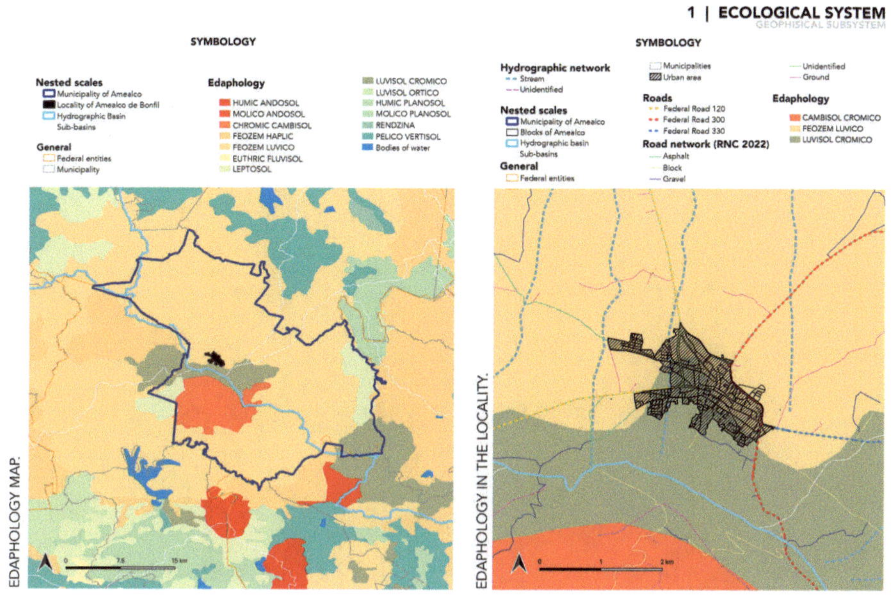

Fig. 6.8 Edaphology of Amealco de Bonfil by basin, municipality and locality. Own elaboration based on the National System of Information on Biodiversity (SNIB) (CONABIO, 2023)

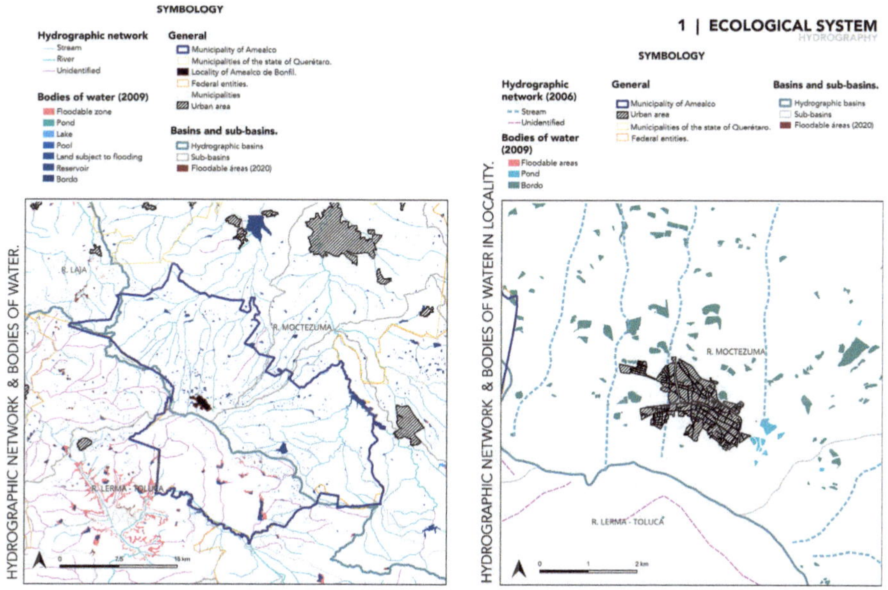

Fig. 6.9 Hydrographic network (2006) by basin, municipality and locality. Own elaboration based on the National System of Information on Biodiversity (SNIB) (CONABIO, 2023)

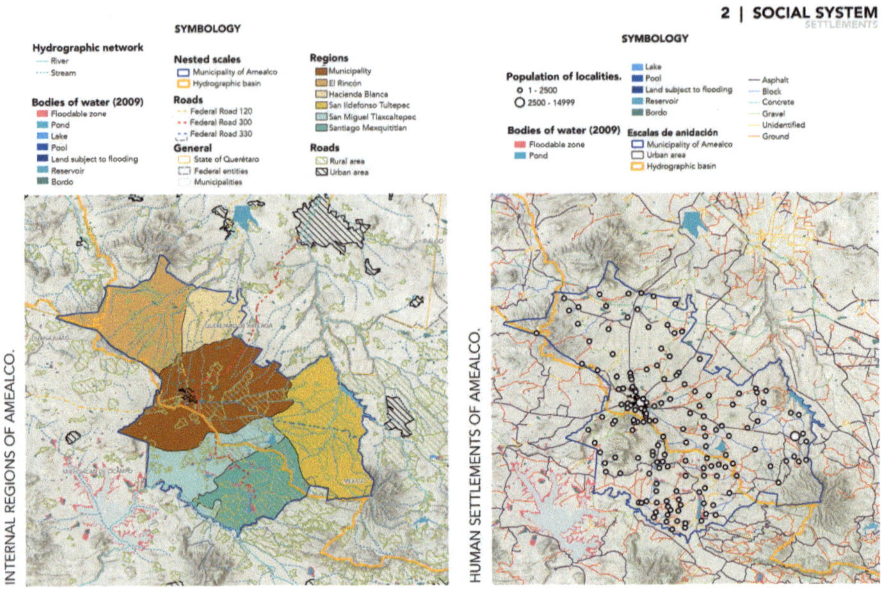

Fig. 6.10 Representation of the internal regions and human settlements within the municipality of Amealco. Own elaboration According to the INEGI 2020 census

population, which becomes a differentiating and characteristic element. The features of the active population, economic activities, people with education, disabilities, etc., are defined, which allow us to have a vision of the complexity of society and its relationship with the economy of the place.

Phase 2b. Ethnography

For anthropology and sociology, the knowledge obtained by applied research is indispensable for the processes of development and transformation of societies, as it provides the qualitative data that are inherent to the realities they study.

From the different tools used, defining elements of the different systems and subsystems are discovered that complement the information and the definition of the systems analyzed from the desk work.

From the exercise of collective memory, it is highlighted how ancestral knowledge is the interrelation between knowledge and wisdom is kept alive through collective biocultural memory, product of thousands of years of interaction between nature and human beings transmitted from generation to generation. Through the tool of Oraliterature, it is discovered how the orality of cultures is kept alive through collective memory; through it, phenomena of interest are reproduced, resignified and updated.

With social cartographies, built through participatory instruments, a practice, a group reflection that through participatory dialogue and pedagogy becomes a collective construction of territorial knowledge, with collective mappings of the territory, the biocultural calendar, the development of actor maps and body mapping. A general view of the organization, scales, dimensions distribution, uses of the territory and its landscapes, in everyday life: geographically, symbolically and productively. It can clarify the power relations of all the key actors of the territorial frameworks, at certain times (as it can change over time). It allows the reflection of the links with bodies and territories, makes evident the transgressions that the territory suffers and how it is lived from the bodies making conscious the place where one lives and the identity of the territories from its people, this is important as it also provides important social, economic and political data, as well as the relationship of the territory-nature and the productive activities of people (Fig. 6.11).

Phase 2c. Systemic Analysis Based on Pattern Reading

Finally, all the information obtained, everything learned through desk work and ethnographic and participatory work, is poured out as defining elements on a gygamap. These defining elements of the systems and subsystems are organized according to the different scales of nested systems to which they correspond, interrelationships are sought, and from the discovery of the patterns of essence, defined by those defining elements that are unique and unrepeatable of the place, the patterns of relationships and nesting, the essence, the vocation and the potential of the place are established and discovered, as well as subsequently the activation nodes on which one could intervene as leverage points of the system, to generate changes and sow the seed for the beginning of processes of co-evolution and regenerative development of the region and the place analyzed and understood, in conjunction, human system and living systems where it is inserted (Fig. 6.12).

Discovery and definition of the essence, vocation and potential of the place

After the work of discovering patterns through systemic analysis from the gygamap, the urban rural character and essence of Amealco is defined, as an identity that is managed between the change and continuity of traditions. Its essence is marked by the access to natural resources in a unique way given its location in the bioregion and access to resources, particularly like water, in addition to having a climate that differentiates it from the rest of Querétaro, where it is inserted. Its main transformation occurred due to its location and the link to an infrastructure that connects it with the rest of the territory, which leads it to have external influences, which keep Amealco in continuous change and evolution, while it has maintained a very special relationship with the conservation of its traditions, marked by the art percentage of the indigenous population of the place. This valued tradition intersects with modernity, as it has always been a place as a central point between some community heads, transforming through their influence, but maintaining its character, its traditions, its customs and ways of life, escaping from neoliberal globalization patterns that mark other places and populations in the region.

This link and connectivity marks a commercial essence of Amealco, but a trade from the use of its natural resources, agricultural and without losing the character and cultural wealth and identity through craftsmanship and gastronomy.

Its relationship with the territory as a central point allows it to dance between change and continuity, the conservation of traditions and modernity and this is

Fig. 6.11 Example of social cartography exercise and collective mapping. Own elaboration

precisely what marks its vocation, from the conservation of its particular characteristics, its education, its culture, its religion, its customs and traditions, its urban dynamics, with the vocation of preserving its cultural essence.

And this that characterizes its vocation, is the key point of its potential, being a central point for trade, from its privileged environmental characteristics regarding the surrounding territory, but from the conservation of its traditions and customs, despite

Fig. 6.12 Excerpt from the section on ecological system within the systemic analysis based on the generated gygamap

recent urban or industrial development with its own life dynamics, from what the place gives it, from the use of its natural resources, from maintaining and preserving a cultural and biocultural aspect of its own.

Activation Nodes

The most important activation nodes ended up related to the living points determined in the first phase of the reading and understanding of the place, which are understood as a good base to generate vital supports for the start of regenerative processes, as points of cohesion for the given dynamics of the place, from the conservation and preservation of the biocultural character of the place, sociocultural and economic, as a crossing point and center.

On these living points, the activation nodes can be understood from or through 3 important axes, which are the trade of own character, from the conservation of its customs, its spirituality and sacred relationship with the place and finally the conservation of its natural resources, which can guide a mutualistic co-evolution of social and economic dynamics.

6.3.3 Discussion of the Results

The implementation and application of the methodological tool of understanding of place from a regenerative approach, leads us to the following discoveries and findings from the reading and discussion of the results:

- It allows to establish a base for the study of the phenomena.
- The study of phenomena is achieved from an integral, interdisciplinary and holistic vision that takes into account the perspective and experience of the inhabitants.
- It allows to ground the real context in which it is studied from the own experience of the researcher and designer.
- It allows the understanding and conceptualization of places from a perspective of living systems through the identification of the central patterns that construct the narrative of a place and a systemic conceptualization.
- It can adapt to the study needs of particular phenomena as soon as the premises, concepts, and base dimensions of the methodology are internalized.
- For the urban–rural study, it is necessary to establish criteria both population-based and by functions and dynamics to characterize a place and proceed to define it through nesting scales.
- Understanding the key concepts and premises of the regenerative paradigm allows establishing a basis for the study of phenomena.
- Technologies (living systems thinking, permaculture and processes of development changes) allow the study of phenomena from an integral, interdisciplinary and holistic vision that takes into account the perspective and experience of the inhabitants.
- A methodology that recognizes the role of the human being, a new mind and a new role, allows grounding the real context in which it is studied from the own experience of the researcher and designer.

6.4 Conclusions

From a point of view of understanding the systemic network and of particular potentials of each place, it is possible to understand the urban–rural phenomenon. The Regenerative Understanding of Place works as a tool in the urban–rural study due to its nature of systemic thinking and transdisciplinary.

There are limitations that can occur when the study tends to a generalization of regions that correspond to diverse contexts. However, studying urban phenomena from regenerative perspectives can give us significant advantages: the opportunity to discover new systemic design processes that advocate for the particularities of each territory and integrate the perspective and experience of its inhabitants, moving away from decision-making in urban planning from blind neoliberal progress.

The proposed methodology allows the understanding and conceptualization of places from a perspective of living systems, which has a methodological flow in which they involve: The definition of the place through nesting scales, a comprehensive evaluation through the identification of the central patterns that construct the narrative of a place, the place seen as a living system that includes the dialogue with stakeholders and a systemic conceptualization with regenerative focus of the discoveries. It can adapt to the study needs of particular phenomena as soon as the premises, concepts, and base dimensions of the methodology are internalized, so it can be in the disposition of the use of tools, methods and technologies that fit the purpose of the study of the particular phenomenon.

For the urban–rural study, it is necessary to establish criteria both population-based and by functions and dynamics to characterize a place and proceed to define it through nesting scales. To acquire the data that allow us a comprehensive evaluation, it is necessary to establish objectives in relation to activities, natural systems and the population.

The adaptation of a methodological tool generally seeks a practical application, so in the study of the urban–rural from a regenerative approach, it remains to evaluate whether the analysis is effective based on a practical case study. In addition, it should be clarified that limitations can occur when the study tends to a generalization of regions that correspond to diverse contexts given the origin of the methodological model foreign to a Mexican reality. Without however studying urban phenomena from regenerative perspectives will give us two significant advantages: understanding literature in a Spanish-speaking and endemic format, as well as the opportunity to discover new systemic design processes that advocate for the particularities of each territory and integrate the perspective and experience of its inhabitants, moving decision-making in urban planning away from blind neoliberal progress.

References

Aguilar, A. G. (2006). Mexico City and its polycentric regional structure. In IG-UNAM, CONACYT (Eds.), *The large agglomerations and their regional periphery*. Experiences in Latin America and Spain (Vol. 1, pp. 115–141). Miguel Ángel Porrúa Editores.

Aguilar Sánchez, G., & Lara Aguazul, C. (2021). Recursos naturales y desarrollo en Amealco, Querétaro.

Alexander, C., Ishikawa, S., & Murray, S. (1977). *A pattern language/a language of patterns*. Oxford University Press, Inc.

Ávila Sánchez, H. (2005). The urban-rural, new territorial expressions? CRIM.

Ávila Sánchez, H. (2009). Peri-urbanization and rural spaces on the outskirts of cities. *Agrarian Studies, 15*, 93–123. ISSN: 1405-2466.

Barsky, A. (2000). Managing the diversity of the peri-urban territory from the complexity of state institutions. In *Implementation of public policies for the sustainability of agriculture on the edges of the metropolitan region of Buenos Aires* (2000–2013).

Bellet, C., Melazzo, E., Sposito, M., & Llop, J. (2015). *Urbanization, production and consumption in medium/intermediate cities*. Edicions de la Universitat de Lleida.

Benard Calva, S. M. (2019). Autoetnografía. In S. Bénard Calva, E. Adams St. Pierre, T. Adams, A. Bochner, M. Burns, K. DeMarrais, C. Ellis, D. Jerz, J. Preissle, C. Rambo Ronai, L. Richardson, J. Tullis (Eds.), *Una metodología cualitativa*. Universidad Autónoma de Aguascalientes.

Benne, B., & Mang, P. (2015). Working regeneratively across scales—Insights from nature applied to the built environment. *Journal of Cleaner Production, 109*, 42–52.

Bolay, J. (2016). *Prosperity and social inequalities: Montes Claros, how to plan an intermediary city in Brazil* (2nd ed., Vol. 4).

Clegg, P. (2012). A practitioner's view of the 'Regenerative Paradigm'. *Building Research & Information, 40*(3), 365–368.

Cole, R. J., Busby, P., Guenther, R., Briney, L., Blaviesciunaite, A., & Alencar, T. (2012). A regenerative design framework: Setting new aspirations and initiating new discussions. *Building Research & Information, 40*(1), 95–111.

Cole, R. J., Oliver, A., & Robinson, J. (2013). Regenerative design, socio-ecological systems and co-evolution. *Building Research & Information, 41*(2), 237–247.

Cotrim, M., & Lara, F. L. (2020). Decolonizing the history of the hybrid housing building in Latin America. The cases of Lima, Caracas, Buenos Aires, Mexico City, Bogota and Sao Paulo. In R.L. Cansino & F.L. Lara (Org.), *Notes on decolonization, architecture and city in the Americas* (Vol. v.1, pp. 279–326).

Dalla Torre, J., Sales, R., Esteves, M., & Ghilardi, M. (2017). The particularities of urban-rural borders in territorial planning. In *Workshop of the Ibero-American Network of Territorial Observation* (5th: 2017: Mendoza, Argentina).

Delgado-Castillo, C. L. (2017). *The analysis of patterns as a tool for defining the essence of a place located in the southern urban periphery of Mexico City, within the framework of regenerative development*. Doctoral Thesis, Pontifical Catholic University of Rio de Janeiro.

Du Plessis, C., & Brandon, P. (2015). An ecological worldview as the basis for a regenerative sustainability paradigm for the built environment. *Journal of Cleaner Production, 109*

Dueñaz Checa, l. A. (2017). Elements for the contemporary analysis of the tension between the urban and the rural. *Virtual Catholic North Magazine, 51*, 272–291. http://revistavirtual.ucn.edu.co/index.php/RevistaUCN/article/view/847/1365

Fernández, P., & de la Vega, S. (2017). *The rural in the urban? Peri-urban localities in the Metropolitan Area of the Valley of Mexico*. Retrieved from http://dle.rae.es/?id=ZcqJYVW

George, P., Guglielmo, R., & Lacoste, B. K. (1967). *Active geography* (p. 414). Barcelona.

Gibbons, L. V. (2020). Regenerative—The new sustainable? *Sustainability, 12*(13), 5483.

González Gómez, C. I., Duering Cufré, E., & Basaldúa Hernández, M. (2012). Current urban morphology in intermediate cities: Santiago de Querétaro, Mexico. In *International Seminar of the Ibero-American Network of Researchers on Globalization and Territory (RII)*, Belo Horizonte, Brazil. October 1–4, 2012

Herrera Torres, H. (2011). Evaluation of municipal performance. Methodological proposal for the semi-urban municipalities of the State of Michoacán.

INEGI. (2013). Mexican Elevation Continuum (CEM).

INEGI. (2020a). Population and Housing Census. INEGI

INEGI. (2020b). National Statistical Directory of Economic Units (DENUE).

INEGI. (2020c). System for the Consultation of Census Information 2020 (SINCE 2020).

INEGI. (2022a). Agricultural Census (CA).

INEGI. (2022b). National Geo-statistical Framework.

Jiménez Ramos, D. (2019). *Community mapping and social cartographies: Creative, pedagogical, intervention and community accompaniment processes for the social management of territories.* Camidabit-The Interns.

Lyle, J. T. (1996). *Regenerative design for sustainable development.* John Wiley & Sons.

Mang, P., & Reed, B. (2012). Designing from place: A regenerative framework and methodology.

Mang, P., & Haggard, B. (2016). *Regenerative development and design: A framework for evolving sustainability.* Wiley.

Mang, P., & Reed, B. (2020). Regenerative development and design. *Sustainable Built Environments,* 115–141.

Martino, D. (2005). Bioregionalism: Introduction to concepts and alternatives for Latin America.

Ornelas Delgado, J. (2000). *The City under Neoliberalism. Population Papers, 6,* 45–69.

Parussini, A. M. J. (2012). The process of suburbanization of the western metropolitan cord of Rosario. *Urban Notebook, 12*(12).

Reed, B. (2007). Shifting from 'sustainability' to regeneration. *Building Research & Information, 35*(6), 674–680.

Restrepo, E. (2018). *Etnografía: alcances, técnicas y éticas.* Universidad Nacional Mayor de San Marcos.

Roggema, R. (2022). *Designing for regeneration.* Springer.

Salazar, G., Irarrázaval, F., & Fonck, M. (2018). Urban transformations and senses of place in the intermediate cities of the Araucanía Region. *AUS, 23,* 4–11.

Svec, P., Berkebile, R., & Todd, J. A. (2012). REGEN: Toward a tool for regenerative thinking. *Building Research & Information,* 40(1), 81–94.

Tafuni, V., & Heß, J. (2019, June). Education for sustainable development as a catalyst and the role of students in the future management of HEIs. In *Proceedings of the keynote at the Bologna Process Conference,* Bologna, Italy (Vol. 24).

Wahl, D. C. (2016). *Designing regenerative cultures.* Ecohabitar

Wals, A. E. (2019). Sustainability-oriented ecologies of learning: A response to systemic global dysfunction. In *Ecologies for learning and practice* (pp. 61–78). Routledge.

Chapter 7
The Environmental Observatory for Mining Projects: A System for the Analysis of Public Environmental Management Information

Kay Bergamini, Piroska Angel, Katherine Mollenhauer, José Ignacio Medina, Vanessa Rugiero, and Javiera Rodríguez

Abstract Access to public environmental management information in Chile suffers from weaknesses such as institutional dispersion of data and asymmetries in access to information by different actors (Civil society, public sector, and private sector). These issues have been particularly acute for the mining industry, a sector with the most investment contribution to the country but also associated with significant socio-environmental conflicts. Recognizing this challenge, the research team proposed and was awarded the FONDEF Environmental Observatory of Mining Projects project. This initiative aimed to develop a pilot environmental management information system focused on mining projects in two regions: Valparaíso and Metropolitana. The system was designed to capture, integrate, and disseminate information, reducing access and knowledge gaps and generating understandable indicators for different types of users. Both the definition of indicators and the construction

K. Bergamini
Institute of Urban and Territorial Studies and Energy Center, Pontificia Universidad Católica de Chile, Santiago, Chile
e-mail: kbergani@uc.cl

P. Angel (✉) · V. Rugiero
Institute of Urban and Territorial Studies, Pontificia Universidad Católica de Chile, Santiago, Chile
e-mail: paangel@uc.cl

V. Rugiero
e-mail: vrugiero@uc.cl

K. Mollenhauer (✉) · J. Rodríguez
School of Design, Pontificia Universidad Católica de Chile, Santiago, Chile
e-mail: kamollenhauer@uc.cl

J. Rodríguez
e-mail: jrodriguezbirke@uc.cl

J. I. Medina
Ministry of Energy, Providencia, Chile
e-mail: jose.medina@uc.cl

M. Vicuña and E. Giorgi (eds.), *Intersections*, Springer Geography,
https://doi.org/10.1007/978-3-031-76402-8_7

of the visualization platform were developed using the co-creation methodology, testing, and validating with potential users. This methodology comprised a robust participatory component to ensure perspective integration from different types of users from the regions analyzed. The co-creation approach facilitated greater accessibility to complex information by exploring ways to present it at different levels of complexity. Furthermore, it enabled screening of new ways to communicate the information, approaching the citizenry looking for a graphic language that is easier to understand and allows the reading and understanding of the data. This article addresses the analytical framework to explain the role of access to environmental information. Then, it presents the methodological framework of the co-creation process and the results of the process. Finally, the conclusions are included.

Keywords Information · Indicators · Environment · Mining · Co-creation

7.1 Introduction

Since the late 20th century, Latin America has experienced an upward trend in mining exploitation, with Chile positioning itself as a global reference in copper production (Valor Minero, 2017), resulting in economic prosperity, such as generating jobs, increasing GDP, and boosting government revenues, among others (Mining Council, 2023; Ministry of Mining, 2022). Indeed, largely thanks to mining, Chile's GDP per capita in the last 40 years increased from 2,540 USD in 1980 to 16,251 USD in 2021 (OECD-UN, 2018; World Bank, 2021).

However, this economic gain has been accompanied by substantial (Alimonda, 2011; Barton et al., 2015; Bebbington, 2012; Bergamini & Dextre, 2022; Engels & Dietz, 2017; Gudynas, 2019; Göbel & Ulloa Cubillos, 2014). According to Schorr (2018), the increase in investments and mega-mining projects generated environmental conflicts related to impacts, such as water use, alteration of ecosystems, and pollution. Nowadays, Chile is among the nations with the most mining-related conflicts in Latin America (Latin American Mining Conflict Observatory, n.d.). On the other hand, the threat posed by climate change calls into question the mining industry, given that the production and reserve of minerals in Chile are located in areas of climate risk and reserve of minerals in Chile are located in areas of climate risk and water stress, its contribution to greenhouse gas (GHG) emissions, its participation in electricity consumption and emissions attributable to the sector (Ministry of Mining 2022; National Society of Mining, Mining Council, Association of Industrial Suppliers of Mining, 2019).

On the other hand, given the global climate and ecological crisis, minerals such as copper, nickel, molybdenum, and lithium are fundamental for the energy transition and the decarbonization of the economy (Bruckmann et al., 2022). In this context and according to World Bank projections, the demand for lithium will increase by 965%. Therefore, Chilean mining faces the challenge of providing these minerals, minimizing environmental impacts, managing waste, protecting biodiversity, establishing relationships with communities, and creating quality jobs (Fundación Encuentros del Futuro et al., 2022).

However, asymmetries in access to environmental management information could exacerbate socio-environmental conflicts. Proper and transparent information management can reduce conflicts and improve companies' environmental and social performance (Cohen, 2007; Vera, 2015). Despite public data availability, its fragmentation and complexity represent significant barriers (Manfredi-Sánchez, 2017; Yáñez & Villatoro, 2005). Information must be presented comprehensibly to a diverse audience, from experts to the public (Gauchi, 2012).

Within the previous scenario, the project funded by the National Agency for Research and Development (ANID) through The Fund for the Promotion of Scientific and Technological Development (Fondef), "Environmental observatory for mining projects: a system for the analysis of public environmental management information," emerged. This project aimed to develop an environmental management information system prototype focused on mining projects in the Valparaíso and Metropolitan regions. This system aims to integrate information, reduce access gaps, and generate understandable indicators for citizens (civil society), public officials, and private entities. The working hypothesis indicates that a system of access to public environmental management information that integrates information available on the websites of the Ministry of the Environment, Environmental Assessment Service, Superintendency of the Environment, and the Environmental Courts of mining projects in the Valparaíso and Metropolitan regions, and delivered to civil society, private agents and public agents through the elaboration of indicators, their visualization and linking to the origin of the data, allows decreasing gaps in access to environmental management information, facilitates the understanding of it and increases the appreciation of the integration.

The case of the Valparaíso and Metropolitan regions are critical, as they present an intense competition for territory, given that they concentrate the country's largest population, develop productive vocations such as agriculture, viticulture and have limited resources of energy and water, and in turn, present a significant mining potential to be developed in the future (Valor Minero, 2017). The above sets a complex scenario and a significant challenge. On the one hand, mining, an activity whose projects are developed in the medium and long term, needs to anticipate the opportunities, risks, uncertainties, and developments that could affect it (Valor Minero, 2017). Simultaneously, it inherently requires its sustainable development, the building of society's trust towards the government (responsibility, reliability, transparency, legal process, ethics, and supervision), and the relationship with the industry in decision-making processes (Valor Minero, 2017). Additionally, it can be estimated from the information on environmental platforms that there are 60 mining projects in both regions; however, according to the SII (2020), there are 28,731 companies in the mining and quarrying sector. It should be noted that, according to CPC & OJO (2020), the judicialized investment in Chile for the mining sector between 2012 and 2019 amounted to 25.618 billion dollars, corresponding to 21 projects (Fig. 7.1).

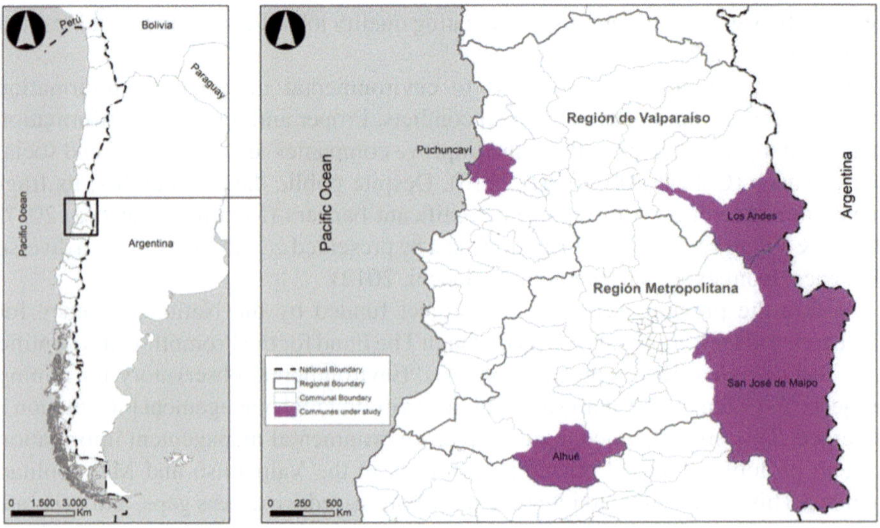

Fig. 7.1 Location of the study area. Valparaíso and Metropolitan Region in Chile. (Own elaboration)

Next, this article presents the conceptual and theoretical framework considered, a synthesis of the project's main results, to end with the discussion, conclusions, and projections given the current context in the subject.

7.2 Theoretical Framework

7.2.1 Access to Environmental Information

Access to environmental information has taken on critical importance in contemporary contexts, especially concerning environmental matters, which has been recognized in various international declarations such as Rio (UN, 1992) and regional ones such as Aarhus (UNECE, 1999) and recently Escazú in Latin America and the Caribbean (ECLAC, 2022). This access plays a crucial role in managing socio-environmental conflicts, companies' environmental and social performance, and informed civic participation (Cohen, 2007; Vera, 2015). However, significant challenges are identified due to the fragmentation of information and lack of understanding, which create barriers for citizens and complicate accountability and informed decision-making (Manfredi-Sánchez, 2017; Yáñez & Villatoro, 2005).

One of the characteristics of environmental information is that it is often expressed in a technical and tangled way in specialized jargon, which requires additional efforts for its democratization and understanding (Botero, 2012; Fuenmayor, 2004). The comprehensible presentation of data and information for a diverse audience is fundamental; as Gauchi (2012) maintains, accessible information is an engine of progress and a strategic axis for common well-being.

In recent years, research on environmental conflicts and inequality in the distribution of their harms has increased, highlighting the need for deliberative spaces and the use of environmental performance indicators to manage impacts (Alonso & Costa, 2002; Niemeijer & De Groot, 2008; Sabatini, 1997; Singh et al., 2012). In this line, Bergamini et al. (2023a) and other experts emphasize how public environmental information improves the performance of companies and fosters informed citizenship, generating resilience and "learning capacity" as crucial elements for an innovative and adaptive society (Jeans et al., 2016).

As mentioned, the Rio Declaration (UN, 1992) already recognized citizen participation as crucial for addressing environmental problems. Therefore, states are responsible for facilitating access to accurate and relevant information, allowing informed and responsible public participation in environmental management (Herrera et al., 2013; Whittaker, 2017).

Specifically, the Escazú Agreement, a milestone in Latin America and the Caribbean, underscores the importance of equality in access to information, promoting regional standards and tools for effective citizen participation in environmental matters (Nalegach, 2019). This agreement aligns with dialogue, transparency, and accountability principles, which are fundamental for environmental governance and conflict prevention.

In this landscape, technology emerges as an essential ally as it broadens access and understanding of environmental information, allowing precise and attractive presentations that encourage active participation (Barceló et al., 2009; Gouveia et al., 2004; Ojeda, 2009; Pérez et al., 2021). However, as Manfredi-Sánchez (2017) warns, having access to information does not guarantee understanding. Technical information must be contextualized so that citizens understand its direct impact on their quality of life.

7.2.2 *Interdisciplinary Work to Address Complex Problems*

One of the main characteristics of the 21st Century is the emergence of problems of great magnitude and complexity that demand an interdisciplinary approach (Vicuña, 2022). Such problems are manifested in various areas, from environmental issues to sociopolitical and economic challenges. The current trend of global development presents unsustainability, and globalization, technological advancement, and demographic increase have intensified the interdependence and complexity of problems, particularly environmental ones (Bergamini & Ángel, 2021; Tiessen, 2007).

In the above context, these global challenges cannot be effectively addressed from a single discipline because, as Max-Neef (2004) indicates, they transcend the individual scope of traditional disciplines.

Interdisciplinarity, according to García (2013), is a way of examining and introducing oneself to complex systems, for which it is necessary to construct a common language and collaborate among actors from various disciplines to understand and address a complex object of the study. In this context, solutions to contemporary problems require a holistic and integral vision that can only be achieved through interdisciplinary work.

7.3 Methodological Framework

For the realization of the project, we applied a methodology based on the triple diamond of service design from the Public Innovation Laboratory (2017), which includes the phases of "discover," "define," "develop," "deliver," "pilot" and "adjust," following five principles: user-centered, co-creative, sequenced, evidenced, and holistic (Stickdorn & Schneider, 2012).

As the primary technique, we used Co-Creation among key actors. This technique allows participants to (a) consider all perspectives of the problem, (b) be empathetic and tolerant to potential errors and iterations, and (c) generate final solutions with more excellent knowledge and wisdom, moving towards a horizontal exchange and the search for knowledge, ideas, and solutions that minimize the risks associated with the design and implementation of new solutions (Public Innovation Laboratory (Lipuc), 2017).

In particular, recent research suggests that co-creation as a participation technique triggers creative processes in which stakeholders create solutions and, therefore, are appropriate in their design and implementation (Mollenhauer et al., 2018; Soto et al., 2021). The techniques used to carry out these processes were surveys, interviews, and online and in-person participatory workshops, considering the COVID-19 pandemic context in which it unfolded (Bergamini et al., 2023b) (Fig. 7.2).

The methodological process proposed in the project based on the co-creation approach considered five stages, as can be seen in the figure and is developed below (Fig. 7.3):

Stage 1, perception of access to public environmental management information

Its purpose was to identify the gaps in access to environmental information, the dimensions of environmental information that interest different types of actors, and the obstacles and facilitators to access it. The process was carried out through 278 face-to-face and self-administered surveys based on an intentional theoretical sampling of both citizens and members of public and private organizations, and 30 semi-structured interviews with key actors (organized civil society interested in environmental issues, private agents (business people with investment projects subject to environmental impact assessment, environmental consultants and mining guild) and public agents linked to mining and environment (ministries, undersecretaries, municipalities).

Fig. 7.2 Photograph of participatory activities carried out in the project context. (Own elaboration)

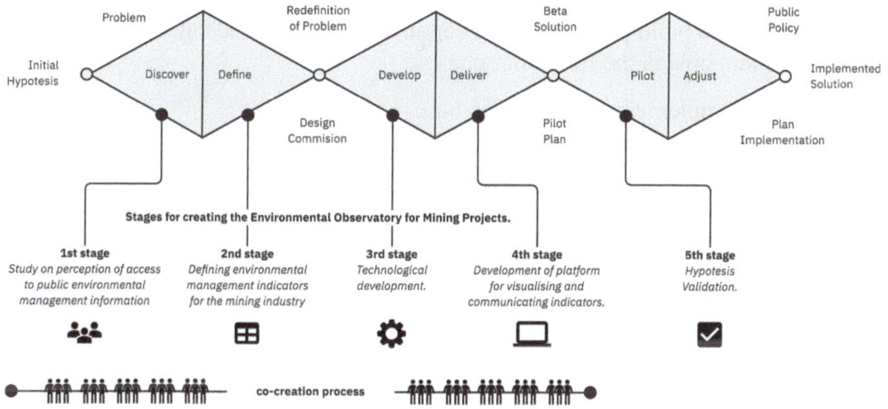

Fig. 7.3 Stages for the construction of the Environmental Observatory of Mining Projects. (Own elaboration)

Stage 2, definition of environmental management indicators of the mining industry

Its purpose was to propose indicators that will be monitored on the platform. The process involved two participatory workshops, feasibility analysis, and indicators ranking.

Workshop 1: Based on the results of the interviews, an activity was created that allowed for the identification and creation of a consensus on dimensions and variables of interest in environmental information focused on mining. A prototype of indicators was generated with 27 online participants.

Workshop 2: The interest and focus of the identified variables were validated, inviting participants to comment on the prototype and its formulation, obtaining prototype 2 of indicators. The activity was conducted online (16 participants) and in person on the field (84 participants).

Subsequently, the feasibility of the indicators was analyzed using the SMART criteria suggested by Schomaker (1997) to select preliminary indicators. This methodology evaluates the quality of the indicators from the point of view of technical feasibility, clarity, possibility of measuring and achieving, and their relevance. The definition of the acronyms was adjusted to contextualize the project.

Finally, the technical feasibility of data collection, the existence of input data, the use of public and reliable data, and the feasibility of automation in data extraction were investigated to have a matrix of indicators (prototype 3) that is usable and understandable.

Workshop 3: Prototype 3 was used after a SMART technical evaluation, inviting participants to rank the indicators. The activity was conducted online (20 participants) and in person on the field (44 participants). As a final result, prototype 4 of environmental management indicators was obtained.

Stage 3, technological development

Its purpose was to build the software to capture, normalize, and calculate indicators. Its development covers the following phases:

- Design and implementation of a database.
- Phase of programmed data extraction and storage in the database
- Data normalization
- Design and development of the calculation engine for the first ten indicators of the feasible ranking to calculate according to the project development times.

Stage 4, development of visualization and communication platform for indicators

Its purpose is to obtain the web page for consulting indicators. It involved iterative testing to adapt information delivery to different types of users, contacting new people, and using samples from previous stages.

The process includes the review of background and references, followed by an internal workshop to establish technical and operational requirements, giving coherence to the construction of the interface for the needfinding with users. Then, wireframes of the platform were created, and a cycle of user testing began, improving the design iteratively.

Three instances of information collection and testing were designed for the interface design:

- Concept workshops with 11 experts in environmental training.
- Workshop with 13 experts from the field of environmental education.
- Quick validation tests of an actionable prototype with 43 people from the public, private sector, and citizens.

The result was an interface designed and tested by different types of users. Subsequently, the platform was programmed to connect the back and front end.

Once the programming was finished, two validation instances were designed: (1) Thorough internal review by project professionals and researchers; (2) Controlled soft launch with people who participated in the co-creation instances, seeking to validate the usability and understanding of the platform and its indicators. The guided navigation consisted of a 1:1 instance of a researcher with a participant, performing tasks and evaluating navigation qualitatively. It was conducted with seven people (3 from the public, two from the private, and two citizens) through the Zoom platform over 30 min.

Stage 5, Hypothesis Validation

It sought to generate evidence to validate the initial hypothesis of the project, which holds that the platform reduces access gaps to environmental management information, facilitates its understanding, and increases the valuation of information integration. The platform was piloted in surveys and navigation processes with users and citizens, public and private agents, contacting municipalities and key actors from the pilot communes of the study. A total of 110 surveys were conducted.

The process had three moments: (i) initial survey to characterize the respondents and to know their vision about environmental information; (ii) presentation of the platform and handling by the users; and (iii) closing survey to evaluate the platform in terms of understanding, navigation, and reliability. Descriptive statistics and the qualitative comparative method (QCA) were applied to analyze the results. Descriptive statistics allowed us to determine patterns and compare results without and with the platform. The QCA identified patterns and concurrences of cases using independent variables: relevance of information, information search, experience in the search, and learnings from the platform.

7.4 Results

7.4.1 Diagnosis of Access to Information

A diagnostic report was generated on access to public information on environmental management that, in its summarized version, can be consulted at (chrome-extension://efaidnbmnnnibpcajpcglclefindmkaj/https://estudiosurbanos.uc.cl/wp-content/uploads/2022/05/Primer-informe-de-avances_Observatorio-Ambiental-2.pdf).

About the initial findings of the research, it can be noted that the need for actors to have an integrated public information system to reduce gaps in access to environmental management information is validated, recognizing its current dispersion and the abundance of technicalities that prevent efficient and useful access for decision-making, citizen participation, and opinion formation based on data interpretation. At the same time, the importance of improving information delivery mechanisms focused on users is highlighted, as well as seeking strategies that allow the democratization of environmental language and positioning itself within the significant challenges of the project.

7.4.2 Environmental Management Indicators for the Mining Industry

In total, 25 indicators were defined and prioritized, as seen in Table 7.1.

These indicators go through a ranking process by the participating actors, where the top 10 in the ranking are selected for development.

7.4.3 Information Capture and Normalization Software

Public environmental management information capture software is developed to allow integration into a database and make its information available. Its development faced significant challenges, as the information available in different public services is in different file formats and contents, typing errors in the data, and incomplete and unstructured databases complicate the processes to standardize and join data, a problem that this software addresses.

This software stored on a server is programmed to capture public environmental information from public service websites through web services and web scraping from portals of the Environmental Assessment Service, Superintendency of the Environment, Ministry of the Environment, the Second Environmental Court, National Geology and Mining Service and General Water Directorate (Fig. 7.4).

7.4.4 Display Interface

We create an environmental information display interface linked to mining projects for the Valparaíso and Metropolitan regions. The platform provides environmental information on the regional, communal, and project scales.

The integration of the platform (data model, back-end, and front-end) allows the search for reliable information from the integration of different sources of public information along with its visualization from graphs and maps that facilitate its understanding. The platform is available at the following link: https://observatorioambientaluc.cl/.

The most relevant elements regarding the effective communication of the platform correspond to the following (Table 7.2 and Fig. 7.5).

Table 7.1 List of environmental management indicators for the mining industry according to dimension and priority ranking. (Own elaboration)

Dimension	Indicators	Ranking
Industrial and mining activities	Jobs associated with the mining sector in the municipality	11
	Jobs associated with projects in the SEA	13
	Operating mining projects in the territory	20
	Approved investment	21
	Investment under evaluation	23
Water	Emissions to water	1
	Water use	4
	Water demand	16
Air	Local mining air emissions	2
	Air emissions that contribute to climate change	3
Biodiversity	Protected areas or areas of conservation interest	15
Environmental management	Environmental compliance	5
	Conflict of mining projects	8
	Environmental management activities in the territory	10
	Average project approval times	14
	Complexity of environmental management of projects	18
	Mining patent	19
	Projects entered into the environmental assessment system	25
Human environment	Occupational safety	17
Waste	Hazardous waste	6
	Non-hazardous waste	12
Soil	Mine closures	2
	Surface and location of tailings	7
	Surface affected by mining activity	9
Territory	Projects in a state of abandonment	22

7.4.5 Hypothesis Validation

Based on a survey conducted with users who tested the platform, the results showed that 63.3% of the participants found it easy or very easy to search for information on it; 98.1% found the information they were looking for; 74.5% would navigate to find environmental information; 53.6% would use it for easy access; and 52.7%

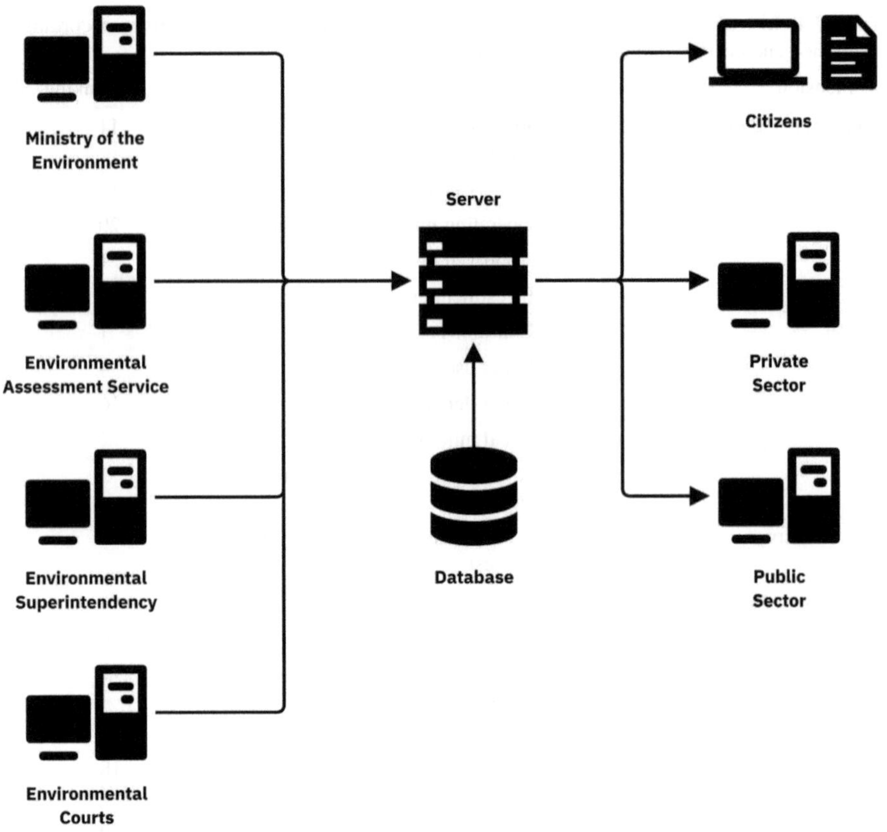

Fig. 7.4 Example scheme of the Information capture and normalization software. (Own elaboration)

because it contains relevant information. 83.9% mentioned that it would be likely or very likely to use the platform in the future.

On the other hand, the QCA analysis showed a consistency of 96% in the formula used and a 95% "raw consist" in the "truth table," explaining the probability of using the platform with the mentioned variables, generating evidence in favor of the hypothesis, demonstrating that the platform reduces gaps in access to environmental management information, facilitates its understanding and increases the appreciation of integration.

Table 7.2 Attributes for effective communication and design elements considered in the developed web platform. (Own elaboration)

Category	Attribute	Web Implementation
Platform navigation	Intuitive	Few "steps" or "clicks" to reach the information of interest Use of colors to highlight key messages Incorporation of "buttons" to facilitate navigation Filters (by commune, region, project, or indicator variables) in order to facilitate the search for information
Usability	Understandable	Texts with simple and little technical writing Incorporation of hyperlinks with complementary information for complex concepts The user manual is available Incorporation of a glossary of terms specific to environmental management Uses simple and commonly recognized graphics by users Incorporate graphics that allow comparing the value of the indicator between regions, communes, and projects Generation of communal reports downloadable in PDF for distribution
	Attractive	Use of graphic resources (maps and graphs) Interactive Link to social networks
	Reliable	Explains the methodology for calculating the indicators Indicates the sources of information for the indicators Only uses validated public data sources Link to data sources used Incorporation of disclaimer and conditions of use of the platform's data

7.5 Discussion and Conclusions

Based on the experience in the Environmental Observatory of Mining Projects, the results confirm the importance of public access to environmental information. Among the findings, the fragmentation of information and its presentation in technically complex formats, similar to what was identified by Yáñez and Villatoro (2005) and Manfredi-Sánchez (2017), continue to represent significant barriers to effective access and understanding by the general public. Furthermore, the experience allowed us to identify that clarity in data presentation, interactivity, and simplification of technical language are critical elements for effective communication.

In the previous context, while regulations have established standards for accessing information, implementation faces obstacles, such as the discrepancy between the availability of information, its accessibility, and comprehensibility for the general public, similar to the challenges pointed out by Botero (2012) and Fuenmayor (2004). This highlights what has been argued in this work concerning the fact that having information available differs from effective access. Therefore, work is needed to 'translate' this information, as warned by Manfredi-Sánchez (2017), allowing

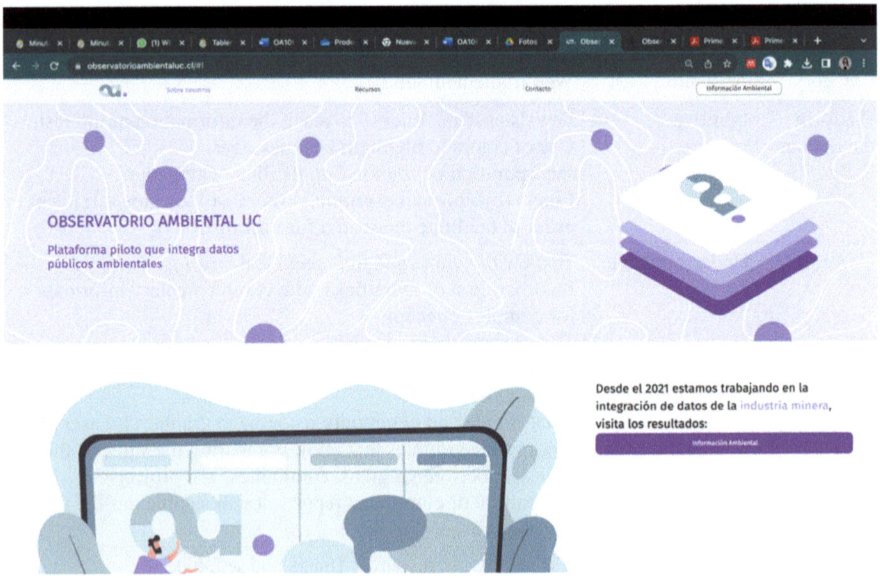

Fig. 7.5 Screenshots of the environmental observatory of Mining Projects. (Own elaboration)

citizens to empower themselves to make informed decisions and actively participate in the socio-environmental dialogue, companies to make better decisions with knowledge of the consequences and the State to propose and manage better public policies.

The complexity of the work carried out, which dealt with environmental themes, design, data, and computer science, among others, could be addressed through the interdisciplinary approach of the team, in line with the emergence of large-scale and complex problems, as pointed out by Vicuña (2022). In the same vein and aligned with ECLAC-UN (2003), this approach allows for a broader, more enriching vision. Also, García (2013) mentions that interdisciplinarity is a way of examining complex systems for which a common language and active collaboration between different disciplines are required, for which the practices carried out, in which various techniques of information gathering, analysis, and development were triangulated, together with work with actors from the public sector, private sector, and civil society, allowing for the completion of a product like the Environmental Observatory of Mining Projects, which helps to advance towards a reduction of the gaps in access to environmental information as mentioned by Cohen (2007) and Vera (2015). Therefore, this interdisciplinary collaboration is expected to result in an innovative solution and one that is sustainable over time.

Finally, this work contributes to the existing literature, offering a practical guide on addressing complex challenges of resolving socio-environmental problems, the incorporation of technology, and access to information through interdisciplinary perspectives in future initiatives, especially in times of the climate and environmental emergency facing the planet.

Funding Research funded by the Fund for the Promotion of Scientific and Technological Development (FONDEF) ID20I10084 'Environmental observatory of mining projects: a system for analyzing public information on 'Environmental management.' and Technological Research Fund IT23I0066 'Environmental observatory of mining projects'.

References

Alimonda, H. (2011). The coloniality of nature. An approach to Latin American political ecology. In H. Alimonda (Coord.), *Colonized nature*. Political Ecology and Mining in Latin America. Bs. As., CLACSO.

Alonso, A., & Costa, V. (2002). Por uma sociologia dos conflitos ambientais no Brasil. In Alimonda, H. (ed), *Ecología política. Nueva Naturaleza, sociedad y utopía*. CLACSO, Buenos Aires.

Barceló, F., Gutiérrez-Pérez, J., & Perales, F. (2009). What tools do information and communication technologies provide for environmental education? *Eureka Journal on Teaching and Dissemination of Sciences, 6*(3), 318–344. ISSN 1697-011X. https://doi.org/10.25267/Rev_Eureka_ensen_divulg_cienc.2009.v6.i3.02

Bebbington, A. (2012). Mining, social movements, and peasant responses. A political ecology of territorial transformations. Lima, IEP–CEPES.

Bergamini, K., & Angel, P. (2021). Conceptos Básicos para la Gestión Ambiental. In Bergamini, K. (eds), Gestión del Cumplimiento Ambiental. Editorial: RIL editores, Santiago de Chile, 565. ISBN 9788418982439.

Bergamini, K., & Dextre, R.-M. (2022). Export of pollution in Chile: Analysis of sanctioning procedures of the superintendency of the environment, 2013–2019. *EURE (Santiago), 48*(145), 1–27. https://doi.org/10.7764/eure.48.145.06

Bergamini, K., Ángel, P., Rugiero, V., Medina, J. I., & Mollenhauer, K. (2023b). Building resilient communities: The environmental observatory for mining projects and climate change indicators. *Sustainability, 2023*(15), 6947. https://doi.org/10.3390/su15086947

Bergamini, K., Mollenhauer, K., Ángel, P., Rugiero, V., Medina, J., & Curillán, C. (2023b). Challenges of the pandemic in the participatory and co-creative process for constructing the environmental observatory of mining projects. In *Suspensions* (pp. 176–184). UC Editions.

Botero, C. (2012). *The right to public information in Latin America*. Commission of the American States; Inter-American Commission on Human Rights.

Bruckmann, M., Barrios, M. A., & Lajtman, T. (2022). The environmental geopolitics of the United States and its northern allies. Implications for Latin America. Chapter 4. Latin America in the geopolitics of the 21st century. The Decline of the United States and the Rise of China. Ed. CLACSO and UNIJUI (pp. 121–180).

Cohen, M. A. (2007). A new environmental management: Risk and the precautionary principle. *Open Space, 16*(2), 209–222.

Decree 2: Approves National Mining Policy 2050. Ministry of Mining. 2023.

ECLAC. (2022). https://repositorio.cepal.org/server/api/core/bitstreams/a6049491-a9ee-4c53-ae7c-a8a17ca9504e/content

Encounters of the Future Foundation et al. (2022). Chile is a world leader in green mining. https://cloudcomisionfuturo.bcn.cl/index.php/s/yc4miHxtgSPPLS4

Engels, B., & Dietz, K. (Eds.). (2017). *Contested extractivist, society, and the state: Struggles over mining and land*. Springer.

Fuenmayor, A. (2004). *The right of citizens to access public information*. UNESCO.

García, R. (2013). Investigación interdisciplinaria de sistemas complejos: lecciones del cambio climático. *INTER DISCIPLINA, 1*(1). https://doi.org/10.22201/ceiich.24485705e.2013.1.46545

Gauchi, V. (2012). Right of access to public information. *Information Methods (MEI), II Era, 3*(5), 167–176.

Göbel, B., & Ulloa Cubillos, A. (Eds.). (2014). *Mining extractivism in Colombia and Latin America.* National University of Colombia, Bogotá, Faculty of Human Sciences.

Gouveia, C., Fonseca, A., Câmara, A., & Ferreira, F. (2004). Promoting the use of environmental data collected by concerned citizens through information and communication technologies. *Journal of Environmental Management, 71*(2), 135–154. ISSN 0301-4797

Gudynas, E. (2019). Development and limits to economic growth: A persistent controversy. In P. García, J. Ordoñez, & R. Munck (Eds.), *Development, society, alternatives* (pp. 252–279). Glasnevin Publishing.

Herrera, A., Moreno, P., & Escobedo, R. (2013). Access to environmental information. *Constitutional Issues, 29*, 219–243.

Internal Revenue Service (SII). (2020). *Company Statistics.* Reviewed on 20/03/20 at http://www.sii.cl/sobre_el_sii/estadisticas_de_empresas.html

Manfredi-Sánchez, J. (2017). Horizons of public information. *The Information Professional., 26*(3), 353–360. https://doi.org/10.3145/epi.2017.may.01

Max-Neef, M. A. (2004). *Fundamentos de la transdisciplinaridad.* Universidad de Cuenca.

Mining Council. (2023). Updated mining figures. https://consejominero.cl/mineria-en-chile/cifras-actualizadas-de-la-mineria/

Mining Value. (2017). *The challenges for the future development of mining in the Central Zone* (p. 44). Mining Value.

Ministry of Mining. (2022). *National Mining Policy.* http://www.politicanacionalminera.cl/

Mollenhauer, K., Figueroa, B., Tello, C., & Wuth, P. (2018). Use of the prototype in the co-creative design of public services. In *FONDART case. Intersections 2016. II Interdisciplinary Congress of Research in Architecture, Design, City and Territory* (pp. 205–222).

Nalegach, C. (2019). Acuerdo de Escazú: desafíos para las decisiones ambientales y oportunidades para los derechos humanos. *Anuario de Derecho Público*, (1), 260–277.

National Society of Mining, Mining Council Association of Industrial Suppliers of Mining. (2019). *Position and commitment of the Chilean mining sector in the context of COP 25.* At: chrome-extension://efaidnbmnnnibpcajpcglclefindmkaj/https://consejominero.cl/wp-content/uploads/2019/09/Cambio-Climatico-Sonami-CM-Aprimin.pdf

Niemeijer, D., & de Groot, R.S. (2008) A Conceptual Framework for Selecting Environmental Indicator Sets. *Ecol. Indic*, 8, 14–25.

Observatory of Mining Conflicts in Latin America (OLCA). (n.d). Mining Conflicts in Latin America. https://mapa.conflictosmineros.net/ocmaldb-v2/

Ojeda, F. (2009). *Environmental education and information and communication technologies: Design, development, and evaluation of a collaborative program in secondary education.* Doctoral Thesis, University of Granada.

Organization for Economic Co-operation and Development/United Nations (OECD/UN). (2018). *Production Transformation Policy Review of Chile: Reaping the benefits of new Frontiers.* OECD Development Pathways. OECD Publishing. https://doi.org/10.1787/9789264288379-en

Pérez, D., Bosque Suárez, R., & Tornés Reyes, E. (2021). Environmental education through information and communication technologies in San Vicente, El Salvador. *Ibero-American Journal Environment & Sustainability, 4*, e056. https://doi.org/10.46380/rias.vol4.e056

Public Innovation Laboratory (Lipuc). (2017). *The co-production of the user in public services* (Vol. 1).

Sabatini, F. (1997). Conflictos ambientales y desarrollo sustentable de las regiones urbanas. *Revista EURE-Revista de Estudios Urbano Regionales*, 23(68).

Schomaker, M. (1997). Development of environmental indicators in UNEP. In *Proceedings of the Land Quality Indicators and Their Use in Sustainable Agriculture and Rural Development, Rome, Italy* (pp. 35–36). Retrieved December 22, 2022, from http://www.fao.org/3/w4745e/w4745e07.htm

Schorr, B. (2018). Unequal opportunities: Companies and states conflict mining in Chile. *Atacameño Studies*, (57), 239–255. https://doi.org/10.4067/S0718-10432018005000601

Singh, R. K., Murty, H. R., Gupta, S. K., & Dikshit, A. K. (2012). An overview of sustainability assessment methodologies. *Ecological indicators*, 15(1), 281–299.

Soto, M., Mollenhauer, K., Sarantou, M., & Miettinen, S. (2021). Building a community through service design and responsiveness to emotions. In T. Seppälä, M. Scranton, & S. Miettinen (Eds.), *Arts-based methods for decolonising participatory research* (pp. 123–145). Routledge. https://doi.org/10.4324/9781003053408

Stickdorn, M., Schneider, J. (2012). *This is Service Design Thonking: Basics, Tools, Cases*. John Wiley and Sons: Hoboken, NJ, USA. ISBN 978-1-118-15630-8.

Tiessen, H. (Ed.) (2007). Communicating global change science to society: an assessment and case studies (Vol. 68). Island Press.

UN. (1992). https://www.un.org/spanish/esa/sustdev/documents/declaracionrio.htm

UNECE. (1999). https://unece.org/DAM/env/pp/documents/cep43s.pdf

United Nations. (1992). *Rio Declaration on Environment and Development*. Preamble.

Vera, A. (2015). Nine answers to ten questions about access to public environmental information. In M. Juliá, R. Salassa, A. Vera, N. Olivares, I. González, M. Ulla, & M. Foradori (Eds.), *Legal research in environmental public policies*.

Vicuña, M. (2022). Territorios del mañana. Modelos predictivos de capacidad de carga para la gestión sustentable de los territorios. In R. Moris, et al. (Ed.), Instituto de Estudios Urbanos y Territoriales UC. Santiago:RIL editores. ISBN: 978-956-01-0907-1.

Whittaker, S. (2017). The right of access to environmental information and legal transplant theory: Lessons from London and Beijing. *Transnational Environmental Law, 6*(3), 509–530. https://doi.org/10.1017/S2047102517000115

The World Bank Annual Report (2021). From Crisis to Green, Resilient, and Inclusive Recovery (Vol. 1 of 3) (English). Washington, D.C.: World Bank Group. http://documents.worldbank.org/curated/en/120541633011500775.

Yañez, M., & Villatoro, P. (2005). Information and communication technology (ICT) and social institutionalism are new technologies. Towards a knowledge-based management. Division of Social Development, ECLAC (83 pp.).

Chapter 8
Huasco Wetland Observatory. Promoting the Environmental and Cultural Values of Urban Coastal Wetlands in the Atacama Desert

Osvaldo Moreno and Ximena Arizaga

Abstract Coastal wetlands in the Atacama Desert constitute important systems that are characterized not only by their landscape beauty and the exceptional biodiversity they harbor, but also by their important role as ecological infrastructure that provides key ecosystem services for sustainable development, resilience, and adaptability to climate change of territories and communities. At the same time, these are understood as the expression of broader and more complex ecological systems, related to regional hydrology and to migratory movements of birds on a continental scale. From a socio-ecological perspective, wetlands are vital ecosystems for human survival, as they represent some of the most productive environments in the world: they are cradles of biological diversity, sources of water and primary productivity on which countless plant and animal species depend for survival. Despite the above, these systems are commonly exposed to severe threats and impacts to their components and functions that affect not only the ecology of wetlands, but also the quality of life for surrounding communities. This chapter examines, based on the case study of the wetland at the mouth of the Huasco river, the approaches and strategies of landscape architecture for enhancing the visibility and conservation of the environmental and cultural values of urban coastal wetlands in the Atacama Desert, implemented through a co-production process with local communities and key stakeholders. The resulting initiative, known as Huasco Wetland Observatory, conceived as a space for mediation between the ecology of the wetland, the local community, and visitors, by activating an in-situ dissemination area that provides scientific and didactic information, thus contributing to the understanding of this important landscape.

Keywords Coastal wetlands · Atacama Desert · Landscape architecture · Co-production · Visibility · Conservation

---

O. Moreno (✉) · X. Arizaga
School of Architecture, Pontificia Universidad Católica de Chile, Santiago, Chile
e-mail: omorenof@uc.cl

X. Arizaga
e-mail: axarizaga@uc.cl

M. Vicuña and E. Giorgi (eds.), *Intersections*, Springer Geography,
https://doi.org/10.1007/978-3-031-76402-8_8

8.1 Introduction

The wetland at the mouth of the Huasco River is the principal coastal wetland in the Atacama Region, situated within the driest desert in the world. It is located immediately north of the city of Huasco, the capital of the province of the same name. This area is characterized by a significant industrial mining development, including an iron pellet production plant, a thermoelectric plant, and a mineral shipping port, which has led to decades of environmental degradation affecting various coastal ecosystems (MMA & CEA, 2017). On the other hand, the intense agricultural activity of the Huasco river valley has also generated effects of degradation and contraction of the wetland, mainly due to the historical occupation of soils for the development of olive plantations and other crops. The city has a population of 11,374 inhabitants (INE—National Institute of Statistics, 2022 projections) and attracts 27,119 tourists annually (Monthly Survey of Tourist Accommodations—INE, 2019), with an average floating population of 695 per day.

During the summer months, such as February, the population in the commune reaches 12,277 people (including residents, tourists, and temporary visitors), representing an 8% increase. This fluctuating population exerts significant pressure on the fragile wetland ecosystem due to unregulated activities and the systematic disposal of waste at its edges. Furthermore, one of the main issues identified in this research is the lack of awareness among the population regarding the wetland's qualities and functions (Moreno & Arizaga, 2022). Despite the highly anthropogenic environment surrounding it, the wetland supports a dynamic and diverse ecology, with water quality suitable for irrigation and conditions appropriate for recreational activities involving direct contact (MMA & CEA, 2017) (Fig. 8.1).

In this context, the purpose of the "Huasco Wetland Observatory"[1] project is to create a space for mediation between the wetland's ecology, the local community, and visitors. This is achieved by establishing an on-site platform for disseminating the wetland's environmental and cultural values, offering scientific and educational information that enhances the visitor experience and contributes to the wetland's visibility and understanding. The Observatory is proposed as a landscape design device that creates a space for exploration and interaction between the community and the captivating wetland landscape. It consists of a central space called "environmental classroom", characterized by a permeable gravel floor that facilitates educational and recreational activities. Around this space, vertical elements called "forests of reeds and cattails" are deployed, which on an enlarged scale represent the qualities of the marsh vegetation fibers found along the edges of the coastal wetlands. Like these natural species, these forests cast dynamic shadows and provide habitats for various forms of life. The Observatory also includes several "interpretive stations" that present information related to hydrology, geomorphology, ecology, and the history of the wetland through a series of horizontal panels.

Considering that the landscape "resides not solely in the object, nor solely in the subject, but in the complex interaction between the two" (Berque, 1994: 5), the

Fig. 8.1 Huasco wetland, located on the coast of the Atacama region, Chile. *Source* Authors, 2022

Observatory establishes a suture of this relationship. It is not, therefore, just a physical and spatial operation, but through the process of creation and implementation of this device, a comprehensive study of the landscape can be pursued, relating the different systems—biotic, abiotic, and anthropic—the various scales and temporalities that constitute the Huasco wetland. Thus, the project operates as a catalyst, revealing the multiple layers of this relationship through the inhabitants themselves, who also become observers. The various activities carried out that seek to challenge instrumentalized participation, aim to make the landscape project of the observatory a "logical procedure through which all the peculiarities of the phenomenon itself are verified, without this being interrupted by experimentation" (Littré Dictionary, 1958 [1873], meaning of "observation").

The following sections will present a synthesis of the study of the Huasco River wetland landscape, including its specific characteristics and ecosystem values. First, the study will be summarized, highlighting the wetland's unique attributes and ecological significance. Next, the methodology for participatory work and co-production will be explained, detailing the activities conducted and the results obtained. Finally, the implementation of the Huasco River Wetland Observatory landscape project will be described.

8.2 Huasco River Wetland: Values of a Water Landscape in the Atacama Desert

Continental and coastal wetlands worldwide cover more than 12.1 million km^2 (RAMSAR, 2018), and in the South American continent, the system of coastal wetlands that extends between the Tumbes river in northern Peru and the Aconcagua valley in central Chile stands out, forming an extensive eco-region of 3,800 km in length, from the semi-arid tropical zone of northern Peru to the semi-arid zone of Norte Chico in Chile (Tabilo et al., 2001). In addition to presenting valuable ecosystems that have sustained human communities for centuries, these wetlands, influenced by the Humboldt Current, serve as critical habitats for migratory bird species that travel to the distant coasts of Canada in the north and Patagonia at the southern tip.

In Chile, coastal wetlands are mainly of the estuarine type, mouths of gorges and rivers, followed by riparian wetlands, lake, marine, and marshy. In the northern regions, which are characterized by an extremely arid climate, the water supply comes from the high plateau with small and sporadic runoff throughout the year and significant increases in the summer. Regarding runoff, the area is characterized by the presence of endorheic type basins, with the basins of the Rio Lluta, Loa, Copiapó, Huasco, Elqui, and Limarí being the main flows that, either by rainfall or snowfall, flow into the ocean. This contribution influences the dynamics of the mouths, being open or closed according to the existing flow and the water extraction processes in each basin. The tides exhibit a micro-tidal regime, leading to the formation of sandbars at river mouths and contributing to the creation of a salt wedge at the mouths of low-slope rivers (Cienfuegos et al., 2012). In this context, the coastal wetlands are located on the coastal plains, close to mouths of gorges and rivers, as well as near surface water sources, whether underground or submarine. Factors such as high solar radiation and the evaporation process have influenced the condition of temporality of the water bodies (MMA, 2023) along with anthropogenic factors related to the construction of upstream reservoirs and water extraction from these wetlands.

In this context, the coastal wetlands south of Copiapó are in semi-arid and desert areas, contrasting with the rest of the environment that surrounds them. They are located within one of the 25 zones "hotspots" with the highest biological diversity in the world. In the Atacama region there are 5 coastal wetlands: the emerging wetland of Chañaral (39.24 ha) induced by the action of the inhabitants (Bresciani et al., 2023), the wetland and Nature Sanctuary of the mouth of the Copiapó River (174 ha), the coastal wetland and Nature Sanctuary Laguna Totoral (16.37 ha), the Wetland and Nature Sanctuary of Carrizal Bajo (46.8 ha) and the wetland of the mouth of the Huasco river (258.9 ha), which is the largest of the five and has been proposed by the municipality, also, to Nature Sanctuary.[2]

The wetland is situated in the estuary of the Huasco river, in the final stretch of the valley before its mouth. In its biotic and abiotic components, the consequences of natural and anthropogenic dynamics developed upstream are evident: to this arrive

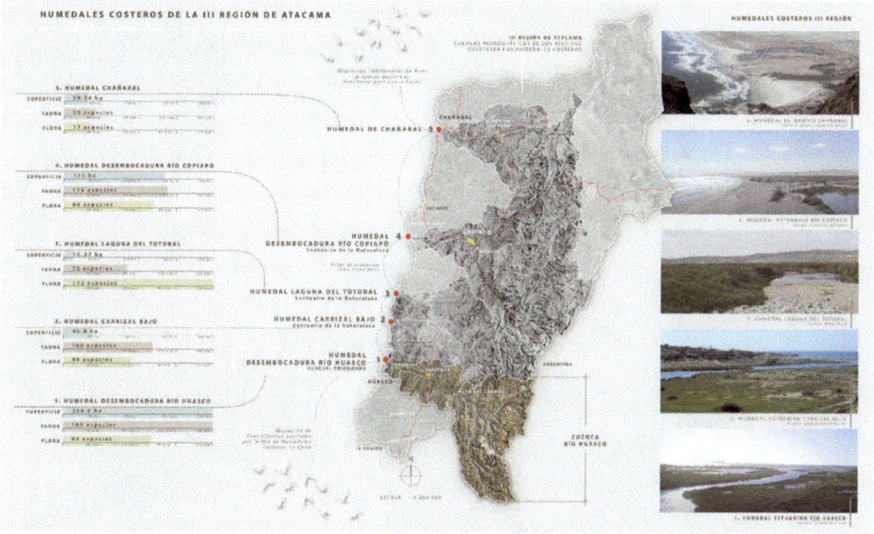

Fig. 8.2 Representation of the coastal wetland system of the Atacama region, Chile. *Source* Own elaboration within the framework of the Huasco Wetland Observatory Project, Promoting the environmental and cultural values of the coastal wetlands of Atacama, 2022.

sediments, organic matter, nutrients, and minerals that descend by the drag of its flow; as well as polluting substances derived from the livestock, agricultural and mining industry present within the basin (Fig. 8.2).

The Huasco valley is primarily characterized by its agricultural use, with predominant olive groves for oil production, grape cultivation, and also horticultural crops. In the upper basin of the Huasco river are located large-scale mining projects that extract gold and copper in addition to medium and small mining operations. South of the city of Huasco is located the industrial mining hub formed by the facilities of the Iron Pellets Plant and the mineral shipping port of the Mining Company of Pacific, next to the Guacolda thermoelectric plant, the largest coal-fired generator in the country. Located 65 km upstream within the valley is the Santa Juana Reservoir, the largest in the Atacama Region, which regulates the flow of the Huasco River. Despite this highly anthropogenic context, the wetland reveals a very dynamic habitat with considerable variation, offering water quality suitable for irrigation and conditions appropriate for recreational activities involving direct contact (MMA & CEA, 2017) (Fig. 8.3).

The wetland develops on fluvial deposit soils and floodplains, which provide a slightly permeable land capable of retaining the hydrological contributions from the Huasco River and maintaining humidity, in addition to providing soils suitable for the development of agriculture. The other landscape units that surround the wetland, such as the dune strip and the beach strip, develop on active and inactive aeolian soils, composed of sands that are characterized by greater permeability, being essential to filter sea water towards the wetland with a reduced salinity percentage. In turn, the

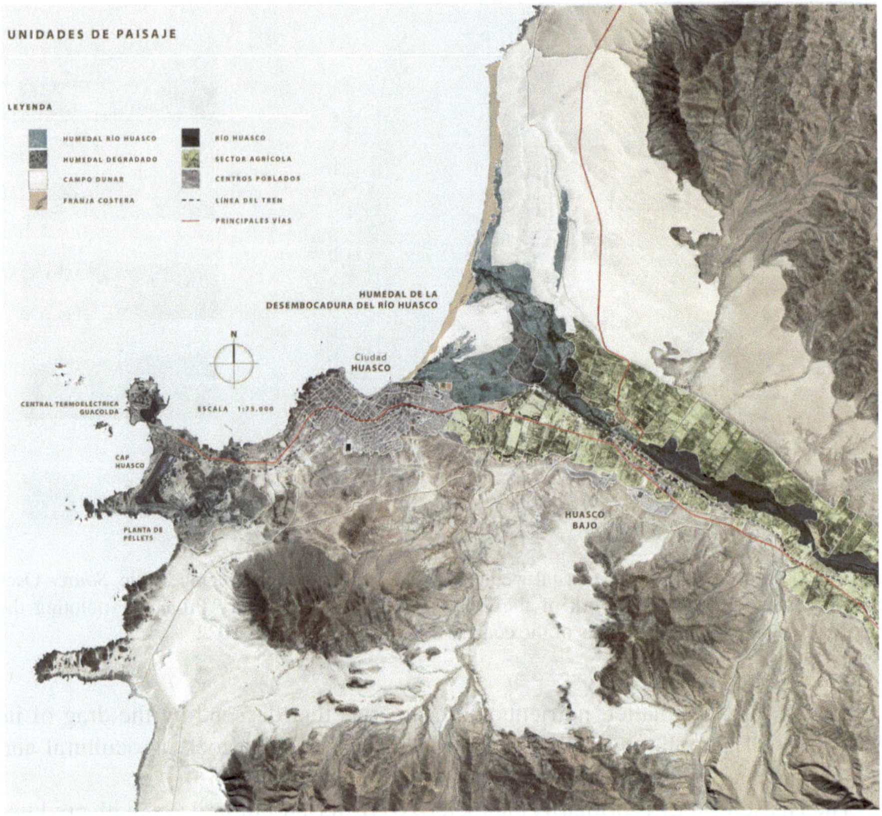

Fig. 8.3 Relationship of the wetland with the coastal landscape units, the valley and the city of Huasco. *Source* Own elaboration within the framework of the Huasco Wetland Observatory Project, Promoting the environmental and cultural values of the coastal wetlands of Atacama

dynamics of the Huasco River mouth is partly influenced by the presence of the dune bar in the coastal area that conditions the hydrological functioning of the wetland, with periods in which the outlet to the sea is obstructed by the dune, particularly in summer. This same dune also acts as a barrier between the wetland and the beach, protecting it from the intrusion of vacationers. Combined with the absence of clearly defined pathways, this natural barrier limits visitation and helps preserve the habitat conditions for the wetland's flora and fauna.

Its flow volume depends on the input from tributaries originating from the Santa Juana Reservoir and the water extraction rights for irrigation and other uses, without considering annual precipitation or ocean currents. By processing these inputs, the wetland has the capacity to purify the river water before it reaches the sea, through phytoremediation processes and sequestration of chemical substances in its subsoil: CO_2, CH_4, Ni, P, among others (MMA & CEA, 2017; Olivares, 2023). In this way, the wetland serves not only as an indicator of the state of the Huasco River basin, but

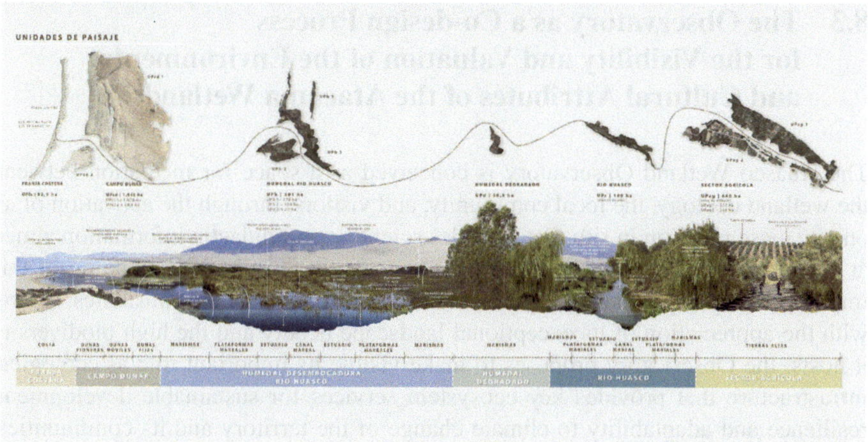

Fig. 8.4 Transect of the wetland and its ecological gradient, encompassing various landscape units. *Source* Own elaboration within the framework of the Huasco Wetland Observatory Project, Promoting the environmental and cultural values of the coastal wetlands of Atacama, 2022

also as a piece of green and blue infrastructure for natural treatment and regulation of river waters, endowing it with significant socio-ecological value in the desert context where it is located (Fig. 8.4).

According to the reviewed data records (MMA & CEA, 2017; Moreno & Arizaga, 2022), the Huasco River wetland has a total richness of 168 species of which 134 species (79.8%) correspond to birds, 16 species (9.5%) to micromammals, 12 species (7.1%) to reptiles, 4 species (2.4%) to macromammals (Guanaco, Culpeo Fox, Chilla Fox and Puma not confirmed in the field campaigns carried out in that context) and 2 species (1.2%) correspond to amphibians. Of these species, 150 are native (91.5%), 14 endemic (8.5%) and four introduced (2.4%), the latter correspond to micromammals. Regarding vegetation, 30 vegetation units have been identified, with a few species dominating the area. The dominance corresponds to perennial herbs followed by shrubs as the most common; and the form with fewer representatives corresponds to succulents. In the field, water bodies could be observed in which the combination of herbaceous plants, the presence of Totora clusters (*Typha angustifolia*) that can reach up to 3 m in height and the colonies of ducks and coots stand out as a characteristic that gives scenic beauty to the wetland.

8.3 The Observatory as a Co-design Process for the Visibility and Valuation of the Environmental and Cultural Attributes of the Atacama Wetlands

The Huasco Wetland Observatory is conceived as a space for mediation between the wetland ecology, the local community, and visitors, through the activation of an on-site dissemination in situ that provides scientific and didactic information aimed at enhancing the visitor experience, thus contributing to the understanding of this important landscape and its various environmental and cultural attributes. Along with the appreciation of its exceptional landscape beauty and the high biodiversity it hosts, the Observatory proposes to make visible its important role as ecological infrastructure that provides key ecosystem services for sustainable development, resilience and adaptability to climate change of the territory and its communities. It also contextualizes the Huasco wetland as part of broader and more complex ecological systems, related to the valley landscape and regional hydrology, with the system of coastal wetlands of the Atacama Desert and with the continental scale bird migration movements.

In this framework, the project proposes that the Observatory is not only the equipment enabled in the physical space built for that purpose, but also the co-design process[3] carried out with various stakeholders in the territory. In this sense, all the actions conducted—such as workshops, meetings, encounters, tours and reviews—contribute to the visibility and understanding of the wetland landscape and, therefore, the observatory is expressed through the activation of the socialization initiatives, discussion and deliberation, which are organized into three phases: (1) the collection and synthesis of scientific information; (2) the development of participation workshops and international seminar[4] with key stakeholders, and; (3) the on-site implementation of the landscape architecture device.

The first activity of the Observatory consisted in collecting the existing information about the Huasco Wetland, which was found mainly, in the application file for the figure of protection of Nature Sanctuary (MMA & CEA, 2017) and in other related studies and research (Moreno & Arizaga, 2022; Olivares, 2023). In this instance, the scientific and technical background is synthesized to transform them into didactic information to be displayed on graphic supports called interpretive panels, which form a sequence of sensitive visibility of the main dynamics and characteristics that assign to the wetland landscape the relevant environmental and cultural values.

Secondly, the information collected and processed that leads to the construction of the series of interpretive panels is presented in 3 collaborative work workshops, allowing for participation and discussion with the community, with representatives from the municipality and public services at the regional level, who systematically participated in the entire process (Fig. 8.5). In each workshop, proposals for improvements to the presented material were generated, as well as new ideas to create additional panels for the project (Figs. 8.6 and 8.7), dedicated—among other topics—to explaining the importance of the wetland in the context of the cultural landscape of the valley, detailing the hydraulic functioning of the wetland in the context of

Fig. 8.5 Exhibition of the workshop 1 sheets. Photography: Huasco Wetland Observatory Project, Promoting the environmental and cultural values of the coastal wetlands of Atacama, 2022

the estuary where it is located, and deepening the understanding of the ecosystem services it provides to the territory.

After the workshops, socialization activities of the Observatory project are conducted on-site with the community, in the border areas between the wetland and the city. Through various tours and technical discussion activities on site (Fig. 8.8), the potential programs for dissemination, education, and recreation that can be included in the project are confirmed. These activities are widely disseminated in the local and regional press and are also highlighted in seminars and scientific meetings where the project has been presented and published. Given the community's interest in having the material used in each of the workshops, which was refined throughout this shared observation, a webpage[5] is enabled to download graphic and audiovisual materials, particularly for those who cannot travel, to learn about the wetland.

Finally, the on-site implementation of the landscape architecture device in the physical space of the Observatory is carried out with the support of the municipality and the active participation of local companies, thus facilitating the quick and efficient construction of the project, which was inaugurated in January 2023, with broad participation from the local community, tourists, and regional and municipal authorities. Its location responds to a consensus vision, aimed at transforming a degraded

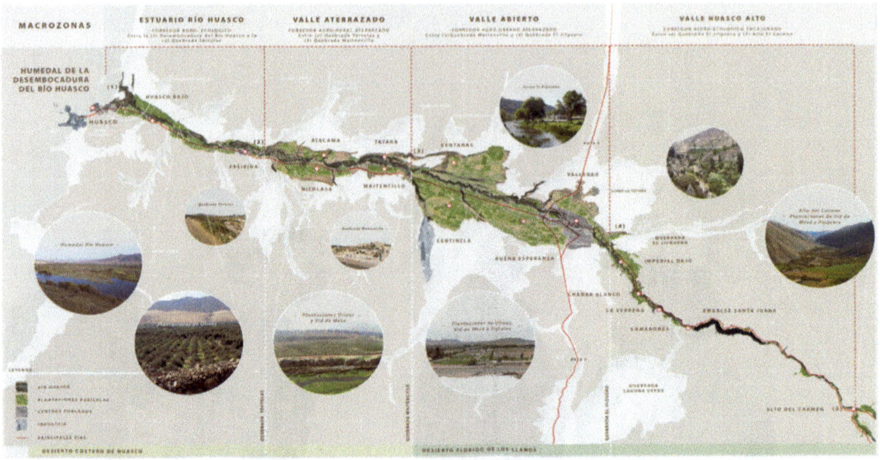

Fig. 8.6 Macro-zones of the Huasco Valley landscape. *Source* Own elaboration within the framework of the Huasco Wetland Observatory Project, Promoting the environmental and cultural values of the coastal wetlands of Atacama, 2022

Fig. 8.7 Biodiversity and metabolic processes of the Huasco wetland. *Source* Own elaboration within the framework of the Huasco Wetland Observatory Project, Promoting the environmental and cultural values of the coastal wetlands of Atacama, 2022

area previously marred by micro-dumps into a memorable public space adjacent to the public access to the beach and coastal promenade (Fig. 8.9).

Fig. 8.8 Wetland tour, August 2022. *Source* Own elaboration within the framework of the Huasco Wetland Observatory Project, Promoting the environmental and cultural values of the coastal wetlands of Atacama, 2022

The Huasco River Wetland Observatory proposes a meeting place to explore and appreciate the multiple environmental and cultural values of this surprising landscape. It features a central area known as the *environmental classroom*, characterized by a permeable gravel floor that supports the development of educational and recreational activities. Around this space, vertical elements called *reed and bulrush forests* are deployed, which on an increased scale represent the qualities of the marsh vegetation fibers that are found at the edges of the coastal wetlands. And just as happens with these species, these forests cast dynamic shadows and provide habitat for various forms of life. The various *interpretive stations* present in the Observatory expose—through a sequence of 14 horizontal panels—content related to hydrology, geomorphology, ecology, and the history of the wetland (Figs. 8.10, 8.11 and 8.12).

8.4 Conclusions and Projections

Wetlands are complex systems that have developed over thousands of years, whose structural components—geological, chemical, and biological—and their processes have managed to establish a level of functioning that allows them to adapt to natural

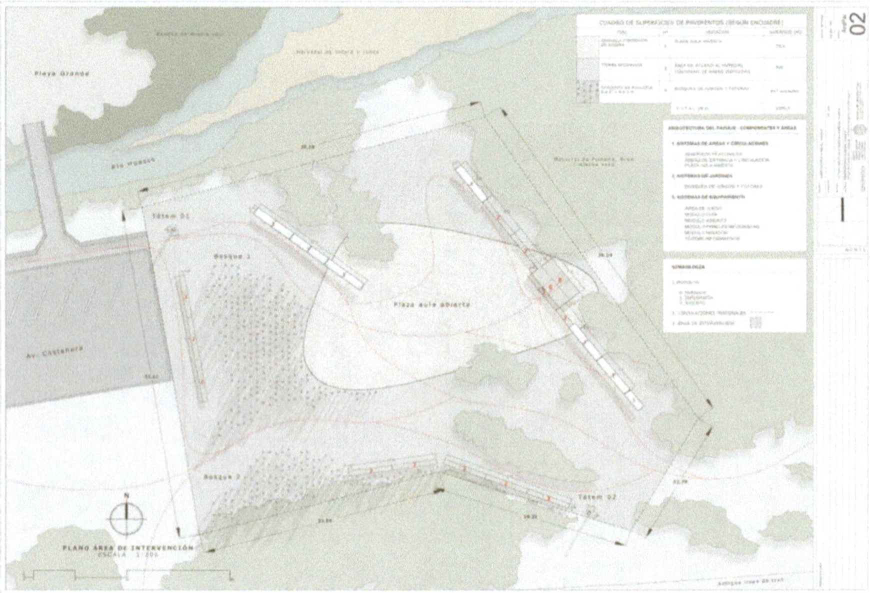

Fig. 8.9 General plan of the Huasco Wetland Observatory, March 2023. *Source* Own elaboration within the framework of the Huasco Wetland Observatory Project, Promoting the environmental and cultural values of the coastal wetlands of Atacama, 2022

Fig. 8.10 Interpretive panels at the Huasco Wetland Observatory, March 2023. Photography: Osvaldo Moreno, 2023

Fig. 8.11 Furniture and Reed Forest at Huasco Wetland Observatory. Photography: Osvaldo Moreno, 2023

Fig. 8.12 General view of Huasco Wetland Observatory. Photography: Osvaldo Moreno, 2023

changes and fulfill a primary ecological function, but at the same time, that makes them unique and fragile systems that currently face threats that put at risk their existence. Knowing and understanding both their importance and their behavior is essential to guarantee active protection by the communities. In the case of the Huasco River Estuary wetland, the residents are especially interested in learning in depth about the functioning of this ecological system and integrating its social and environmental benefits into the city's framework.

The wetland, in this sense, is recognized as a landscape that collects the aspirations for quality of life and a possibility of materializing them, highlighting the environmental and cultural attributes that can support sustainable development and enhance the resilience of the territory. Consequently, the observatory has the potential to *establish a global, integrated, non-sectorized approach to the territory that connects the interrelationships between the elements and processes of ecological, economic, social, and cultural nature* (Tarroja, 2006).

Through the experience of visiting the Observatory, people—and especially children—will be able to understand the relevance of these systems, complementing the direct perception of their landscape with the didactic information displayed on the interpretive panels, expanding the experience and learning. Additionally, through the possibilities of use and programming of this public space, the local community and various stakeholders will be able to carry out cultural, educational, and recreational activities aimed at stimulating the knowledge, appreciation, and conservation of the Huasco wetland and, in general, the coastal wetlands of Atacama.

End Notes

1. Project "Observatory of Coastal Urban Wetlands of Atacama: promoting of its environmental and cultural values" (Code CP21-E014) funded by the National Public Science Competition: Regional Public Spaces for Dissemination and Socialization of Knowledge, from the Ministry of Science Technology Knowledge and Innovation, developed between the years 2021 and 2023.
2. The Sanctuaries of nature determine a category of natural heritage recognized by the State of Chile. According to the Council of Monuments of Chile, all those

terrestrial or marine sites that offer special possibilities for geological, paleontological, zoological, botanical or ecological studies, or that possess natural formations, whose conservation is of interest to science or to the State, are Sanctuaries of Nature.
3. https://www.monumentos.gob.cl/monumentos/definicion/santuarios-de-la-naturaleza.
4. Co-design can be defined as an open and collaborative creative process that involves the users of a product, service or organization to generate their own solutions through a guided search instance that generates a shared view based on the understanding of all key actors, resulting in a design object based on collective learning. This process of joint search makes the resulting solutions generate a higher degree of affective appropriation and, therefore, of greater social validation by those who have been protagonists of its gestation (Britton, 2017; Trischler et al., 2018).
5. Seminar: "Challenges for the Rehabilitation and Conservation of Coastal Urban Wetlands in Contexts of Water Scarcity. Case Study: Huasco, Atacama Region" Held on August 11, 2022, and Funded by the Vice-rectorate of research of the Pontifical Catholic University of Chile. Participating institutions: Pontifical Catholic University of Chile, University of Antofagasta, Polytechnic University of Madrid, CONAF Atacama, Illustrious Municipality of Huasco.

References

Arizaga, X., Moreno, O., Palacios, S., & Gómez, T. (2021). The landscape as a strategic resource: Design of a tourist route in the Province of Chañaral, Atacama region, Chile. In K. Benamara, P. Bachimon, & M. El Boukhari (Eds.), *L'impact sociétal de l'aménagement touristique, Acceptabilité, prise en compte, recevabilité et long terme* (pp. 157–186). Harmattan Editions. ISBN: 978-2-343-23280-5.

Berque, A. (1994). Cinq propositions pour une théorie du paysage. In A. Berque, M. Conan, P. Donnadieu, B. Lassus, & A. Roger (Eds.) Editions Champ Vallon.

Bildstein, K. L., Bancroft, G. T., Dugan, D. H., Gordon, R. M., Edwin, R. M., Noel, E., Payne, L. X., & Senner, S. E. (1991). Approaches to the conservation of coastal wetlands in the Western Hemisphere. *Wilson Bulletin, 103*(2), 218–254.

Bresciani, L., Moreno, O., Arizaga X., Valdés J., Marambio-Alfaro, Y., Bravo-Naranjo, V., Rivera, M., Vasquez, A., & Quevedo, G. (2023). *Study Analysis of systems of the urban coastal area of Chañaral*. Final Report, Ministry of Housing and Urban Planning, Atacama Region.

Britton, G. (2017). *Co-design and social innovation connections, tensions and opportunities*. Routledge.

Cienfuegos, R., Campino, J. R., Gironás, J., Almar, R., & Villagrán, M. (2012). River mouths and coastal lagoons in the central area of Chile. In M.J. Fariña & A. Camaño (Eds.), *Coastal wetlands of Chile* (pp. 21–66). UC Editions.

INE—National Institute of Statistic. (2019). *Monthly tourist accommodation survey*, No. 255. Statistical Bulletin of the National Institute of Statistics (INE). Chile.

Littré, E. (1958). *Dictionary of the French language*. Paris, L. Hachette, 1873–1874.

MMA, Ministry of the Environment. (2023). *Aquatic and continental ecosystems*. https://humedaleschile.mma.gob.cl/ecosistemas/humedales/

MMA, Ministry of the Environment & CEA, Center for Applied Ecology. (2017). *Characterization of the coastal wetland estuary of the Huasco river for its nomination as a Nature Sanctuary.* Final Report.

Moreno, O., & Arizaga, X. (2022). Rehabilitation and conservation of urban coastal wetlands in contexts of water vulnerability: Case study. *Huasco Wetland Urban Research Notebooks, 0*(143), 138–158. https://doi.org/10.20868/ciur.2022.143.5002

Olivares, C. (2023). *Urban coastal wetlands as Socio-ecological Infrastructures for the resilience of territories in Sacrifice Zones: Compensation and environmental rehabilitation park of the Huasco River mouth wetland.* Thesis to opt for the academic degree of Master in Landscape Architecture. Pontifical Catholic University of Chile.

RAMSAR Convention Report. (2018). *13th Meeting of the Conference of the Contracting Parties to the Ramsar Convention on Wetlands.* COP13, Dubai.

Tabilo, E., Jorge, R., & Mondaca, V. (2001). Water birds in coastal wetlands of the Coquimbo region, Chile. *Chilean Ornithology Bulletin, 8*, 13–17.

Tarroja, R. (2006), Territorial transformations and social valuation of the landscape. In R. Mata & M. Domènech (Eds.), *The landscape and the management of the territory. Incorporation of landscape criteria in land use planning and urban planning.* Diputació de Barcelona Editions.

Trischler, J., Pervan, S. J., Kelly, S. J., & Scott, D. R. (2018). The value of codesign: The effect of customer involvement in service design teams. *Journal of Service Research, 21*(1), 75–100. https://doi.org/10.1177/1094670517714060

Chapter 9
Evaluation Criteria for Resilient Civic Design for Response Capacity in Residential Communities Affected by the San Ramón Fault, in Santiago de Chile

Jorge Inzulza Contardo

Abstract In recent decades, Santiago de Chile has seen a substantial increase in urbanized areas along its eastern edge and, in particular, in the foothills of the Andes Mountains, where the San Ramón Fault is located. Although this cortical geological feature is ancient, its presence in urban development debates is more recent, generating a new seismic risk scenario that challenges sustainability agendas and urban planning tools. Using the undeniable attraction of the natural environment, the panoramic view of the city, and a healthier environment, housing projects have been promoted that coexist with this seismic risk, something that seems to have been overlooked by communal and metropolitan planning instruments. The aim of this article is to explore specific evaluation criteria for resilient civic design in two residential communities in the commune of Peñalolén, affected by the San Ramón Fault, to determine their response capacity to seismic risk. To this end, a mixed approach is applied through a matrix of urban resilience indicators that considers four dimensions of action—natural, built, social, and normative—to identify critical situations in the studied sectors, including analysis of official data, mapping techniques and interviews with residents. It is concluded that contrasting urban planning experiences in extreme seismic risk conditions is illustrative, both to permeate the operational layers of risk management and communal and metropolitan regulations, and, thereby, to consider the Andean foothills in a resilient manner, recognizing the San Ramón Fault as a new baseline criterion for 21st century urbanism.

This work contains results that are part of the article Inzulza et al. (2023). Resilient civic design evaluation criteria for response capacity within a seismic risk context in Santiago, Chile. The case of residential communities affected by the San Ramón fault. International Journal of Disaster Risk Reduction 87, https://doi.org/https://doi.org/10.1016/j.ijdrr.2023.103597

J. Inzulza Contardo (✉)
Departamento de Urbanismo, Facultad de Arquitectura y Urbanismo, Universidad de Chile, Santiago, Chile
e-mail: jinzulza@uchilefau.cl

M. Vicuña and E. Giorgi (eds.), *Intersections*, Springer Geography,
https://doi.org/10.1007/978-3-031-76402-8_9

Keywords Urban resilience · Civic design · Seismic risk · San Ramón fault

9.1 Introduction

In recent decades, concern has increased worldwide about the number of socio-natural disasters that have occurred and their threats in inhabited areas (Audefroy, 2003). Latin American countries have been affected on numerous occasions by natural events that have impacted their populations, the environment, and urban planning (Lay, 2015).

Even though these events are not new, the accelerated urbanization of risk areas has been more exponential in recent decades and, with it, more and more settlements susceptible to the occurrence of socio-natural disasters are being generated. Chile is no exception and, as part of the Pacific Ring of Fire, it is constantly exposed to a high seismic risk, being impacted by the strongest earthquakes recorded in the world, in 1960 and 2010 (Ruíz & Madariaga, 2018).

Santiago de Chile has seen a considerable increase in urbanized areas along its eastern edge and, in particular, in the foothills of the Andes Mountains, where the San Ramón Fault is located, covering seven communes (Vitacura, Las Condes, La Reina, Peñalolén, La Florida, Puente Alto, and Pirque) with a total extension of 50 km (Easton et al., 2018, 2022). Using the undeniable attraction of the natural environment, the panoramic view of the city, and a healthier environment, housing projects, mainly of the gated community type, have been built that coexist with this seismic risk, something that seems to have been overlooked by communal and metropolitan planning instruments (Inzulza et al., 2021, 2022).

This article explores criteria for resilient civic design in residential communities affected by the San Ramón Fault, in order to evaluate their response capacities to seismic risk. Two communities affected by the San Ramón Fault have been studied using the matrix of Resilient Civic Design Evaluation (Inzulza et al., 2023) which is complemented by official data, mapping techniques, and interviews with residents. This analysis allows for the delivery of results that can be considered in the Andean foothills in a resilient manner, recognizing the San Ramón Fault as a new baseline criterion for 21st-century urbanism.

9.2 Resilient Civic Design

Urban planning and design constitute an appropriate response strategy for rebuilding cities and towns (Parnell et al., 2009), if their basic postulates and development stages are considered to form neighborhoods, including the civic attitude of their residents and their cultural practices or assets (Horn et al., 2011).

In turn, urbanism and resilience have become increasingly interrelated (UN-HABITAT, 2018). Urbanism, for its part, seeks its pragmatic side through urban design and, prior to this, in civic design (Adshead, 1910), while resilience has transformed into the performance of urban regeneration and reconstruction strategies of recent decades (Allan & Bryant, 2011). This, not only to face post-disaster events, but as a form of permanent action, transferred to their communities (Cutter et al., 2008).

These concepts are also linked with what is expressed in the 2030 Agenda for Sustainable Development in a broad sense and, in particular, with what is established in its ninth goal which seeks to "build resilient infrastructures, promote inclusive and sustainable industrialization and foster innovation" (United Nations, 2018: 43). In this way, it is possible to predict that urban design and resilience are decisively harvested in official agendas and manifestos to promote a sustainable quality of life with a systemic view of risk territories that, in addition, exhibit high vulnerability at present.

Urban resilience and the extent of its contribution in seismic risk contexts are supported by a growing body of literature (Koren & Rus, 2019). However, it remains largely unexplored how to analyze the existing capacity of the urban environment to respond to a seismic event under the umbrella of resilient urban and civic design.

We define resilient civic design as a discipline that attempts to respond to the multi-layer problem in the inhabited territory with the aim of promoting community resilience through interventions driven by urban design (Landry, 2017) and thus become a facilitator of more empowered communities in decision-making (Monna & Auricchio, 2020).

Operationally, resilient civic design acts as an interface area shaped by four areas: the natural, the built, the social, and the regulatory (Fig. 9.1). Specifically, there is a relationship between the natural and the built when the built environment develops according to natural resources and sustainable open spaces; and this relationship includes a suitable social environment with empowered people organized in communities to make decisions about their risk territories (Moser, 2010).

These environments can be better associated in the Resilient Civic Design Evaluation Matrix proposed by Inzulza et al. (2023), as action areas that include specific resilience indicators and evaluation criteria (Table 9.1).

The natural area is the basal system that possesses the territorial conditions including the threats and risks, to which the built and social environment must adapt. Within its evaluation criteria are the threats and risks from geology, geomorphology, river networks, tree systems, and proximity to a geological fault rupture strip.

The built area includes the sustainability approach which is defined as the way in which the territory is occupied seeking harmony with the natural environment. The evaluation criteria include both the response capacity of the built space in the event of a seismic event and the response capacity of open spaces that include safety zones and evacuation routes.

The social sphere is associated with the concept of civility to measure how citizens become part of the relationship between the spatial and natural environment through participatory community, decision-making, and knowledge of the urban and natural territory. Evaluation criteria include education and knowledge of the fault, practices related to seismic risk, and local-level organization.

Fig. 9.1 The concept of resilient civic design. *Source* Inzulza et al. (2023: 3)

Table 9.1 Resilient civic design evaluation matrix

Dimensions	Resilience indicators	Evaluation criteria
Natural scope	Territorial conditions including threats and risks, to which the built and social environment must adapt	Threats and risks from geology, geomorphology, river networks, tree systems, and proximity to geological fault rupture strip
Built scope	Sustainability understood as the way in which the built territory is occupied in harmony with the natural environment	Exposure and response capacityy of buildings, critical infrastructure, built space, and open spaces to a seismic event
Social scope	Civility understood as citizenship becomes part of the relationship between the natural and built environment with decision-making	Knowledge and education about geological faults and practices related to seismic risk with organizations and community worktables
Normative scope	Instruments of change with mechanisms that manage, and handle threats and risks allow promoting resilience from design and urban planning	Urban norms from intercommunal regulatory plans (PRMS, PRIS) and communal (PRCs) such as uses, densities, land surfaces, distances, heights, and environmental impact assessment

Source Inzulza et al. (2023: 4)

These three areas of action—natural, built, and social—are related to the normative sphere that includes change strategies and management mechanisms to promote resilience from urban planning design (Tierney & Bruneau, 2007). Within this sphere, urban norms from intercommunal, communal, and sectional regulatory plans are crucial, including land uses, densities, land surfaces, distances, building heights, and environmental impact assessment.

In this way, it should be achieved that the built environment develops according to natural resources and sustainable open spaces; and this relationship includes a suitable social environment with empowered people and communities organized to make decisions about their threatened territories.

9.3 Methodological Design

The present research applies a mixed approach that is part of the Resilient Civic Design Evaluation Matrix (Inzulza et al., 2023), which considers 4 areas of action—natural, built, social, and normative—to identify critical risk situations. The commune of Peñalolén has been chosen, as it is an area with high risk due to having a larger built-up area with high housing density on or near the San Ramón Fault (Garrido, 2021).

Specifically, two residential communities affected by the San Ramón Fault (Fig. 9.2) were considered, which represent two residential typologies characteristic of the foothills of Santiago but contrasted in their morphology regarding the relationship of the houses with the public space and their socioeconomic composition. The first of these is the San Tadeo and Olga Leiva Community, made up of a traditional residential neighborhood of approximately 34,000 inhabitants (INE, 2017) and an area of 100.84 Ha, located less than 400 m from the buffer of the San Ramón Fault.[1] The second community called Club de Campo is a closed condominium residential complex with 1,100 inhabitants, an area of 31.81 Ha, and located on the buffer of the San Ramón Fault.

9.3.1 Data Collection for the Natural Sphere

A base cartography was carried out using an OpenStreetMap image drawn with the ARCGIS 10.1 software. This base map contains the buffer of the San Ramón Fault through the topographic survey carried out with "Light Detection and Range (LIDAR)", as a remote sensing method that uses light in the form of a pulsed laser to measure ranges (variable distances) to the Earth. Specifically, the section of the commune of Peñalolén is presented, highlighting the analyzed areas of the San Tadeo and Olga Leiva, and Club de Campo residential communities.

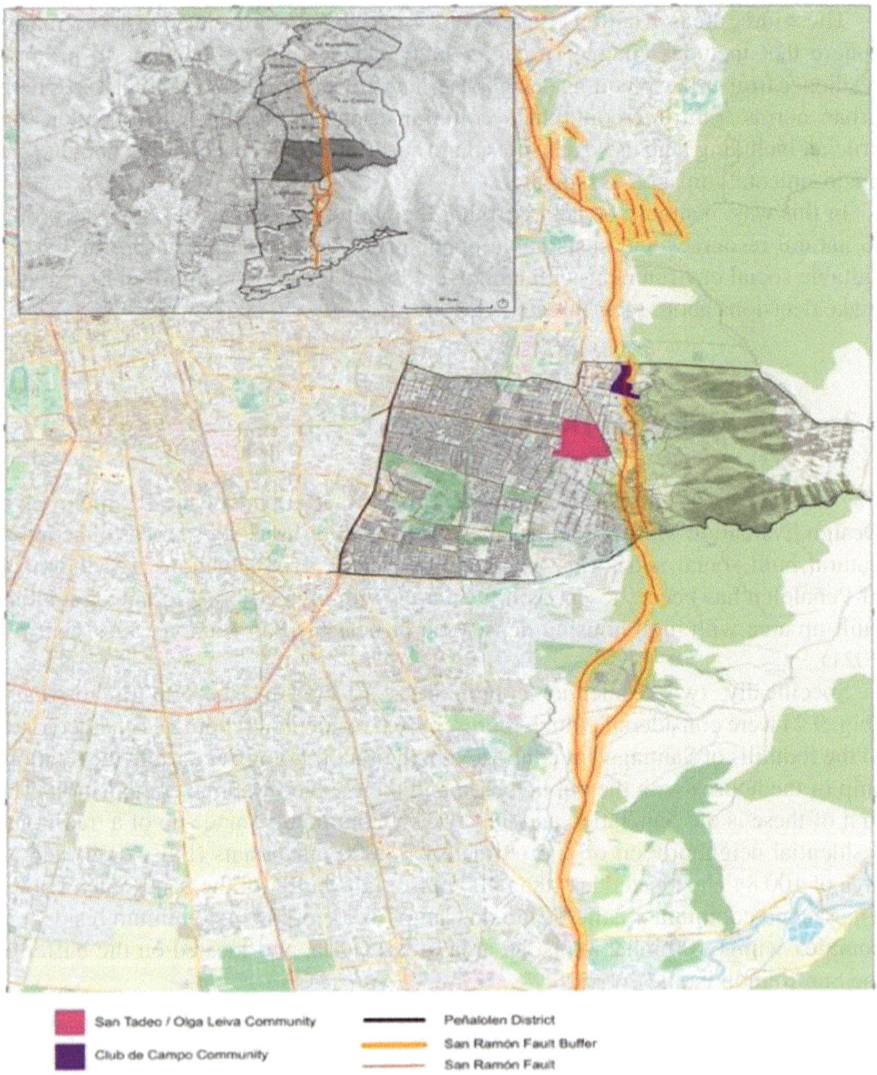

Fig. 9.2 Location of San Tadeo and Olga Leiva communities and Club de Campo. *Source* Inzulza et al. (2023: 6)

9.3.2 Data Collection for the Built Sphere

Data on the current conditions of the San Tadeo and Olga Leiva community were collected during September 2021, and that of the Club de Campo community, during December 2020. Although the collected data were based on the Resilient Civic Design Evaluation Matrix, the evaluations of some variables were carried out differently considering the scale and architectural design of these studied areas. As the San

Tadeo and Olga Leiva community is part of a city expansion, the analysis of the typology of housing in its different existing typologies. On the contrary, as the Club de Campo area was built as a gated community, the analysis contemplating a single typology that covers the entire sector under study.

9.3.3 Data Collection for the Social Scope

Semi-structured interviews were applied to both residential communities: 5 interviews with inhabitants for San Tadeo and Olga Leiva were conducted, between 21-10-2021 and 20-01-2022, as part of a master's thesis (Garrido, 2020); and 5 interviews with inhabitants for Club de Campo, between 18-12-2020 and 28-12-2020, as part of an undergraduate dissertation (Mattei, 2021). The interviews focused on two aspects: knowledge about the FSR and practices related to seismic risk (at a private and community level). A selection of representative textual quotes related to the variables of interest was developed to interpret the community's response capacity to seismic risk.

9.4 Resilient Communities of the Foothills?

The application of the action scopes in the residential communities San Tadeo and Olga Leiva, and Club de Campo, located in the foothills of Santiago, allows obtaining types of responses that challenge the resilient capacity of their communities.

9.4.1 Natural Scope

In the case of San Tadeo and Olga Leiva and as a natural consequence of its location, there is a risk environment derived from the distance between the San Ramón Fault and this community under study, with less than 400 m (Fig. 9.3). This is part of the consequences of a weak urban planning system that has not considered the seismic risk of what it means to be located in the foothills. In addition, more than 34,000 inhabitants of this community would be directly affected in case the San Ramón Fault is activated.

In the case of Club de Campo, the proximity to the San Ramón Fault is especially critical, as this community is located directly on its buffer or rupture zone, which crosses the center of this residential area. In addition, being a gated community type, there is a single main entrance for all the houses, which increases the exposure to geological risk.

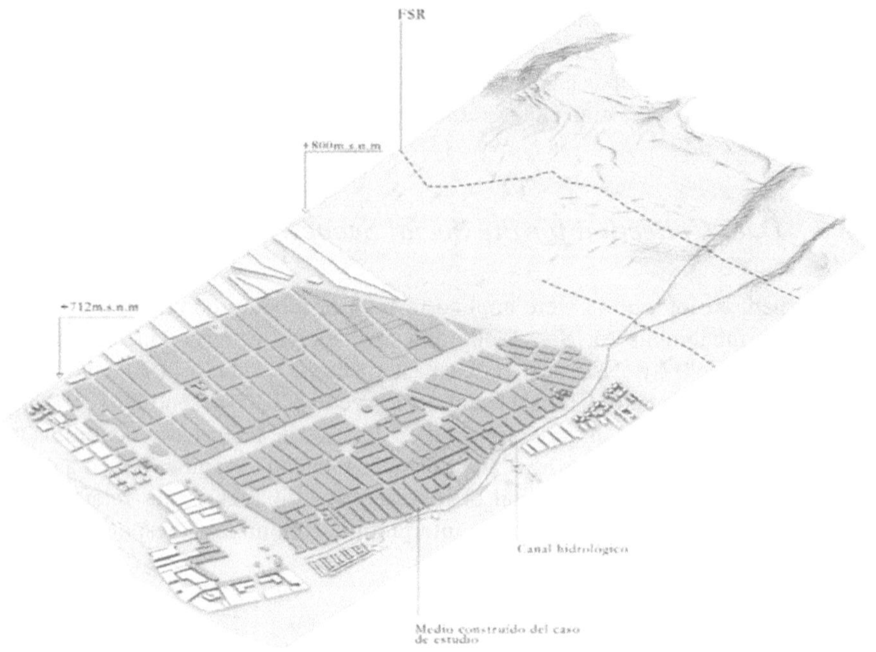

Fig. 9.3 San Tadeo and Olga Leiva condominium in the foothills of the mountain range. *Source* Mattei (2021: 29)

9.4.2 Built Scope

The analysis of San Tadeo and Olga Leiva allows identifying the degree of response to seismic risk in different elements that structure this residential community. It is observed that the degree of response varies, with housing density and safety areas contributing to a lower degree of response in the event of a detonation of the San Ramón Fault.

For the Club de Campo community, the extensions made in the houses of this gated community have modified the originally built building type, affecting the resilience of the original houses (Fig. 9.4). Regarding safety zones, in general terms, the distribution of green areas as safety zones is adequate, as there is a wide green area with good response capacity to seismic risk and easy access for the community. At the same time, there are smaller areas that serve as safe zones for the north and south sectors of the condominium.

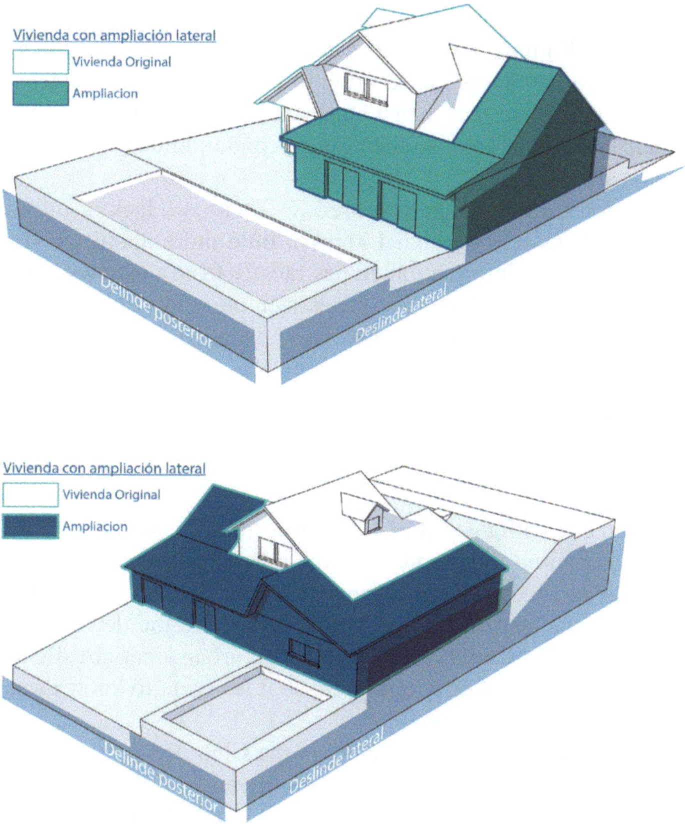

Fig. 9.4 Expansion typologies of houses in Club de Campo condominium. *Source* Garrido (2020: 90)

9.4.3 Social Scope

Although the community of San Tadeo and Olga Leiva knows about the existence of the San Ramón Fault and its risks, and mainly from its geological origin, it does not have information on how to act in case of a seismic event. They also do not have a level of organization, or are very supportive of each other, although they think that younger families could act better as they did in the earthquake of 27/F 2010:

"My young neighbors, I think they know well how to act, because when the 2010 earthquake occurred, everyone was in their home, all the neighbors were asking how they were, and the younger ones were asking if there was any problem… [To be better organized] everything depends on how we all know each other from childhood, we all play outside, so apart from being neighbors here we are like family with the people" (Valeria, Resident).

In the case of Club de Campo, in general terms there is knowledge about the existence of the San Ramón Fault and its geological origin; however, they do not feel that as a community they are prepared to act in the event of a high-intensity seismic event. They also recognize that their homes are in the rupture zone which covers more than their residential area, so their entire condominium would be affected in the event of an earthquake.

"Currently I don't think we are prepared … There is a lack of information. As a person I can perhaps tell you that I know a little more about the situation, but not as a community. The information comes mostly from the news [on television]. We are here in Peñalolén and we have never seen or, at least I, initiatives from the municipality on earthquake issues or evacuation plans" (Laura, Resident).

9.5 Discussion

Three aspects can be highlighted (Fig. 9.5). First, there is the importance of residential communities exposed to seismic risk which is part of the natural environment. Today more than ever, due to climate change, it is crucial to know the types of threats that communities face, especially in urban areas, where the density of population exposed to risk is higher (CNID, 2016). Almost all urban areas are affected by territorial limitations and risks of different types, so it is urgent to know these territorial limitations and plan appropriately.

In the studied cases and, more critically in the Club de Campo community, it is possible to understand how exposed it is to the activation of the San Ramón Fault. It can be verified that there is poor urban planning that puts this residential area at risk, and it demonstrates the importance of rethinking how to proceed in the future with

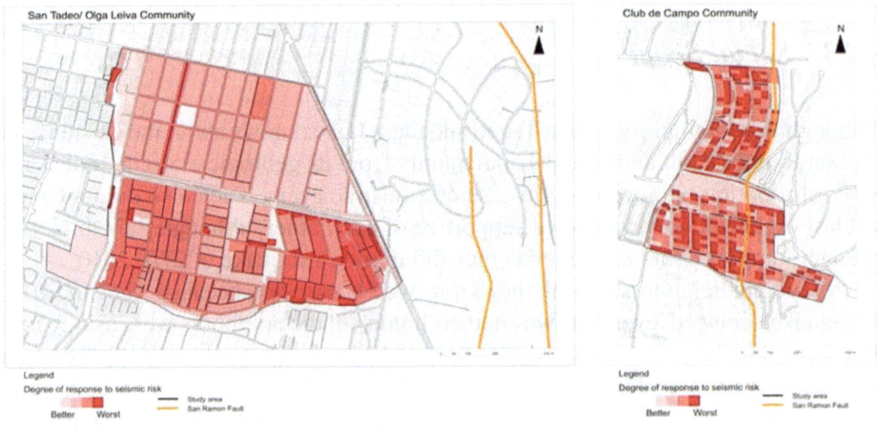

Fig. 9.5 Degree of response to seismic risk of San Tadeo and Olga Leiva and Club de Campo communities. *Source* Inzulza et al. (2023: 11)

the areas that are still unbuilt along the San Ramón Fault (mainly in the southern sector).

Secondly, it is important to reflect on the role of the built environment in the capacity to respond to seismic risk (built environment). Resilient civic design defines a way of using the territory in harmony with the natural environment and the basis of sustainable development (Espinosa et al., 2013). From the analysis of variables associated with the built environment, it is possible to know the urban design elements that affect the communities' response capacity to seismic risk, such as population density, available green areas as safe areas, the type of housing, and escape routes.

The studied communities show a disconnection between the built environment and the natural environment, not only because they are residential areas located on the San Ramón Fault, but because they were not developed including a type of response to seismic risk. San Tadeo and Olga Leiva shows how the lack of a system of green areas and open public spaces in high-density areas can significantly reduce their response to a seismic event. The search for urban design strategies in highly seismic areas then requires a proposal on types of parks and green areas for an adequate response.

Finally, it is crucial to highlight the role of communities as urban actors and decision-makers (social environment). The capacity to respond to seismic risk cannot be assessed from a resilient civic design approach without knowing the degree of participation and education of the affected communities. Therefore, it is necessary to ensure that these communities have a high degree of civility, that is, with residents aware of their environment and capable of making decisions about their territory.

The analyzed cases are illustrative to reflect on the need for the formation of an informed and participative community at the local level. While there is organization and knowledge of seismic risk at the family level, there is a lack of organization on the part of the community, especially at the municipal level. Organized communities can respond better to critical events, with strong social cohesion and active action mechanisms (Adger et al., 2005), which are more resilient at the local level.

9.6 Conclusion

The recognition of threat types at planning levels with a resilient civic design approach is especially significant in highly urbanized seismic risk contexts like the one analyzed in the foothills of Santiago. The San Ramón Fault represents a new seismic risk scenario that challenges sustainability agendas and urban planning instruments.

The study of the residential communities San Tadeo and Olga Leiva and Club de Campo, as two characteristic typologies of the foothills of Santiago, but contrasted in their morphology, has allowed the application of the resilient civic design evaluation matrix (Inzulza et al., 2023), with the idea that it can be evaluated and replicated in other communes that present similar housing complexes exposed to this type of seismic risk. It is established that the proper habitation in the foothills depends on specific indicators to evaluate such as population density, surface area of public spaces, slope, presence of anthropic elements, vegetation, and distance and height of nearby houses, among the most relevant.

In this way, the results of this multi-layer analysis can be understood as a first approximation to understand the response capacity of these areas and communities to seismic risk. Thus, this first approach to evaluating resilient civic design and its areas of action is an invitation to continue exploring new methods of approach to understand the relationships between natural, built, and social environments. They can then be improved so that their relationship is increasingly sustainable (natural and built environment), conscious (natural and social environment), and participative (social and built environment).

In addition, the analysis of these two communities can contribute to other similar experiences where those green areas that play a key role as safe open spaces and other factors that are relevant for the planning of resilient cities such as the building typology, densities, and evacuation routes. In this sense, the role of local authorities and municipalities in the transfer of civic assets and risk management with a thorough knowledge of their environments is crucial.

For the above, it is recommended to continue reviewing resilient civic design strategies with new indicators and variables that can better describe the singularities of urban settlements and territories affected by geological faults, landslides, and other types of natural threats. To the extent that these natural elements can be adequately included in ecological manifestos, urban agendas, and design manuals, the impacts of earthquakes like the San Ramón Fault will be part of an active built natural environment that will allow us to live in a more sustainable way.

End Notes

1. The buffer of the San Ramón Fault is the rupture area formed by 300 m wide and that runs 50 kms from the foothills of Santiago (from the commune of Vitacura in the north to those of Pirque in the south), and is measured from the axis of its trace, 100 m to the east, and 200 m to the west.

Funding ANID/FONDECYT/ Regular 1230350.

References

Adger, W., Hughes, T., Folke, C., Carpenter, S., & Rockstrom, J. (2005). Social-ecological resilience to coastal disasters. *Science, 309*, 1036–1039. https://doi.org/10.1126/science.1112122

Adshead, S. D. (1910). An introduction to the study of civic design. *The Town Planning Review, 1*(1), 3–17. https://doi.org/10.3828/tpr.1.1.k25421150511g571

Allan, P., & Bryant, M. (2011). Resilience as a framework for urbanism and recovery. *Journal of Landscape Architecture, 6*(2), 34–45. https://doi.org/10.1080/18626033.2011.9723453

Audefroy, J. (2003). The problem of disasters in urban habitat in Latin America. *Invi, 47*, 52–71. https://doi.org/10.5354/0718-8358.2003.62233

CNID, National Council for Innovation for Development. (2016). Towards a Chile resilient to disasters: An opportunity. National research, development and innovation strategy for a Chile resilient to natural disasters. CREDEN. https://www.cr2.cl/wp-content/uploads/2017/01/INFORME-DESASTRES-NATURALES.pdf

Garrido, D. (2020). Physical-social assets and urban resilience in a context of seismic risk by the San Ramón Fault (FSR). The case of the Club de Campo sector in Peñalolén, Chile. Master's Thesis in Urbanism, Faculty of Architecture and Urbanism, University of Chile.

Cutter, S., Barnes, L., Berry, M., Burton, C., Evans, E., Tate, E., & Webb, J. (2008). A place-based model for understanding community resilience to natural disasters. *Global Environmental Change, 18*(4), 598–606. https://doi.org/10.1016/j.gloenvcha.2008.07.013

Easton, G., Boroschek, R., Rebolledo, S., Inzulza, J., Ejsmentewicz, D., Vergara, P., & Giessen, E. (2022). *Policy brief: The San Ramón Fault and the sustainability of the Santiago foothills: Recommendations for public policy.* Vice-rectorate of Research and Development. University of Chile. https://repositorio.uchile.cl/handle/2250/183864

Easton, G., Inzulza, J., Pérez, S., Ejsmentewicz, Y., & Jiménez, C. (2018). Failed urbanization? The San Ramón Fault as a new scenario of seismic risk and the sustainability of Santiago, Chile. *Journal of Urbanism, 38*, 1–20. https://doi.org/10.5354/0717-5051.2018.48216

Espinosa, M. Prieto, I., & Peña, J. (2013). Green areas and outdoor spaces as an evacuation option in case of earthquakes. *Science on Your PC, 4*, 58–74. https://www.redalyc.org/articulo.oa?id=181332462004

Horn, P., Stein, A., & Moser, C. (2011). *MDGs and assets, Briefing Paper*. Global Urban Research Centre, The University of Manchester. http://hummedia.manchester.ac.uk/institutes/mui/gurg/working_papers/briefingpapers/GURC_BP1_March2011.pdf

Inzulza, J., Cerda, C., Garrido, D., & Mattei, R. (2023). Resilient civic design evaluation criteria for response capacity within a seismic risk context in Santiago, Chile. The case of residential communities affected by the San Ramón fault. *International Journal of Disaster Risk Reduction, 87*. https://doi.org/10.1016/j.ijdrr.2023.103597

Inzulza, J., Curihuinca, M., Pérez, S., & Easton, G. (2022). Revealing seismic risk in the foothills of Santiago, Chile. Multicriteria analysis for determining vulnerability in the San Ramón Fault (FSR). *Revista de Geografía Norte Grande, 81*, 305–330. http://revistanortegrande.uc.cl/index.php/RGNG/article/view/22431

Inzulza, J., Gatica, P., Easton, G., & Pérez, S. (2021). Resilient urban design in the foothills in Santiago? Contrast of community scenarios with seismic risk facing the San Ramón Fault. *Revista Urbano, 43*, 96–107. http://revistas.ubiobio.cl/index.php/RU/article/view/4529/4018

INE, National Institute of Statistics. (2017). Synthesis of Results. CENSUS 2017. http://www.censo2017.cl/descargas/home/sintesis-de-resultados-censo2017.pdf

Koren, D., & Rus, C. (2019). The potential of open space for enhancing urban seismic resilience: A literature review. *Sustainability, 11*(21), 5942. https://doi.org/10.3390/su11215942

Landry, C. (2017). *The Civic City in a Nomadic World.* Nai010 Publishers, Rotterdam.

Lay, T. (2015). The surge of great earthquakes from 2004 to 2014. *Earth and Planetary Science Letters, 409*, 133–146. https://doi.org/10.1016/j.epsl.2014.10.047

Mattei, R. (2021). *Morphological effects on open public spaces in a geological risk context: The San Ramón Fault (FSR) in the northeastern sector of Peñalolén 2017–2021.* Research Seminar, Faculty of Architecture and Urbanism, University of Chile.

Monna, V., & Auricchio, V. (2020). Building the Civic City: Civic design as enabler of resilient communities. *Constructii: Journal of Civil Engineering Research, 12*, 166–170. https://www.researchgate.net/publication/347910197

Moser, C. (2010). *Neighborhood people, extraordinary lives. Assets and poverty reduction in Guayaquil, 1978–2004.* Sur Professionals.

United Nations. (2018). *Agenda 2030 and the sustainable development goals an opportunity for Latin America and the Caribbean.* Economic Commission for Latin America and the Caribbean ECLAC.

Parnell, S., Pieterse, E., & Watson, V. (2009). Planning for cities in the global South: An African research agenda for sustainable human settlements. In H. Blanco & M. Alberti (Eds.), *Shaken, shrinking, hot, impoverished and informal: Emerging research agendas in planning. Progress in Planning, 72*, 195–250. https://doi.org/10.1016/j.progress.2009.09.001

Ruiz, S., & Madariaga, R. (2018). Historical and recent large megathrust earthquakes in Chile. *Tectonophysics, 733*, 37–56. https://doi.org/10.1016/j.tecto.2018.01.015

Tierney, K., & Bruneau, M. (2007). Conceptualizing and measuring resilience: A key to disaster loss reduction. *TR News, 250*, 14–17. https://onlinepubs.trb.org/onlinepubs/trnews/trnews250_p14-17.pdf

UN-HABITAT. (2018). *Building urban economic resilience during and after COVID-19.* https://urbanresiliencehub.org/economicresilience/

Chapter 10
Adaptability of Architectural Forms for Irregular Buildings Located in the Flood-Prone Coastal Zone of Laguna Verde, Valparaíso

Felipe Igualt, Abraham Vallejos Arias, and Diego Apablaza Díaz

Abstract Coastal floods have recorded a high recurrence and destructive impact in the central area of Chile in recent years. In areas of irregular occupation, the impact on residential infrastructure is visibly greater. However, it is observed that in the context of irregularity there are faster adaptation responses to floods than in consolidated urban areas. These adaptations are mostly intuitive and lack professional judgment. To establish which type of building adapts best to flood conditions in the context of irregular occupation within the central area of Chile, the case of Laguna Verde, Valparaíso, was studied. Here it was detected that adaptation occurs from transformations of the houses generally from the cube, which is the most governable figure culturally for the residents. Tests were developed with scale models to identify a structural form that allows supporting a progressive house. Subsequently, a prototype of a residential module was built, taking care to make a simple, replicable, affordable system that can be self-built. The experience allowed guiding an initial approach toward the design of progressive houses adapted for areas of irregular occupation in the process of settlement.

Keywords Coastal floods · Adaptation · Vulnerability · Resilience · Laguna Verde

F. Igualt (✉) · A. Vallejos Arias · D. Apablaza Díaz
Escuela de Arquitectura y Diseño, Pontificia Universidad Católica de Valparaíso, Valparaíso, Chile
e-mail: figualt@ead.cl

A. Vallejos Arias
e-mail: a.vallejos@ead.cl

D. Apablaza Díaz
e-mail: diegoapablaza@hotmail.com

M. Vicuña and E. Giorgi (eds.), *Intersections*, Springer Geography,
https://doi.org/10.1007/978-3-031-76402-8_10

10.1 Introduction

10.1.1 Irregular Occupation and Coastal Floods

Irregular settlement is a phenomenon that has progressively increased over the last decades in Chile, especially in recent years. It is a little-studied phenomenon that today shows negative consequences for both people and the environment. In Chile, there are 113,887 families currently living in irregular settlement, of which 27.2% are located in the region of Valparaíso. This is the region where the largest number of families living in camps at the national level is found. Only in the conurbation between Valparaíso, Viña del Mar, and Concón, 4 large concentrations of camps are visualized, each of which is subject to 3 types of risks: fires, mass removal, and floods (CETCH, 2023).

Although they are mostly concentrated on the outskirts of cities, in the Greater Valparaíso GV there are also irregular settlements located in the coastal area. Examples of these are the Las Gaviotas Camp and La Isla de Concón in the south of the commune of Quintero and El Campamento Otaegui in Laguna Verde, Valparaíso, which have been pointed out as the most vulnerable settlements in the GV in the event of a tsunami (Contreras-López et al., 2019). This type of informal settlement represents for some families the only opportunity to access a home, due to the difficulties that exist to access one from formal mechanisms. These are collective occupations located on unprotected lands, close to a consolidated urban center, where a high environmental deterioration associated with the lifestyle of its occupants is recorded (Villamar & Aguilar, 2020).

The consolidation of irregular occupation in the GV has led to the creation of highly vulnerable scenarios in the face of the effects of extreme events, among which coastal floods stand out. These can occur both from coastal events such as tsunamis and storm surges, as well as from river floods. In the GV there is a long history of flood events from the aforementioned sources, which have generated both loss of infrastructure and housing. Among these, the meteo-tsunami of August 8, 2015, stands out, which caused considerable losses on the coasts of the GV (Winckler et al., 2017). The low settlement of Laguna Verde was the most affected residential sector, with more than 50 homes flooded.

10.1.2 Laguna Verde

The irregular settlement located in the lower area of Laguna Verde (Fig. 10.1) corresponds to an area administered by the Municipality of Valparaíso. This entity aimed to allocate the coastal strip for the construction of a beach resort. However, the failure of this initiative led to the abandonment and obsolescence of planning in this area, resulting in a series of irregular appropriations of the coastal territory, where today most residents do not have property titles, which prevents them from obtaining

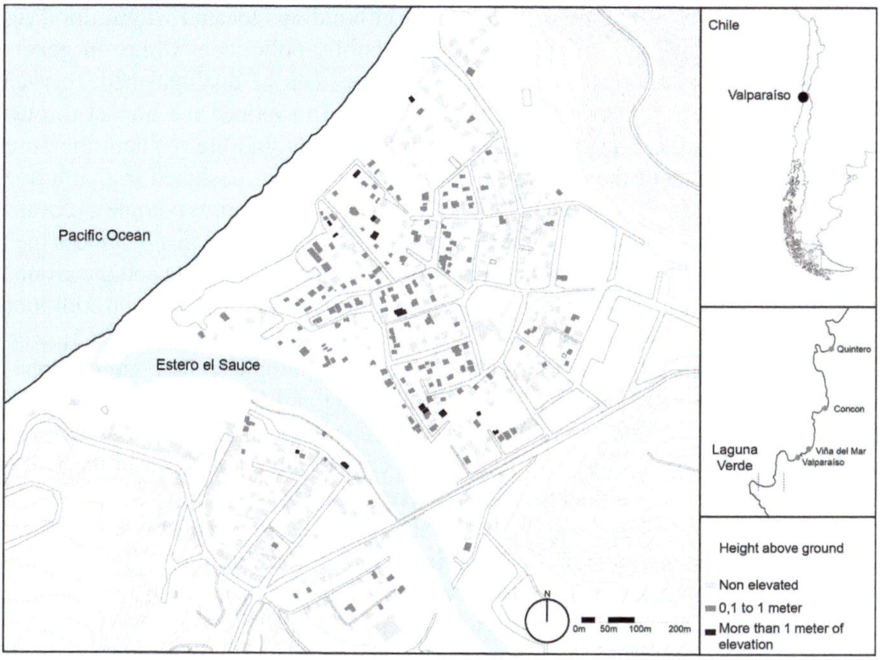

Fig. 10.1 Adaptation of houses in the flood-prone area of Laguna Verde

subsidies or grants for housing improvement. This settlement is geographically characterized by being located adjacent to Laguna Verde beach, as well as on the edge of the Estero El Sauce. Because of this, it is prone to flooding from both sources. Despite this, it records a constant increase in homes in flood-prone areas. The Municipality of Valparaíso has considered a strategy of settling this area, so accommodation adaptation (raising the house above the flood height) is considered as a possibility to reduce exposure to floods.

10.2 Adaptability

Adaptability is an essential concept within the resilience cycle (Rosati et al., 2015). It is used to point out possible remediation strategies in a situation of negative vulnerability, aiming to improve responses to future disturbances, reducing damage and allowing a faster recovery (Rosati et al., 2015; Walker et al., 2004). In this context, adaptability to coastal floods is understood as the ability of a settlement to reinterpret the territory and infrastructure after being affected by a destructive event. Adaptability will be conditioned to the response capacity of the local community, with various variables through which it can be developed, such as urban layout, structural typologies, and use of materials.

The adaptability of the architectural form in buildings located within flood risk areas has been slow and scarcely guided by public policies in Chile. In general terms, two main categories of adaptation measures can be distinguished: resistant and resilient, where the resistant ones are designed to reduce the impact through elements that block the arrival of water to the building and the resilient measures allow the interaction of the building with the water, but are designed so that it does not collapse and has short recovery times. Among the resilient typologies, elevated buildings stand out. These are located on a supporting structure that distances them from the ground, allowing the passage of flows between the building and the ground. It is estimated that elevated buildings that can withstand a tsunami have an additional cost between 10 and 20% over a conventional building (FEMA P-646, 2012).

In Chile, after the 2010 tsunami, elevated housing typologies called "tsunami resistant" were incorporated by the Ministry of Housing and Urban Planning MINVU. These correspond to two-story houses of 54 m^2 supported on columns that raise the house 2 m above the ground, implying a considerable reduction in the risk of flooding. However, in various localities where these houses were built, the owners closed the first level to increase the living area of the house (Martínez et al., 2017), returning to the initial vulnerability situation. This context highlights the difficulties that adaptation will bring about traditional ways of living, visualizing a problematic of habitability in adapted homes. This is primarily evident in the spatial configuration of the stilt space, ensuring it is neither perceived nor treated as a residual space.

According to Siders (2019), there are three main strategies for adaptation for buildings exposed to floods: (a) Resist: through the incorporation of hard works, such as breakwaters and dikes to block or mitigate the flow of flooding; (b) Relocate: involves moving the building to a place free of exposure to floods; (c) Accommodate: the building is raised above the known flood height, supported by a structural framework beneath. Each of these strategies has advantages and impacts, both economic and on the natural environment. In the case of Laguna Verde, it is observed that, due to the recurrence of flood events, there is a high number of homes that have been adapted through accommodation. This is because, being an irregular settlement, there is no investment by the local government for major flood mitigation works. Relocation is not possible due to resident's lack of resources to acquire land in risk-free areas. Other approaches, such as nature-based adaptation (Vuik et al., 2016), has been ignored, mainly due to a lack of awareness of the mitigating function of the Estero El Sauce wetland, affecting the capacity for natural mitigation and damaging the coastal ecosystem.

10.3 Diagnosis of Adaptability in Laguna Verde

In order to understand the trends of accommodation adaptation in the flood-prone area of Laguna Verde, an analysis was conducted on 565 buildings, the majority of which were residential. The analysis consisted of identifying those homes that have been raised or raised their first level in response to the exposure to floods that affect the ground on which they are located. In addition, the construction systems used to raise the buildings were recorded, verifying the prevailing trends.

Figure 10.1 shows the studied area, as well as the results obtained. Of the total 565 buildings, 384 (68%) correspond to conventional housing typologies, which are not raised above ground level. On the other hand, 166 homes (29.4%) show adaptation by accommodation, rising from 0.1 to 1 m above ground level. Finally, 15 homes (2.6%), located in more exposed areas, rise to a height from 1 to 3m. These results allow us to affirm that, despite being an irregular settlement, there are instances of adaptation of the architectural form by the residents for the exposure condition that affects the homes.

Regarding the construction trends in the buildings that were raised, Fig. 10.2 shows some examples of buildings adapted by accommodation in Laguna Verde. It is observed that 156 homes (92.9%) use lightweight materials for their construction, such as wood or simple partitioning, representing the major trend in relation to the adaptation of homes by accommodation in Laguna Verde. A number of 7 homes (4.2%) use a solid construction system of reinforced concrete or reinforced masonry and 4 homes (2.4%) use a steel system for the support structure. Only 1 home (1.2%) uses a mixed system of steel with wood. Those homes that adapt with solid structural systems mainly correspond to formal homes that account for a greater investment than those irregular ones. Another construction trend observed in the area is the prevalence of self-built structures, where the landowners or occupants are the ones who build the building with recycled or reused materials, progressively (growth by cubic units according to the need and availability of resources) considering simple and standard measures for the construction of the building.

It is important to note that the response of the residents through the accommodation of homes, raising them above the known heights of flooding, is intuitive and lacks, for the most part, professional judgment. A raised house built of wood, despite reducing the exposure to flood flow, is still highly vulnerable due to the use of a lightweight construction system. If the flood is turbulent, the possibility of collapse remains high. In Chile, although there is no current regulation that governs these types of buildings, the entry into force of the NCh3363 standard (INN, 2015) could significantly contribute to improving the standard of buildings in areas exposed to coastal floods.

Taking into consideration the data presented above, a constructive proposal for adaptation by accommodation should ideally contemplate a structure that is composed of parts and allows the reuse of materials, recognizing the cultural reality of self-construction. For this reason, reinforced concrete is discarded due to its difficulty in reuse, as is wood, which fails to provide the minimum resistance required against hydrostatic and hydrodynamic loads. In contrast, metal structures are presented as a suitable material for flood contexts, despite the fact that the data collection shows a low use of this material, which is due to the high initial cost, but which, according to De Paris and Lopes (2018), will mean a lower subsequent cost in the intervention of the construction system.

Fig. 10.2 Typologies of adapted houses in Laguna Verde. **a** Partially elevated on a concrete structure, **b** partially elevated on a lightweight wooden structure, **c** located on fill, and **d** elevated on a steel structure

10.4 Development of a Progressive Housing Model Adapted to Flood Conditions

In a second instance of this study, in order to evaluate a structural form that allows the adaptation of progressive homes by accommodation, tests were carried out with scale models on a seismic oscillations table. This allowed to qualitatively analyze a structural form to determine an efficient form in the use of material and resistance to the increase of loads on it. Based on the trends described above, the structural form raises 3 m a cubic space made of wood measuring 3 × 3 × 3 m. The models were built with 2 mm model sticks, at a 1:10 scale, adding a surface of 2 mm white cardboard, which functions as a floor and allows to place the weights with which the resistance of each of the models will be obtained. Tests were carried out on eleven different models. Of these, three are the models that stand out corresponding to Model 07 (inverted pyramid), Model 10 (3-sided pyramidal structure), and Model 11 (rhomboidal pillars). Figure 10.3 shows the three structural models with greater efficiency. This efficiency derives directly from the load that each one managed to support, since the most efficient model (Model 07) supported a greater weight than the others before the collapse.

Model 11, composed of four rhomboidal pillars, stands out for being a two-level model that, unlike the previous ones, seeks constructive ease, through the elaboration of coplanar elements. For this, a proposal of a prism of 3 m wide and long, with a height of 6 m, was developed, which corresponds to two levels of 9 m^2 each. This structure is reinforced by diagonals that extend 1 m outside the prism and then

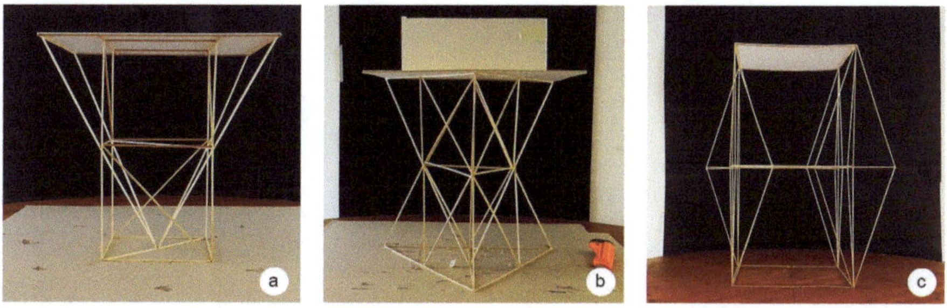

Fig. 10.3 Structural models with higher efficiency. **a** Model 07: inverted pyramid, **b** Model 10: three-sided pyramidal structure, and **c** Model 11: four rhomboidal pillars

integrate at the top vertex; this operation is repeated on each of the edges of the prism. In the tests, the deformation is restricted to the diagonals, avoiding the deformation of the vertical elements. Thus the time that elapses between deformation and fracture is prolonged, allowing to predict the point of collapse, which occurs when the verticals start to deform, because the diagonals reached their maximum resistance.

From these results, three structural conditions are obtained for the elaboration of a progressive structural form, which are detailed below.

1. Horizontal segmentation should be avoided, as this causes the module to work in parts, increasing the breaking points, rigidifying the structure, without allowing the flexibility of its parts.
2. It should avoid the oscillation of vertical elements, as the onset of deformations in these elements leads to a loss of rigidity, requiring external components to transmit and absorb the resulting forces.
3. It should consist of recognizable modular elements, since, in this way, the inhabitants will be able to reconstruct each one of its elements, prioritizing those that are composed of coplanar pieces, thus discarding the construction of complex parts that can potentially increase inaccuracies.

10.5 Construction of a Progressive Housing Prototype for Flood Adaptation

With the aim of prototyping a progressive modular proposal that incorporates adaptation by accommodation—and thereby revealing the possible difficulties that may arise in the construction of this building system—Model 11 was chosen for a construction development. Adjustments were made to the structural form, seeking to reduce the use of materials in its construction. Figure 10.4 shows the evolution of the structural form to define the final form used in the prototype. The adjustments allowed the essence of the proposal to be maintained, in addition to responding to the three structural requirements of the previous stage. For this, two modifications were made: the first corresponds to the use of a different profile for the diagonals, which is smaller

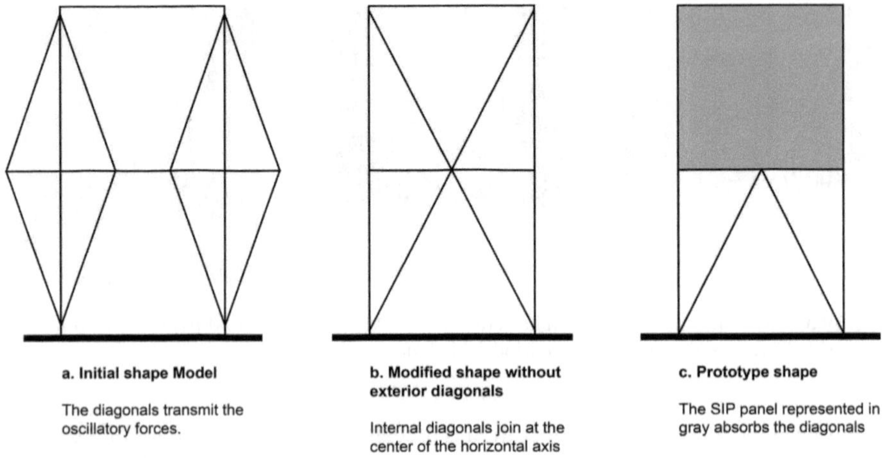

Fig. 10.4 Evolution of the structural shape, from tests on scale models to the prototype: **a** initial shape Model 11, **b** modified shape without exterior diagonals, and **c** prototype shape: the SIP panel represented in gray absorbs the diagonals

and open, thus being able to visualize the deformation primarily in these pieces; the second modification consists of the elimination of the external reinforcements, as this way the virtual surface required is reduced, from 16 to 9 m^2.

Then the materiality of the prototype was defined, following the guidelines of the NCh3363 standard. A structure of steel profiles of 80 and 100 mm of 3 mm thickness was selected, with which the entire support structure and upper structure of the prototype were developed. 86 mm SIP panels were used for the enclosures, such as floor and walls, but not for the roof, which is built with trusses, OSB boards, and zinc sheets. The prototype was developed in three stages in order to test the compatibility of the materials and the selected fastening systems. The materials were reused, to verify their ability to be disassembled and reassembled again.

Figure 10.5 shows the stages of prototyping developed in this study. In the first prototype, a corner of the module was developed, with the aim of verifying the feasibility of the joints between the concrete (foundation), steel (pillar, brackets, and beams), and the SIP panel (floor). It should be noted that the mobile foundations that were developed respond to the location where the prototype was built (concrete floor with no risk of flooding) and are not suitable for use in a flood zone. For these areas, a specific calculation of foundations is required, which will mainly depend on the composition of the soil. The second prototype consisted of the construction of the lower support structure, without considering the elevated level. This stage aimed to develop a metal structure that does not require welding, but only assemblies, allowing a quick assembly and disassembly. A bracket-shaped fitting was chosen for manufacture, which is joined by 3/8″ bolts supporting the weight of the beam, both by compression, when placed under the beam, as by traction, when placed on the beam.

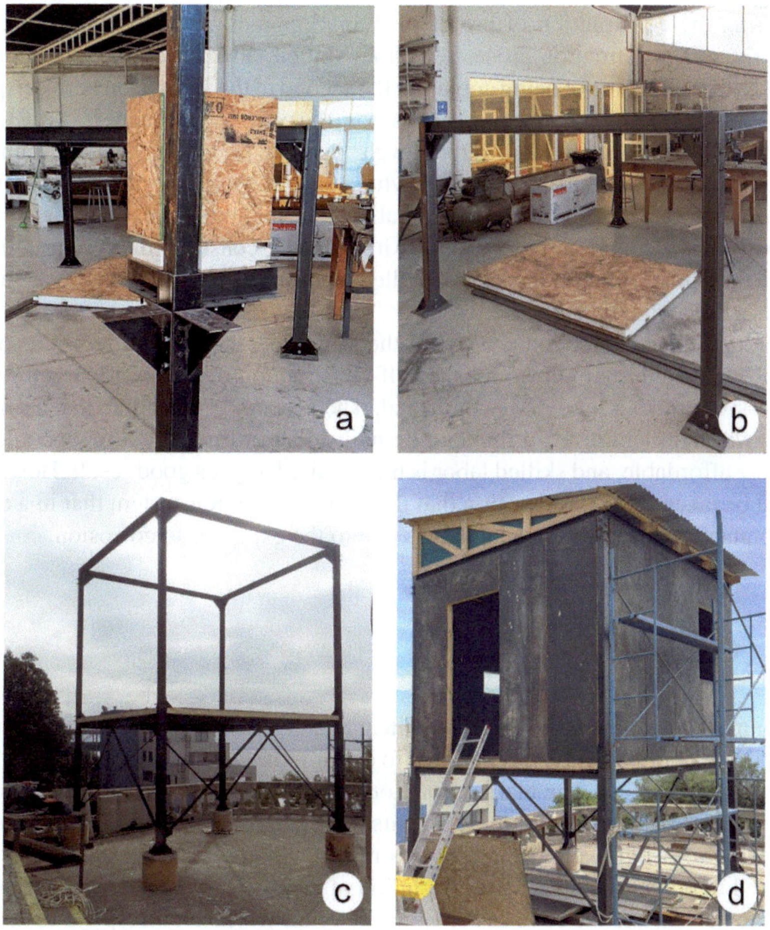

Fig. 10.5 Stages of the prototyping process. **a** Prototype 1: feasibility verification for joints of different materials, **b** Prototype 2: assembled support structure, **c** Prototype 3: total projected structure, and **d** Prototype 3: total closed module

For the third prototype, some of the elements manufactured in the previous stage were reused, verifying the feasibility of disassembling and reassembling, increasing the height from 1.0 to 2.2 m in the support structure. The cuts of the SIP panel were also standardized to 61cm (half of the board), which allowed to minimize the number of cuts, facilitating the constructability (previously cuts were made in various dimensions). The reduction in the size of the parts also allowed decreasing the weight of transport, allowing a single person to be able to transport each piece and at the same time reduce the amount of waste material. To make the fit between the 3-m metal pieces (2.8 m inside) and the pieces of SIP of 0.61 m, accommodation pieces were made, which are smaller SIP panels that allow the closure of each facade.

During the construction of prototype 5 inaccuracies were revealed, which can affect the self-construction process (without skilled labor): the first corresponds to the faces of the pillars not being coplanar; the second to the separation between pillars being different from 2.8 m; the third arises from the possible unevenness of the terrain, for which the brackets are at different levels; just like previously, the unevenness of the floor generates a fourth inaccuracy, where the pillars are not plumb; the fifth and last inaccuracy corresponds to the fact that the interior angles are different from 90°, and with that a square is not formed. Although the construction system aimed to reduce these inaccuracies, the lack of skilled labor in the construction process makes it highly likely to encounter them.

Through the construction process of the prototype, the possibility of creating an assembled system that allows for the self-construction of a support structure for a progressive housing module is confirmed. The process is considered successful as it achieves a structural form that can be built in short times by a few people. The costs are affordable, and skilled labor is not required to get a good result. However, it should be taken into consideration that this is a construction system that in a coastal environment requires high maintenance due to the exposure to corrosion.

10.6 Conclusions

In Laguna Verde, despite being a settlement characterized by low governability and high degree of informality, it is possible to confirm the presence of buildings adapted to the risk of flooding. The main adaptation strategy that was identified in this study corresponds to accommodation. Residents have used various construction systems, especially with lightweight materials to raise their homes above the flood marks recorded in previous events. The adaptation of homes is mainly carried out in areas closer to the sea and the edge of the El Sauce estuary, which correspond to the areas most exposed to the risk of flooding. These locations mostly correspond to those of greater precarity, where the houses reflect a low investment and poor quality of materials. Most of the houses are raised to a slight height, which would not prevent a significant flood, corresponding to an intense storm surge or a possible tsunami, like the scenario shown by the CITSU Laguna Verde chart (SHOA 2017).

This is a reactive adaptation in response to events that have previously affected a house. Adaptation occurs mainly from self-construction, using recycled materials where there is no application of regulations or professional criteria for risk mitigation. The growth in these houses is progressive; it occurs in parts according to the need and availability of resources, generally using cubic forms in 3 × 3 m modules. It was observed that this form is the one that allows a greater control by the residents, which is consistent with the mode of growth in informal sectors of Valparaíso.

In the central zone of Chile, the use of materials adapted to floods, such as steel and concrete, is expensive and complex. For this reason, for the manufacture of an adapted prototype it was chosen to work with prefabricated systems and then assemble them on site. The modular systems allow greater flexibility in design and adapt better to the needs of the residents. The incorporation of a supporting structure made with a triangulated structure allows incorporating rigidity to the structure and at the same time make it difficult to close its lower part. From a minimum module of 3 × 3 m, it is possible to project groupings of modules that compose a more resistant building, as shown in Fig. 10.6.

Regarding the possibilities for improvements to the construction process, from the developed prototype, the following recommendations are established:

1. The structural brackets that link the beams with the columns should be made in pairs of adjacent faces on the pillar; in this way, a greater precision in the leveling of the floor is guaranteed.
2. The structural brackets should be installed under the beam, not because of the efforts made, but because, being on the beam, it requires several cuts in the SIP

Fig. 10.6 Progressivity and grouping of projected houses from the prototyped adapted module. The lower image simulates the progressive growth of a house. The upper image shows a settlement of progressive houses adapted to flood conditions

panel, which slows down the work, in addition to generating a greater number of thermal bridges.

3. The openings should account for the furniture that can be installed inside, for which multiple module options must be developed, which respond directly to the space they occupy.
4. The metal elements should be permanently coated with anticorrosive paint. These should include an assembly piece or sealant in the metal fittings, to prevent internal corrosion.

Acknowledgements This study was funded by ANID FONDECYT Iniciación 11200300. The School of Architecture and Design e.ad of the Pontificia Universidad Católica de Valparaíso PUCV provided the facilities for the construction of the prototype.

References

CETCH TECHO Chile Study Center. (2023). The ongoing emergency in the camps of Chile: Exposure to natural threats and vulnerability. National Report. TECHO Chile, Study Center Reports. Retrieved October 14, 2023, from https://ceschile.org/wp-content/uploads/2023/08/Riesgos-campamentos_compressed.pdf

Contreras-López, M., Araya, P., Figueroa-Sterquel, R., et al. (2019). Assessment of tsunami vulnerability for the tourism sector in Valparaíso, Chile. *REDER, 3*(1), 5–23.

De Paris, S., & Lopes, C. (2018). Housing flexibility problem: Review of recent limitations and solutions. *Frontiers of Architectural Research, 7*(1), 80–91. https://doi.org/10.1016/j.foar.2017.11.004

FEMA P-646. (2012). *Guidelines for design of structures for vertical evacuation from tsunamis* (2nd Ed.). Federal Emergency Management Agency.

INN. (2015). *NCh3363 structural design buildings in areas at risk of flooding by tsunami or seiche.* National Institute of Standardization, Santiago de Chile.

Martínez, C., Rojas, O., Villagra, P., et al. (2017). Risk factors and perceived restoration in a town destroyed by the 2010 Chile tsunami. *Natural Hazards & Earth System Sciences, 17*(5).

Rosati, J., Touzinsky, K., & Lillycrop, W. (2015). Quantifying coastal system resilience for the US Army Corps of Engineers. *Environment Systems and Decisions, 35*, 1–13. https://doi.org/10.1007/s10669-015-9548-3

Siders, A. R. (2019). Managed retreat in the United States. *One Earth, 1*(2), 216–225.

SHOA Hydrographic and Oceanographic Service of the Chilean Navy. (2017). Tsunami Inundation Charts (CITSU). Retrieved October 14, 2023, from http://www.shoa.cl/php/citsu.php

Villamar, I. R., & Aguilar, A. G. (2020). Environmental deterioration and irregular urban occupation in the Lacustrine Zone of Xochimilco, Mexico City. *Journal of Latin American Geography, 19*(4), 43–67.

Vuik, V., Jonkman, S. N., Borsje, B. W., et al. (2016). Nature-based flood protection: The efficiency of vegetated foreshores for reducing wave loads on coastal dikes. *Coastal Engineering, 116*, 42–56.

Walker, B., Holling, C. S., Carpenter, S. R., et al. (2004). Resilience, adaptability and transformability in social–ecological systems. *Ecology and Society, 9*(2).

Winckler, P., Contreras-López, M., Campos-Caba, R., et al. (2017). The storm of August 8, 2015, in the regions of Valparaíso and Coquimbo, Central Chile. *Latin American Journal of Aquatic Research, 45*(4), 622–648.

Part III
Right to Housing, To the City and Services: Alternatives for Territorial Inclusion and Social Integration

Chapter 11
Nutritious Cities: An Exploration of Food Environments in Five Mexican Cities

Aleksandra Krstikj and Edgar Alan Egurrola Hernández

Abstract This chapter explores food environments in five Mexican cities: Mexico City, Monterrey, Guadalajara, Queretaro, and San Luis Potosi. The food environment refers to the physical, economic, and socio-cultural conditions that shape people's interactions with the food system. In this work, we analyze the role of informal street markets in food environments. Informal street markets are a common feature in the cities of Mexico. Little research is available on informal markets since these markets are not considered by the official census of economic units. We gathered data on informal markets through field surveys carried out from September to December 2022. We developed a framework to analyze the accessibility, affordability, and acceptability of fresh food offered in informal markets to shed light on the conditions that shape purchasing and consumption decisions. We concluded that informal markets reduce nutritional vulnerability, especially in the urban periphery. This work provides a base for planning decisions that target the improvement of access to a healthy diet in informal settings.

Keywords Food environments · Informal markets · Nutrition · Fresh food · Mexico

11.1 Food Environments in Mexico

The food environment refers to the physical, economic, political, and socio-cultural conditions that shape people's interaction with the food system and shape decisions about food acquisition and consumption (Reyes-Puente et al., 2022). Based on the

A. Krstikj (✉) · E. A. E. Hernández
School of Architecture, Art and Design, Tecnologico de Monterrey, Lopez Mateos, Mexico
e-mail: sandra.krstik@tec.mx

E. A. E. Hernández
e-mail: edgar.egurrola@tec.mx

E. A. E. Hernández
School of Architecture, Art and Design, Tecnologico de Monterrey, Guadalajara, Mexico

M. Vicuña and E. Giorgi (eds.), *Intersections*, Springer Geography,
https://doi.org/10.1007/978-3-031-76402-8_11

analysis of factors that determine the accessibility, affordability, and acceptability of unprocessed foods, this work explores the role of the Mexican informal market—*tianguis*—in the food environment and evaluates its contribution to decreasing food access vulnerability in five cities, i.e., Mexico City (CDMX), Guadalajara, Monterrey, Querétaro, and San Luis Potosí (SLP).

Recently, several studies have emerged that evaluate the proximity of food outlets and the quality of the food environment in Mexico (Pineda et al., 2021; Reyes-Puente et al., 2022). However, these studies focus mainly on fixed markets, convenience stores, and grocery stores, not including the *tianguis*, which is one of the most visited food outlets by Mexicans. Considering that only 6–10% of products sold in supermarkets are unprocessed foods (Gonzalez Arellano & Capron, 2020)—such as fruits, vegetables, and seeds—their role in the nutritious food environment is limited. Not considering the informal market in the food environments of Mexican cities with a high degree of informality can bring about biased results on evaluation of community nutritional access.

This research raises the question: what is the role of *tianguis* in the food environments in Mexico? Do the *tianguis* reduce inequality in access to fresh food and, if so, to what extent? Current planning does not consider the *tianguis*; thus their role in the food environment is unclear. The benefit of the study is to support local decision-making in promoting urban health and food security, especially in areas with a high degree of informality.

11.1.1 The Tianguis of Mexico

Traditional food markets have been key development poles of our cities, in both economic and social sense (Krstikj & Koura, 2013). They are not only a place for food retail but also a place of encounter, enjoyment, and socio-cultural recreation that makes them the heart of a neighborhood (Istijabatul et al., 2017). The traditional street market is an open or partially open commercial complex with vending stalls (fixed or movable) organized in rows. They are a familiar retail space in neighborhoods from Latin America to southeast Asia that offers affordable access to local, fresh, and healthy foods (Zhong et al., 2020). Even in large cities, such as Mexico City, Taipei, and Singapore, they are a common urban element. They have been found to create freshness through their sensoria, atmosphere, and trust between food vendors and consumers (Zhong et al., 2020) as a key attraction point that is lacking in supermarkets. In the US, a comparison can be drawn to farmers market, although those markets have higher prices compared to supermarket food and are usually not present in underserved neighborhoods (Larsen & Gilliland, 2009).

In Mexico, the *tianguis* originated in pre-Hispanic times as a producer market-on-wheels that offers fresh and prepared foods, as well as artisan products (Linares & Bye, 2016). Surprisingly, the *tianguis* in Mexico have resisted the global "retail revolution" that predicted the rapid decline of small-scale retail shops and traditional markets in favor of supermarket chains (Zhong et al., 2020). Today, the *tianguis* in

Fig. 11.1 Photo of one stall selling fresh food in the tianguis on Avenida el Colli in Guadalajara (by authors, taken October 1, 2022, at 10 am)

Mexico offer food, cosmetics, clothes, tools, and many other products (Fig. 11.1). The percentage of fresh food sold in the *tianguis* differs from city to city and from market to market.

Although informal, the *tianguis* are regulated by local governments to some degree. Local governments are aware of their existence and know exactly which day and where they appear. There is usually a representative from the community of street vendors that negotiates their function with local authorities. In the municipality of Atizapan de Zaragoza, we found from the vendors that they pay a monthly fee to the local government for using the public street (approximately 10 US dollars per stall per month). Unfortunately, there is no reliable data of the number and location of *tianguis*, type of fresh food sold, or prices of food. This is a problem when evaluating social disparities in access to fresh food sold in these markets, considering only the fixed and censused markets can show a partial picture of the food environment.

11.1.2 Case Studies

This study focused on five metropolitan areas, i.e., CDMX, Monterrey, Guadalajara, Queretaro, and SLP (Table 11.1).

CDMX is the capital and central urban node of the country in geographic and economical terms. The rapid urban expansion of the past 20 years has generated a heterogeneous, car-centric, and fragmented periphery, while the affluent inner-city zones experienced population loss (Gonzalez Arellano & Capron, 2020). To compare the different conditions between inner-city neighborhoods and the metropolitan periphery, we conducted field surveys in the central municipality of Benito Juarez and in the peripheral industrial municipality of Atizapan de Zaragoza 23 km north

Table 11.1 Case studies general information. Official census data for 2020, National Statistics Bureau of Mexico

City	Urban area (km^2)	Population (in millions)	Density (people/ km^2)	GDP/capita (m × n)	Working poverty	Number of market stalls*
CDMX	1,494	9.21	6,164	316,761	2.8	1,367
Atizapan	93	0.52	5,591	85,184	/	46
Monterrey	958	1.14	1,189	225,862	0.6	287
Guadalajara	151	1.38	9,139	133,857	2.3	450
Queretaro	363	0.79	2,189	160,301	3.1	143
SLP	385	0.82	2 140	119 143	2.5	184

* Data gathered from field surveys, municipality registers, and open sources

of Benito Juarez. Atizapan has the lowest number of *tianguis* per capita in the five case studies (1 *tiangui* per 11,304 people).

Monterrey is the industrial powerhouse of Mexico due to its strong economic and cultural relationships with the United States. As a reaction to the different development policies passed by federal governments in the twentieth century and with the signing of the North American Free Trade Agreement in 1993, Monterrey experienced rapid urban sprawl and car-oriented development in its peripheral zones (Carpio et al., 2021). Its metropolitan area is nearly the same size as that of CDMX but is significantly less dense. Monterrey has low agricultural output in which hydric stress recently became a severe problem (Aguilar Barajas et al., 2016). Monterrey holds the second place in the number of *tianguis* per capita in the five case studies (1 *tiangui* per 3,972 people).

Guadalajara is in the eastern Pacific zone. With its conducive climate and soil, Guadalajara has some of the richest farmlands in Mexico. Guadalajara had been an agricultural hub since the seventeenth century and, despite recent problems with urban expansion, has continued to hold a leading place in agricultural production in Mexico (Secretaría de Agricultura y Desarrollo Rural de México: https://www.gob.mx/agricultura/articulos/cinco-estados-con-cinco-estrellas-en-produccion-agricola). Nevertheless, the agricultural communities located at the urban periphery are currently sites of urban sprawl and impoverishment (Barleta et al., 2020). Guadalajara has the highest number of *tianguis* per capita in the five case studies (1 *tiangui* per 3,066 people).

Querétaro and SLP are mid-sized developing cities in the central zone of Mexico. Both cities are undergoing rapid car-centric urban sprawl (Moreno-Mata, 2018; Obregón-Biosca et al., 2016). As important nodes on the axis between CDMX and Monterrey, both cities are developing industrial and transport hubs (Moreno-Mata, 2018). Querétaro has 1 *market stall* per 5,524 people, while SLP has 1 *market stall* per 4,456 people.

The working poverty rate in Mexico is assessed as the percentage of the population with a labor income less than the monetary value of the food basket (ITLP, *I. S. labor poverty*). Monterrey has the lowest working poverty rate, as well as the highest GDP/capita ratio. On the other hand, Queretaro has the highest working poverty rate, while Guadalajara has the lowest GDP/capita ratio.

11.2 Important Indicators in Food Environments

The data on the location of *tianguis* was obtained through different sources. For CDMX, a shapefile was published on the city government web-platform in 2021 that georeferenced the *tianguis* in the city's administrative zone. The metropolitan area is not covered in the data.

CDMX is the only city in Mexico that has georeferenced the locations of *tianguis*. In other cities, usually the municipal office has a register of locations of *tianguis* with addresses but without coordinates, which sometimes also contains information about weekdays when they open, and the number of stalls. Our team had to verify the location data and georeferenced each market in Atizapan de Zaragoza, Guadalajara, and Querétaro. For the *tianguis* in Monterrey and SLP, we relied on open web maps published by citizens. The maps of *tianguis* in Monterrey and SLP were also later checked with field surveys to verify the data.

The field surveys were conducted during a three-month period from September to November 2022. We visited several markets in each city and documented the conditions with photography and videos (Table 11.2). We collected data on the percentage of stalls that sold fresh and unprocessed food. Further, we collected data on the prices of common and non-seasonal fruits and vegetables. We also collected several interviews with customers and sellers.

In Fig. 11.2, we present the location of *tianguis* in each case study in relation to the marginalization index by neighborhood estimated by the National Commission of Population of Mexico in 2020 (CONAPO, 2020). The map also shows the location of *tianguis* where we collected the data shown in Table 11.2.

Previous studies have focused on disparities in access, including the existence of food deserts and the neighborhood characteristics associated with them (Ginsburg et al., 2019; Walker et al., 2010; Wetherill & Gray, 2015). Others explored the effects of food deserts on diet and obesity as a result (Cooksey-Stowers et al., 2017; Liese et al., 2018). Most use the method of Caspi et al. (2012) to explore food environments in 5 dimensions:

- Availability: adequacy of the supply of healthy foods; measurements: store presence, shelf space, product availability.
- Accessibility: the location of the food supply and the ease of getting to that place; measurements: distance and travel time.
- Affordability: prices and people's perceptions of value in relation to cost; measures: audits of specific food stores or regional price indices.

Table 11.2 Description of the materials gathered through field surveys of tianguis

City	Address/street	Period	Obtained data
Atizapan de Zaragoza	Cuautitlán Izcalli, Campanario	September 5, 2022	26-min video two interviews
	Lago de Guadalupe, Margarita Maza	September 25, 2022	15-min video one interview
Monterrey	del Chorro	September 23, 2022	15-min video two interviews
	Centro Histórico	September 24, 2022	9-min video one interview
	Florida	September 24, 2022	8-min video
Guadalajara	el Colli	October 1, 2022	32-min video two interviews
	from the Agricultural	October 1, 2022	10-min video one interview
SLP	from Saturdays, av Jesus Yuren	October 8, 2022	35-min video two interviews
	Saturday, Av Real to Guanajuato	October 8, 2022	6-min video
	Mexico Avenue, San Felipe de la Cruz	October 8, 2022	12-min video one interview
	Garibaldi	October 9, 2022	3-min video
Queretaro	the Tepetate	October 9, 2022	15-min video two interviews
	Lomas de Casa Blanca	October 9, 2022	12-min video one interview
CDMX	New York street, Naples House of the World Worker, Narvarte	November 21, 2022	20-min video two interviews 30-min video One interview

- Acceptability: people's attitudes about the attributes of their local food environment, and whether or not the given supply of products meets their personal standards.
- Measures: survey participants; store auditors assign food quality scores to products.
- Accommodation: how well outlets accept and adapt to the needs of residents; measurements: store hours and accepted payment types.

In our research, we focus on measures of accessibility, affordability, and acceptability as previous literature found evidence linking these indicators to consumption patterns (Bridle-Fitzpatrick, 2015; Krstikj et al., 2022; Turner et al., 2021).

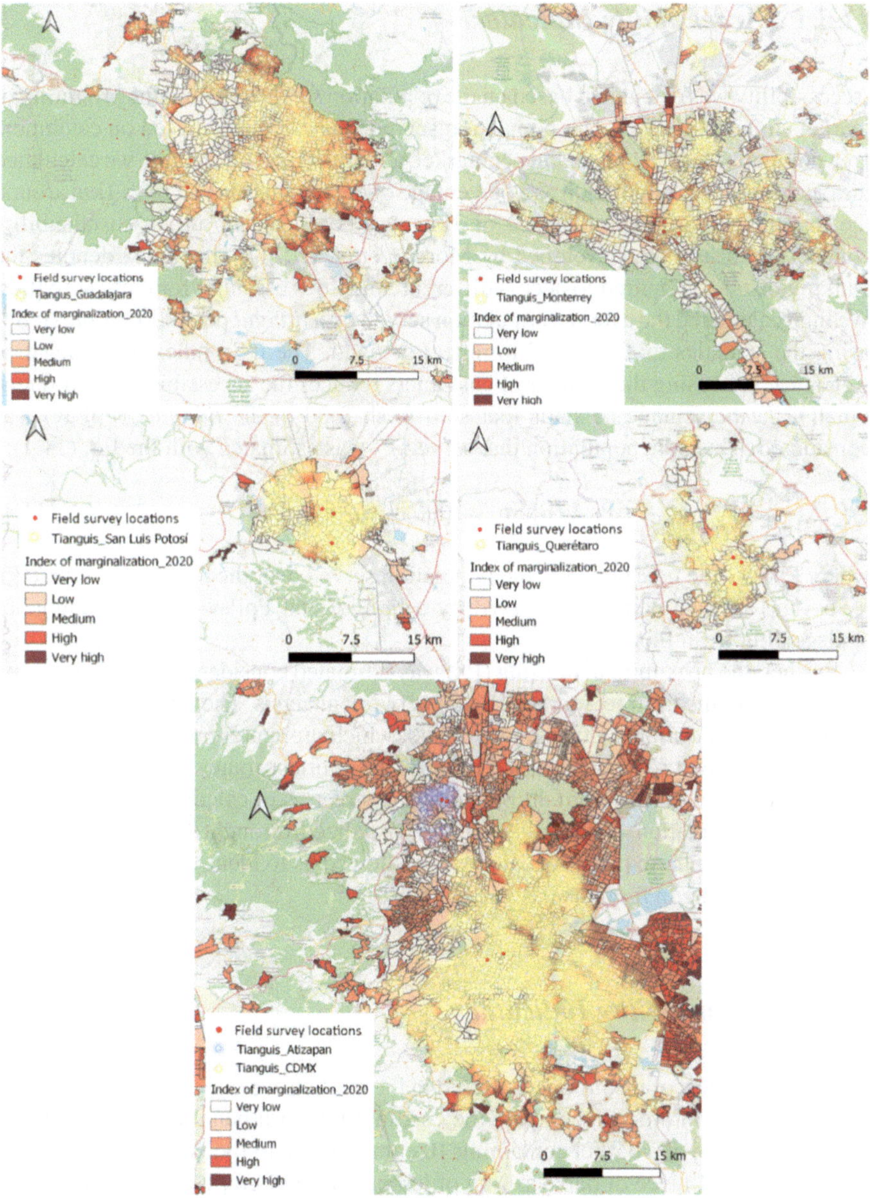

Fig. 11.2 Map of the location of tianguis in the cities (white dots), their proximity buffer space (yellow glow), and field survey locations (red dots) (by authors)

11.2.1 Accessibility Indicator

Accessibility was calculated based on the population served (Ps) and the proximity of markets to housing (Px). The population served was calculated based on catchment area population, mean of demand per market, and fresh food offer. First, we calculated the catchment area population as a percentage of the population that lives 1km around the tianguis (PpSp). The distance of 1km was based on the urban sustainability concept of *15-min city* developed by Moreno et al. (2021). This index denotes the walking distance required to access different amenities essential for daily activities. We then calculated the average population served by a *tiangui* per 1000 people (μ'Sp) using the statistics tool in QGIS to reflect demand. The μ'Sp index has a negative sign since increasing the number of customers reduces the accessibility of products. Finally, the percentage of stalls that sold fresh food in the *tianguis* (PpPav) was considered. Thus, the population that served Ps was estimated with the Eq. (11.1):

$$\mathrm{Ps} = \mathrm{PpSp} - 1000/\mu'\mathrm{Sp} + \mathrm{PpPav} \tag{11.1}$$

where PpSp—the percentage of the population served over the total population (%); μ'Sp—the average population served by a *tiangui* and PpPav—the percentage of stalls selling fresh food (%).

Further, the proximity to housing (Px) was estimated considering the spatial proximity of the home to street markets, using network analysis. The layout of roads was used as the connections between two points, which are represented by the intersections between two roads and, based on this, the shortest routes that a person must travel to reach a market from all points of the street were calculated: the city. The aggregate information of the cities can be visualized through the average time in minutes to reach these points of sale, which allowed us to identify the value that the practice of selling in *tianguis* has in each city.

11.2.2 Affordability Indicator

The prices of most of the products offered in the markets were recorded through field work; however, in order to compare the cost in different urban areas, we calculated the average price of a basket containing 10 commonly sold foods—1 kg of bananas, 1 kg of apples, 1 piece of papaya, 1 kg of lemons, 1 kg of grapes, 1 piece of lettuce, 1 kg of tomatoes, 1 kg of carrots, 1 kg of potatoes, and 1 kg of onion. With this method, we could avoid biases due to seasonal and regional differences in the supply of fruits and vegetables, as well as imported goods that may be outliers in terms of higher prices. Then the price of the basic basket was compared with the prices of the same products registered by the Federal Consumer Protection Agency or PROFECO. PROFECO collects information from fixed and formal points of sale, for which the comparison serves to establish price differences between the formal and informal sectors (Asq1). Finally, the price of the basic basket was multiplied by 4 weeks, and it was crossed with the minimum wage declared by the National Minimum Wage Commission or CONASAMI. Thus, the percentage of minimum wage necessary for the acquisition of basic fruits and vegetables in a month (Asq2) was also calculated. The final affordability index (Asq) was estimated by reducing Asq2 from Asq1, as in Eq. (11.2):

$$\mathrm{Asq} = \mathrm{Asq1} - \mathrm{Asq2} \tag{11.2}$$

11.2.3 Acceptability Indicator

The population for the interviews was chosen randomly from people who were willing to chat. We acknowledge the fact that the sample is by no means statistically significant, but only illustrative of the socio-cultural context in the informal market. Thus, the questions in the interview were open-ended, providing an opportunity for the interviewees to express information important to them. Nevertheless, three key questions formed the structure of the interview:

1. Do you live near this market and how do you commute here?
2. Do you always buy your fresh food from this market, and if yes, how often do you visit it?
3. How do you compare the quality and price of the fresh food in this market with formal supermarkets?
4. The questions for the sellers were the following:
5. Do you live near this market and how do you commute here?
6. How long have you been selling food here?
7. Do you produce the food you sell, or do you re-sell it?

11.3 The Role of the Informal Market in the Food Environment

The results showed that CDMX has the best spatial agreement between *tianguis* distribution and population, where 71% of the total population lives within 1 km of a *tiangui*. In Guadalajara 57% of the population is served by these markets, while in Atizapán and San Luis Potosí only 53%. The population served is smaller in Querétaro (44%), while Monterrey has the worst result for population served (29%) (Table 11.3). The index of the served population (Ps), considering the products offered and demand, confirms that the best conditions are in CDMX while the worst are in Monterrey.

The accessibility analysis shows that the cities of Querétaro and Monterrey have the lowest accessibility to the market with more than 35 min on average (Fig. 11.3).

Table 11.3 Assessment of the population served by the tianguis

Area	PpSp	PpPav	1000/μ′Sp	Ps
CDMX	0.71	0.6	0.3687	0.94
Atizapan	0.53	0.5	0.4042	0.63
Monterrey	0.2	0.1	0.3120	0.08
Guadalajara	0.57	0.6	0.3162	0.85
Queretaro	0.44	0.4	0.2599	0.58
SLP	0.53	0.4	0.3272	0.60

This means that there is low connectivity between housing and markets. The cities with the greatest connectivity are CDMX and Guadalajara, which meet the 15-min requirement. Also, we zoomed into individual cities to see the distribution of proximity to *street markets* compared to other points of sale—supermarkets, fixed markets, and groceries, as well as evaluating the proximity to these spaces against the coverage of educational equipment, commerce, recreational activities, and community services. In this comparison, we can see that both the *street markets* and the food sales spaces in CDMX have a wide coverage in the territory, although the quality of the food sold may vary from one to another (Fig. 11.4, CDMX). However, the optimal coverage of amenities is contained in specific areas of the city, which allows us to infer that the coverage of the markets is not limited by the level of urban consolidation of the community it serves.

The affordability analysis discovered that the lowest price in the *street markets* is found in the metropolitan periphery of Atizapan, while the highest is in the capital. This result differs significantly from the data from the formal sector where the cheapest foods were found in Guadalajara and the most expensive in SLP. Comparing the cost of the *street markets* with the minimum wage, we find that most of the salary is paid for food in CDMX while the least in the Atizapan periphery, which can be identified in the final index Asq (Table 11.4).

From the interviews in the *markets,* we concluded that people who buy food there are usually locals who live in proximity or same neighborhood where the *market* is located. The commute is mostly by walk or by bicycle to the market. They said that they visit the *market* at least once a week. They found the prices of fresh food lower in the *market* which confirms our analysis. In general, people were satisfied with the quality of food offered in the *market* and said that they do not see any difference between food sold in the *market* and supermarkets. From two interviews with locals in Monterrey, we found that people who have cars (and cars are a priority in a car-centric city) prefer to buy food in one of the three *Supply Centers*—wholesale markets—or in fixed supermarkets since parking is available there. They noted that in Monterrey informal markets are usually flea markets for second-hand clothes or tools and hinted that cultural factors—such as the dominant processed and meat-based diet of the north—could be a factor as well. During our field survey in San Luis Potosi and Queretaro, we noted that the average age of sellers of fresh food in the *market* is higher. In several interviews there we were informed that those sellers have local small

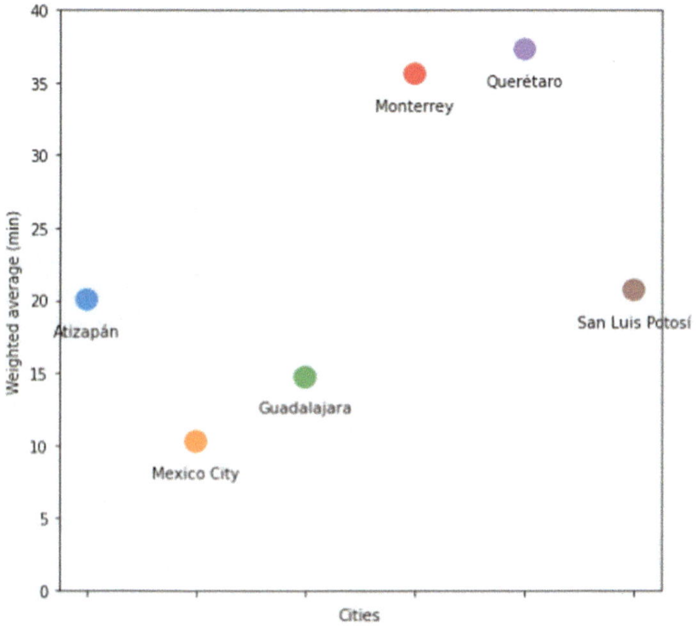

Fig. 11.3 Estimation of accessibility of street markets per city, in minutes of walking (by authors)

urban gardens and produce the food themselves, which contributes significantly to cutting middlemen in the food supply chain and reducing prices. It seems a remanent practice of historical land uses in those cities before the rapid industrialization of the last few decades where small urban gardens were very common. Nevertheless, the fresh food supply in big and densely populated cities like Mexico City depends almost entirely on complicated food supply chains. This asks for a serious revision and policy control of food supply chains since the current trend could only exacerbate the hollowing of the inner-city zones.

11.4 Conclusions

This research is a preliminary exploration of the role of the informal market in food environments in Mexico. In previous literature, we found that the metropolitan periphery is more vulnerable in access to nutrition considering the scarcity of sales points and long distances. This was confirmed by our analysis of accessibility and connectivity in cities. Additionally, cities that are experiencing rapid urban expansion along with car-dependent planning are also vulnerable in terms of access, such as Monterrey and Querétaro. Food availability also seems to be related to food culture and agricultural practices, where cities such as Guadalajara and the metropolitan periphery in Atizapan almost reach the same level of service provision as the capital,

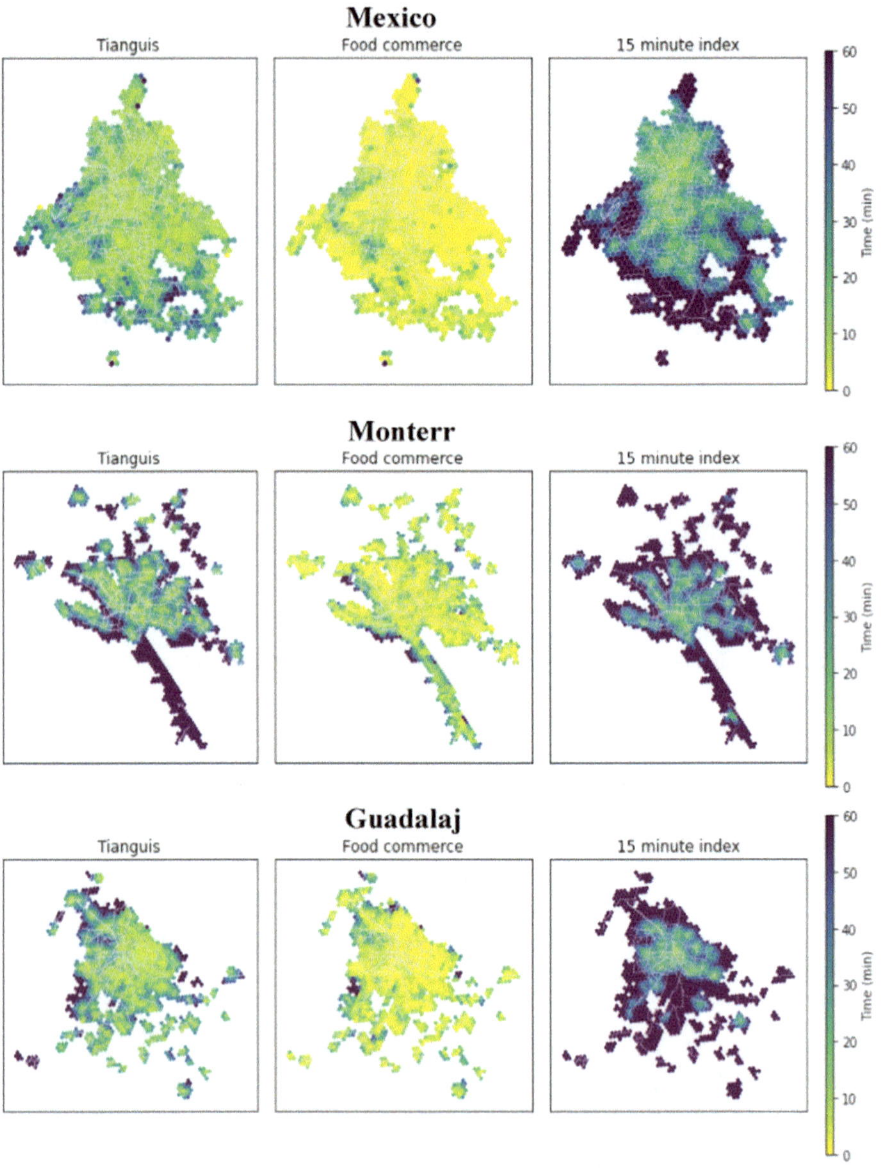

Fig. 11.4 Comparison of accessibility to tianguis (left), formal food outlets (middle), and other services expressed as a 15-min index (right) in Mexico City (CDMX), Monterrey and Guadalajara (by authors)

Table 11.4 Assessment of the affordability of fresh food in the street markets

City	Price* food basket street markets	Price* food basket formal outlets PROFECO	Asq1 (+)	Asq2 (−)	Asq
CDMX	258	310	20%	19%	0.003
Atizapan	156	310	97%	12%	0.853
Monterrey	223	303	36%	17%	0.187
Guadalajara	165	297	79%	12%	0.668
Queretaro	180	310	72%	14%	0.579
SLP	190	317	67%	15%	0.5219

* In MXN pesos

while Monterrey shows surprisingly low availability conditions. Finally, the study showed that street markets fill the void left by the lack of official urban nutritional planning by providing an affordable food option for the most vulnerable populations. Although informality can often be a drawback for the planner, we should not underestimate the resilient nature of this informal sector and its evident contribution to urban health and promote public policies that protect informal markets.

The appearance of *tianguis* in areas where formal outlets do exist, especially in cities where *tianguis* have a high Nutritive index such as Guadalajara, Mexico City, and its metropolitan periphery, means that the actual access to fresh food is significantly higher than shown by previous studies that rely on census reported outlets only. It is important to promote future studies to better clarify the ingestion of fresh food in these zones considering the alternative informal market as an important contributor to availability and accessibility of food. Thus, the informal market is filling a significant gap left by official planning in bringing fresh food closer to vulnerable communities, although its precise contribution is still unknown. With this preliminary study, we identified the relevance of *tianguis* as a means of bringing fresh food closer to the population and highlighted the importance of considering these informal spaces when studying access to food in different communities.

References

Aguilar Barajas, I., Sisto, N. P., & Ramírez, A. (2016). Water governance in Mexico: An overview and the case of the Monterrey metropolitan area. Multilevel governance in the urban context. *Innovative Governance of Large Urban Systems, 2*(1), 2–7. https://iglus.org/wp-content/uploads/2019/05/IGLUS-Quarterly-Vol-2-Issue-1.pdf

Barleta, L., Carrillo, M., Frank, Z., & Steiner, E. (2020). Ejidos, urbanization, and the production of inequality in formerly agricultural lands, Guadalajara, Mexico, 1975–2020. *Land, 9*(12), Article 526. https://doi.org/10.3390/land9120526

Bridle-Fitzpatrick, S. (2015). Food deserts or food swamps?: A mixed-methods study of local food environments in a Mexican city. *Social Science and Medicine, 142*, 202–213. https://doi.org/10.1016/j.socscimed.2015.08.010

Carpio, A., Ponce-Lopez, R., & Lozano-García, D. F. (2021). Urban form, land use, and cover change and their impact on carbon emissions in the Monterrey metropolitan area, Mexico. Urban Climate 39, Article 100947. https://doi.org/10.1016/j.uclim.2021.100947

Caspi, C. E., Sorensen, G., Subramanian, S. V., & Kawachi, I. (2012). The local food environment and diet: a systematic review. *Health & Place, 18*(5), 1172–87. https://doi.org/10.1016/j.healthplace.2012.05.006. Epub 2012 May 31. PMID: 22717379; PMCID: PMC3684395.

CONAPO or the National Commission of Population of Mexico. (2020). https://www.gob.mx/conapo/documentos/indices-de-marginacion-2020-284372

Cooksey-Stowers, K., Schwartz, M. B., & Brownell, K. D. (2017). Food swamps predict obesity rates better than food deserts in the United States. *International Journal of Environmental Research and Public Health, 14*(11), 1366. https://doi.org/10.3390/ijerph14111366

Ginsburg, Z. A., Bryan, A. D., Rubinstein, E. B., et al. (2019). Unreliable and difficult-to-access food for those in need: A qualitative and quantitative study of urban food pantries. *Journal of Community Health, 44*, 16–31. https://doi.org/10.1007/s10900-018-0549-2

Gonzalez Arellano, S., & Capron, G. (2020). No. 22: Inclusive Growth and Informal Food Vending in Mexico City, Mexico. Hungry Cities Partnership, 46. https://scholars.wlu.ca/hcp/46

Istijabatul, A., Bambang, S., & Wisnu, P. (2017). Spatial variety and distribution of traditional markets in Surakarta as potential factors in improving spatial-based management. *Geoplanning: Journal of Geomatics and Planning, 4*(1), 63–74.

Krstikj, A., & Koura, H. (2013). Identifying the significance of historic place in Skopje's Old Bazaar—based on analysis of facility's location and primary road in the Ottoman urban strategy. *Journal of Architecture and Planning (Transactions of AIJ), 78*(686), 829–835.

Krstikj, A., Lukoseviciute, G., Boyes, C., & Contreras Ruiz Esparza, M. G. (2022). Nutritious landscapes: The food environments in the metropolitan peripheries of Mexico. In E. Giorgi, T. Cattaneo, A.M. Flores Herrera, & V.d.S. Aceves Tarango (Eds.), *Design for vulnerable communities*. The Urban Book Series. Springer. https://doi.org/10.1007/978-3-030-96866-3_20

Larsen, K., & Gilliland, J. (2009). A farmers' market in a food desert: Evaluating impacts on the price and availability of healthy food. *Health & Place, 15*(4), 1158–1162.

Liese, A. D., Lamichhane, A. P., Garzia, S. C. A., Puett, R. C., Porter, D. E., Dabelea, D., D'Agostino, R. B., Standiford, D., & Liu, L. (2018). Neighborhood characteristics, food deserts, rurality, and type 2 diabetes in youth: Findings from a case-control study. *Health & Place, 50*, 81–88. https://doi.org/10.1016/j.healthplace.2018.01.004

Linares, E., & Bye, R. (2016). *Traditional markets in Mesoamerica: A mosaic of history and traditions, 151–177*. Springer.

Moreno-Mata, A. (2018). Urban sprawl, environmental justice and equity in the access to green spaces in the metropolitan area of San Luis Potosí, Mexico. In: W. Leal Filho, R. Noyola-Cherpitel, P. Medellín-Milán, & V. Ruiz Vargas (Eds.) *Sustainable development research and practice in Mexico and selected Latin American Countries*. World Sustainability Series. Springer. https://doi.org/10.1007/978-3-319-70560-6_32

Moreno, C., Allam, Z., Chabaud, D., Gall, C., & Pratlong, F. (2021). Introducing the "15-Minute City": Sustainability, resilience and place identity in future post-pandemic cities. *Smart Cities, 4*(1), 93–111. https://doi.org/10.3390/smartcities4010006

Obregón-Biosca, S. A., Romero-Navarrete, J. A., Mendoza-Sanchez, J. F., & Betanzo-Quezada, E. (2016). Impact of mobility induced by urban sprawl: Case study of the Querétaro metropolitan area. *Journal of Urban Planning and Development, 142*(2). https://doi.org/10.1061/(ASCE)UP.1943-5444.0000295

Pineda, E., Brunner, E. J., Llewellyn, C. H., et al. (2021). The retail food environment and its association with body mass index in Mexico. *International Journal of Obesity, 45*, 1215–1228. https://doi.org/10.1038/s41366-021-00760-2

Reyes-Puente, A. L., Peña-Portilla, D. G., Alcalá-Reyes, S., Rodríguez-Bustos, L., & Núñez, J. M. (2022). Changes in food environment patterns in the metropolitan area of the valley of Mexico,

2010–2020. *International Journal of Environmental Research and Public Health, 19*, 8960. https://doi.org/10.3390/ijerph19158960

Turner, G., Green, R., Alae-Carew, C., & Dangour, A. D. (2021). The association of dimensions of fruit and vegetable access in the retail food environment with consumption; A systematic review. *Global Food Security, 29*, 100528. https://doi.org/10.1016/j.gfs.2021.100528

Walker, R. E., Keane, C. R., & Burke, J. G. (2010). Disparities and access to healthy food in the United States: A review of food deserts literature. *Health & Place, 16*(5), 876–884. https://doi.org/10.1016/j.healthplace.2010.04.013

Wetherill, M. S., & Gray, K. A. (2015). Farmers' markets and the local food environment: Identifying perceived accessibility barriers for SNAP consumers receiving temporary assistance for needy families (TANF) in an urban Oklahoma community. *Journal of Nutrition Education and Behavior, 47*(2), 127-133.e1. https://doi.org/10.1016/j.jneb.2014.12.008

Zhong, S., Crang, M., & Zeng, G. (2020). Constructing freshness: The vitality of wet markets in urban China. *Agriculture and Human Values, 37*(1), 175–185.

Chapter 12
PICTOS: A Service Focused on Cognitive Accessibility for Navigation and Evaluation of Services in Chile

Katherine Exss, Herbert Spencer González, Vanessa Vega Córdova, Izaskun Álvarez-Aguado, and Renée Rodo Iunnissi

Abstract Universal accessibility is a topic of recent interest in the field of disability in Chile. Through public policies, the State has worked to guarantee mainly physical and sensory accessibility of spaces and public services. However, cognitive accessibility is still scarcely addressed despite its relevance and impact on the inclusion of people in society. Cognitive accessibility ensures that information is easy to understand by all people, regardless of their previous knowledge or intellectual abilities. In this context, since 2018, a web service (PICTOS) has been co-designed which provides support to ensure cognitive accessibility in the use of services. From inclusive research, with the active participation of an advisory group composed of adults with intellectual disability, a pictographic system and an information architecture have been created aimed at promoting and improving the cognitive accessibility of public and private services. Recently, progress has been made in the implementation of the support in various cities of the country, and in the transfer of the cognitive accessibility model for digital services.

Keywords Cognitive accessibility · Intellectual disability · Services · Supports · Pictograms

K. Exss · V. Vega Córdova
Millennium Institute for Care Research (MICARE), Santiago, Chile
e-mail: vanessa.vega@pucv.cl

K. Exss (✉) · H. Spencer González · R. Rodo Iunnissi
Pontificia Universidad Católica de Valparaíso, Valparaíso, Chile
e-mail: kexss@ead.cl

H. Spencer González
e-mail: hspencer@ead.cl

R. Rodo Iunnissi
e-mail: reneerodo@ead.cl

I. Álvarez-Aguado
Universidad de las Américas, Mexico City, Mexico
e-mail: ialvareza@udla.cl

M. Vicuña and E. Giorgi (eds.), *Intersections*, Springer Geography,
https://doi.org/10.1007/978-3-031-76402-8_12

12.1 Introduction

In Chile, universal accessibility has become a topic of recent interest in the field of disability. As a result of current public policies, which are manifested in laws such as 20.422 (Ley Chile, 2010) and 21.015 (Ley Chile, 2017), the State has gradually been working to guarantee the physical and sensory accessibility of public spaces and services. However, cognitive accessibility (CA) has received little attention in our country and remains a practically unknown phenomenon, despite being an indispensable requirement to ensure the principle of universal accessibility (Plena Inclusión, n.d.).

Cognitive accessibility focuses on making information easy or intuitive to understand by anyone, regardless of their intellectual abilities (Brusilovsky, 2018). In this sense, it is known that Intellectual Disability (ID) is a disorder that includes limitations of intellectual functioning as well as adaptive behavior in the conceptual, social and practical domains (Schalock et al., 2021), but with the appropriate support, it is possible that they perform satisfactorily in daily life activities (Asociación Americana de Psiquiatría, 2014). It should be mentioned that the current definition of disability recognizes that it is not only an individual characteristic or condition, but the result of the interaction between the person with a disability and the social, cultural, environmental and structural barriers that can limit their participation in society (Schalock, 2009). This approach coincides with the social model of disability, which emphasizes the elimination of barriers imposed by the built or designed environment, and the promotion of inclusion and equality of opportunities for people with disabilities (Naciones Unidas, 2006).

Among the topics addressed in discussions about disability, independent living emerges as one of the main interests and demands of communities and is promoted by important movements led by people with disabilities, such as the Independent Life Movement (Morris, 1994). In this sense, it is considered that the development of an independent life constitutes a human and civil right that allows people with disabilities to make decisions associated with control over their own lives (Atack et al., 2019). This interest has also been recognized in the local Chilean context as a primary need, which is not exempt from obstacles, such as the difficulty in being economically independent or the lack of accessibility in the workplace, in the city and its services (Jarpa-Azagra et al., 2021; Vega Córdova et al., 2023).

Cognitive accessibility in services is one of the fundamental pieces to facilitate the full inclusion of people and promote independent living (Blackman et al., 2003; Peterson et al., 2022). As a result of the pandemic and the forced digitalization of services, it has been observed that digital media offer transactions equivalent in relevance and complexity to physical media, requiring an expansion of the right to services, regardless of the channel used by people. However, the sense of Web spatiality does not literally coincide with physical spaces. The challenges around accessibility that physical services face closely involve the tasks of architects and designers, while the challenge of guiding and allowing users to find their way in the hypertextual space touches the interest of information architecture and user experience (UX).

Accessibility seems to have been at the center of the interests of Web services from the beginning. Evidence of this is the famous phrase from the director and inventor of the World Wide Web, Tim Berners-Lee where he mentions that the power of the Web is in its universality, being a central aspect of the access of any person, regardless of disabilities (W3C, 1997). In this line, international guidelines have recommended a series of requirements mainly associated with physical and sensory disability, focusing on the perception and operation of the Web (W3C, 2019). It is only in recent years that requirements for cognitive accessibility have begun to gain a more relevant space in international recommendations. In Chile, the standards and guidelines established to guide Web accessibility have replicated international recommendations (Senadis, 2016, 2022), developing technical standards that detail the requirements for digital public services (Ley Chile, 2015). However, there are no effective sanctions for public services that do not meet the standard, which causes important institutions for all citizens, such as the Civil Registry, to maintain a digital service with significant Web accessibility shortcomings, which are recordable in online evaluation services (for example, Accessibility Checker https://www.accessibilitychecker.org/). Additionally, current standards exclude municipalities, universities, state companies and private services, restricting their scope and effectiveness.

From the challenge of creating supports to ensure cognitive accessibility, and promoting independent living, since 2018 work has been done on the co-design of a Web application (PICTOS) to strengthen the cognitive accessibility of services. Within the framework of inclusive and interdisciplinary research, with the active participation of adults with ID, a pictographic system has been created and an information architecture aimed at promoting and improving the understanding and proper use of physical services (Exss Cid et al., 2022) and recently, also digital services.

12.2 Methodological Approach

Within the paradigm of inclusive research, this work has incorporated the participation of an advisory panel (AP) composed of adults with ID from the Valparaíso region, along with a team of experts from the disciplines of design, special education and sociology. The AP participates as "experts by experience" (Álvarez-Aguado et al., 2021; Sanders & Stappers, 2008) contributing with their personal knowledge of what it means to be a person with ID in Chile.

The AP that accompanied the first years of conceptualization of the support, was made up of seven women and four men, between 22 and 54 years old. The selection criteria for the participants included having a diagnosis of mild or moderate ID, with oral communication skills, being of legal age (over 18 years in Chile) and actively participating in support institutions for independent living in Valparaíso region. The selected people had no previous experience in research topics.

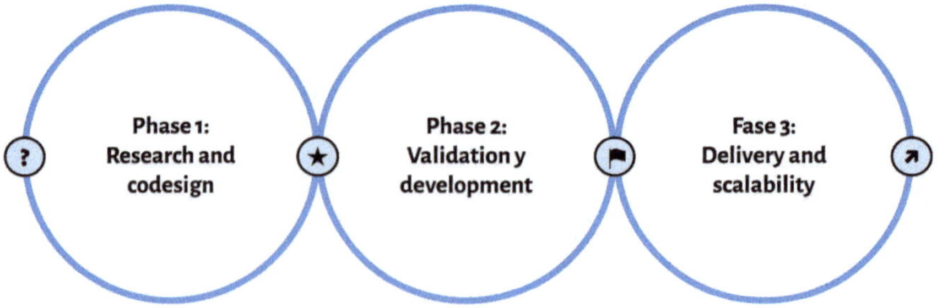

Fig. 12.1 Inclusive research process, development and implementation of PICTOS. *Source* Own elaboration

The inclusive research process was conceived as a project-based research through design. This implied the adaptation and design of various methods to be cognitively accessible with the ultimate goal of facilitating the participation of all voices. Some examples of these adaptations are elicitation notebooks to promote equal participation in individual interviews, field research probes to explore services and identify accessibility problems, and ideation workshops where generative co-design was stimulated for the conceptualization of design proposals (Spencer González et al., 2020).

This approach allowed the AP to be involved in research and subsequent developments relevant to their own interests, enabling a transition in their roles: From subjects of study, to collaborators and co-designers (Vega et al., 2022). It should be noted that all stages of R&D were designed for enhancing the full participation of the AP.

Under this regime, the work was structured in the following phases that have been developed over years of joint work (Fig. 12.1):

1. Research and co-design: This phase focused on collecting the research interests of the AP around issues of cognitive accessibility in city services, to subsequently co-design a support that responds to these concerns.
2. Validation and development: The second phase allowed for various validations of the proposed concept and the development of a functional prototype based on data collection in which the AP also participated.
3. Implementation and scalability: The third phase aimed to insert the support proposal in real public and private services and in designing the scalability of the proposal towards digital services.

12.3 Results

12.3.1 Research and Co-design: A Support for Inclusion in Cities

The process initiated in 2018 allowed the co-design of the Web service PICTOS, which seeks to develop and implement a methodology and tools to improve cognitive accessibility in various environments and services, both public and private. The ultimate goal of the service is to facilitate the inclusion and participation of people with intellectual disabilities in society, while at the same time raising awareness and promoting the adoption of good practices in terms of cognitive accessibility with a universal design approach.

Co-designing with the AP allowed us to propose that services can be understood as a set of tasks or transactions to be performed in a specific place, where each transaction consists of steps, and each step in turn, can be represented as an instruction written in plain language accompanied by a pictogram. This information architecture model allowed us to generalize and scale the proposal towards the PICTOS application that subsequently moved to the validation and development stage. From collaborative and inclusive work, a pictographic system and an information architecture aimed at promoting and improving the cognitive accessibility of services, with orientation to physical public services (Fig. 12.2). The pictographic system was proposed as an open repository, in constant update and growth (https://pictogramas.pictos.cl/), where each pictogram is built from three visual layers: Space, User Actions and Landmarks (Exss Cid et al., 2022). This layered composition allows its expansion to different sectors and applications, projecting the growth of the support. Along with this, a syntax was defined to compose the pictograms and texts through three elements:

1. The verb or action to be performed
2. The element with which one interacts
3. And the space that contains it.

With this, it was sought to establish a direct correspondence between the linguistic space and the visual space. The icon that accompanies the instruction arose from the need to emphasize both the action as the element or landmark to ensure the clarity of the instruction.

The process of research and co-design required multiple adaptations and delivered learnings that can be transferred to future collaborative experiences with groups with ID. Below are five central learnings from this stage:

- Research activities with the AP must be focused and declared in easy read or plain language depending on the context.
- The role of support people or facilitators of the sessions is fundamental for the integration of each participant with ID, in environments of respect, seriousness and commitment to research.

Fig. 12.2 Co-design process of pictograms composed of visual layers with the advisory panel of adults with intellectual disability. *Source* Own elaboration

- It is advisable to provide multiple ways for eliciting from people with ID. This can consider oral expression, in writing or through concepts or responses previously printed or illustrated for the occasion (in the form of post-its, for example).
- The use of pictograms and supporting illustrations is encouraged, however, infantilization in the visual language used in presentations and printed material should be avoided.
- When using toolkits and other generative design tools it is preferable to work with concrete materials and shapes instead of abstract ones.

12.3.2 Validation and Development: Design Heuristics for Cognitive Accessibility

The validation and development phase allowed the creation of a functional prototype of the PICTOS Web application (https://www.pictos.cl/). The application considers accessible navigation and consolidates the proposal of initial support co-designed, through three communication channels: audio, text, and pictograms (Fig. 12.3). The participation of people with ID in this phase allowed early validation of interface design proposals and prioritizing content through categories that proved relevant to the AP. The PICTOS pilot integrates two functionalities:

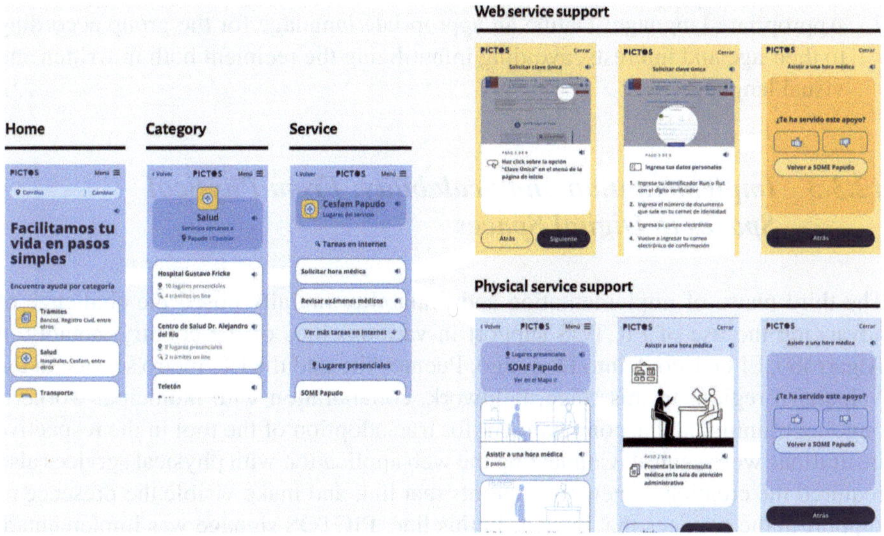

Fig. 12.3 Screens of the PICTOS Web application, with support flow for face-to-face services and web services. *Source* Own elaboration

- Support to ensure cognitive accessibility: Support system through step-by-step pictograms for the tasks or transactions of a physical service, accompanied by audio and easy read text.
- Cognitive accessibility evaluation: Cognitive accessibility evaluation system through a comprehension scale based on the information available in physical services.

Throughout the validation and development process, the following principles were identified that proved useful when developing cognitively accessible prototypes. Seven design heuristics for cognitive accessibility are proposed:

1. Sequentiality: Organize complex tasks into clear and short sequential steps, avoiding delivering more information than necessary.
2. Progression: Define steps in a logical and continuous progression that always allows you to locate yourself in the current task and anticipate the next step in the process.
3. Focus: Maintain a clear and visually hierarchical focus of attention in activities or screens, avoiding parallel distractions.
4. Low density: Decrease the density of visual information. Redundant elements, which are often considered favorable in terms of usability, can be detrimental to constructing a clear meaning for the user.
5. Purpose: Make the purposes transparent at all times, explaining the reasons for the actions that are taken
6. Quality and consistency: Offer simple interfaces and navigation structures, ideally with a maximum of 2 or 3 levels of depth.

7. Appropriate language: Define an appropriate language for the group according to their age and interests, avoiding infantilizing the recipient both in written and visual language.

12.3.3 Implementation and Scalability: From Physical Spaces to Digital Spaces

The third phase of implementation and scalability initially posed the challenge of advancing the use of PICTOS support in various cities of the country, notably in Algarrobo, El Quisco, Santo Domingo, Puente Alto and the Efe metro service in the Valparaíso region. In this stage, fieldwork, collaboration with municipal workers, ongoing training, and accompaniment for true adoption of the tool in the respective institutions were central. Connecting the web application with physical services also required the creation of new touchpoints that link and make visible the presence of support in the services that have it. In this line, PICTOS signage was implemented, which, through a QR code, links to the tasks available for the in-person service (Fig. 12.4).

On the other hand, this phase also allowed us to explore and experiment with the transfer of the PICTOS cognitive accessibility model for digital services. In this context, where national web services present difficulties to be understood by users, PICTOS aims to act as a support in web navigation on computers, accompanying users from their mobile phones. The web support is based on the information architecture provided by the digital services and develops a guide that accompanies the user's actions through page jumps and interactions in more complex spaces such as forms or transactions. PICTOS Web presents plain language textual information and screenshots that highlight the interaction zones of each step.

In this new scenario, both interactions coexist in the same virtual space, so there is a risk of overcomplicating the user experience. To address this problem, web support in two incremental phases was put forward (Fig. 12.5), which allows to validate and develop this functionality in an agile manner. The first phase consists of PICTOS Web as external support, residing on the phone as a second screen and representing the simplest case technologically (referred to as fallback). The second phase advances towards a parallel support format through a browser extension, which, upon identifying that the site has supports, offers a sidebar to navigate the tasks. This second alternative is thought of as a secondary phase due to its greater technical and development complexity. For both cases, the step-by-step logic is maintained, which allows to locate and guide the person within a process, and incorporating pictographic supports, now oriented to interface elements such as buttons, dropdowns, forms, among others.

Fig. 12.4 Support implemented through contextual signage and Web application in the city of Algarrobo, Chile. *Source* Own elaboration

12.3.4 Limitations and Projections

The insertion of the PICTOS support system in national face-to-face services has had great growth in recent years. The evaluation of its effectiveness is one of the pending aspects that is currently under development. This evaluation will serve to quantify and qualify the way in which PICTOS impacts the cognitive accessibility of services. It is expected that the results will also open new research questions that will allow us to continue contributing to the study of the transactional accessibility of services.

The transfer of the PICTOS support model to digital services is a process that is underway and requires various adaptations due to differences in the sense of spatiality and administration of processes. Firstly, digital services update their interactions and contents more frequently than urban services. In this sense, there is a risk that the support becomes outdated quickly. Additionally, it is common that during web navigation, external services are integrated into the flows of a website, which could complicate the definition of the beginning and end of a task within a service. Webpay

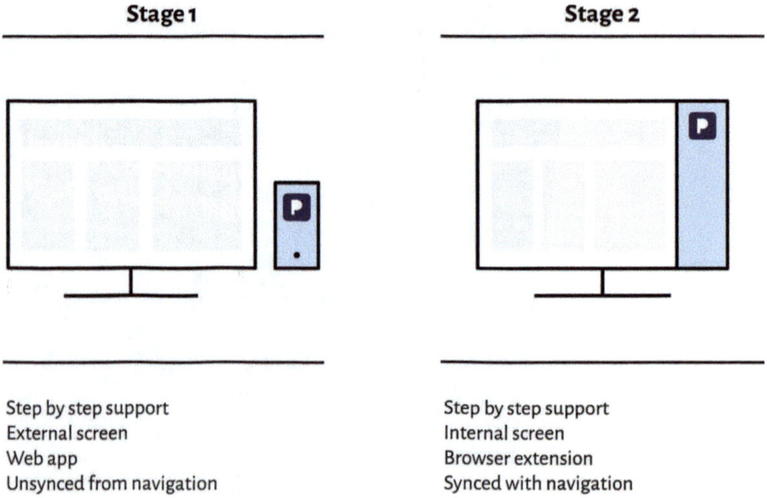

Fig. 12.5 Development phases of the web support with stage 1 (application from the mobile, external to the screen) and stage 2 (browser extension that coexists on the same screen). *Source* Own elaboration

as a payment system, or confirmations or validations through the use of email are examples of the above. These aspects are still in the process of discussion, as the beta version of the web support is released, but it is anticipated that they will need to be observed and evaluated later to guide the verification of the effectiveness of the support for digital services.

It is projected for the continuity of the development of PICTOS the growth in terms of adoption of the tool in national services, especially in regions, that begin to periodically evaluate their cognitive accessibility, while offering cross-cutting supports for their physical services. The web support can be tested more broadly as it is effectively implemented and published in beta state during 2024. The next steps of the project also point towards the internationalization of the tool, posing new challenges associated with language, translation, interpretation of pictograms and availability in other countries.

Through these years of interdisciplinary and inclusive research and development, in constant collaboration with communities with ID and through the creation of PICTOS, it has been sought to promote the culture of accessibility and facilitate the adoption of these supports by public and private institutions. It is proposed that according to the social model of disability, the inclusion and social integration of people with disabilities depend directly on architecture and design, which is an opportunity to question disciplinary tasks and turn them towards the full inclusion of our services and cities.

Acknowledgements This work is framed within the project FONDEF IT21I0065 "PICTOS: Service for the navigation and evaluation of cognitive accessibility of services in Chile" and corresponds to the continuity of a project Senadis 13514-2019 "Visual supports for the full inclusion of people with intellectual disabilities in public services in Chile". The process described in this article represents a period of six years where many collaborators have participated. In particular, we want to thank the members of the advisory group and the municipal teams that since 2022 have joined the initiative to create cognitively accessible services. This project has the bioethics consent of the PUCV and the authorization of all participants for the use of their photographs in academic publications.

References

Álvarez-Aguado, I., Carrasco Delgado, P., González Carrasco, F., Exss Cid, K., Spencer González, H., & Vega Córdova, V. (2021). *Expertos por Experiencia. El proceso de investigación inclusiva: Un desafío para los nuevos tiempos*. Ediciones Universitarias de Valparaíso.

Asociación Americana de Psiquiatría. (2014). Manual Diagnóstico y estadístico de los Trastoprnos Mentales (DSM-5®), 5a Ed. ed. Editorial Médica Panamericana, Arlington.

Atack, L., Verseghy, J., Maher, J., Herie, M., Poirier, M., MacNeil, F., & Grimley, M. (2019). An emerging model for semi-independent living for persons with intellectual and developmental disabilities. *Inclusion, 7*, 57–70. https://doi.org/10.1352/2326-6988-7.1.57

Blackman, T., Mitchell, L., Burton, E., Jenks, M., Parsons, M., Raman, S., & Williams, K. (2003). The accessibility of public spaces for people with dementia: A new priority for the "open city." *Disability & Society, 18*, 357–371. https://doi.org/10.1080/0968759032000052914

Brusilovsky, B. (2018). *Índice de accesibilidad cognitiva: Consideraciones para el diseño.* Entinema.

Exss Cid, K. E., Spencer González, H., Vega Córdova, V., Jarpa Azagra, M., Álvarez-Aguado, I., Pastén Bernales, & von Unger Martínez, M. I. (2022). Investigación inclusiva y codiseño: Cocreación de un sistema de apoyo tecnológico para la discapacidad intelectual. Rev. 180 49, 95–106. https://doi.org/10.32995/rev180.Num-49.(2021).art-866

Jarpa-Azagra, M., Vega-Córdova, V., Álvarez-Aguado, I., Spencer, H., & Exss, K. (2021). Posicionamiento discursivo de adultos con discapacidad intelectual frente a los desafíos de una vida independiente. Raled 21. https://doi.org/10.35956/v.21.n2.2021.p.160-182

Ley Chile. (2017). LEY 21015 Incentiva la inclusión de personas con discapacidad al mundo laboral.

Ley Chile. (2015). Decreto 1 Aprueba Norma Técnica Sobre Sistemas y Sitios Web de los Órganos de la Adminisitración del Estado [WWW Document]. www.bcn.cl/leychile. https://www.bcn.cl/leychile. Accessed 10 Feb 23

Ley Chile, del C. (2010). LEY 20422 Establece normas sobre igualdad de oportunidades de inclusión social de personas con discapacidad.

Morris, J. (1994). Community care or independent living? *Critical Social Policy, 14*, 24–45. https://doi.org/10.1177/026101839401404002

Naciones Unidas. (2006). Convención Sobre los Derechos de las Personas con Discapacidad.

Peterson, C. M., Birkeland, R. W., Barsel, S., Statz, T. L., Gaugler, J. E., & Finlay, J. M. (2022). "Sick with stress": perspectives on airport travel from persons living with dementia and their travel companions. Disability & Society. https://doi.org/10.1080/09687599.2022.2070060

Plena Inclusión. (n.d.) Accesibilidad universal [WWW Document]. Plena Inclusión. Retrieved September 28, 2023, from https://www.plenainclusion.org/discapacidad-intelectual/recurso/accesibilidad-universal/

Sanders, E. B.-N., & Stappers, P. J. (2008). Co-creation and the new landscapes of design. *CoDesign, 4*(1), 5–18. https://doi.org/10.1080/15710880701875068

Schalock, R. L. (2009). la nueva definición de discapacidad intelectual, apoyos individuales y resultados personales. Siglo Cero 40.

Schalock, R. L., Luckasson, R., & Tassé, M. J. (2021). *Discapacidad Intelectual: Definición, diagnóstico, clasificación y sistemas de apoyo U.S. Spanish Translation* (12th ed.). AAIDD.

Senadis. (2022). Manual Accesibilidad Web—Guía Técnica para la Implementación de Sitios Web Accesibles [WWW Document]. Retrieved February 10, 2023, from https://kitdigital.gob.cl/archivos/insumos/nuevos/Manual%20Accesibilidad%20Web.pdf

Senadis. (2016). Accesibilidad Web en Chile—Guía Técnica para la Implementación de Sitios Web Accesibles.

Spencer González, H., Vega Córdova, V., Exss Cid, K., Jarpa Azagra, M., & Álvarez-Aguado, I. (2020). Including intellectual disability in participatory design processes: Methodological adaptations and supports. In *Proceedings of the 16th Participatory Design Conference 2020—* Participation(s) Otherwise—Volume 1. Presented at the PDC '20: Participatory Design Conference 2020—Participation Otherwise, ACM, Manizales Colombia (pp. 55–63). https://doi.org/10.1145/3385010.3385023

Vega Córdova, V., Álvarez-Aguado, I., González Carrasco, F., Spencer González, H., Jarpa Azagra, M., & Exss Cid, K. (2023). Experiencias de personas adultas con discapacidad intelectual sobre el derecho a la vida independiente: Un estudio de caso. *Interdiscip. Rev. Psicol. Cienc. Afines, 40*, 299–318. https://doi.org/10.16888/interd.2023.40.2.18

Vega, V., Álvarez-Aguado, I., Jarpa, M., Spencer, H., Exss, K., & Muñoz, R. (2022). Capítulo 7. Definiendo roles en investigaciones inclusivas desde una experiencia en Chile. In *Investigación Inclusiva Diez Años de Trabajo Con El Consejo Asesor Del Grupo de Investigación En Diversidad de La Universitat de Girona*. Girona, España: Universitat de Girona-Servei de Publicacions.

W3C. (2019). How to meet WCAG (Quickref Reference) [WWW Document]. Retrieved February 10, 2023, from https://www.w3.org/WAI/WCAG21/quickref/

W3C. (1997). World Wide Web Consortium launches International Program Office for Web Accessibility Initiative [WWW Document]. W3C. Retrieved February 10, 2023, from https://www.w3.org/press-releases/1997/ipo-announce/

Chapter 13
Walking in Deprived Neighborhoods: Understanding the Role of the Built Environment

Cristhian Figueroa-Martínez

Abstract There is abundant evidence demonstrating that the built environment influences human behavior, people's travel patterns, and in particular, their willingness to make walking trips. However, there is little clarity about the impacts that the built environment has on walks that occur in deprived neighborhoods. The few studies that have addressed the matter indicate that the built environment is of little relevance in contexts of social deprivation since other phenomena (such as insecurity) and the lack of resources are more important, making walking an activity that must be carried out regardless of external conditions. This chapter seeks to delve into that discussion and explore the ways in which the built environment affects the walks that occur in deprived neighborhoods. Thus, the results of 42 semi-structured interviews are reported in which residents of 3 deprived neighborhoods in Santiago de Chile were invited to talk about the built environment and the difficulties they face when walking. The results indicate that the built environment influences the walk of the deprived neighborhoods of Santiago in at least three ways. The urban structure reinforces boundaries and dichotomies, restricting the area that is considered safe and, therefore, walkable. The deterioration of the built environment can turn walking into an unpleasant practice; while poor pedestrian infrastructure prevents the realization of recreational activities and transforms walking into a permanent reminder of the profound inequalities of society and the few opportunities they have to integrate into it.

Keywords Built environment · Walking · Deprived neighborhoods · Qualitative methods · Santiago de Chile

---

C. Figueroa-Martínez (✉)
Departamento de Planificación y Ordenamiento Territorial, Universidad Tecnológica Metropolitana y Centro de Desarrollo Urbano Sustentable, CEDEUS, Santiago, Chile
e-mail: cafigueroa@utem.cl

M. Vicuña and E. Giorgi (eds.), *Intersections*, Springer Geography,
https://doi.org/10.1007/978-3-031-76402-8_13

13.1 Introduction

There is a broad consensus that the built environment influences walking. Compiled in numerous systematic reviews (Cervero & Kockelman, 1997; Ewing & Cervero, 2010; Wang & Yang, 2019), the abundant evidence indicates that dense neighborhoods, with a high mixture of land uses, with pedestrian-friendly public spaces, well connected to public transport, to the services and jobs available in the city promote sustainable travel patterns and, particularly, walking. Recent works, however, suggest that this prolific body of literature has tended to minimize the relevance of the socio-cultural context (Næss, 2015) and has made invisible the influence that the built environment could have on groups that, lacking resources, have no other alternative but to walk (Benediktsson, 2018; Bostock, 2001; Figueroa et al., 2019). To this is added that the vast majority of the works have been carried out in places where walking is a potential alternative to decrease the disproportionate use given to the car. Very little is known about what happens in regions (e.g., Latin America) where walking is a widespread practice (Figueroa et al., 2019). Exploring the ways in which the built environment affects the walks that occur in the deprived neighborhoods of Santiago de Chile, this work seeks to contribute to those discussions.

The chapter reports the results of a qualitative research in which residents of three deprived neighborhoods in Santiago de Chile were invited to participate in a semi-structured interview and reflect on: the characteristics of the built environment of their neighborhoods, the socio-spatial practices that occur in their public spaces and the way in which both affect their walks. With the analysis of those interviews, this chapter seeks to contribute to the literature that understands walking as a complex activity (Lorimer, 2016); a key practice for the reproduction of capitals of different natures and the exercise of citizenship (Middleton, 2018; Rybråten et al., 2019; Spinney et al., 2015) and a daily habit that evolves over time according to personal trajectories and the socio-cultural context (Ingold, 2011; Koskela & Pain, 2000; McFarlane, 2011). Likewise, it seeks to contribute to an emerging body of work that indicates that the built environment influences people's movement patterns in combination with socio-cultural factors (Banerjee & Hine, 2014; Næss, 2015, 2016; Stefansdottir et al., 2019) and can transform the walks of those who live in deprived neighborhoods into unpleasant practices (Benediktsson, 2018; Bostock, 2001; Figueroa et al., 2019).

Studying Santiago de Chile, the chapter also seeks to collaborate with the Latin American discussion and, more specifically, with the understanding of the role that the built environment plays in Latin American metropolises which, as the literature indicates (Caldeira, 2017; Gasparini & Cruces, 2021), are characterized by deep inequalities and clear patterns of segregation. It is expected to contribute to a growing number of Latin American studies that have investigated the links between the built environment and walking and that, among other findings, highlight the relevance of the quality of pedestrian infrastructure (Arellana et al., 2020; Gómez et al., 2010; Hermida et al., 2019; Herrmann-Lunecke et al., 2021; Valenzuela-Montes & Talavera-García, 2015; Vergel-Tovar & Rodríguez, 2018). In parallel, it is expected to contribute to studies on the walks that vulnerable people (e.g., children, adolescents, women, elderly people) undertake, which, in recent decades, have significantly increased in Latin America and, particularly, in Chile (Figueroa, 2023; Figueroa & Forray, 2015; Figueroa et al., 2019; Herrmann-Lunecke et al., 2022; Salazar, 2021; Waintrub et al., 2024).

Excluding this introduction, the chapter is structured in three sections. In the following section (Sect. 13.2) details of the methods outlined in the previous paragraphs are provided. Organized into three subsections, Sect. 13.3 contains the main findings of the research that address the influences on walking of the urban structure of neighborhoods (Sect. 13.3.1), the level of deterioration of public space (Sect. 13.3.2) and pedestrian infrastructure (Sect. 13.3.3). Finally, Sect. 13.4 contains a synthesis of the research, and the main conclusions reached.

13.2 Methods and Materials

To explore the ways in which the built environment affects the walks that occur in the deprived neighborhoods of Santiago de Chile, the chapter reports the results of forty-two semi-structured interviews in which people were invited to talk about (i) the trajectory of the places where they live, (ii) the affection and attachment they feel towards their communities and their neighborhoods and (iii) the socio-spatial practices that they and other people carry out in the streets of their neighborhoods. Additionally, they were asked about (iv) what they believe will happen to their neighborhoods in the future and (v) the socio-spatial practices they believe occur on streets located in other places in Santiago.

The interviews were guided by a questionnaire that contained general questions to open the conversation and more specific ones to delve into topics of interest that the participants might not naturally touch on. All the interviews were recorded in audio, transcribed and pseudo-anonymized (i.e., removal of direct identifiers) and then imported into the qualitative analysis software *atlas.ti*. Using tools from the software, the transcriptions were reviewed in detail (i.e., persistent observation), isolating significant fragments and assigning them descriptive labels (or codes) that reflected their content. The fragments that received similar codes were then visualized simultaneously in order to identify narratives, patterns and omissions in the responses given by individuals. All this analysis was carried out by a researcher and then discussed with others in triangulation meetings.

The participants, on the other hand, were recruited in situ, directly inviting residents to participate in the research. Those who accepted the invitation were interviewed and, at the end of the activity, asked for the contact details of people who might be interested in participating in the research. Potential participants were contacted by the researcher, interviewed according to their availability and, following a snowball scheme, also asked for the details of other people who might be interested in participating. Following this mixed recruitment scheme, 42 participants were reached. Twenty-five participants were women (60% of the total) and 17 men (the remaining

40%), and they averaged forty-four years old. All were also residents of three deprived neighborhoods which, in order not to reproduce and amplify the stigmas that currently afflict them, will henceforth be named only with letters. As Fig. 13.1 shows, the three neighborhoods are located on the southern outskirts of Santiago de Chile and well represent the changing relationship between the State and civil society. In detail:

Case A: A neighborhood built in the 1960s, under the umbrella of incremental policies that sought to encourage progress through citizen participation and the strengthening of "the public" (Ministerio de Vivienda y Urbanismo [MINVU], 2004). It is a neighborhood of self-built houses, whose abundant public space was slowly consolidated by the community and the intermittent action of the State.

Case B: A sector built during the military dictatorship (1973–1990), in the 1980s by policies that, seeking to alleviate a rampant housing deficit, promoted the entry of private agents into sectors that, like social housing, traditionally were in the hands of the State (MINVU, 2004). It is a neighborhood of attached houses that have been expanded to occupy most of the small lots on which they are located.

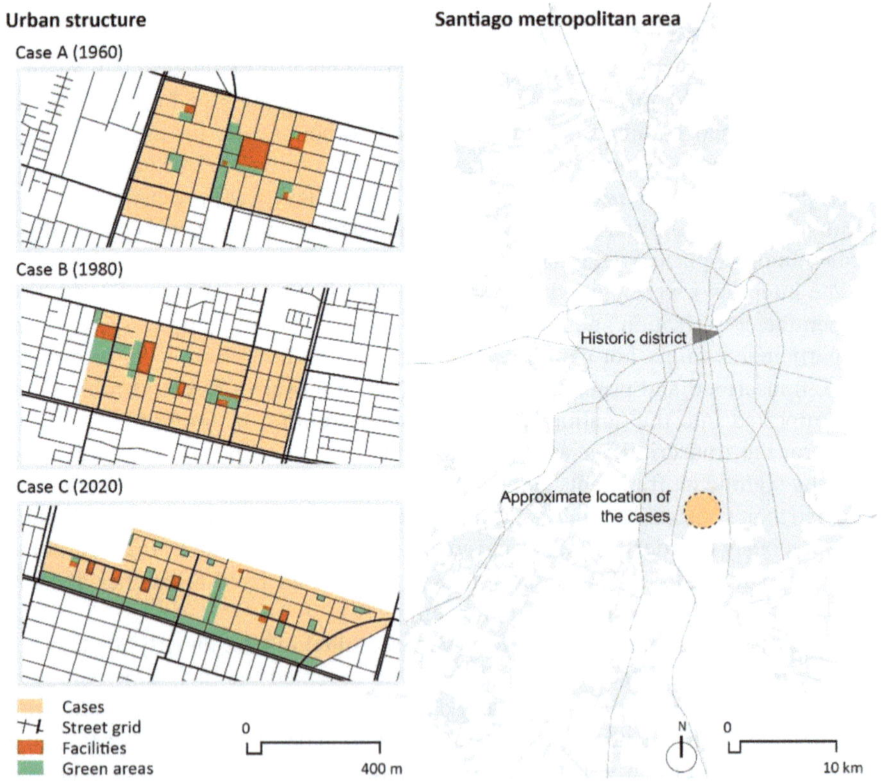

Fig. 13.1 Location and urban structure of the cases. *Source* Author

Public space is scarce, most of the streets are narrow passages with shared roads, while the few avenues have narrow sidewalks (less than a meter in some cases).

Case C: A neighborhood that began to be inhabited at the end of 2020 and was produced by recent policies that seek to generate well-equipped housing complexes with a social mix. It is a sector of semi-detached houses arranged around squares or community facilities (Servicio de Vivienda y Urbanismo [SERVIU], 2018). At the time of writing this chapter, this neighborhood has only received families coming from other deprived territories. The land that will host higher income families remains empty.

The three neighborhoods also have intertwined trajectories. In Chile, it is common for houses in neighborhoods produced under the umbrella of social housing policies to be expanded to accommodate family units that have not been able to access a permanent solution. The phenomenon, known locally as "allegamiento", means that most of the inhabitants of the socioeconomically vulnerable outskirts of Santiago have links in neighborhoods produced by social housing policies different from the one that built the place where they currently reside. Thus, many residents of Case B lived at some period of their lives in a neighborhood like Case A. Likewise, most of the inhabitants of Case C come from places with characteristics similar to those of the older cases. All participants were highly familiar with the dynamics of neighborhoods different from their own. Many of their responses were, in fact, comparisons between what they experience daily in the present and what happened in their old neighborhood or observed when they visit friends or relatives.

The next sections contain the main findings of the analysis of the interviews just described. To exemplify significant themes, the sections contain photographs and plans of the case studies and small fragments of the interviews accompanied by general identifiers (i.e., gender, age range, neighborhood of residence of the participant). Finally, all the protocols described in this section were approved by the Ethical Scientific Committee of Social Sciences, Arts and Humanities and the Committee of Safety in Research of the Pontifical Catholic University of Chile.

13.3 Main Findings

All the people interviewed walk regularly through their neighborhoods, either to stock up at local shops and street markets, reach public transport stops, or access health and education facilities. Most, however, consider their neighborhoods to be not very walkable. Those who reside in the oldest neighborhood (Case A) feel unprotected when walking. The community that actively participated in the process of consolidation was weakened, and the public space deteriorated due to a lack of investment. Residents of the sector built during the dictatorship (Case B) pointed out that the implementation of urban regeneration programs—about a decade ago—significantly improved the neighborhood, but added that those efforts have not been enough to rearticulate communities devastated by urban violence (e.g., drug trafficking), solve

13.3.2 Deterioration

As various works suggest (Figueroa et al., 2019; Herrmann-Lunecke *et* al., 2022), the level of cleanliness of the streets emerged on numerous occasions as an aspect that plays a fundamental role in the experiences that people have while moving on foot. Walking through places where garbage and debris are observed causes discomfort. On the contrary, walking through clean places causes some joy, as it shows that those who live in the surroundings are interested in the well-being of others and in the future of the neighborhood. The analysis of the collected data showed that cleaning the streets daily is seen as a way of showing consideration for the other people who live or pass through the neighborhood and, more broadly, one of the few mechanisms that communities have to counteract the widespread absence of the State, and the general deterioration suffered by their neighborhoods due to lack of investment:

> I find that we have a lot of inequality because in well-off neighborhoods this [level of deterioration] is not seen. The issue of a sidewalk in poor condition, you don't see streets or main avenues with these famous holes, because they [well-off municipalities] improve them *right away* [sic]. We, on the other hand, have to wait for years. (woman, 51 years old, Case A)

Most of the participants from the older neighborhoods, and especially those who resided in Case A, indicated that public space has gradually deteriorated due to a lack of investment. Sidewalks fractured by the passage of time, cracked by tree roots, or uneven due to poorly executed repairs are part of the daily landscape and are affecting the daily life of people who, for various reasons, have greater difficulties walking. Several older participants or those with disabilities, for example, indicated that they do not feel comfortable walking on the sidewalks of their neighborhoods. Despite the risk that it implies, some walk on the roads because the surfaces are smoother and better maintained; while other participants have suffered falls and, out of fear, have preferred to stay in their homes and not go out to the streets of their neighborhoods:

> It's hard to walk... a lot, especially for people who are sick, who can't walk, who walk with a cane [...]. It's hard for me because I... walk a lot... but I tend to lean, I lose my balance. That's why I always walk arm in arm [with someone] or with a cane that I use for support.
>
> The streets don't help at all, the other day I tripped while turning, I almost hit my head on a metal pole. (woman, 80 years old, Case A)

Coinciding with the findings of several authors (Benediktsson, 2018; Bostock, 2001; Figueroa, 2023), the deterioration of the pedestrian space has also transformed the walks that occur in the older neighborhoods into a permanent reminder of the abandonment which they find themselves. Many people indicated that the State and its institutions "appear" sporadically in their neighborhoods and do not manage to reverse the general deterioration they observe when they walk. These narratives of abandonment even permeated the walks of the participants in the recently inaugurated case (Case C). The neighborhood has sidewalks in good condition, ramps at the corners and well-marked pedestrian crossings, however, the vast majority of residents walk with the concern that what they observe will not remain in good condition for long. They believe that deterioration is the "inevitable destiny" of all the complexes built by Chilean housing policies, as in their places of origin they experienced similar processes and the State, which was present throughout the process of consolidating the new neighborhood, has already left them alone.[2]

13.3.3 Pedestrian Infrastructure

It was common for participants to indicate that it is not pleasant to walk through the streets of their neighborhoods. In the two older neighborhoods (cases A and B) it was pointed out that the streets are barely adequate for walking (Fig. 13.2). The sidewalks are narrow, often deteriorated, dirty, plagued with obstacles (e.g., streetlight poles), lack amenities and vegetation and do not allow a series of recreational activities (e.g., strolling) and sports (e.g., jogging). Those activities were, on repeated occasions, portrayed as "privileges" of other neighborhoods and, especially, of those inhabited by higher income groups. As previously indicated, the pedestrian infrastructure of Case C has a better standard, but, from the participants' point of view, it still does not reach that of other neighborhoods in the city. Urban furniture is present only in some green areas and the existing vegetation is young and will require years of care to consolidate. The space of the streets is also unevenly distributed; occupied by disproportionately wide roads that, on the one hand, allow vehicles to reach high speeds and, on the other, leave pedestrians cornered in marginal spaces (Fig. 13.2).

When shown images of streets from other places in Santiago, a significant number of participants highlighted the width of the sidewalks and how places with these characteristics allow walking in pairs, groups or accompanying children. Many people also noticed the width of the furnishing zones and emphasized how this helped them feel more protected from vehicular traffic and its externalities (e.g., noise, bad smell). It was also highlighted how wide furnishing zones enriched the landscape and the walk, as they allowed vegetation to grow (e.g., trees, gardens) and urban furniture to be installed (e.g., benches, trash cans). Several people also noticed details of the images; sidewalks covered by colored tiles, underground wiring and lanterns that directly illuminated the sidewalks and that, from their perspective, evidenced the "care" with which the public spaces of other neighborhoods in Santiago are designed:

> I like [the place in] the photo, because it looks orderly. You can see nature, shade for the summer. You can see the streets clean, orderly. The care of the garden already tells you that it is not… it is not around here… [it is] middle class… upwards (Man, 53 years old, Case B).

Participants from the three neighborhoods indicated that the public spaces of the neighborhoods studied were not thought out, nor cared for, with care. This lack is felt, as what the participants see while walking through the streets of their neighborhoods reflects the deep inequalities that Santiago suffers from and reminds them that they have fewer opportunities than others to progress.

Fig. 13.2 Pedestrian infrastructure of the case studies. *Source* Author

13.4 Conclusions

The purpose of this work was to explore the ways in which the built environment affects the walks that occur in deprived neighborhoods. In this regard, the narratives of the participants showed that walking is an activity they regularly perform and that is influenced by the built environment in, at least three, ways. The **urban form** of the neighborhoods can be an obstacle to building links, reinforcing boundaries (e.g., us/them) and dichotomies (e.g., familiar/unknown) and modifying the extent of what is considered walkable. The **deterioration** of public space has transformed walking into a constant reminder of the absence of the State and the neglect in which the places they inhabit have remained. While the standard of sidewalks and the rest of the **pedestrian infrastructure** reflected the deep inequalities from which Santiago suffers, further impoverishing walking. The results also showed that the built environment does not seem to play a relevant role in the reproduction of the fear that accompanies the walks that the residents of deprived neighborhoods make, as improving or intervening tended not to modify the predisposition they had to go out on the street in places and times that are considered dangerous.

The findings of this work suggest that urban design can make significant contributions to public policies. The planning of new neighborhoods should avoid urban structures where spaces of limited connectivity (e.g., cul-de-sacs) that promote isolation abound. Efforts should be redoubled to keep public space in good condition and prevent deterioration from becoming an obstacle capable of immobilizing those people who have reduced mobility (e.g., older people). In addition, the pedestrian spaces of deprived neighborhoods should be reconfigured, expanding sidewalks and furnishing zones and incorporating vegetation and furniture in order to reduce the gap that separates them from what is observed in more affluent areas of the city. The results show that the built environment of Santiago's deprived neighborhoods needs to be improved, not only to facilitate walking but also to create a more inclusive city.

The study had a series of limitations. The three neighborhoods studied are located in areas of Santiago where there are few opportunities to come into contact with higher income groups. It would be of great interest to contrast the narratives detected with those articulated by residents of deprived neighborhoods located in socio-economically mixed areas. The participants were recruited through a snowball scheme, which could have underrepresented the voices of some groups of people and affected the generalization of the results. During the analysis, a series of differences were detected according to the gender and age of the participants that were not explored in depth in this study. In this research, a series of strategies were used to give reliability to the data collection (e.g., *prolonged engagement*) and the subsequent analysis (e.g., persistent observation of data, triangulation). Finally, throughout the chapter, fragments of the interviews were integrated and a series of data from the case studies were provided so that the reader can judge the transferability of the findings.

Acknowledgements The author thanks the support for the research provided by CEDEUS, ANID FONDAP 1523A0004.

Notes

1. Second half of the year 2022.
2. The residents of Case C claimed on many occasions that the institutions abandoned them when the neighborhood began to be inhabited. Several indicated that problems that emerged immediately after the inauguration have not been resolved despite constant complaints; while others said that the institutions no longer respond to them.

References

Arellana, J., Saltarín, M., Larrañaga, A. M., Álvarez, V., & Henao, C. A. (2020). Urban walkability considering pedestrians' perceptions of the built environment: A 10-year review and a case study in a medium-sized city in Latin America. *Transport Reviews, 40*(2), 183–203. https://doi.org/10.1080/01441647.2019.1703842

Banerjee, U., & Hine, J. (2014). Identifying the underlying constructs linking urban form and travel behaviour using a grounded theory approach. *International Journal of Environmental Science and Technology, 11*(8), 2217–2232. https://doi.org/10.1007/s13762-014-0585-0

Benediktsson, M. (2018). Where inequality takes place: A programmatic argument for urban sociology. *City & Community, 17*(2), 394–417. https://doi.org/10.1111/cico.12302

Bostock, L. (2001). Pathways of disadvantage? Walking as a mode of transport among low-income mothers. *Health & Social Care in the Community, 9*(1), 11–18. https://doi.org/10.1046/j.1365-2524.2001.00275.x

Caldeira, T. (2017). Peripheral urbanization: Autoconstruction, transversal logics, and politics in cities of the global south. *Environment and Planning D: Society and Space, 35*(1), 3–20.

Cervero, R., & Kockelman, K. (1997). Travel demand and the 3Ds: Density, diversity, and design. *T Ransportation Research Part D: Transport and Environment, 2*(3), 199–219. https://doi.org/10.1016/S1361-9209(99)00032-2

Ewing, R., & Cervero, R. (2010). Travel and the built environment: A meta-analysis. *Journal of the American Planning Association, 76*(3), 265–294. https://doi.org/10.1080/01944361003766766

Figueroa, C., & Forray, R. (2015). Movilidad femenina: Los reveses de la utopía socio-espacial en las poblaciones de Santiago de Chile. *Revista de Estudios Sociales, 54*, 52–67. https://doi.org/10.7440/res54.2015.04

Figueroa, C., Hodgson, F., Mullen, C., & Timms, P. (2019). Walking through deprived neighbourhoods: Meanings and constructions behind the attributes of the built environment. *Travel Behaviour and Society, 16*, 171–181. https://doi.org/10.1016/j.tbs.2019.05.006

Figueroa, C. (2023). Entre lo social y lo construido: Comprendiendo la caminata en barrios vulnerables de Santiago. *Revista de Urbanismo.* https://doi.org/10.5354/0717-5051.2023.6941

Gasparini, L., & Cruces, G. (2021). The changing picture of inequality in Latin America: Evidence for three decades. UNDP LAC Working Paper No. 01. Background Paper for the UNDP LAC Regional Human Development Report 2021.

Gómez, L., Parra, D., Buchner, D., Brownson, R., Sarmiento, O., Pinzón, J., Ardila, M., Moreno, J., Serrato, M., & Lobelo, F. (2010). Built environment attributes and walking patterns among the elderly population in Bogotá. *American Journal of Preventive Medicine, 38*(6), 592–599. https://doi.org/10.1016/j.amepre.2010.02.005

Hermida, C., Cordero, M., & Orellana, D. (2019). Analysis of the influence of urban built environment on pedestrian flow in an intermediate-sized city in the Andes of Ecuador. *International Journal of Sustainable Transportation, 13*(10), 777–787. https://doi.org/10.1080/15568318.2018.151444

Herrmann-Lunecke, M., Figueroa, C., & Vejares, P. (2021). Caminata y vejez: explorando el espacio público peatonal de las personas mayores en los instrumentos de planificación urbana en Chile. Urbe. Revista Brasileira de Gestão Urbana, 13.

Herrmann-Lunecke, M., Figueroa-Martínez, C., Parra, F., & Mora, R. (2022). The disabling city: Older persons walking in central neighbourhoods of Santiago de Chile. *Sustainability, 14*(17), 11085. https://doi.org/10.3390/su141711085

Ingold, T. (2011). *Being alive: Essays on movement, knowledge and description.* Routledge.

Koskela, H., & Pain, R. (2000). Revisiting fear and place: Women's fear of attack and the built environment. *Geoforum, 31*(2), 269–280. https://doi.org/10.1016/S0016-7185(99)00033-0

Lorimer, H. (2016). Walking: New forms and spaces for studies of pedestrianism. In *Geographies of mobilities: Practices, spaces, subjects* (pp. 19–33). Routledge.

McFarlane, C. (2011). *Learning the city: Knowledge and translocal assemblage.* John Wiley & Sons.

Middleton, J. (2018). The socialities of everyday urban walking and the 'right to the city.' *Urban Studies, 55*(2), 296–315.
Ministerio de Vivienda y Urbanismo [MINVU]. (2004). Un siglo de políticas de vivienda y barrio . Santiago, Chile: MINVU.
Næss, P. (2015). Built environment, causality and travel. *Transport Reviews, 35*(3), 275–291. https://doi.org/10.1080/01441647.2015.1017751
Næss, P. (2016). Built environment, causality and urban planning. *Planning Theory & Practice, 17*(1), 52–71.
Rybråten, S., Skår, M., & Nordh, H. (2019). The phenomenon of walking: Diverse and dynamic. *Landscape Research, 44*(1), 62–74. https://doi.org/10.1080/01426397.2017.1400527
Salazar, A. (2021). Experiencias de movilidad cotidiana de mujeres migrantes: Explorando el barrio La Chimba desde una perspectiva feminista. *INVI, 36*(102), 127–156.
Servicio de Vivienda y Urbanismo [SERVIU]. (2018). Informe de balance de gestión regional 2014–2018. Santiago, Chile: SERVIU.
Spinney, J., Aldred, R., & Brown, K. (2015). Geographies of citizenship and everyday (im)mobility. *Geoforum, 64*, 325–332. https://doi.org/10.1016/j.geoforum.2015.04.013
Stefansdottir, H., Næss, P., & Ihlebæk, C. (2019). Built environment, non-motorized travel and overall physical activity. *Travel Behaviour and Society, 16*, 201–213. https://doi.org/10.1016/j.tbs.2018.08.004
Valenzuela-Montes, L., & Talavera-García, R. (2015). Environments of pedestrian mobility: A review of approaches, factors and conditions. *Eure (Santiago), 41*(123), 5–27. https://doi.org/10.4067/S0250-71612015000300001
Vergel-Tovar, C., & Rodríguez, D. (2018). The ridership performance of the built environment for BRT systems: Evidence from Latin America. *Journal of Transport Geography, 73*, 172–184. https://doi.org/10.1016/j.jtrangeo.2018.06.018
Waintrub, N, Jones, P., & Tyler, N. (2024). Children and young people's unaccompanied mobility: The role of the built and social environments in an unequal Latin American Metropolis. *Children's Geography,* 1–21. https://doi.org/10.1080/14733285.2024.2386244
Wang, H., & Yang, Y. (2019). Neighbourhood walkability: A review and bibliometric analysis. *Cities, 93*, 43–61. https://doi.org/10.1016/j.cities.2019.04.015

Chapter 14
Separate and Unequal Childhoods: Residential Segregation and Neighborhood Inequalities of Children in Chile

Quentin Ramond

Abstract This study examines differences in residential segregation between children and adults in Chile, as well as its consequences for the inequalities that both groups experience in their neighborhood contexts. I link individual-level census data with a series of databases related to neighborhood characteristics at the tract level. First, I use segregation indices to measure segregation levels along age and socioeconomic lines. Results indicate that segregation is higher for children compared to adults, both in terms of evenness and exposure. Specifically, children from socially advantaged backgrounds have higher levels of spatial separation and isolation than any other group. Second, results of a linear regression model with city fixed effects show that children live in neighborhoods that are significantly more unequal than those of adults regarding two key characteristics: exposure to violence and school test score performance. These results are stable across all cities under study, although with varying magnitudes. Overall, this article highlights how age and socioeconomic factors interact in the spatial sorting process and specifies its relationship with contextual inequalities that crucially affect the distribution of life opportunities, especially among children. I close with a discussion of the challenges that these findings raise for social cohesion.

14.1 Introduction

Residential segregation—that is, the uneven distribution of socioeconomic or ethnic groups across neighborhoods—is a persistent feature of large cities, not only in Chile where this study is based (Agostini et al., 2016), but also in other Latin American countries (Fernández-de-Córdova et al., 2021; Marques & França, 2020) and more egalitarian societies across Europe (Musterd et al., 2017). This trend raises concerns because segregation sorts people into unequal neighborhoods, which affects their

Q. Ramond (✉)
Centre for Economics and Social Policy (CEAS), Universidad Mayor, Santiago, Chile
e-mail: quentin.ramond@umayor.cl

M. Vicuña and E. Giorgi (eds.), *Intersections*, Springer Geography,
https://doi.org/10.1007/978-3-031-76402-8_14

opportunities and outcomes, especially for young people. Growing up in deprived neighborhoods negatively affects children's health (Minh et al., 2017), cognitive skills (Sharkey & Elwert, 2011), educational attainment (Kauppinen, 2007), and neighborhood outcomes during adulthood (McAvay, 2020). Conversely, living in advantaged neighborhoods yields positive effects on these dimensions (Troost et al., 2023). Scholars have identified six mechanisms through which neighborhoods affect people's life chances (Galster & Sharkey, 2017): (1) the quality of institutional resources (Nieuwenhuis & Xu, 2021); (2) local social organization (norms, values, social control, and social cohesion); (3) collective socialization processes, in particular adult role models (Ainsworth, 2010); (4) interaction with peers; (5) perceived local inequalities (also known as relative deprivation); (6) exposure to unequal conditions, including contaminants (Crowder & Downey, 2010) and violence (Caudillo & Torche, 2014). In general, living in an impoverished neighborhood is associated with greater levels of social isolation, lower social organization, poor access to good quality services, and greater exposure to contaminants and violence, which have negative consequences.

In this context, scholars have devoted growing attention to differences in segregation levels between children and adults. They have found that income and racial segregation are higher among children than adults (Bernelius & Vilkama, 2019; Owens, 2016, 2017; Sabater & Catney, 2019). Several pathways along which this may occur have been suggested, including the higher economic inequality among households with children and variation in residential preferences, especially greater concerns about the characteristics of local schools when parents choose where to live (Cuddy et al., 2020; Goldstein & Hastings, 2019; Ramond & Oberti, 2022). However, these valuable contributions have generally focused on a limited number of countries where student's allocation to school is based on place of residence, such as the US, Finland, and the UK (with the exception of Boterman, 2020). Therefore, the findings of these studies may not be directly applicable to other contexts where parents have greater autonomy to choose schools for their children, like Chile. Moreover, past research has primarily focused on how neighborhood ethnic-racial composition (Goyette et al., 2014) and the characteristics of the local school options (Kauppinen et al., 2021; Owens, 2016) influence residential selection and segregation processes among children and adults. Therefore, research into whether and how residential segregation is associated with gaps in exposure to other contextual resources that crucially shape individual well-being and outcomes remains limited.

Against this background, the goal of this study is to examine the socioeconomic residential segregation of children and adults as well as its consequences for the inequalities they experience in their neighborhood contexts. It makes two contributions to the literature. First, the chapter focuses on Chile, a Latin American country with high levels of economic inequality and an educational system of completely free school choice. As school district or catchment area boundaries are a well-identified mechanism that explains differences in residential segregation between children and adults, I ask whether this gap remains in choice-based systems. Second, I analyze the relationship between residential segregation and several forms of inequalities that children and adults experience in their neighborhoods. That is, the study aims to uncover the degree to which residential segregation either constrains or enhances access to multiple contextual resources differently for children and adults. I address the following research questions:

1. How does socioeconomic residential segregation vary between children and adults?
2. Is residential segregation of adults and children associated with inequalities in local school options? What other dimensions of neighborhood inequalities are associated with the segregation of children and adults?

The chapter begins with a presentation of the data, measurement, and methods used in the analysis. Results of segregation indices and linear regression models are then reported, followed by a discussion and a conclusion.

14.2 Data and Methods

14.2.1 Data and Sample

I use individual-level data from the 2017 Chilean Census provided by the National Institute of Statistics (INE). Residential location is known at the census tract level, my definition of the neighborhood. A census tract is an administrative-territorial unit defined by the INE with relatively stable boundaries and the average number of residents is 3000. Studies indicate that tract-based measure of residential segregation has limitations, because it does not capture variations in the scale of segregation (Lee et al., 2018). Moreover, this approach conceives a neighborhood as a fixed and bounded location, while not considering surrounding tracts and assuming that all residents within a tract effectively share space, no matter whether they live in the middle of the tract or at its edges. Despite these limitations, I use the census tract for data availability reasons and because, in the Chilean context, it matches relatively well with people's perceptions of their neighborhoods (Link & Valenzuela 2018).

I link census data with neighborhood information produced by the Centre for Territorial Intelligence (Adolfo Ibañez University). I also use geocoded data on school test scores (Ministry of Education) and data on healthcare facilities (Ministry of Health).

I analyze the 19 largest Chilean cities with at least 100,000 residents. I remove old individuals (aged 65 and over), because the spatiality of aging has been the focus of several previous studies (Sabater et al., 2017) and because older people have very specific housing and neighborhood preferences and constraints that could bias my estimates (Mulliner et al., 2020). My sample consists of 9.9 million individuals, nearly 57% of the total population, distributed across 3,403 neighborhoods.

14.2.2 Measurement

This study uses individual-level and neighborhood-level variables. At the individual level, I distinguish between two age groups: children, who are up to seventeen years old, and adults, who are aged between eighteen and sixty-five years old. Socioeconomic status (SES) is based on the highest level of education obtained. I distinguish between four SES groups. Low-SES individuals have primary education, medium–low SES comprise people with secondary education, medium–high SES individuals have vocational education, and high-SES people have reached university education. To determine children's SES, I consider the education level of the household's head. Additional analyses (not presented here) were carried out using the highest level of education attained by either the household head or the spouse in the case of two-parent households to measure children's SES. The results do not change significantly because, in most cases, the household's head has the highest level of education. Table 14.1 presents the sample by age and SES.

It is worth noting that years of education do not perfectly capture social class, which is generally measured using occupation schemes (Erikson & Goldthorpe, 1992) or by considering the intersection of cultural, social, and economic resources (Savage, 2015). However, years of education, as a form of institutionalized cultural capital (Bourdieu, 1984), significantly shape people's life opportunities. In post-industrial societies, individuals with higher education levels are in a better position to participate in the labor market competition and to mobilize politically, while less educated people often face higher economic uncertainties, professional and political marginalization. There is also a growing interest among scholars to study spatial sorting processes by education groups. Research carried out in Europe (Boterman et al., 2021) and in Latin America (Ariza & Solís, 2009; Garreton et al., 2020; Peña, 2012) finds that segregation between education groups has increased in recent years. In some contexts, individuals with university education are even more segregated than the highest income group.

Neighborhood context is defined using a set of eight indicators that capture the mechanisms through which the residential neighborhood potentially affects individual outcomes. First, I use two variables to describe the potential pool of negative adult role models in individuals' neighborhood: the share of unemployed people above age 18 and the proportion of female-headed single-parent households. I measure the quality of environmental conditions with three variables that indicate the

Table 14.1 The sample

	Children	Adults	All
Low-SES	19.3	14.2	15.8
Medium–Low SES	47.8	46.6	46.3
Medium–High SES	17.1	23.1	21.8
High-SES	15.8	16.1	16.1

Source Own calculations based on INE Census, 2017

proportion of poor-quality housing, a public space quality index, and the proportion of a census tract's land area covered by green space. I created a composite index to measure exposure to violence which includes drug trafficking, homicide, robbery, and fights. Finally, I characterize local school options by measuring the academic performance of the schools located within a 1-km radius around the centroids of each census tract, based on standardized average mathematic test scores in 4th, 8th, and 10th grades. Similarly, I calculate proximity to health facilities by measuring the ratio between the number of residents and the number of available seats in healthcare centers and hospitals located within a 1-km radius.

14.2.3 Analytical Strategy

The analysis begins with estimating segregation levels of children and adults between neighborhoods within cities. I measure evenness with the Information Theory Index *H* (Theil & Finizza, 1971). *H* measures the difference between the average socioeconomic diversity (or entropy) within neighborhoods and the socioeconomic diversity of their city. Intuitively, it measures the extent to which neighborhoods are, on average, less diverse than their city. The index varies between 0 and 1. A value of 0 indicates no segregation—each neighborhood has the same socioeconomic composition as the city as a whole—and a value of 1 indicates complete segregation—each neighborhood is socially homogeneous. I estimate multigroup *H* and binary *H* between low- and high-SES groups. In addition, I measure exposure using interaction and isolation indices (Massey & Denton, 1988). They also vary between 0 and 1. The interaction index estimates the exposure of a typical member of group x to individuals of group y in their neighborhood, while the isolation index estimates the exposure of members of group x to other members of the same group x. For both evenness and exposure indices, I present unweighted average estimates of *H* across the 19 largest Chilean cities. I report statistical significance using a two-tailed test when discussing differences in segregation levels between children and adults.

To address the second research question, I perform linear regression models with city-fixed effects predicting residential segregation between neighborhoods. I run separate models for children and adults. The unit of analysis is the census tract. My dependent variable is the local Information Theory Index estimated in each census tract i and city j separately for children and adults. This local measure of segregation estimates the extent to which the neighborhood entropy is reduced below the city's entropy, standardized by the city's entropy. To distinguish between privileged and disadvantaged neighborhoods, I calculate a polarized version of the local Information Theory Index by multiplying its values by -1 in predominantly low-SES areas (low and middle-low). Values close to -1 are associated with a higher concentration of lower socioeconomic groups, while values close to 1 are associated with a higher concentration of advantaged groups. Values close to 0 indicate socioeconomic diversity. All independent variables (i.e., the eight neighborhood characteristics) are standardized so that each has a mean of zero and a standard deviation of one. Intuitively, a positive coefficient indicates that an independent variable is more prevalent in advantaged neighborhoods, whereas a negative coefficient indicates that an independent variable is more prevalent in disadvantaged neighborhoods.

14.3 Results

14.3.1 Segregation Levels of Children and Adults

I begin with the analysis of how segregation in terms of evenness differs for children and adults. Table 14.2 shows the averaged estimates of *H* between neighborhoods among children and adults within the 19 largest Chilean cities. Multigroup *H* indicates that the segregation of children is significantly higher compared to adults. On average, *H* reaches 0.129 for children and 0.084 for adults. This means that the residential neighborhoods of children are 12.9% less diverse than they would be if there were no segregation, while this proportion is only 8.4% among adults, a significant difference of 54.4%. Binary *H* provides estimates of segregation levels between specific social groups. Results show that segregation between high- and low-SES, between high-SES and all other groups as well as between low-SES and all other groups is stronger among children than adults. These differences are statistically significant.

Gaps in segregation levels by age group are particularly marked regarding the separation of high-SES children from all other groups. High-SES children's *H* index exceeds by 67% that of high-SES adults. The relative difference by age among low-SES groups is only 27.5%. Moreover, the average segregation of high-SES children is more than two times higher than the segregation of low-SES children. It is also worth

Table 14.2 Average levels of segregation of children and adults (H)

	Children	Adults	Difference	Δ%
	(1)	(2)	(1–2)	
Multigroup	0.129	0.084	0.046***	54.4
	(0.03)	(0.02)	(0.01)	
High-SES/All other groups	0.231	0.138	0.093***	66.8
	(0.05)	(0.03)	(0.02)	
Low-SES/All other groups	0.104	0.081	0.022**	27.5
	(0.02)	(0.02)	(0.01)	
High-SES/Low-SES	0.446	0.333	0.113***	33.9
	(0.08)	(0.06)	(0.02)	

Source Own calculations based on INE Census, 2017. Averages are unweighted. *$p < 0.05$; **$p < 0.01$; ***$p < 0.001$ (two-tailed test). Standard deviations in parenthesis noting that segregation between low- and high-SES children is significantly higher than segregation between the same adult groups. Together, these findings indicate that the spatial concentration of high-SES children exceeds that of any other group, including high-SES adults.

Segregation is high in absolute terms. Although comparison with other studies is methodologically challenging, research carried out in Mexico finds that *H* between adults from low- and highly educated groups is 0.28 (Peña, 2012, p. 83) which is significantly lower than in Chilean cities. My estimates show that the levels of segregation of children and adults in Chile are closer to those of racial segregation in the US (Owens, 2017).

Across all cities, gaps in segregation levels between children and adults are significant, and high-SES children are more strongly segregated than any other group. Beyond these common patterns, Santiago, the capital city, shows higher levels of segregation. Cities in the northern regions of the country have lower levels of segregation, which could be explained by lower levels of economic inequality compared to southern cities (Brevis, 2020).

Next, I use interaction and isolation indices to examine whether segregation in terms of exposure differs for children and adults. Note that I estimate children's and adults' exposure to their own age group. Figure 14.1 shows the average neighborhood socioeconomic composition of low- and high-SES children and adults. On average, low-SES children live in neighborhoods where 28% of children also come from low-SES backgrounds, whereas low-SES adults live in neighborhoods that are only 21% low-SES. Low-SES children also encounter fewer medium–high and high-SES children in their neighborhoods compared to adults (−6 and −2 percentage points, respectively).

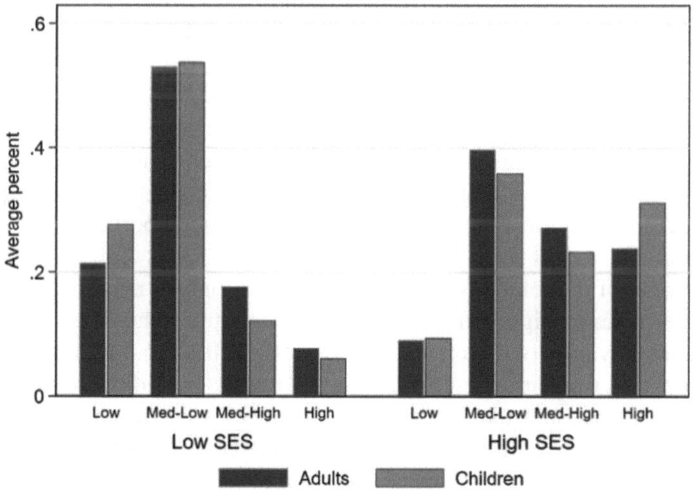

Fig. 14.1 Neighborhood socioeconomic composition of children and adults. *Source* Own calculations based on INE Census, 2017. Unweighted averages

On the other hand, high-SES children experience more advantaged neighborhoods than high-SES adults. The average high-SES child lives in neighborhoods where 31% of children come from high-SES backgrounds, whereas the average high-SES adult lives in neighborhoods where 24% of adults come from high-SES backgrounds. Gaps between high-SES adults' and children's neighborhood composition are smaller when considering exposure to other SES groups. These patterns are found in all cities under study.

Collectively, these findings indicate that children and adults of the same SES live in neighborhoods with different compositions: low-SES children experience higher levels of social disadvantage than low-SES adults, while high-SES children experience higher levels of social advantage than high-SES adults. These patterns may have important consequences regarding the kind of peers with whom children and adults interact, with potentially far-reaching consequences for the unequal distribution of life opportunities. I delve further into these issues in the conclusion.

14.3.2 Segregation and Neighborhood Inequalities of Children and Adults

Results thus far reveal that residential segregation is higher among children than among adults, which is mostly linked to the greater spatial separation and isolation of socially advantaged children. The following regression analysis examines the extent to which residential segregation is associated with neighborhood inequalities among children and adults. To allow for a more straightforward interpretation of the results from the linear regression models, Fig. 14.2 shows the regression coefficients and 95% confidence intervals for every independent variable.

The results indicate a statistically significant relationship between the concentration of disadvantaged individuals, unemployment rate, the share of female-headed single-parent households, the share of low-quality housing, exposure to violence and drug trafficking. Conversely, a higher degree of neighborhood socioeconomic advantage is positively correlated with access to better-quality public spaces and proximity to higher-performing schools. The presence of green spaces and proximity to health facilities are not significantly associated with neighborhood socioeconomic composition.

The strength of the relationship between segregation and neighborhood inequalities is systematically higher among children than adults. This is evidenced by the higher and lower values of regression coefficients for children. Importantly, the coefficients for children depart strongly from those of adults regarding two key variables: exposure to violence and the quality of the local schools. Among children, a one-standard-deviation increase in local schools' test score performance is associated with a 0.05-point increase in the segregation index, against 0.03 point for adults. This indicates that, for children, residential segregation is associated with stronger inequalities in exposure to violence and in the quality of the local school options

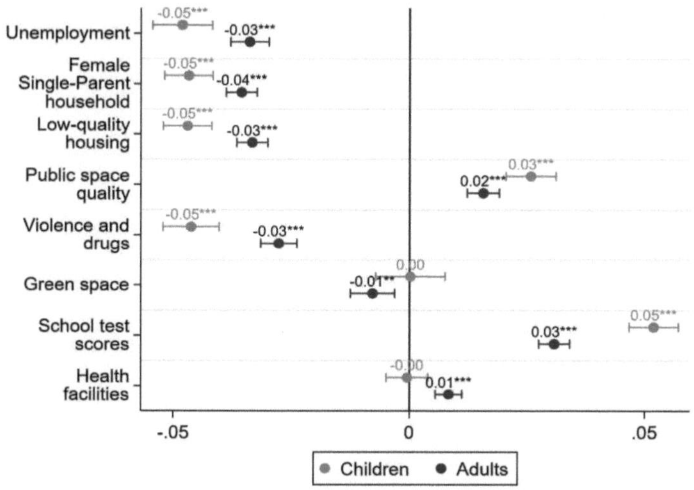

Fig. 14.2 Linear regression with city fixed effects predicting local residential segregation (H). *Source* Own calculations based on INE Census, 2017, CIT-UAI, Ministry of Education, Ministry of Health

compared to adults. High-SES children tend to cluster in neighborhoods with better schools, more than their adult counterparts. Low-SES children are exposed to higher levels of violence in their neighborhood spaces, even more so than low-SES adults.

To unpack further the differences between children and adults, Fig. 14.3 illustrates the predicted values of H across different levels of exposure to violence and average standardized school test scores of nearby schools. The slopes indicate the strength of the relationship between segregation and contextual inequalities such that the steeper the slope, the stronger the link between segregation and neighborhood characteristics. Children show the steepest slopes, indicating a stronger relationship between neighborhood SES composition, exposure to violence, and local school test scores. Exposure to violence is more strongly concentrated in socially disadvantaged neighborhoods among children. Conversely, proximity to high-performing schools mostly benefits homogeneous advantaged areas, especially among children. Exposure to violence and proximity to high-performing schools occur in more diverse neighborhoods for adults.

14.4 Conclusion and Discussion

The aim of this study has been to examine the socioeconomic residential segregation of children and adults, as well as its consequences for the inequalities that both groups experience in their neighborhood contexts in Chile. I relied on segregation indices to examine segregation levels along age and socioeconomic lines. Moreover,

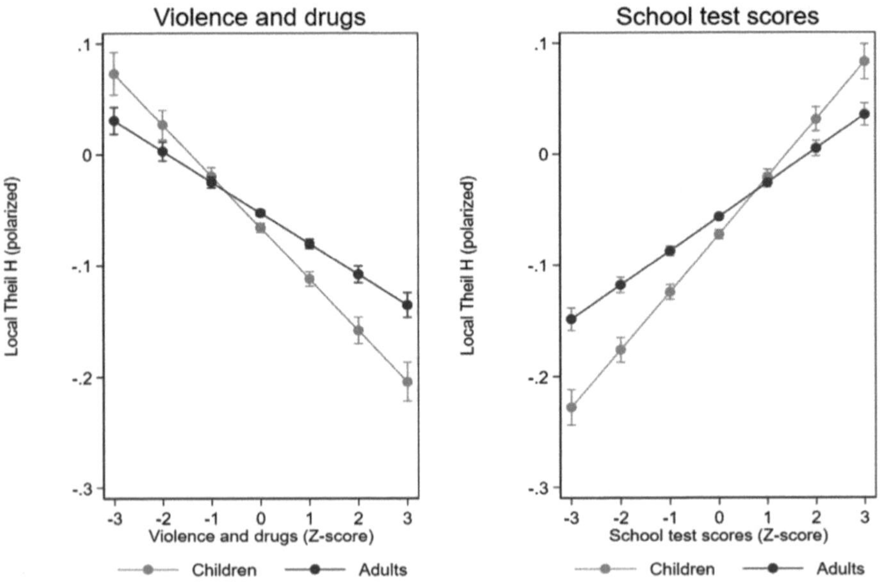

Fig. 14.3 Segregation, exposure to violence, and school test scores. Predicted estimates with 95% confidence intervals. *Source* Own calculations based on INE Census, 2017, CIT-UAI, Ministry of Education, Ministry of Health

I performed linear regression models to measure the association between segregation and several dimensions of neighborhood inequalities.

Results indicate that children are less evenly distributed across neighborhoods than adults. This difference is largely driven by the greater spatial separation of high-SES children from other socioeconomic groups. Moreover, children live in less diverse neighborhoods than adults. High-SES children are exposed to more high-SES peers than high-SES adults, while low-SES children experience higher exposure to low-SES peers than low-SES adults. These patterns are observed across all cities under study, although with varying magnitudes. Residential segregation contributes to sorting children and adults into unequal neighborhoods, but important differences emerge between both groups.

Children experience stronger neighborhood inequalities than adults in their residential neighborhoods. Specifically, residential segregation is associated with higher inequalities in local school options among children, although the Chilean educational system provides families with a large autonomy in the school choice process. Previous works carried out in Chile have shown that high-SES families incorporate schools into their residential decisions as part of their social reproduction strategy (González-Espejo et al., 2022). The present study adds to these works by showing that high-SES parents probably use their resources to cluster in socially advantaged neighborhoods with better schools and less violence. Conversely, low-SES families with children experience strong constraints on the housing market that prevent them from living in neighborhoods that ensure their children's well-being and access to opportunities.

Overall, this study highlights the significant interactions between age and socioeconomic resources in the spatial sorting process and exposure to contextual inequalities at the neighborhood level. The higher degree of segregation and neighborhood inequalities experienced by children indicates that they are less likely to establish everyday social interactions with peers from other social backgrounds in their neighborhood. This pattern raises concerns for the future of social cohesion in Chile for, at least, two reasons. First, these segmented social networks forged during childhood could contribute to sustaining inequality in access to social capital that crucially influences children's future opportunities (Kaztman & Retamoso, 2009). Second, experiencing greater segregation reduces exposure to and interactions with people from other parts of the social structure, which, in the long run, could undermine tolerance to social diversity (Sabatini et al., 2012) and impede the emergence of cross-class coalitions in support of collective goods over private alternatives (Préteceille, 2006). Therefore, implementing public policies that address children's specific housing, location, and infrastructure needs is necessary to create more equal childhoods and foster social cohesion throughout Chile. For instance, housing policies may prioritize allocating social housing in advantaged areas to disadvantaged families with children, while educational policies may strive for a better spatial distribution of attractive school options (Ramond, 2023).

Acknowledgements I am grateful to Pablo González for helpful feedback on earlier versions.

References

Agostini, C. A., Hojman, D., Román, A., & Valenzuela, L. (2016). Income residential segregation in Greater Santiago, 1992–2002: A robust estimate. *EURE Santiago, 42*, 159–184.

Ainsworth, J. W. (2010). Does the race of neighborhood role models matter? Collective socialization effects on educational achievement. *Urban Education, 45*, 401–423.

Ariza, M., & Solís, P. (2009). Socioeconomic dynamics and spatial segregation in three metropolitan areas of Mexico, 1990 and 2000. *Sociological Studies El Col Mexico, 27*, 171–209.

Bernelius, V., & Vilkama, K. (2019). Pupils on the move: School catchment area segregation and residential mobility of urban families. *Urban Studies, 56*, 3095–3116.

Boterman, W. (2020), Intersections of class, ethnicity and age: Social segregation of children in the metropolitan region of Amsterdam. In *Handbook of urban segregation* (pp. 200–218). Edward Elgar Publishing.

Boterman, W. R., Musterd, S., & Manting, D. (2021). Multiple dimensions of residential segregation. The case of the metropolitan area of Amsterdam. *Urban Geography, 42*, 481–506.

Bourdieu, P. (1984). *Distinction: A social critique of the judgement of taste.* Harvard University Press.

Brevis, M. (2020). The dynamics of inequality in Chile: A regional perspective. *Review of Economic Analysis, 35*, 91–133.

Caudillo, M. L., & Torche, F. (2014). Exposure to local homicides and early educational achievement in Mexico. *Sociology of Education, 87*, 89–105.

Crowder, K., & Downey, L. (2010). Interneighborhood migration, race, and environmental hazards: Modeling microlevel processes of environmental inequality. *American Journal of Sociology, 115*, 1110–1149.

Cuddy, M., Krysan, M., & Lewis, A. (2020). Choosing homes without choosing schools? How urban parents navigate decisions about neighborhoods and school choice. *Journal of Urban Affairs, 42*, 1180–1201.

Erikson, R., & Goldthorpe, J. H. (1992). *The constant flux: A study of class mobility in industrial societies*. Oxford University Press.

Fernández-de-Córdova, G., Moschella, P., & Fernández-Maldonado, A. M. (2021) Changes in spatial inequality and residential segregation in Metropolitan Lima. In M. Van Ham, T. Tammaru, R. Ubarevičienė, & H. Janssen (Eds.), *Urban socio-economic segregation and income inequality: A global perspective* (pp. 471–490). Springer.

Galster, G., & Sharkey, P. (2017). Spatial foundations of inequality: A conceptual model and empirical overview. *RSF Russell Sage Foundation Journal of Social Science, 3*, 1–33.

Garreton, M., Basauri, A., & Valenzuela, L. (2020). Exploring the correlation between city size and residential segregation: Comparing Chilean cities with spatially unbiased indexes. *Environment and Urbanization, 32*, 569–588.

Goldstein, A., & Hastings, O. (2019). Buying In: Positional competition, schools, income inequality, and housing consumption. *Sociological Science, 6*, 416–445.

González-Espejo, F., Astroza, S., & Hurtubia, R. (2022). On the relation between school and residential location choice: Evidence of heterogeneous strategies from Santiago de Chile. *Journal of Transport Geography, 102*, 103359.

Goyette, K., Iceland, J., & Weininger, E. (2014). Moving for the kids: Examining the influence of children on white residential segregation. *City & Community, 13*, 158–178.

Kauppinen, T. M. (2007). Neighborhood effects in a European city: Secondary education of young people in Helsinki. *Social Science Research, 36*, 421–444.

Kauppinen, T. M., van Ham, M., & Bernelius, V. (2021). Understanding the effects of school catchment areas and households with children in ethnic residential segregation. *Housing Studies*, 1–25. Spatial Demography.

Kaztman, R., & Retamoso, A. (2009). Residential segregation in Montevideo: Challenges to educational equality. In B. R. Roberts & R. H. Wilson (Eds.), *Urban segregation and governance in the Americas* (pp. 97–119). Palgrave Macmillan US.

Lee, B. A., Farrell, C. R., & Reardon, S. F., et al. (2019). From census tracts to local environments: An egocentric approach to neighborhood racial change. *Spatial Demography, 7*, 1–26.

Link, F., & Valenzuela, F. (2018). *The structure of socio-residential density in the metropolitan area of Santiago*. Institute of Urban and Territorial Studies UC, Working Papers of the IEUT, N° 3.

Marques, E., & França, D. (2020). Segregation by class and race in São Paulo. In *Handbook of urban segregation* (pp. 36–54). Edward Elgar Publishing.

Massey, D. S., & Denton, N. A. (1988). The dimensions of residential segregation. *Social Forces, 67*, 281–315.

McAvay, H. (2020). Socioeconomic status and long-term exposure to disadvantaged neighbourhoods in France. *Urban Studies, 57*, 2663–2680.

Minh, A., Muhajarine, N., Janus, M., et al. (2017). A review of neighborhood effects and early child development: How, where, and for whom, do neighborhoods matter? *Health & Place, 46*, 155–174.

Mulliner, E., Riley, M., & Maliene, V. (2020). Older people's preferences for housing and environment characteristics. *Sustainability, 12*, 5723.

Musterd, S., Marcińczak, S., van Ham, M., & Tammaru, T. (2017). Socioeconomic segregation in European capital cities. Increasing separation between poor and rich. *Urban Geography, 38*, 1062–1083.

Nieuwenhuis, J., & Xu, J. (2021). Residential segregation and unequal access to schools. *Social Inclusion, 9*.

Owens, A. (2016). Inequality in children's contexts: Income segregation of households with and without children. *American Sociological Review, 81*, 549–574.
Owens, A. (2017). Racial residential segregation of school-age children and adults: The role of schooling as a segregating force. *RSF Russell Sage Foundation Journal of Social Science, 3*, 63–80.
Peña, L. S. (2012). Living increasingly separated? A multigroup analysis of residential segregation in Mexico City, 1990–2005. *Estudios Demográficos y Urbanos, 27*, 57–93.
Préteceille, E. (2006). Segregation against social cohesion: The Parisian metropolis. In H. Lagrange (Ed.), *The test of inequalities, Presses Universitaires de France* (pp. 195–246). Presses Universitaires de France.
Ramond, Q., & Oberti, M. (2022). Housing tenure and educational opportunity in the Paris metropolitan area. *Housing Studies, 37*, 1079–1099.
Ramond, Q. (2023). Infancias separadas y desiguales. Segregación residencial y desigualdad espacial entreniños/as en Chile. *Temas de la Agenda Pública, 18*(163), 1–19. Centro de Políticas Públicas UC, Chile.
Sabater, A., & Catney, G. (2019). Unpacking summary measures of ethnic residential segregation using an age group and age cohort perspective. *European Journal of Population, 35*, 161–189.
Sabater, A., Graham, E., & Finney, N. (2017). The spatialities of ageing: Evidencing increasing spatial polarisation between older and younger adults in England and Wales. *Demographic Research, 36*, 731–744.
Sabatini, F., Rasse, A., Mora, P., & Brain, I. (2012). Is residential integration possible in Chilean cities?: Disposition of the middle and high groups to integration with groups of popular extraction. *EURE Santiago, 38*, 159–194.
Savage, M. (2015). *Social class in the 21st century*. Penguin UK.
Sharkey, P., & Elwert, F. (2011). The legacy of disadvantage: Multigenerational neighborhood effects on cognitive ability. *American Journal of Sociology, 116*, 1934–1981.
Theil, H., & Finizza, A. J. (1971). A note on the measurement of racial integration of schools by means of informational concepts. *Journal of Mathematical Sociology, 1*, 187–194.
Troost, A. A., Ham, M. van, & Manley, D. J. (2023) Neighbourhood effects on educational attainment. What matters more: Exposure to poverty or exposure to affluence? *PLOS ONE, 18*, e0281928.

Part IV
Gender Focus and Ethics of Care: New Disciplinary and Projectual Perspectives in Design, Architecture and Urbanism

Chapter 15
Public Health as Public Space: Feminist Design for Spaces of Care, Encounter and Collaboration Between State and Citizenship Around Well-Being

Patricia Sánchez-Hidalgo

Abstract Nowadays, at the San Rafael Family Health Center (Cesfam), territory and Health System converge in two simultaneous interaction spaces. The first are the waiting and stay areas within the building that have incorporated libraries, interior gardens, and free-use spaces for patients, neighbors and officials, favoring the encounter between them. The second is a more intangible space built through the coordinated work between a group of Cesfam workers—called Family Executives—and the Social Leaders of the sector that expand the monitoring capacity of the Cesfam and build a bridge between the socio-health needs of the neighborhood and the Health Center. The description of this case constitutes an inspiration for the form that a feminist design of public health spaces could take that welcome and enhance their role as a moment of coexistence between State and Citizenship, and materialize an accessible space from care, attention and encounter. The case was addressed through a series of field work sessions during 2021 and 2022 that included observation and description of the waiting and transit spaces inside the Health Center, interviews and conversations with workers and patients, and the subsequent analysis and translation of what was observed into design recommendations with a feminist approach. Finally, in addition to the potential of the Cesfam San Rafael as a benchmark for public space that fosters an effective encounter between state and territory, the work opens the space to reflect on the capacity of design as a discipline to materialize political approaches and counter-hegemonic narratives that can be experienced corporally by people to push for a transformation in the way of conceiving and inhabiting these spaces from a feminist perspective.

Keywords Public space · Public health · Primary care · Feminist design

P. Sánchez-Hidalgo (✉)
School of Design, Pontificia Universidad Católica de Chile, Santiago, Chile
e-mail: pdsanche@uc.cl

M. Vicuña and E. Giorgi (eds.), *Intersections*, Springer Geography,
https://doi.org/10.1007/978-3-031-76402-8_15

15.1 Introduction

The Cesfam San Rafael is a public health center located in the La Pintana district, Santiago de Chile, with a multidimensional poverty index well above the average of the rest of the city (Congressional Library, 2021).[1] This center—inaugurated in 2007—is part of the Metropolitan Health Service South East network[2] (SSMSO) and serves about 31,000 people annually under a care model named by its former director as "innovating towards the dignity of people", where they try to understand and embrace the context of the territory to cover not only the biomedical aspects of people's health but all those that may influence their well-being. Under this model, various meeting and support spaces have been created within the premises and the position of Family Executives has been created, officials who work in the monitoring and accompaniment of neighbors through constant coordination with the local Neighborhood Leaders.

The forms of relationship and organization present in this health center resemble what could result from the incorporation of a feminist perspective into design, incorporating the discourses and practices of subjects normally ignored by the supposedly neutral and universal canon of the discipline (Zambrini & Flesler, 2017).

From my situated position as a feminist female designer and researcher, I present this text as part of a research conducted under a qualitative and feminist approach, which puts at the center above all the perspectives and ways of living of the women involved in the daily operation of Cesfam as a "significant indicator of the «reality» against which hypotheses must be contrasted" (Harding, 1987, p. 21), contributing to the deprivatization of their stories, giving them the place they deserve within the analysis of culture (Castañeda, 2012).

This from a Research for Design methodology, where the tools and techniques already existing in the social sciences were used to inquire about a topic involved in the development of a potential design artifact (Roggema, 2016), in this case, to understand the context of Cesfam San Rafael to inspire and inform a series of design recommendations for Public Health Spaces with a feminist approach. The data were collected through semi-structured interviews and observation of activities and facilities within the Cesfam. For systematization and analysis, dense descriptions of the operation and inhabitation of the place were generated from the written and audiovisual records of the field visits. Finally, the categories of collaboration, encounter and care were created as axes to translate the descriptions into design recommendations and make the step from descriptive to prescriptive and define the design actions that should be carried out to enhance, reduce or eliminate the influence of the critical factors (Mollenhauer et al., 2020) involved in the situations that shape these three categories.

As an evidence-based study, I embrace the challenge that gender issues be recognized and legitimized and then integrated into the public agenda for their subsequent solution (Muñoz Rojas, 2021). Thus, in the final reflection, I will reaffirm my beliefs about the potential of design for the reduction of the implementation gaps of the public gender policies but also the contradiction that my recommendations present by affirming many of the design recommendations in the work and commitment of the women who today already sustain the operation of the place, perpetuating its sustenance in the affective bond and care (Rojas Lasch, 2018) that have been historically a female gender mandate.

15.2 Public Health Spaces as Public Spaces

For the analysis of the different spaces of the Cesfam San Rafael, I start from the idea that public spaces are composed of a physical dimension but also a mental and social one (Lefevre, 1991) that overlap and intersect at the same time. Delgado (2007, 2011) adds that these should be conceived as places for everyone and where relationships between those who coincide are produced. Despite this integral role of public space in the production of society (Massey, 2012), it presents frustrated access by women, children, older people and people with functional diversities—historically relegated to the private and domestic space—who need to extend their practices from private space for optimal inhabiting of the public (Jirón, 2007). Therefore, studying and welcoming this diversity of ways of life is a contribution towards justice with geographical implications (Soto, 2012).

To define Public Health—and its spaces of waiting, attention and contact with citizens—as Public Space, I will consider these as places to which we all have legal access and that act as mediators between people and broad structures and social institutions (Jirón, 2007) that can occur both in open spaces and interiors that embrace the same logic of street or parks despite not being equally exposed (Soto, 2012). Based on the above, the waiting spaces of health centers constitute a meeting point between the State and Citizenship with the potential to strengthen this relationship, being an overlap of the extension of the boundaries of both.

These spaces must move out of the category of non-places (Augé, 2000) or spaces exclusively for transit between A and B and strengthen as spaces to inhabit freely (Musset, 2015). This is in line with Delgado's (2007) observation about places like shops, kindergartens or school gates that remain predominantly female territories from which most women cognitively organize their urban morphology. If we consider health spaces as everyday spaces where people act and develop, the value of studying them lies in examining the invisible and problematic aspects ignored in decision-making because they are so ingrained in daily life (Jirón, 2007).

But the idea of Public Space suggests the existence of an antagonist that would be Private Space. To address this dichotomy, the fields of discussion of the public–private exposed by Rabotnikof (1998) and the different definitions and problems that each one raises for the analysis of Public Health Spaces are recovered:

1. From a political-economic point of view, the public is associated with the state and collective benefit while the private with the market or privatized.
2. In terms of civic and participatory tradition, "the public" is the sphere of participation, what belongs to everyone and "the political".

3. In the category of social life, the criterion of exposure and visibility prevails. The public understood it as a space of fluid sociability (Rabotnikof, 1998), of interaction in sight of all. In contrast, the private corresponds to the intimate sphere, which is less visible and that is outside of public interference.

In the eyes of feminist literature, the unequal designation of men and women to the different spheres that the public–private classification proposes under the idea that "the personal is political" is criticized. This shows that the conceptual pair is far from referring to two "ontologically differentiated" areas and that, as Soto (2009) establishes, "they are not univocal, they do not form a single binary opposition, on the contrary, there is a network of oppositions that are associated with it, identifying problems, criticisms and diverse and often divergent senses." (p. 54). Therefore, the waiting and attention spaces belonging to public health services are contradictory in the eyes of the public–private dichotomy for being an interior space—therefore, of low visibility -, but protected by the State as a public entity that should promote social life and citizen participation.

15.3 A Feminist Perspective on the Design of Public Health Spaces

To understand the potential of incorporating a feminist perspective into the design of public spaces that materialize a relationship of encounter, care and collaboration between the State and Citizenship, we must consider the implementation gap in the material aspect of public policies that hinders us from advancing in programs and devices that do not perpetuate unequal gender relations. Health spaces are in this sense an opportunity to strengthen the relationship between the State and Citizenship, however, institutional guidelines around access and enjoyment of these spaces coexist in practice with various class and gender biases that do not allow them to be appropriated or inhabited by people for purposes other than transit, transmitting a clear point of view on who has the right to what (Anzorena, 2017). This directly undermines state policies of gender mainstreaming by not incorporating this perspective into the material aspect of the delivery of public goods and services (Muñoz Rojas, 2021).

15.3.1 Current Guidelines and Gaps in the Design of Public Spaces

As Lahera (2005) exposes, the design and implementation of Public Health Spaces should be taken as part of the essential and endogenous tasks of the public sector. In terms of gender mainstreaming, it becomes complex to design public spaces with a gender perspective because their projection and construction are in the hands of state actors different from those who lead the gender agenda (Muñoz Rojas, 2021). In this case, the guidelines are mainly delivered by the Ministry of Health with an expected bias in health care provision.

Ministerial documents associated with the design of waiting rooms and health infrastructure such as the 'Design Guide for Medium Complexity Hospital Establishments' (Ministry of Health, 2019) and the document 'Technical Guidelines for Design of Preliminary Projects of Complex Hospitals' (Ministry of Health, 2017) propose that waiting spaces be open and flexible to potential emergency scenarios. In contrast, the 'National Health Plan for Health Objectives by 2020' (Ministry of Health, 2016) proposes "equity in health as a state responsibility, seeking to reduce unjust inequalities" but does not detail how this could be materialized in Public Health Spaces. I think it's important to point out that health spaces—and public spaces in generalare designed by those who hold power and seek to promote and preserve what Delgado (2011) describes as "civility", that is, what the elite considers as good practices in public space. Jirón (2007) adds to this diagnosis by stating that those who think about public spaces do not incorporate the way in which people live in the city because they are based on abstract models that do not consider the recognition and feedback of the everyday. This design that does not consider the gender and class biases in the management of institutions or in the way in which public goods and services are delivered (Muñoz Rojas, 2021) results in a city aggressive towards women, dissidents, girls and boys, elderly people, people with disabilities, etc. who experience daily situations of discomfort, limitation and even aggression that feed their agoraphobia (Lindón, 2006) or fear of circulating in public space.

15.3.2 Feminist Design for the Public

I consider design a critical and reflective discipline capable of responding to the complexities of people's thinking and cultural production through mediating (Bentz & Franzato, 2017). In its vocation to know and transform at the same time (Sánchez & Hermansen, 2021), design must open to project and materially represent a diversity of narratives of possible futures to be communicated through its tangibility to increase the reach and effectiveness of these alternative life ideas (Dunne & Raby, 2013; Heidaripour & Forlano, 2018). However, as a discipline born and established in Europe within the modern paradigm of progress and development, design has adopted a "neutral" and "objective" character in line with the historical moment in which it was consolidated.

To combat this, I adhere to the idea of designing based on an ethical analysis and a critical perspective of established norms, aware of how these norms, values and other social constructs are produced and reproduced in society through design and interactions (Öhrling et al., 2018). Therefore, a feminist design perspective would not only question "for whom" it is designing to focus solely on women, but rather "from where" it does it and what ideas it promotes and perpetuates in its practice (Sánchez & Neira, 2021). As a political horizon, I advocate for stopping designing to remedy the problems of the model and start prototyping alternative models that put life and not capital at the center (Carrasco & Recio, 2014).

15.4 The Inhabiting of Cesfam San Rafael

Regarding the Cesfam San Rafael as a case study, it should be mentioned that the period studied corresponds to the COVID-19 pandemic where, aligned with its care model, the center did not stop attending even during the period of health restrictions, focusing on chronic and mental health patients. In addition, as a strategy to not lose contact with the community, they joined community activities such as communal pots and took advantage of virtual platforms like Zoom to hold meetings and workshops with the Neighborhood Leaders.

In relation to the spaces where the Health System and territory converge, there is an overlap between the tangibles ones like the building, waiting rooms and transit places and those spaces built through the network of work between the Family Executives and the Neighborhood Leaders of the territory that is operationalized in active follow-up and the expansion of the monitoring capacity of patients, covering more areas of well-being than those corresponding to the physiological.

Next, I present some of the situations and interactions that distinguish the operation of the Cesfam San Rafael and that help to define the categories of encounter, care and collaboration that will inspire the feminist design recommendations for health spaces.

15.4.1 Roots and Trust to Build a Place of Well-Being

The respect and recognition of the organization dynamics of the territory in which this health center is inserted, along with the territorial inscription of the staff –where a large percentage currently live or have lived there in the neighborhood—have been key to generating roots in San Rafael. The staff transcends the sense of territoriality that they have built over the years (Lindón, 2006) and apply it to their workspace, helping to transform this health space into a place to feel attended and cared for as neighbors. This rooting can be considered an advantage for state capacities that seek to build agreements and synergies among the staff to achieve better results (Ricardo, 2016) based on life agreements common to adhere to.

Another important aspect has been the strategies of building trust between the State and Citizenship through the transparency of the working conditions and fluid and constant communication. This aligns with Lahera's (2005) approach on the way in which public institutions must be prepared to explain what is being done, in order to make it understood that public policies seek the common good even if at the point we are at it does not seem evident.

The adaptation of spaces to the changing needs of the territory is also perceived as an act of attention by patients and their companions. Although the flexibility of these spaces is considered an essential characteristic from the health perspective that requests to have open areas and adaptable to an extraordinary flow of patients (Ministry of Health, 2017) it is also important from a feminist geography approach that proposes to have a perspective of openness towards an unpredictable and unexpected future (Massey, 2012) and the joint construction from the needs of those who inhabit the spaces. As an example, one of the adaptations of the Cesfam to the sanitary measures for the containment of COVID-19 to the local resources and culture of the territory was the installation of an outdoor sink at the entrance of the building to promote hand washing despite the resistance to the use and carrying of hand sanitizer.

15.4.2 The Network of Family Executives and Neighborhood Leaders as a Foundation for Well-Being

The role of family executives was created to heal the relationship between the health center and the territory through a closer, more fluid, and transparent relationship with these officials as mediators. They were also considered an indispensable profile for a family health model based on prevention and the ability to anticipate and accompany the well-being of individuals. The routine of these officials mainly involves scheduling medical appointments and tracking patients both over the phone and in person through home visits that allow them to understand the living conditions that may be influencing their well-being. The supportive role they play has proven essential for many "mom-grandmothers" who serve as primary caregivers and need to feel that they are not alone in this situation, as well as for older people who do not have other support networks.

The tracking not only focuses on physiological aspects, but also on mood states, relationships with family and neighbors, and even working conditions. This form of accompaniment contrasts with the monitoring proposed by traditional and impersonal institutional mechanisms that rely on quantitative indicators to measure family achievements that grant access to state benefits (Rojas Lasch, 2018). However, it is essential to problematize how they are adding tasks and overloading the working days of the women behind these functions.

On their part, Neighborhood Leaders amplify the voice of the Cesfam through their credibility as community leaders. An example is the coordination around the COVID-19 vaccination processes where the flow of people varies according to the information circulating through neighborhood WhatsApp groups about service hours and the number of people waiting. Another way in which the Cesfam has managed to extend and expand its reach in the territory has been through first aid training for some of these leaders, which has not only made them ambassadors but also a first line of care and monitoring in cases of medical emergencies.

Despite its success, this network of working women is also an expression of the "neoliberalization of the social" for being a precarious and budgetarily restricted network but at the same time highly moral and subjective in its motivations (Rojas Lasch, 2018) that leads them to contribute with their own resources as part of an emotional commitment to the work. Thus, these women embody and replicate unequal gender stereotypes and relationships in their role as caregivers from affection and sacrifice according to gender subordination mandates (Lamas, 2007) installed from their childhood as activities inherent to their female condition.

15.5 Feminist Perspective Design Recommendations for Health Spaces

From the analysis of the ways of inhabiting the San Rafael Cesfam, the categories of Care, Collaboration, and Encounter emerge to group the activities and interactions that constitute this health space in a place marked by the affective relationships that are produced there and inspire the subsequent design recommendations.

The Care category is reflected in being attentive and making a fluid and constant follow-up of the needs that impact the well-being of the neighbors. This care extends to domestic spaces in home visits, health attention in the neighborhood and other informal support spaces within the Cesfam facilities. The neighbors perceive that the Cesfam is effectively attentive to their needs, despite the occasional disagreements with the care staff that usually achieve a broad resolution of their concerns as an act of care.

The Encounter category refers to the ability of Cesfam spaces to promote the construction of relationships between patients, staff, and neighbors and to highlight social life as an influential factor in people's well-being. This mainly occurs in the corridors, open spaces, and high-traffic areas of the Cesfam, but it does not only refer to spontaneous encounters but also to scheduled instances.

Finally, the Collaboration category refers to the Cesfam as a space co-constructed between the State and Citizenship that fosters belonging and rooting to the health space and the spaces it extends through. This collaboration is mainly observed in the coordinated work of the Family Executives with the key actors of the territory and the contribution of the neighbors to the production and maintenance of the common spaces.

15.5.1 Feminist Perspective Design Recommendations for Health Spaces

Below are a series of feminist design recommendations to build public health spaces that promote encounter, care, and collaboration as forms of relationship.

1. Think of waiting rooms as places for encounter, with a layout that allows people to look at each other and be attentive to the conditions or needs of others. This layout should be flexible not only to health contingencies as proposed by ministerial guidelines, but above all to the diverse and changing ways of inhabiting by workers, patients, companions, and neighbors.
2. Formalize a system of patient registration that allows a broader tracking and vision of the factors affecting their well-being. Despite the benefits of having this information at hand, there is no current way to formally register it on the medical records platform available, leaving its record and use to the judgement of the different workers.
3. Incorporate collective use spaces for community well-being, both for autonomous neighborhood groups and for activities proposed and mediated by the Cesfam. Also, consider virtual spaces when there are conditions that hinder the displacement of its participants due to domestic commitments or territorial contingencies.
4. Open the spaces of the Health Center to the participation of neighbors in its planning, production, or maintenance to precipitate the construction of the common in a tangible way.
5. Promote spaces for transparency and greater territorial reach through health ambassadors who, just as Neighborhood Leaders do today, can contribute to the functioning of the Cesfam by alerting about possible risk situations for the well-being of people.

15.6 Conclusions

Based on the above, I think it is important to reflect on the material dimension of the exclusion that women suffer through devices that mediate and emphasize unequal gender relations. These devices are often forgotten by equality and gender mainstreaming policies that perpetuate the exclusion and marginalization of women from decision-making spaces, economic and social development (Muñoz Rojas, 2021) despite aiming to improve their quality of life. I, therefore, invite you to carefully analyze how the physical space of public institutions like the Cesfam San Rafael replicates or amplifies the pre-existing dynamics in the territory. Here is an opportunity for the State to reduce the implementation gaps of gender policies, through the incorporation of a feminist perspective when thinking about its material dimension.

Therefore, design when practiced from a critical and feminist perspective could be an ally to mediate the programs and spaces of Public Health as a materialization of the different proposed care models. In the same line, design opens the possibility that the questioning is not only verbal, but also through letting emerge a way of inhabiting the Public Health Spaces that resists the imposed arrangements and incorporates people as participants of this design and redesign process from the recognition of their daily practices.

Faced with the proposal of a feminist design that gives importance to the experiences and needs of historically marginalized groups such as women, there will always be arguments about the "dangers" of building biased futures. But in my opinion, the greatest risk of what is proposed in this text is to continue perpetuating the gender mandate that falls on women around care tasks (Rojas Lasch, 2018). My own design recommendations lack a broad consideration of the people—mostly women—involved in its implementation and are mostly actions that, without a transformation of the current living and working conditions of the female workers, will continue to rest its success on them personally.

This awareness of the living conditions of the women involved, along with the intention to transform them considering local cases such as the Cesfam San Rafael, could be a possible path towards the projection of more diverse lifestyles focused on well-being in favor of society in general and women in particular (Harding, 1987). The problem then lies in how we overcome a social structure that constantly tests the limits of the ethics and vocation ocf these women when they are pushed to extreme situations where their intervention and involvement are the core of the institution's performance. The work being done in San Rafael could be—in addition to inspiring—another example of public practices that mobilizes and promotes solidarity under neoliberal conditions, creating the feeling of a lifestyle not so atomized or isolated (Rojas Lasch, 2018) but doesn't really align with the search for horizontality, collaboration, and unbiased distribution of tasks that feminisms invite us to.

Notes

1. La Pintana has a multidimensional poverty rate of 32.74%, while the average of the Metropolitan Region and the country is 20% according to the CASEN 2017 survey.
2. The SSMSO corresponds to the South East area of Santiago, comprising the communes of Puente Alto, La Florida, San Ramón, La Granja, La Pintana, San José de Maipo and Pirque; two of these communes, Puente Alto and La Florida, have the most populous populations in the country.

References

Anzorena, C. (2017). Feminist readings for the theoretical and empirical analysis of public policies. In A. De Oto & M. Alvarado (Eds.), *Methodologies in context: Interventions from a feminist, postcolonial, Latin American perspective* (pp. 63–82). CLACSO.

Augé, M. (2000). *Non-places, Spaces of anonymity: An anthropology of supermodernity*. M. Mizraji (Trans.). (5th ed.). Gedisa.

Bentz, I., & Franzato, C. (2017). The relationship between strategic design and metadesign as defined by the levels of knowledge of design. *Strategic Design Research Journal, 10*(2), 134–143. https://doi.org/10.4013/sdrj.2017.102.06

Carrasco, C., & Recio, A. (2014). From measured time to lived times. *Critical Economy Journal, 17*, 82–97.

Castañeda, M., et al. (2012). Feminist ethnography in feminist research. In N. Blasquez (Ed.), *Epistemology, methodology and social representations* (pp. 217–238). Debate and Reflection Collection. National Autonomous University of Mexico.
Delgado, M. (2011). *Public space as ideology*. Catarata.
Delgado, M. (2007). The woman from the street. In *Moving societies* (pp. 224–261). Anagrama.
Dunne, A., & Raby, F. (2013). *Speculative everything*. The MIT Press.
Harding, S. (1987) Is there a feminist method? G. Bernal (Trans.). Indiana Press University.
Heidaripour, M., & Forlano, L. (2018). Formgiving to feminist futures as design activism. In *Design research society 2018 catalyst* (pp. 518–528). DRS.
Jirón, P. (2007). Gender implications in the experiences of urban daily mobility in Santiago de Chile. *Venezuelan Journal of Women's Studies, 12*(29), 173–198. Caracas.
Lahera, E. (2005, October 18–21). From saying to doing: How to implement policies? [lecture]. X CLAD Congress. Santiago, Chile.
Lamas, M. (2007). Gender, development and feminism in Latin America. Iberoamerican Thought n°0, 133–152.
Lefevre, H. (1991) The production of space. In D. Nicholson-Smith (Trans.). Ed Blackwell.
Library of the National Congress of Chile. (2021). La Pintana: 2021 Community Report. https://www.bcn.cl/siit/reportescomunales/reporpdf.html?anno=2021&idcom=13112
Lindon, A. (2006). Territoriality and gender: An approach from spatial subjectivity. In P. Ramírez & M. Aguilar (Eds.), *Thinking and inhabiting the city. Affectivity, memory and meaning in contemporary urban space* (pp. 13–31). Anthropos, Autonomous Metropolitan University.
Massey, D. (2012). The philosophy and politics of spatiality: Some considerations. In *A global sense of place* (pp. 156–181). Icaria editorial.
Metropolitan Health Service South East. (n.d.). Communes of the network. https://redsalud.ssmso.cl/communes-of-the-network/
Ministry of Health. (2016). National Health Plan for Health Objectives by 2020.
Ministry of Health. (2017). Technical guidelines for design of complex hospital preliminary projects.
Ministry of Health. (2019). Design guide for medium complexity hospital establishments.
Mollenhauer, K., Moraga, E., & Bernasconi, R. (2020). Instrumental model for complex projects. *Poliedro Academic Value Proposal. Urban Territorial Logbook, 30*(II), 127–140.
Muñoz Rojas, C. (2021). From policies to management: Implementation gaps and state capacities for gender equality. In *Treaty of genders, rights and justice. Public policies and multidiscipline* (L. Pautassi coord.). Rubinzal Culzoni Publishers.
Musset, A. (2015). From waiting places to waiting territories. A new dimension of social geography? *Documents d'Anàlisi Geogràfica, 61*(2), 305–324. https://doi.org/10.5565/rev/dag.315
Öhrling, T., Normark, J., & Wikberg Nilsson, Å. (2018). Norm creativity in student design projects—One approach to creating sustainable societies. International conference of engineering and product design education. [lecture] London: Dyson School of Design Engineering, Imperial College of London.
Rabotnikof, N. (1998) Public/Private, Feminist Debate 18.
Ricardo, E. (2016). State capacity, capacity gaps and institutional strengthening. *Political Analysis, 87*, 60–76.
Roggema, R. (2016). Research by design: Proposition for a methodological approach. *Urban Science, 1*(2). 10.3390.
Rojas Lasch, C. (2018) Affection and care: Pillar of neoliberal social policy. *Polis, Latin American Journal, 49*, 127–149.
Sánchez, P., & Hermansen, P. (2021). Recognize (ourselves) and transform (ourselves): Systematization of the co-design experience of Murales Minga workshops for the Dr. Exequiel González Cortés Hospital.
Sánchez, P., & Neira, A. (2021, March 8). Our experience of Feminist Design Encounters. Patricia Sánchez Medium. https://patriciaisidora.medium.com/our-experience-of-feminist-design-meetings-bba7bf940b9b

Soto, P. (2012). Women's fear of violence in Mexico City. A matter of spatial justice. *Invi Magazine* Nº 75, *27*, 145–169.

Soto, P. (2009) The public and the private in the city. *Time House Magazine*, UNAM.

Zambrini, L., & Flesler, G. (2017) Gender perspective and design: Deconstructing the neutrality of typography and clothing. *Inclusion Magazine, 4*. Special Issue, 11–22.

Chapter 16
Sensing the City. Co-designing Urban Space with Deaf Children and Their Caregivers

Carolina Carrasco Pizarro

Abstract This study addresses a relevant, but often overlooked issue: the inclusion of deaf children and their caregivers in public space. Although efforts have been made to address the needs of people with motor and visual disabilities in urban environments, the difficulties experienced by this group of people in public spaces have received little attention. In early childhood, language development and socialization are critical aspects. For children with hearing difficulties, reducing the gaps in their development is imperative. This study is part of the applied research project "Sensing the City", which is being developed in collaboration with the school for deaf children Ana Bruzone 197, Montevideo Innovation Laboratory, ARUP and the Bernard Van Leer Foundation. Its main objective is to improve the accessibility and inclusion of these children and their caregivers in public spaces. To achieve this, experimental prototypes have been created in co-design with the affected group and have been arranged in the public space. This has allowed to explore in a situated way the difficulties and opportunities they face when moving and communicating in the street and the neighborhood. The project is not limited to identifying barriers but also seeks to open new perspectives for the development of a city that takes into account all senses. The group of deaf children has been actively involved in co-design workshops, which has facilitated the contribution of their ideas and experiences. The data has been collected through analysis of previous studies, participant observation in workshops and evaluation of the prototypes through non-intrusive observation and semi-structured interviews. The results of this study reveal the need to define urban artifacts as prototypes that explore the improvement of visual communication and the creation of safe spaces for play that promote communication through sign language. In addition, the incorporation of other senses, such as smell and touch, to stimulate exploration and play, which expands the opportunities for active and social learning of this group of people, has been examined. In summary, the "Sensing the City" project represents an innovative initiative that effectively addresses the accessibility

C. Carrasco Pizarro (✉)
Urban Studies and Architecture, NGO Espacio Lúdico, Pontificia Universidad Católica de Chile, Santiago, Chile
e-mail: ccarrascop@gmail.com

M. Vicuña and E. Giorgi (eds.), *Intersections*, Springer Geography,
https://doi.org/10.1007/978-3-031-76402-8_16

and inclusion of deaf children and their caregivers in public spaces. This collaborative and co-design-based approach contributes to envisioning how communication and movement gaps in public space could be reduced, thus improving the quality of life of this group and their caregivers.

Keywords Co-design · Urban experimentation · Deaf space · Inclusive design · Public space · Childhood

16.1 The Challenge of Urban Design to Facilitate Communication and Social Inclusion of Deaf Children in Their Urban Environment

Urban design has tended to overlook certain sensory difficulties, in particular, deafness, and has not systematically addressed these issues in the configuration of public spaces (Harold, 2013). This limited approach has resulted in a significant restriction in the ability of deaf people, in general, to use and move around in urban environments. This limitation becomes even more critical in early childhood, a period in which language development should be actively stimulated, along with the promotion of social learning. These deficiencies have accentuated the gaps in the comprehensive development of deaf people. The present study is located in the city of Montevideo and is linked to the deaf school number 197 "Ana Bruzone". This educational institution plays a fundamental role in the lives of deaf children, offering care and educational programs that include specific methodologies for communication and education for the benefit of the deaf community.

Research in the field of deafness is inherently interdisciplinary. In the field of architecture and planning, studies on deafness aim to improve the control that individuals have over the spaces in which they reside (Dirksen et al., 2010, p. 219).

The project proposes to address the following question: What lessons can we draw from experimentation on communication in public spaces with deaf children and their relationship in stimulation, communication, and social relations? Faced with this challenge, the NGO Espacio Lúdico has conceived the applied research project "Sensing the City" in collaboration with Arup, Urban 95 of the Bernard Van Leer Foundation, and the Intendencia of Montevideo, within the framework of the evaluation of the "Proximity of Care" methodology.

In this way, through the creation of experimental prototypes designed in collaboration with the affected group and deployed in public space, a situated exploration of the difficulties and opportunities that these individuals face when moving and communicating in the urban environment is carried out.

From this, we have been able to learn that the socio-spatial experience of deaf childhood and their caregivers requires specific care in terms of space design, road signage, and sound, while it is important to allow models of urban artifacts that promote inclusive communication between deaf and hearing children, as well as between adults and infants. Similarly, mobile furniture is shown as an option to facilitate the visibility of gestures and forms that are basic in sign language and body language.

16.2 Care and Singularities

16.2.1 New Challenges in City Design

For a long time, design and urban planning have overlooked the singularities of a broad group of individuals who experience urban space in diverse ways. The "Critical Care Theory" has played a relevant role in highlighting this problem and promoting its integration into the fields of architecture, design, and urbanism (Fitz et al., 2019).

Feminist thinkers, like Donna Haraway (1988), urge us to consider the ethics and politics of care, taking into account the differences and particularities of people (Carli & Frediani, 2021; Ferreri, 2021). The ethics of care in urbanism also recognizes the inequity in decision-making in urban planning. According to Tronto and Fisher (1990), the act of caring requires both the ability of caregivers, as capacity factors (time, material resources, skills, and knowledge), as well as the specific conditions to carry out this work. In this context, the city must allow caregivers to perform their tasks in the best possible way. However, in the case of deaf people, the design of the experience in the city has not taken these needs into account.

The challenge then calls for reducing the gap related to the factors that limit the capacity of caregivers, both in terms of material resources and in the ability to support and educate those who face the care of deaf children.

16.2.2 Public Space Design in "Deaf Space" Studies

The concept of "DeafSpace" (Byrd, 2017; Gulliver, 2006; Harold, 2013) has been coined as a research approach in architecture and urbanism, with the purpose of promoting a reevaluation of conventional approaches in space design, emphasizing the need to consider other senses when hearing is not present.

In this context, the design of space and stimuli related to vision, touch, and smell play a fundamental role. However, to date, the city has not addressed these considerations adequately, which could hinder the full inclusion and accessibility of people in public space, particularly deaf children. This lack of attention is especially disturbing in the case of all children, as childhood represents a critical stage in which they are expected to fully develop their social and cognitive potential. Aspects such as language development, social interaction, and motor development are fundamental for promoting communication and language (Iverson, 2010).

On the other hand, it has been observed that motor development in early childhood has a considerable influence on the language development of children (Iverson, 2010). This area of study continues to evolve, based on the hypothesis of the importance of considering the dynamic relationship between motor development and language in child development. This underscores the need for a more detailed exploration of the design of inclusive urban spaces that promote the comprehensive development of children, addressing their multiple dimensions.

16.2.3 The Emerging Relevance of Urban Experimentation

Urban experimentation has become a valuable tool for addressing the growing urban challenges of the twenty-first century. This approach, characterized by its iterative and exploratory focus, has emerged as an innovative alternative in city management (Eneqvist et al., 2021; Evans et al., 2018; Montero, 2017; Wendler, 2015). It allows urban planners and designers to creatively explore possible urban futures (Carrasco et al., 2021; Carrasco Pizarro et al., 2022; Evans, 2011; Evans et al., 2016; Jaworski, 2022; Sengers et al., 2016).

Urban experimentation has recently gained relevance and promises more democratic processes while socializing the planning process and promoting the inclusion of diverse perspectives. However, there is still much to learn in this field. To address urban challenges, it is essential to involve key actors, including groups that are often overlooked, such as children, girls, and deaf caregivers. This approach not only enriches decision-making but also addresses the issue of legitimacy in urban governance, as the problem is understood from and alongside its fundamental actors.

The implementation of urban experimentation faces challenges in the search for situated urbanism, especially in the context of Latin America. This approach seeks to address the particularities of each territory and requires a creative effort in the urban design process. However, hegemonic conventions in decision-making, the lack of methodologies to promote situated knowledge, and the absence of recognition by decision-makers can underestimate the competencies and real possibilities of this type of urbanism in the construction of the living environment.

The inherent duality of urban planning is essential to understand the challenges of urban planning. On the one hand, it is perceived as a process led by experts with technical knowledge and specific tools; on the other hand, it is conceived as a social struggle and a mobilization in search of justice and opportunities. This reveals that urban planning encompasses both technical expertise and participation in a process of public debate and transformation of the public sphere and its symbols.

Neil Brenner, in his work "New Urban Spaces: Urban Theory and the Scale of Question" (Brenner, 2019) underscores the importance of a theoretically reflective process-based approach in urban studies. Brenner criticizes the prevailing tendency to consider cities and urban spaces as obvious and concrete places in research, which he considers a naive view. Instead, he argues that the "urban is not something given or transparent but requires a process of theoretical abstraction".

This theoretical abstraction implies that understanding what constitutes an urban space and how it functions goes beyond observing its physical characteristics. It is necessary to consider how people perceive and experience that space, how they relate to each other in it, and how they construct mental and conceptual representations of the urban space. These theories and abstractions are not mere backdrops in urban research; they are the very basis of how urban planners approach their studies. They determine the objectives of the research, how data is collected and analyzed, and how solutions or interventions are proposed in urban environments.

Critical reflexivity in urban studies is essential for questioning these underlying theories, connecting them with the historical and geographical context in which they are applied, and effectively understanding the constantly changing urban challenges, which could contribute to better governance and planning of cities.

Urban experimentation stands to facilitate this dialogue and more inclusive decision-making. It allows citizens to test new ideas and approaches in their urban environments and actively contribute to the transformation of the city according to their shared visions.

16.2.4 Study Objectives

To address the challenge for the design of urban space in line with promoting the use of senses beyond the auditory, as well as improving the perception and experience of living for deaf people, especially children and their caregivers, the following objectives are defined to probe:

1. Understand the Spatial Experience of Deaf People: The first objective is to acquire a deep understanding of the spatial experience of deaf people in the urban environment. This involves identifying the barriers, challenges, and opportunities they face in terms of communication, stimulation, and safety in public spaces.
2. Develop Innovative Spatial Strategies: The second objective focuses on creating innovative spatial strategies that enhance the experience of deaf people in the urban environment. This includes proposing alternative communication strategies that do not depend on oral/auditory language and introducing new ways of perceiving and experiencing space.
3. Evaluate the Effectiveness of the Proposed Artifacts: The third objective is to evaluate the effectiveness of the artifacts developed in the co-design stage. This involves testing and carefully observing how these artifacts contribute to improving communication, stimulation, and safety for deaf people in public spaces. In addition, feedback from deaf people is sought to perfect the artifacts.
4. Address comprehensively the challenge of improving the experience of deaf people in urban space, promoting inclusion, participation, and quality of life in the city. The results of this methodology will serve as a basis for future research and projects focused on creating more inclusive and accessible urban environments.

16.2.5 Methodology

Based on these objectives, an applied research methodology is developed based on experimental methods, focusing on the evaluation of strategies to improve the spatial experience of deaf people in the urban environment, with an emphasis on communication, stimulation, and safety in public space. The methodology was developed in three consecutive stages to achieve the specific objectives of the study:

Stage 1: Understanding General Aspects of the Spatial Experience of Deaf People. In this initial phase, interviews and surveys were conducted with deaf people to gain a deep understanding of their experiences in public spaces. In addition, a review of previous studies related to the space of deaf people was carried out to enrich our understanding of this context. The main objective of this stage was to identify the main barriers and challenges that deaf people face in urban spaces.

Stage 2: Co-design of Rest and Stimulation Prototypes (Figs. 16.1 and 16.2). In the second stage, an active collaboration was established with deaf children. The purpose was to jointly design urban prototypes that seek to improve communication, stimulation, and safety conditions in public space. During this process, the participation of children was encouraged, recognizing their unique perspective and direct experience. Creativity and innovation were promoted in the search for practical and effective solutions by providing the co-design workshop with materials to test ideas and discuss them within the group.

Stage 3: Evaluation of the Proposed Prototypes (Figs. 16.4, 16.5, 16.6, 16.7 and 16.8). The third stage consisted of the evaluation of the prototypes developed in the previous stage. An activity was organized to intervene in public space and test these artifacts. For this, an ethnographic study process was developed based on observation and perception surveys to determine the effect of these artifacts on improving the experience of deaf people in public spaces. This evaluation process was carried out in collaboration with deaf people, capturing their feelings and conjectures, which will form an essential part of the feedback for the final implementation evaluation of the project.

16.3 Results and Projections

16.3.1 The Expectation and Evaluation of Caregivers

To understand the needs and expectations of caregivers of deaf children in relation to activities outside the home, a focused online survey was conducted. The results of this survey were oriented towards the identification of various challenges that caregivers face in public spaces, and they expressed the following expectations and criticisms.

Fig. 16.1 Co-design workshop group 2 with deaf children from school 197 Ana Bruzone, Montevideo. Prototype artifact of communication in public space. *Source* NGO Espacio Lúdico

First, caregivers emphasized the importance of promoting eye contact between deaf children and their peers, as well as in meeting spaces, highlighting the relevance of environments that facilitate visual communication.

A second aspect highlighted in the survey was the need to reduce levels of environmental noise and acoustic pollution in public spaces. This is especially important for children with cochlear implants, as excessive noise can be overwhelming and hinder their participation in daily activities.

Also, the incorporation of other forms of communication in public spaces was proposed, preferably of a visual nature and based on graphic mechanisms, to facilitate the interaction of deaf people, pointing to the importance of adapting the urban environment to meet the communication needs of this group.

The importance of providing appropriate alerts about dangers and care in public spaces to deaf people was also emphasized, highlighting the need for accessible signage that ensures the safety of everyone.

They highlight the need to raise society's awareness of the importance of considering the dimension of deafness in civic education and in the design of urban space. This points to the promotion of greater inclusion and awareness of the experiences of deaf people in urban life.

Fig. 16.2 Co-design workshop group 2 with deaf children from school 197 Ana Bruzone, Montevideo. Prototype artifact of communication in public space. *Source* NGO Espacio Lúdico

The creation of spaces enriched with flowering plants is proposed, considering the sense of smell and sight to offer a multisensory experience in public spaces, which would contribute to greater stimulation and participation.

Finally, the incorporation of educational technology, such as QR codes and informative screens, was suggested, which educates and raises awareness about deafness, providing accessible information in urban spaces. These proposals highlight the importance of using technology as an educational and awareness-raising tool.

In this way, we can see that the expectations and criticisms of caregivers shed light on various areas of improvement in public spaces to promote the inclusion and accessibility of deaf people. These results provide a valuable basis as key inputs for the development of the co-design workshop developed later.

16.3.2 Ideas Co-designed by the Group of Deaf Children

The co-design workshop represented a significant step in the project, as its central objective was to capture the perspectives, experiences and expectations of a group

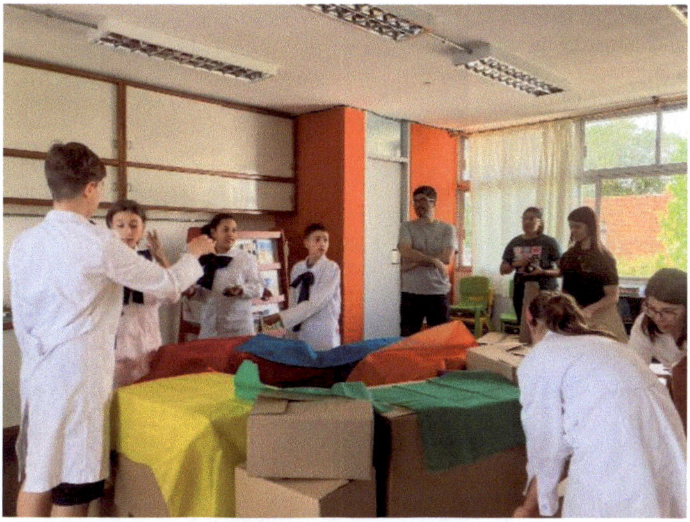

Fig. 16.3 Co-design workshop group 1 with deaf children from school 197 Ana Bruzone, Montevideo. *Source* NGO Espacio Lúdico

of deaf children, being the essential focus group in the research. This workshop was organized into two groups, each with a specific purpose:

In the first group of children, they worked on the development of a prototype type of furniture for public space (Fig. 16.3). At the beginning of the workshop, the purpose of designing for rest, promoting play and group interaction was proposed. The results revealed the importance of physically linking the participants, through the creation of a central space that allowed interaction and play, promoting that the furniture could be flexible in its spatial configuration, facilitating movements that had a response in a frontal way among people. In addition, various heights are integrated to better visually confront the playgroup.

On the other hand, the second group of children focused on the design of an artifact that promoted communication through play (Figs. 16.1 and 16.2), whose main challenge was to promote multilingual interaction and contribute to the learning of sign language. In this context, the participants emphasized the importance of visual communication and proposed the inclusion of transparencies and spaces that allowed the use of the panel on both sides, which would facilitate visual interaction among deaf children. In addition, the need to include elements related to touch and smell was highlighted, suggesting the incorporation of smell boxes as part of the multisensory experience. The exploration of vibrations as part of the construction of the artifact was also identified as a valuable feature.

The co-design workshop stood out for its ability to actively involve deaf children in the design process, promoting a space to express opinions, needs, and desires in a creative way. The results obtained in this workshop reveal several forms devised from the children's experience with their urban space, which would promote inclusion and

Fig. 16.4 Rest artifact tested with the deaf community in the Portugal Square, Montevideo. *Source* NGO Espacio Lúdico

multisensory stimulation in the public space, and which are taken directly into the design of the prototype to be tested in the public space.

16.3.3 Implementation of Prototypes and Experimentation in Public Space

The implementation of the final prototypes, conceived from the ideas presented by the deaf children in the Co-Design workshop, attracted the active participation of various actors, including residents of the sector, organized groups from the deaf community, families linked to School 197 Ana Bruzzone, as well as representatives of the local government. This collaboration was oriented towards the promotion and education of all individuals involved in the planning of public spaces, with a special focus on the public space associated with the deaf school.

The urban intervention covered a series of objectives, highlighting education, awareness, and the promotion of the multisensory approach in the city's culture. However, in this article, we will focus on the evaluation of the prototypes that emerged

Fig. 16.5 Rest artifact tested with the deaf community in Portugal Square, Montevideo. *Source* NGO Espacio Lúdico

Fig. 16.6 Rest artifact tested with the deaf community in Portugal Square, Montevideo. *Source* NGO Espacio Lúdico

Fig. 16.7 Stimulation and play artifact tested in the Plaza Portugal, Montevideo. *Source* NGO Espacio Lúdico

from this co-design process. Attention will be directed specifically towards the analysis of the functionality and effectiveness of these urban artifacts, examining their ability to improve the experience of deaf children in public space.

Regarding the prototypes submitted for evaluation, it was observed that the panel (Figs. 16.7 and 16.8) stood out as an extremely attractive element both for deaf children and for those who did not have this condition. The versatility of the panel in offering a wide range of sensory experiences made it a component of great interest. The participants explored and enjoyed the various sensory forms provided by the panel, which generated a high degree of fascination among them.

It is essential to highlight that the panel was used both by deaf children and by those who can hear, which emphasizes its universal appeal and its ability to captivate

Fig. 16.8 Stimulation and play artifact tested in the Plaza Portugal, Montevideo. *Source* NGO Espacio Lúdico

a diverse audience. Its double-sided design allowed the children to connect with each other through play, driving interaction and communication effectively.

The graphic component that taught sign language was particularly impactful. It not only motivated adults and children's listeners to learn some concepts of this language, but it also became a form of play and communication between caregivers and deaf children. This aspect played a significant role in raising awareness about the importance of sign language and its integration into public space.

The inclusion of colors and smells on the panel was received with enthusiasm, and children were observed actively participating in the exploration and interaction with these sensory elements. This positive response emphasizes the relevance of considering multiple senses in the design of urban artifacts, with the purpose of promoting inclusion and stimulating multisensory experiences in public space.

On the other hand, non-participant observation of the bench reveals various facets of its use and its influence on social interaction.

Firstly, the bench is not limited to its conventional function as a seat but becomes a playful space for children, who use it as a play area by crossing between its structures and climbing it (Fig. 16.6). This aspect highlights the versatility of urban furniture as an active element in child development.

When the bench is under adult supervision, specifically sports monitors, it is observed that it facilitates horizontal and visual communication between the interacting children (NN) and adults (NA). This dynamic reflects the importance of public space design in promoting intergenerational interactions and in creating an inclusive environment (Fig. 16.5).

In addition to its recreational function, the bench presents itself as a space that stimulates multiple sensory awakenings in childhood. Its spatial configuration not only allows direct communication between individuals but also delineates an area conducive to the articulation of group dynamics in a safe and controlled environment (Fig. 16.6).

16.4 Conclusion

In summary, the implementation of the prototypes designed in collaboration with deaf children has yielded novel results in terms of improving their experience in public spaces. These artifacts, conceived to promote visual communication and multisensory stimulation, have proven their effectiveness throughout various stages of the project, from the initial interviews to the experimentation in Plaza Portugal.

However, it is important to acknowledge that the process encountered some challenges. The original planning, which foresaw the occupation of public space for several days, was hindered by problems with public administration and logistical limitations of the project team, which led to the display and evaluation of the prototypes being concentrated on a single weekend day in Plaza Portugal.

A notable aspect that emerged from this experience is the interest aroused in other schools that cater to children with various disabilities, such as autism. This suggests the potential to extend and adapt this co-design methodology to other people facing access and communication challenges in public spaces.

In addition, discussions held with participants revealed the expectation that projects of this nature can contribute to the creation of more inclusive spaces, where communication is intuitive and based on all the senses of the body. This multidimensional approach would not only benefit children with disabilities but would also enrich the experience of all children in a more accessible and diverse urban environment.

The process continues its development, and it is hoped that the transformation of public space will not only benefit deaf children but will also open the door to other people with characteristics that have been traditionally overlooked in urban design.

Acknowledgements Intendencia of Montevideo and the Participation Montevideo Lab department. To the Deaf CINDE community of Montevideo, deaf school 197 Ana Bruzzone. To the ARUP team, Sara Candiracci and Dasha Moschonas. To the Bernard Van Leer Foundation. Especially to the team of professionals from the NGO Espacio Lúdico Montevideo, Noelia Botana and Sofía Alonso and the CEO José Gómez. All of them made this project possible.

References

Brenner, N. (2019). New urban spaces: Urban theory and the scale question. *Oxford University Press*. https://doi.org/10.1093/oso/9780190627188.001.0001

Byrd, T. (2017). Deaf Space. In J. Boys (Ed.), *Disability, space, architecture. The reader* (pp. 241–246). Routledge.

Carli, B., & Frediani, A. (2021). Situated perspectives on the city: A reflection on scaling participation through design. *Environment and Urbanization, 33*, 376–395. https://doi.org/10.1177/09562478211028066

Carrasco, C., Gómez, J., & Egger, T. (2021). How to engage citizens in low-digitalized neighbourhoods during social distancing? CUBOX: A creative analog-digital solution for ludic interaction How Place Matters? Leadership, Governance & Public Administration, England/online.

Carrasco Pizarro, C., Rivera Marchevsky, M., Acosta Scandela, L., David, B., & Lazo de la Cerda, M. (2022). Potential implications in the methodological application of design thinking in urban design processes. Analysis from the designer's experience. *Journal of Urbanism*, 75–94.

Dirksen, H., Bauman, L., & Murray, J. J. (2010). Deaf studies in the 21st century: "Deaf-gain" and the future of human diversity. In M. Marschark & P. Spencer (Eds.), *The Oxford handbook of deaf studies, language, and education* (pp. 246–260). Oxford University Press.

Eneqvist, E., Algehed, J., Jensen, C., & Karvonen, A. (2021). Legitimacy in municipal experimental governance: Questioning the public good in urban innovation practices. *European Planning Studies, 30*, 1–19. https://doi.org/10.1080/09654313.2021.2015749

Evans, J. (2011). Resilience, ecology and adaptation in the experimental city. *Transactions of the Institute of British Geographers, 36*, 223–237. https://doi.org/10.1111/j.1475-5661.2010.00420.x

Evans, J., Bulkeley, H., Voytenko, Y., McCormick, K., & Curtis, S. K. (2018). Circulating experiments: Urban living labs and the politics of sustainability.

Evans, J., Karvonen, A., & Raven, R. (2016). *The experimental city*. Routledge.

Ferreri, M. (2021). The permanence of temporary urbanism: Normalising precarity in austerity London. Amsterdam University Press. https://doi.org/10.1017/9789048535828

Fitz, A., Krasny, E., & Wien, A. (2019). Critical care architecture and urbanism for a broken planet. The MIT Press. https://doi.org/10.7551/mitpress/12273.001.0001

Gulliver, M. (2006). Introduction to deaf space.

Haraway, D. (1988). Situated knowledges: The science question in feminism and the privilege of partial perspective. *Feminist Studies, 14*(3), 575–599. https://doi.org/10.2307/3178066

Harold, G. (2013). Reconsidering sound and the city: Asserting the right to the deaf-friendly city. *Environment and Planning d: Society and Space, 31*(5), 846–862. https://doi.org/10.1068/d3310

Iverson, J. M. (2010). Developing language in a developing body: The relationship between motor development and language development. *Journal of Child Language, 37*(2), 229–261. https://doi.org/10.1017/S0305000909990432

Jaworski, P. (2022). Experimental urbanism. https://experimentalurbanism.com/index.php/2020/12/

Montero, S. (2017). Worlding Bogotá's ciclovía: From urban experiment to international "best practice." *Latin American Perspectives, 44*(2), 111–131.

Sengers, F., Berkhout, F., Wieczorek, A. J., & Raven, R. (2016). Experimenting in the city. In J. Evans, A. Karvonen, & R. Raven (Eds.), *The experimental city* (pp. 36–52). Routledge.
Tronto, J. C., & Fisher, B. (1990). Toward a feminist theory of caring. In E. Abel, & M. Nelson (Eds.), *Circles of care* (pp. 36–54). SUNY Press.
Wendler, J. (2015). Experimental urbanism: Grassroots alternatives as spaces of learning and innovation in the city. University of Manchester.

Chapter 17
Openly Caring: Integrating Sustainability and Inclusion in Urban Education

Karen Hinojosa Hinojosa and Nélida Astrid Escobedo Ruíz

Abstract The challenges that cities currently face have opened reflections on the competencies needed by those who manage them, highlighting the importance of Education for Sustainable Development. At the same time, the lack of diversity in pedagogy in urban planning has been questioned, due to its lack of representation of both the student body and the needs of citizens. Despite efforts and advances towards a decolonial, intersectional education in relation to race and disability, the incorporation of the feminist care framework in urban planning pedagogical agendas is still limited or relegated to elective courses. Using a case study approach, we analyze the results of the implementation of a mandatory urban design course for a postgraduate degree in architecture and urban design at a private university in Mexico. Sustainability, inclusion, and early childhood needs are integrated as theoretical content and are focused through activities, reflection, and feedback to students. The spatial and written integration of sustainability and inclusion in their final projects, as well as in student perception surveys, was coded and analyzed. The findings suggest that most students openly engaged with sustainability and inclusion, with varying levels of integration in the projects and their professional practice. Our findings emphasize how, despite decades of academic reflection on the importance of education for sustainable development and inclusion, spatial expressions of gender issues as an integral part of sustainability and inclusion can still be perceived as strange by urban design and planning students.

Keywords Inclusive cities · Education for sustainable development · Caring cities · Early childhood · Educational innovation

K. Hinojosa Hinojosa (✉) · N. A. Escobedo Ruíz
School of Architecture, Art and Design, Tecnologico de Monterrey, Monterrey, México
e-mail: khinojosa@tec.mx

N. A. Escobedo Ruíz
e-mail: nescobedo@tec.mx

N. A. Escobedo Ruíz
Center for the Future of Cities, Tecnologico de Monterrey, Monterrey, México

M. Vicuña and E. Giorgi (eds.), *Intersections*, Springer Geography,
https://doi.org/10.1007/978-3-031-76402-8_17

17.1 Introduction

Currently, cities around the world, and specifically in Latin America, face significant challenges linked to social inequity, which are exacerbated by the effects of climate change. These interrelated challenges require immediate and strategic attention from highly trained experts in the field of sustainability. It is essential that these professional profiles have a precise understanding of how these impacts can manifest differently in different segments of the population, that is, they have a gender and inclusion perspective. This understanding is essential in the context of decision-making at the governmental, organizational, and community levels, with the purpose of developing visions, objectives, and strategies suitable to address this complex problem.

The modern concept of sustainability has its roots in environmental concerns but has evolved to encompass social and economic aspects. Initially, the focus was on the preservation and protection of the environment and biodiversity to ensure that future generations could enjoy the same natural resources as present generations (Brundtland, 1987). Sustainability as a goal is an idealization of the relationship between nature and society within certain reference systems (Salas-Zapata & Ortiz-Muñoz, 2019). A fundamental criticism of this concept of sustainability focuses on the lack of clarity about what is really sought to be sustained. If present and future generations sustain the same values, systems, and basic strategies that constitute the very roots of the problems that sustainability seeks to address, these will continue or increase (Sen, 2013; Trainer, 1990).

Despite the operational difficulties of this polysemic concept, sustainability today plays a fundamental role in international agendas that guide the actions of governments, the private sector, and society in general. The vision of a sustainable world is embodied in 17 goals that range from promoting peace to creating equal opportunities for all people while preserving the health of our planet.

In the context of an increasingly urban world seeking sustainability, gender equality emerges as a goal in itself but also as a cross-cutting element, as it not only represents a fundamental human right, but also reflects a critical reality: women experience the world in a differentiated way and face greater vulnerability to the impacts of global challenges such as climate change, migrations, and poverty. Therefore, any consideration of alternatives to the development model must be rooted in the active promotion of gender equality (Navarro-Martínez, 2017).

17.1.1 Addressing Sustainability and Inclusion in an Integrated Way in the Caring City

To face the great challenges that cities face, professionalizing urban managers in sustainability and inclusion is key (Beebeejaun, 2022; Mulder, 2017; Sami et al., 2022; Sweet & Etienne, 2011; Terashima & Clark, 2021; Wesely & Allen, 2019;

Wiek et al., 2015), although often these topics are addressed in a dissociated way. The intersectional view that arises from feminism and critical race theory (Crenshaw, 2017) provides an integrative framework for these concepts in the caring city (Campusano & Schiavo, 2021).

As cities are the places with the highest population concentration, the study of urban dynamics from a gender perspective, allows us to understand how inequalities are produced and reproduced in space (Valdivia, 2018). According to feminist theory, the current urban model is based on the segregation of spaces, which are assigned based on the sexual division of labor (de Madariaga, 2004). Under this system, in the public sphere, productive activities, paid work are prioritized, and in the private sphere, reproductive activities or unpaid work, are mainly carried out by women (Muxí Martínez et al., 2011).

Consequently, there are large residential areas, far from basic facilities and workplaces, devoid of proximity services and with poor quality in streets and parks for socializing, moving around and spreading safely. In addition, this urban model, consistent with the capitalist system, is expansive and predatory of natural systems (Borja & Muxí, 2003; Marcuse, 2009; Muxí-Martínez, 2020).

The caring city provides a conceptual framework for developing policies and design strategies that allow the inclusion of marginalized groups and sustainable life on the planet (Valdivia, 2018). The caring city claims the right to differentiated functions of the life cycle of people (Rico & Segovia, 2017) in which there is illness, death, vulnerability, especially in childhood and maturity (Busquets, 2019). Despite the relevance of addressing these concepts in an integrated way, the incorporation of the feminist framework in the pedagogical agendas of urban planning is still limited or relegated to elective courses (Pérez Cano et al., 2019).

17.1.2 Childhood and Early Childhood in the Urban Context

According to United Nations statistics, the greatest population growth in the world happens in cities. In 2021, 50% of the global population lived in cities, and in the next 50 years, this trend will continue to increase. Although cities provide benefits for people, such as greater job opportunities, education, and access to health, the supply of these services and possibilities is still not distributed or accessible to a significant part of the urban population. Added to this, the effects of climate change pose even greater challenges, especially for vulnerable groups such as children, migrants, people in poverty, women, people with disabilities, and older adults (UN-Habitat, 2022).

Within these vulnerable groups, children represent a significant proportion of the total number of people living on the planet. More than 30% of the global population by 2023 will be under 18 years old and 650 million people will be under 5 years old (UNICEF, 2023a, 2023b). Of this number, approximately half live in precarious settlements, in conditions of poverty, and with insufficient coverage of basic services to grow and live fully (UNICEF, 2017). Despite the fact that since 1989 the

Convention on the Rights of the Child establishes the foundations of survival, development, protection, and participation rights, the challenges in cities pose difficulties in ensuring the fulfillment of these rights.

From the point of view of the physical environment, there are multiple vulnerabilities and challenges for children in cities. Inevitably, people are born into a material space and its quality and proximity determine access to medical services, educational spaces without pollution, play areas, and social connections. Despite this relevance, these physical and material aspects in which children grow are rarely incorporated into theories and research on development (Evans, 2006).

More precisely, the vulnerabilities of children in the environments are related to the lack of access to basic services, such as drinking water, electricity, and sanitary drainage. They are also related to the quality and accessibility of local facilities such as schools, nurseries, markets, cultural spaces, and clinics. Housing and neighborhoods, likewise, are key to a full life, because it is here where initial care takes place and the first point of contact for children with their families and caregivers. For example, houses with poor material quality, construction, or without property titles, represent a constant physical risk for their inhabitants, in addition to the psychological stress that this may mean for families. Quality public spaces are also fundamental, mainly because streets allow movement through the city, and also because parks and squares are places to play, exercise, learn, build skills, and have contact with nature and other people (Woodhead & Brooker, 2012).

Within the population group of children, early childhood is a stage that is generally invisible (Ibid). This stage begins in gestation, up to around 5 years of age (UNICEF, 2023a, 2023b). During this period, the human body is in a crucial moment of development that will establish the foundations for the rest of the life cycle. Neuroscience research indicates that in the first 36 months of life, the human brain builds its foundations, reaching a peak of more than 1 million neural connections per minute, an activity that does not repeat for the rest of life (Harvard Center on the Developing Child, 2017). Due to this importance, investing and ensuring a good quality of life in these early years of life is socially more efficient.

Children from 0 to 5 years old and caregivers have specific spatial needs that must be carefully considered in the design of urban spaces. First, it is essential to provide safe and risk-free environments for people at this early age, as they are in a stage of development where exploration and play are fundamental for their learning and healthy growth. This implies the creation of safe play areas, elimination of dangerous obstacles, and ensuring that not only the play areas but the urban spaces where their daily life takes place are adequately supervised with enough eyes on the street (Jacobs, 2020).

In addition, it is crucial to take into account the mobility needs of young girls and boys. Urban spaces must be accessible and friendly for baby strollers, as well as for children to crawl or walk safely. This includes the availability of ramps, suitable sidewalks, and the minimization of architectural barriers. No less important is to consider the sensory and cognitive stimulation of children in the design of urban spaces. This implies the inclusion of elements such as sensory play areas, green areas with diverse vegetation, and social interaction places where children can learn and actively develop. The spatial needs of Children aged 0–5 in urban spaces focus on safety, accessibility, and appropriate stimulation for their comprehensive development.

Finally, in addition to this, the global crisis resulting from climate change will have significant effects on cities and particularly disadvantaged groups such as children. The impacts will have effects in a differentiated manner, depending on the social, economic, and environmental context in which the children are found (Sheffield & Landrigan, 2011). Various measures must be taken, depending on the specific impacts in a given place. For example, in events of drought and extreme heat, it is necessary to add to this list sufficient areas of shade and cooling, given that the thermoregulation capacities of children are less efficient than those of most adults.

17.2 Methodology

Using a case study approach, the perception of students about their learning in a mandatory course with a theoretical-practical approach for a postgraduate degree in architecture and urban design at a private university in Mexico is analyzed. This information from the course Cities for Childhood: Towards Caring Cities is collected and coded from four sources:

- The digital presentations with the students' proposals
- Their comments in the Class Videos
- A final reflection on the collaborative platform Miro at the end of the course (N = 20)
- Follow-up evaluation six months after the end of the course (n = 13).

The responses are coded using discourse analysis methodology, identifying the integration of the key concepts of early childhood inclusion and sustainability, as well as opportunities and challenges in the learning process.

17.3 Results

The results of the analyzed discourses, proposals, class videos, final reflection, and 6-month follow-up reflection showed the integration of sustainability and early childhood inclusion in urban environments in significant ways, and are presented in this section, after a brief introduction to the context of the case study.

17.3.1 Description of the Course, Work Format, and Site of the Academic Projects

The course aimed to reflect and generate proposals for the design of public spaces, within the conceptual framework of a caring, sustainable, and inclusive city. Through various reading materials and dialogue with experts in childhood, landscaping, and public space, the students generated conceptual proposals for projects that facilitate sustainable, healthy, and fair cities prioritizing the care of early childhood and caregivers.

This academic project arises from the alliance with the FEMSA Foundation, a private organization, that has within its strategic work areas the area of Social Investment in Early Childhood and which since 2019, has collaborated with the Monterrey Institute of Technology in the design of public spaces, implementation of participatory processes, and generation of a network of allies in the territory of the Campana-Altamira Initiative. This initiative emerged in 2014, in an area near the central campus of the Monterrey Institute of Technology, where 42% of its inhabitants have food insecurity and 12% are in extreme poverty (Comprehensive Plan for the Development of the Campana-Altamira Polygon, 2016).

The structure of the course is divided into three moments. The first stage consisted of the analysis of various diagnostic documents generated in the territory of the Campana-Altamira Initiative, in particular those that include socioeconomic data and that define urban policies and projects, mainly the Master Plan Resilient Territories for the Bell and the Partial Program District Bell Altamira. Students were asked to analyze the inclusion of the perspective of early childhood and caregivers in these documents. The second stage consisted of developing a strategy for intervening in public spaces in the territory, in a group scheme, with the aim of jointly arriving at a series of articulated projects to improve the conditions of public space in the area. Finally, in the third stage, conceptual proposals were generated for pilot projects that would translate the strategic framework into specific interventions. The course's work format was hybrid, with weekly classes combining theoretical content and project workshop sessions, in which the professors guided the formulation and development of the projects.

The faculty includes various professions and specialties. For this course, the main professors have experience in urban design and planning focused on early childhood. The participation of Professor Julieta Treviño Sherk, a landscaper from North Carolina State University, and Professor Diana Rodriguez, an architect professor at the University of the Andes in Colombia, with experience in public service focused on services for children, was counted on.

Fig. 17.1 Section of park in urban stream during drought time, conceptualizing natural spaces as play areas

17.3.2 Analysis of the Proposals

The proposals had flexibility in terms of the format of presentation, approach, and location of the interventions, which allows a greater range of expressiveness in the proposals than what we traditionally observe. Despite the diversity, all highlighted concepts such as playfulness, safety, vegetation, and the senses. Factors that favor biodiversity, like gentle hills and plants attractive to pollinators (Fig. 17.1), were common, as they are also opportunities for early childhood to have "adventure and multisensory free play" in the words of one of the students.

17.3.3 Class Videos

The students highlighted how normalized it is in an adult-centric society to reject the need for movement, noise, play, and occupying space in childhoods. Despite the positive reception and the openness to the gender perspective, comments also emerged like "I have learned a lot, but I suddenly feel like I'm in a parenting class." The course also triggered a lot of reflection and discussion about the students' own childhoods, who shared experiences like the loss of independent mobility during childhood compared to their parents' generation: "my mom and her siblings would walk from the center of Guadalupe to Fundidora to go to school at Adolfo Prieto, at five or six years old… alone, and I was a child with zero free play. Also, the issue of gender had an impact. It's quite impressive how the city has altered these dynamics".

The professors took advantage of these comments to return the conversation to what our spatial disciplines compete to facilitate healthy development.

17.3.4 Final Reflection

The final reflections were organized around three axes: objectives, achievements, and lessons learned. The students internalized the pedagogical objectives in ways that were expressed in their proposals. The author of a proposal that incorporated drawings and texts, like a story (Fig. 17.2) expressed the objective as "Childhood: question it, remember it, reinterpret it, use it as a basis to reincorporate it into current cities".

Fig. 17.2 Sketch of an urban corner in the child city proposal

In the achievements, sensitization, previous ignorance of the topic, and practical application were the most recurring themes. In the lessons learned and proposals for improvement, the short duration of the course was the most common theme, as it was taught during the summer over twelve sessions twice a week. The need for citizen participation was also highlighted as an area of opportunity "This subject requires more time than a summer, not only because of the speed, but also because of the enrichment of the people and especially the children, I missed that contact and comments from citizen participation".

17.3.5 Follow-Up Evaluation 6 Months Later

Half a year after the conclusion of the course, the students referred to having transferred the concepts to their subsequent workshops and to professional practice. The importance of caregivers and children in public spaces as an indicator of urban health were two of the most significant learnings mentioned, and more than half of the students reiterated the recommendation to extend the duration of the course to a trimester. Although not all students responded to this follow-up evaluation, among those who answered, 85% reported having already applied the learnings from this course in their professional practice or in subsequent courses within the academic program, a positive result.

17.4 Discussion and Conclusions

Although there is diversity in the strategies followed to achieve the integration of the gender perspective, sustainability, and the inclusion of children in urban proposals, there is great interest in the subject among postgraduate students. Our findings emphasize the importance of individuality and reflective practice when considering how students engage and respond to topics that are expanding the boundaries of urban planning practice and pedagogy. The persistence of comments about limited time in the analyzed discourses is evidence of a student interest that can be motivating for academia in similar programs and contexts in Latin America.

The focus of this course is closely linked to our research interests, which we articulate with our teaching from the paradigm of situated knowledge. It is important to highlight that the planning and development of the course have been built from a knowledge base rooted in experience. This focus on situated knowledge also emphasizes the importance of considering the context and local realities when addressing issues of gender, sustainability, and the inclusion of children in urban planning.

It is worth noting that, since the initial elaboration of this research and the teaching of the course, a second edition is under development. In this new edition, the focus has been adjusted towards a more practical application in a particularly illustrative urban context due to its extremes in terms of challenges and vulnerabilities. This specific context is an informal settlement that is crossed by a stream.

This adjustment in focus reflects an evolution in the understanding of sustainability in the urban context, moving from a broader concept to a more specific and adapted application to the particular circumstances of this city development model. This choice allows for more effectively addressing the specific challenges and dynamics present in this community and aligns with the need to find practical and contextually relevant solutions to improve the quality of life of those living in vulnerable urban environments.

Acknowledgements We thank the FEMSA Foundation for its support of early childhood through the resources and time invested in the course Cities for Children, Towards Caring Cities. We also thank architect Diana Rodríguez from the University of the Andes in Colombia and landscape architect Julieta Trevino Sherk from North Carolina State University, professors who taught with us the first edition of this course and whose contributions were extremely enriching. Likewise, we acknowledge the support provided by the team of the Campana-Altamira Initiative, particularly Tania Arrambide and Víctor González. Also, Rubén Segovia, Director of the Master in Architecture and Urban Design program at the School of Architecture, Art, and Design at Tec de Monterrey for his continuous support and for being an early advocate for this course to happen. Finally, we thank Sofía Bojórquez, Eduardo Garza, Fernanda Muñoz, Adrián Segovia, Saúl Caballero, Kevin Castellanos, Carolina Leos, Paulina Quintero, Azucena Cabuto, Ilse Cancino, Alejandro Sánchez, Brandon Vargas, Leticia Esquer, Nurith Gómez, Luis Neri, David Tapia, Óscar Cardozo, Rodrigo González, Alejandro Meza, and Elsa Romero, students of this course, for their feedback that allows us to advance in our understanding of how we can design more equitable and sustainable urban environments.

References

Beebeejaun, Y. (2022). Race, gender, and positionality in urban planning research. *International Journal of Qualitative Methods, 21*. https://doi.org/10.1177/16094069211066167

Borja, J., & Muxí, Z. (2003). Public space: City and citizenship.

Brundtland, G. H. (1987). Our common future—Call for action. *Environmental Conservation, 14*(4), 291–294.

Busquets, M. (2019). Discovering the ethical importance of care. *Folia Humanística, 12*, 20–39.

Campusano, M. C. V., & Schiavo, L. (2021). Ecology of care from a feminist perspective. *Revista Caminhos-Revista De Ciências Da Religião, 19*(2), 354–371.

Comprehensive Plan for the Development of the Campana-Altamira Polygon. (2016). Campana-Altamira Initiative. Retrieved October 15, 2023, from https://www.scribd.com/document/426149958/Plan-Integral-Poligono-Campana-Altamira-FINAL

Crenshaw, K. W. (2017). *On intersectionality: Essential writings*. The New Press.

de Madariaga, I. S. (2004). Infrastructures for everyday life and quality of life. *Ciudades: Revista Del Instituto Universitario de Urbanística de La Universidad de Valladolid, 8*, 101–133.

Evans, G. W. (2006). Child development and the physical environment. *Annual Review of Psychology, 57*, 423–451. https://doi.org/10.1146/annurev.psych.57.102904.190057

Harvard Center on the Developing Child. (2017). Key concepts: Brain architecture. Retrieved October 14, 2023, from https://developingchild.harvard.edu/science/key-concepts/

Jacobs, J. (2020). Death and life of great cities. Capitán Swing Libros.

Marcuse, P. (2009). From critical urban theory to the right to the city. *City, 13*(2–3), 185–197.

Mulder, M. (2017). Competence and the alignment of education and work. In M. Mulder (Ed.), *Competence-based vocational and professional education: Bridging the worlds of work and education* (pp. 229–251). Springer International Publishing. https://doi.org/10.1007/978-3-319-41713-4_11

Muxí Martínez, Z., Casanovas, R., Ciocoletto, A., Fonseca, M., & Gutiérrez Valdivia, B. (2011). What does the gender perspective contribute to urban planning?

Muxí-Martínez, Z. (2020). Applying gender perspective to urban planning and architecture. Experiences at regional and municipal level in Catalonia | Application of the gender perspective to urban planning and architecture. Experiences at regional and municipal level in Catalonia. *Ciudad y Territorio Estudios Territoriales, 52* (203), 57–70. https://doi.org/10.37230/CyTET.2020.203.05

Navarro-Martínez, V. (2017). Playgrounds: The educational importance of outdoor space. *Revista Internacional de Educación Para La Justicia Social (RIEJS).*

Pérez Cano, M. T., Mosquera Adell, E., Navas Carrillo, D., Mosquera Pérez, C., & Navarro De Pablos, F. J. (2019). The perspective of gender in the teaching of urban planning in Spain. A critical review after the implementation of the Bologna process in the public university Spanish. *Education and Social and Cultural Transformation.*

Rico, M. N., & Segovia, O. (2017). Who cares in the city?: Contributions for urban policies of equality. ECLAC.

Salas-Zapata, W. A., & Ortiz-Muñoz, S. M. (2019). Analysis of meanings of the concept of sustainability. *Sustainable Development, 27*(1), 153–161.

Sami, N., Lall, R., Anand, G., & Anand, S. (2022). Rethinking planning education for urban equality: Higher education as a site for change. *Environment and Urbanization, 34*(2), 413–429.

Sen, A. (2013). The ends and means of sustainability. *Journal of Human Development and Capabilities, 14*(1), 6–20.

Sheffield, P., & Landrigan, P. (2011). Global climate change and children's health: Threats and strategies for prevention. *Environmental Health Perspectives, 119*(3), 291–298.

Sweet, E. L., & Etienne, H. F. (2011). Commentary: Diversity in urban planning education and practice. *Journal of Planning Education and Research, 31*(3), 332–339. https://doi.org/10.1177/0739456X11414715

Terashima, M., & Clark, K. (2021). The precarious absence of disability perspectives in planning research. *Urban Planning, 6*(1), 120–132. https://doi.org/10.17645/up.v6i1.3612

Trainer, T. (1990). A rejection of the Brundtland report. *IFDA Dossier, 77*, 71–84.

UN-Habitat. (2022). World cities report 2022. Retrieved October 13, 2023, from https://unhabitat.org/sites/default/files/2022/06/wcr_2022.pdf

UNICEF. (2023). UNICEF data: Monitoring the situation of children and women. Retrieved October 14, 2023, from https://data.unicef.org/how-many/how-many-children-under-18-are-in-the-world/

UNICEF. (2023). Early childhood development index. Retrieved October 14, 2023, from https://data.unicef.org/resources/early-childhood-development-index-2030-ecdi2030/

UNICEF. (2017). Shaping urbanization for children. A handbook on child-responsive urban planning. Retrieved October 14, 2023, from https://www.unicef.org/media/47616/file/UNICEF_Shaping_urbanization_for_children_handbook_2018.pdf

Valdivia, B. (2018). From androcentric urbanism to the caring city. *Habitat and Society, 11.*

Wesely, J., & Allen, A. (2019). De-colonising planning education? Exploring the geographies of urban planning education networks. *Urban Planning, 4*(4), 139–151. https://doi.org/10.17645/up.v4i4.2200

Wiek, A., Bernstein, M., Foley, R., Cohen, M., Forrest, N., Kuzdas, C., Kay, B., & Withycombe Keeler, L. (2015). Operationalising competencies in higher education for sustainable development. In M. Barth, G. Michelsen, M. Rieckmann, I. Thomas (Eds.), *Handbook of higher education for sustainable development* (pp. 241–260).

Woodhead, M., & Brooker, L. (Eds.). (2012). *Healthy environments. Early childhood in focus, 8.* The Open University.

Chapter 18
Urban Proximity: A Methodological Approach from a Gender Perspective

Alessandra Cireddu, Verónica Livier Díaz Núñez, and Dulce Esmeralda García Ruíz

Abstract The study presented is part of a broader research on women, city, and everyday life funded by the National Council of Humanities, Sciences, and Technologies (CONAHCYT) of Mexico and the Observatory of Cities of Tecnologico de Monterrey, carried out in collaboration between these three institutions: Universidad de Guadalajara, Tecnologico de Monterrey, and Universidad Autonoma de Guadalajara. When we talk about cities from a gender perspective, proximity is an indispensable value. Women have the difficult task of reconciling the role and times of care with their productive work and the rest of their daily activities. Thinking about a reorganization of urban space from a neighborhood scale means supporting care tasks, favoring the creation of support networks and making the use of time more efficient, which would also allow a more equitable distribution of these tasks, leading to the construction of more inclusive cities and societies. The crises that cross contemporary times, accentuated by the recent global pandemic, have increased reflection on these issues: "the quarter-hour city", a concept coined by Carlos Moreno in 2015 and taken up by Mayor Anne Hidalgo as an urban proposal for the city of Paris has become the foundation of a new paradigm of sustainability and quality of life. The objective of this chapter is to argue and propose the construction of a methodology for qualitative analysis at the neighborhood scale that allows measuring proximity from a gender perspective. As part of the results and future applications of this study, the possibility of allowing new mappings of the territory through qualitative measurements of proximity is envisaged.

Keywords Proximity · Gender · Everyday life

A. Cireddu (✉)
School of Architecture, Art and Design, Tecnologico de Monterrey, Monterrey, México
e-mail: acireddu@tec.mx

V. L. Díaz Núñez
Universidad de Guadalajara, Guadalajara, México
e-mail: veronica.diaz01@academicos.udg.mx

D. E. García Ruíz
Universidad Autónoma de Guadalajara, Zapopan, México
e-mail: dulce.garcia@edu.uag.mx

M. Vicuña and E. Giorgi (eds.), *Intersections*, Springer Geography,
https://doi.org/10.1007/978-3-031-76402-8_18

18.1 Introduction

When we talk about the importance of urban proximity from a gender perspective, it is necessary to understand that there are differences in the use of the city according to assigned roles. The "universal" displacement focused on production and the quantitative planning of urban transport according to the logic of "peak hours" makes care tasks invisible (Jirón, 2017; Sánchez de Madariaga, 2009), therefore excluding the female gender in its intersection with other characteristics such as income, age, family type, among others (Punt6, 2021). According to CEPAL data showing the use of time by men and women in nine Latin American metropolises, although the hours that men dedicate to productive work are higher than those of women, the female sex dedicates more total weekly hours to work, the differential being the hours dedicated to reproductive work (Scuro & Vaca-Trigo, 2017). The disparities increase as the socioeconomic level decreases in the case of women, while the weekly working hours of men are not subject to large variations (Scuro & Vaca-Trigo, 2017), which underlines the intersectionality of gender and its impact on urban mobility. Care tasks are carried out at home, but also outside of it and are the ones that determine the main differences in urban mobility patterns between men and women. Care mobility is what makes visible the chain of tasks of everyday life in urban journeys (Sánchez de Madariaga, 2009), showing that, compared to men, women move closer to their homes, make more use of public transport, have polygonal routes with multiple stops and travel accompanied by more people (Ciocoletto and Punt 6, 2014; Jirón, 2017). Understanding and addressing urban proximity from a gender perspective becomes a critical challenge in the search for more inclusive cities that recognize and address gender disparities in mobility patterns and daily life. This article aims to explore the importance of proximity from a gender perspective and address how to effectively measure and analyze this concept in the context of urban planning. It should be mentioned that the article is part of a broader research within which various empirical qualitative studies have been carried out to measure proximity in the Metropolitan Area of Guadalajara (Mexico). However, the central objective of this chapter does not lie in the presentation of empirical data, but in the contribution to methodological construction. The aim is to set the foundations for the development of a methodology that, based on the lessons learned from previous research, can be applied to different urban contexts. In this sense, the objective of this chapter is to outline criteria for measuring proximity in a comprehensive way and with a gender approach. These criteria will not only be based on the collection of quantitative data but also on the qualitative analysis of people's experiences in their urban environment. The intention is to provide a methodological basis that allows evaluating proximity considering different experiences in the city, contributing to the existing literature on urban proximity by providing a methodological approach that can be used as a conceptual framework to include personal experiences in the study of urban proximity.

18.2 Defining Proximity: A Multidimensional Approach

Proximity in an urban environment is not limited solely to the physical distance between housing and essential services but becomes a multidimensional concept that encompasses quantitative and qualitative aspects, especially when approached from a gender perspective. In this sense, measuring proximity with a gender approach becomes a complex but fundamental challenge for inclusive urban planning. Some of the dimensions involved in defining this concept are the accessibility of routes, safety (real and perceived), and the quality and human scale of public spaces and streets. These dimensions are directly related to spatial issues that can limit mobility: uphill, potholes, closed streets, blind walls, difficult access stairs, are some of the obstacles that limit people's autonomy in public space, therefore reducing the real proximity between places. Other dimensions can also be identified that are related to factors that become very important when talking about proximity for women. Care mobility, linked to tasks of social reproduction, as defined by Inés Sánchez de Madariaga (2009, 2013), inserts different dynamics and logics in urban transfers, which are based on daily needs. Therefore, other barriers are identified, less visible in urban or spatial terms, but equally significant. These barriers can be of an economic or technological nature or related to the personal skills of each person; there are also organizational limitations that can manifest in terms of how daily activities are structured and managed at home and outside (Jirón, 2017). The intersectionality of gender, which depends on age and socioeconomic level among other factors, further amplifies these barriers, which impacts women to a greater extent, limiting the real proximity of their networks of daily places. In general, women value proximity to services more than men, as men tend to compensate for the lack of nearby activities in a neighborhood by traveling greater distances, while women tend to limit their participation in activities, especially if they are leisure activities (Gil Solá & Vilhelmson, 2022; Næss, 2008), confirming that time constraints are more important for the female gender. A multidimensional approach to urban proximity from a gender perspective is therefore necessary because it allows considering not only the physical aspects of urban space but also makes visible the experiences and barriers that women face in their search for proximity. Proximity becomes an essential concept that must be addressed in inclusive urban planning, recognizing the diversity of needs and challenges that people face in their daily lives in the city.

18.3 Measuring Proximity: Methodological Approach

Given all the above, proximity is an essential urban quality for rethinking inclusive urban environments capable of responding to the needs of different groups of people and inhabitants. It is also clear that proximity is a complex multidimensional concept that highlights the importance of understanding urban routes not only as a matter of transportation, but as a fundamental aspect of urban life that is intrinsically linked with the social, economic, and gender dynamics of the city.

But how do we measure proximity? Moreno (2020) introduces the concept of the 15-min city, based on the idea that a truly livable and sustainable metropolis should be within reach of all its inhabitants, offering a wide range of functions and services within a 15-min walk or bike ride from their homes. The nearby city is a "living" city that must respond to seven functions: living, working, shopping, accessing health, education, culture, and resting in harmony with nature.

To quantify and understand urban proximity, the City Observatory of Tecnológico de Monterrey conducted a study in the main Mexican cities during the COVID-19 pandemic. This study used a quantitative approach by building a proximity index that was based on the calculation of the actual distance between places, rather than a radius distance. The distance to supermarkets, hospitals, and pharmacies was calculated, reflecting the accessibility to essential services during the pandemic emergency. The study was subsequently expanded to include the idea of the "fifteen-minute city", increasing the type of places which, in the second stage, included schools, community services, commerce, and entertainment. These measurements allowed the creation of territorial maps in which areas with different levels of proximity are highlighted, and visually represented by color spots.[1] Although these maps are a valuable tool for assessing quantitative aspects of proximity, they cannot fully capture all the nuances and factors that affect the use of urban space, as specific proximity criteria related to the daily needs of the population are left out.

Since the 1960s, Jane Jacobs has advocated for a nearby, diverse, vital, and vibrant city (Jacobs, 1961), qualities taken up by the Punt6 collective in the guide "Spaces for Everyday Life" (Ciocoletto & Punt6, 2014), where a methodology is proposed to audit the quality of urban environments with a gender perspective, measuring five qualities (proximity, diversity, vitality, autonomy, and representativeness), from which a series of indicators are derived that allow evaluating the different spaces that house everyday activities: urban facilities, public spaces, and the neighborhood's daily network. Among the proposed qualities, proximity is a kind of criterion that encompasses all the others, as a prior condition that must characterize the space under a gender approach. Proximity is defined from three different scales: neighborhood, district, and super-district which correspond to a radius of, respectively, 5, 10, and 20 min of walking, which correspond to distances of 300 m, 600–800 m, and 1300–1500 m, where ideally all daily needs should be met.

Empirical studies carried out in the city of Guadalajara, Mexico, by the authors of this chapter (Cireddu, 2017; Díaz & Cireddu, 2018) show that, when this condition of spatio-temporal proximity is not met, people adopt strategies to meet their daily needs outside the traditional limits of proximity, leading to a redefinition of the very notion of proximity. This reconfiguration is based on lived experiences, constructed in the realm of imaginaries and representations, that go beyond the restrictions of physical space. The case study of Huentitán el Alto, a no central historic neighborhood in the city of Guadalajara, provides a significant example of how the proximity dynamics vary according to the characteristics of the urban environment and different types of inhabitants (Díaz & Cireddu, 2018). In this context of a traditional neighborhood, where new housing projects have recently been introduced in the form of gated communities in urban voids, the analysis demonstrates the existence of two types of proximity. In the case of the inhabitants of the gated communities, an "aspirational" proximity is observed, where people carry out their daily lives in places that are distant from home but are easily accessible by car (Díaz & Cireddu, 2018). These places offer better services and represent the desired urban environment to live in. This phenomenon highlights how individual mobility and car accessibility can alter the perception of proximity, as also shown in the case study of suburban gated communities (Cireddu, 2017).

In contrast, for the inhabitants of the open traditional neighborhood in Huentitán el Alto, a limited geographical proximity is identified (Díaz & Cireddu, 2018). Although the neighborhood has services, shops, and facilities, proximity is hindered by physical barriers such as poor sidewalks, a deficient urban image, perceptions of insecurity, and insufficient lighting, among other factors. These limitations in accessibility and quality of the urban environment affect the perception of proximity, creating a reality in which residents experience reduced proximity despite the presence of mixed uses in the neighborhood.

These findings highlight the importance of considering not only the physical location of services and equipment, but also the elements that affect people's daily experience in their urban environment. The perception of proximity is shaped by a complex interaction between physical, social, and emotional factors, underlining the need to address these elements in urban planning to ensure inclusive and equitable cities. Thus, when we talk about a nearby city, it is not enough to talk about mixed uses, dense or compact city, it is also necessary to integrate people's experiences in terms of accessibility, security (both real and perceived), through the inclusion of tools that allow the evaluation of urban environments from personal experiences and perceptions that allow identifying and mapping specific problems in relation to physical space (for example, lack of lighting or public bathrooms) as well as deeper vulnerabilities. As already mentioned, the concept of urban proximity is not limited exclusively to physical space but encompasses a set of interrelated dimensions that influence the accessibility and quality of life of the inhabitants in an urban environment. These dimensions mainly include economic, social, and internal organizational dynamics aspects within households, which are also influenced by individual skills and access to technology. Economic resources play a crucial role in the ability to take advantage of proximity in the city. Financial limitations can restrict housing and transportation choices. For example, a person with limited income may be forced to live in areas where housing is more affordable but further away from their daily destinations. Likewise, the purchase of a private vehicle or even the costs of public transportation can be prohibitive for some people, limiting their mobility and access to essential services. Social and cultural interactions can also influence the perception and use of urban proximity. For example, security concerns and the perception of the environment can vary according to gender, age, and other demographic factors. The fact that people avoid areas they consider unsafe limits their autonomy and therefore their proximity to certain places. The structure and responsibilities of households play an important role in defining proximity. It is evident that single mothers or families with small children have specific care needs that influence their mobility and daily routes. Tasks such as caring for an elderly person, or a newborn can restrict transportation options and the time-distances one is willing to travel.

Based on all the above, the methodological approach proposed for the analysis of urban environments with a gender focus is based on three steps: the delimitation of a certain study perimeter whose fundamental requirement is the presence of housing; the identification of a physical proximity network that encompasses transportation options, mixed-use with the presence of commerce, equipment, and public space; the in-depth diagnosis from a gender perspective that integrates other parameters and layers that allow defining a proximity capable of supporting daily life in all its spheres (the productive; the reproductive or care; the personal or own; the social or political), including the organizational dynamics of households that are intrinsically linked to individual decisions and the management of activities and movements in the urban environment. As shown in Fig. 18.1, this methodological approach integrates different types of criteria, each of which provides a unique and complementary perspective, necessary to evaluate urban proximity in all its complexity.

18.3.1 Basic Criteria for Proximity Analysis

This category focuses on mainly quantitative criteria that can be measured using data and geographic information systems, which allow defining levels of proximity in extensive areas and highlighting areas with greater or lesser accessibility. The basic criteria are fundamental to understanding the physical structure of proximity and ensuring efficient access to services and resources. The criteria include:

- Presence of mixed uses: The existence of a mix of uses in an urban environment, combining housing with educational, health, sports, and cultural services, as well as commerce, is considered essential for urban proximity. This feature promotes a more efficient daily life by bringing people's needs closer to their homes.
- Presence of public space: The availability of public areas, such as parks, children's play areas, and rest places, contributes to the quality of life and proximity, with access to green and recreational spaces being a significant factor in ensuring the well-being of a community.
- Presence of different types of transport: This topic focuses on aspects related to mobility, addressing as specific criteria the following: 1. Possibility of pedestrian routes: evaluates the existence of safe and accessible pedestrian routes, essential to promote walking mobility. 2. Presence of public transport: refers to the proximity of public transport stops and the availability of efficient bus or train routes. 3. Public bicycle stations: evaluates the presence of a public bicycle service. 4. Bike lanes: considers the presence of exclusive lanes that promote the use of the bicycle as a sustainable means of transport.

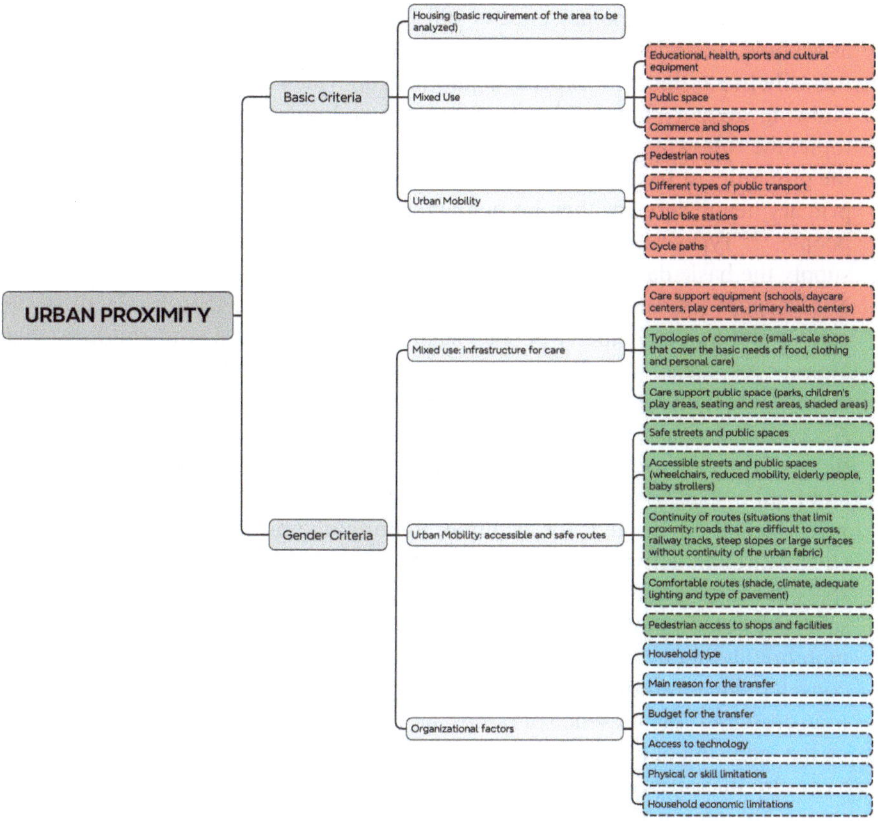

Fig. 18.1 Summary of criteria for the analysis of proximity in urban areas. In red are shown the criteria that can be mapped quantitatively; in green the qualitative criteria that can be mapped with observation and field visits; in blue qualitative criteria that can only be measured from personal experience. The three types of criteria are fundamental for the construction of a comprehensive proximity measurement methodology

18.3.2 Criteria for Proximity Analysis with a Gender Perspective

This category focuses on mainly qualitative criteria related to the gender perspective, which are collected through techniques such as direct observation, perception surveys, or in-depth interviews. The main feature of these criteria is that they allow capturing the daily experience of people in the urban environment, delving into what the urban space should be like to ensure gender equality and meet specific needs in the different spheres of daily life, especially the care sphere, significant for the female gender and forgotten in traditional urban planning. The criteria of this second category are detailed below.

- Care infrastructure: the presence of infrastructure that supports care tasks, such as schools, nurseries, parks, and small-scale shops, among others, is fundamental for daily life as this type of urban equipment alleviates the reproductive workload allowing greater participation in economic and social activities. In this broad topic, the following criteria are evaluated: 1. Care support equipment: evaluates the proximity of facilities such as schools, nurseries, playrooms, care centers, and primary health services that are fundamental for the care of children and elderly people. 2. Types of commerce: refers to the presence of small-scale shops that supply the basic daily needs for food, clothing, personal care, and home care. 3. Public space for care: evaluates the existence of public spaces that facilitate the care of children and elderly people, such as parks equipped with children's play areas or rest areas, with the presence of trees, benches, and shaded areas, as well as adequate lighting that allows safe use.
- Accessible and safe routes: the focus is on mobility and displacements, defining the quality of the routes through the following specific criteria: 1. Safety in streets and public spaces: factors that influence the safety of the space, both real and perceived, are evaluated; for example, lighting, the presence of activities in the streets through active ground floors, open facades that look out onto the public space through balconies and windows, and other elements that contribute to a safe urban environment. The feeling of safety is a determining factor in women's mobility and their autonomous participation in urban life. 2. Accessibility: it is considered in terms of urban design, including the width of sidewalks, safe pedestrian crossings, the presence of ramps, and the quality of pavements, with a focus on the inclusion of people with reduced mobility, wheelchairs, and baby strollers, to ensure that all people can access services and public spaces. 3. Continuity of routes: the presence of obstacles that limit proximity, such as difficult-to-cross roads, railway tracks, steep slopes, or areas without continuity in the urban fabric, is evaluated. These obstacles can affect accessibility and the comfort of displacements, as well as the autonomy with which people move. 4. Comfort of routes: aspects that influence the quality of displacements in terms of the presence of shade, adequate lighting, type of pavement, rest areas, cleanliness, urban image are considered. 5. Pedestrian access to shops and facilities: this indicator evaluates whether shops and facilities are accessible by walking and to what extent they meet the needs of the community. The "human scale" in the accessibility to basic services is a key factor for urban proximity.
- Organizational factors: for an urban analysis with a gender perspective, we will have to take into account all those organizational and intersectional factors that are not directly related to space, but to people and their specific needs. Thus, for example, a park may meet proximity requirements at the two previous levels, however, for someone it may be difficult to access because that person is a single mother who has to organize and financially maintain a household of three children of different ages and does not find the time to take them to the park, even though it meets some proximity criteria. Organizational factors include 1. Type of household: the composition of the household and family responsibilities can influence people's ability to access nearby services. For example, a single mother

or a family with newborn children may face additional challenges in balancing childcare with the rest of the daily tasks. 2. Purpose of the transfer: the main reason for the transfer informs the decisions about mobility and can increase or decrease travel times and the perception of proximity. It is not the same to move to go to work or to the supermarket or to take the children to school. 3. Budget for the transfer: the availability of economic resources for a particular move can limit transportation options. 4. Access to technology: people who have access to technology, for example through a mobile phone or a computer, can more easily plan their trips or shared transportation, in addition to being able to more effectively organize and share care tasks, for example. 5. Physical or skill limitations: considered physical or skill limitations, such as the inability to ride a bicycle or drive a car. 6. Economic limitations: economic limitations of households, such as the inability to pay for public transportation or the lack of a car, affect their proximity.

18.4 In Conclusion

From all the above, it becomes evident the importance of reviewing urban environments through a gender perspective that incorporates people's daily needs and the barriers they face in their search for proximity. This approach highlights the need to understand urban mobility not only as a matter of transportation, but as a fundamental aspect of urban life that is intrinsically linked to the social, economic, and gender dynamics of the city.

The reflections presented in this chapter allow the construction of some general criteria for the review of urban environments through the lens of gender, which involves the use of ethnographic techniques and a qualitative construction of the analysis from personal experience and can be useful to inform the construction of more detailed methodologies. The main challenge we face in constructing an urban diagnostic methodology with a gender focus lies in the need to collect a large amount of qualitative and experiential data efficiently and in a reasonable time frame. This data, fundamental to fully understanding people's experiences in the urban environment and addressing gender issues that influence their mobility and accessibility in the city, needs to be represented through large-scale mappings that can inform urban planning decisions.

As part of the results and future applications of this study, we perceive the possibility of generating urban maps through qualitative measurements of proximity using digital tools. For example, a mobile phone application that allows the collection of data on routes and individual perceptions on a large scale, generates qualitative maps that are valuable for incorporating information about urban experience and care in decision-making.

This article is a call to action for the importance of addressing proximity from a gender perspective in urban planning to be recognized. Through a multidimensional approach, the integration of digital tools, and putting in value urban experience, we can move towards more inclusive cities that reflect the needs and desires of all the people who inhabit them.

Acknowledgements Consejo Nacional de Humanidades, Ciencias y Tecnologías de México; Observatorio de Ciudades Tecnológico de Monterrey.

Note

1. Maps and data available at https://observatoriodeciudades.mx.

References

Ciocoletto, A., & Punt 6 (2014). Espacios para la vida cotidiana. Auditoría de calidad urbana con perspectiva de género. Barcelona.

Cireddu, A. (2017). *Urbanizaciones residenciales cerradas y habitar: análisis de las dinámicas cotidianas en la Zona Real, Área Metropolitana de Guadalajara, México*. Doctoral Thesis, UdeG, Guadalajara.

Díaz, V., & Cireddu, A. (2018). La proximidad urbana a partir del análisis de la vida cotidiana con perspectiva de género en el barrio de Huentitán el Alto, Guadalajara (México). Quid 16, n. 11.

Gil Solá, A., & Vilhelmson, B. (2022). To choose, or not to choose, a nearby activity option: Understanding the gendered role of proximity in urban settings. *Journal of Transport Geography, 99*.

Jacobs, J. (1961). *The death and life of great American cities*. Random House.

Jirón, P. (2017). Planificación urbana y del transporte a partir de relaciones de interdependencia y movilidad del cuidado. In O. Segovia, & M. N. Rico (Eds.), ¿Quién cuida en la ciudad? Aportes para políticas urbanas de igualdad (pp. 405–431). CEPAL.

Moreno, C. (2020). *La revolución de la proximidad. De la ciudad global a la ciudad de los quince minutos*. Alianza Editorial.

Næss, P. (2008). Gender differences in the influences of urban structure on daily travel. In T. Cresswell (Ed.) *Gendered mobilities*. Routledge.

Punt 6. (2021). *Movilidad cotidiana con perspectiva de género*. Buenos Aires: CAF

Sánchez de Madariaga, I. (2013). Mobility of care: Introducing new concepts in urban transport. In I. Sánchez Madariaga, & M. Roberts (Eds.), *Fair shared cities: The impact of gender planning in Europe*. Routledge.

Sánchez de Madariaga, I. (2009). Vivienda, movilidad y urbanismo para la igualdad en la diversidad: ciudades, género y dependencia. *Ciudad y territorio estudios territoriales, XLI* (161–162), 581–597.

Scuro, L., & Vaca-Trigo, I. (2017) La distribución del tiempo en el análisis de las desigualdades en las ciudades de América Latina. In O. Segovia, & M. N. Rico (Eds.), ¿Quién cuida en la ciudad? Aportes para políticas urbanas de igualdad (pp. 117–148). CEPAL.

Chapter 19
House on Kings Road. The House of Pauline Gibling Schindler

Andrea Castro-Marcucci

Abstract The research focus in the events that occurred during the design and construction process of the Schindler house, originally attributed to the Viennese architect R. M. Schindler in Los Angeles between 1919 and 1925. It aims to give visibility to the participation of his wife, the composer, writer and American political activist Pauline Gibling, who played an important role in the development of the ideas that gave rise to the house, while actively participating during its construction and in the meetings that began around 1923. The house went down in history as the Schindler house, overlooking the significant participation of Gibling in the outline of the way of living that gave rise to its spaces, the development of the design, the management of the financing of the construction, and her effort to drive the project to its completion. This house brought together important cultural actors of the time (politicians, artists and architects) who supported avant-garde thinking and were part of the materialization of the emerging modernity of the twentieth century. The study of the case of the house on Kings Road has allowed, in addition, to make a review of the definition of the authorship of the architecture project during the modern movement, a period in which the figure of the architect was built individually, omitting the historiography the intervention of other important actors like Gibling Schindler, who promoted, built and were decisive in the foundation of modernity.

Keywords Pauline gibling · Modern architecture · Avant-garde · Collaborations · R. M. Schindler

19.1 Introduction

The historiography of modern architecture has been told, in most cases, from the vision of the architect as an individual figure, without considering their collaborators (Colomina, 2018). However, when reviewing the reference architects of the modern movement in Europe and America, there are figures who accompanied, contributed

A. Castro-Marcucci (✉)
Facultad de Arquitectura, Universidad ORT Uruguay, Montevideo, Uruguay
e-mail: castro_ac@ort.edu.uy

M. Vicuña and E. Giorgi (eds.), *Intersections*, Springer Geography,
https://doi.org/10.1007/978-3-031-76402-8_19

and collaborated on its development. In recent decades, these collaborators have been the subject of multiple research, their stories have begun to be known and this has allowed to make sense of changes, adaptations and even identify the different moments of the trajectory of modern architects. This is possible to observe in the itineraries of Le Corbusier, Frank Lloyd Wright, Mies van der Rohe, Richard Neutra, among many others.

The work aims to make visible one of these collaborations omitted in the history of architectural modernity, that of Pauline Gibling (Schindler) composer, designer and activist American political, who was the wife of the Viennese architect R. M. Schindler between 1919 and 1927 and who went down in history as a muse of artists of the time like John Cage (Hines, 2019). However, in the history of Californian modernity, the tireless work that Gibling carried out for the dissemination of the generation of architects who settled in southern California from 1920 until 1950 has been downplayed.

Gibling was a vivid writer, her letters have become a determining source of information and have allowed to demonstrate the influence she had on Schindler's projects (Parra-Martínez, 2012), from the approach with clients, the financing of the works and the subsequent publication in local magazines. The information existing about her trajectory had to be extracted from sources referred to the architects R. M. Schindler, Richard Neutra, Frank Lloyd Wright and later in the correspondence of the composer John Cage, and the letters published by R. Sweeney and Thomas S. Hines, as there is no single archive. This research was carried out, mainly by reading between the lines about facts that were not possible to explain only from the documented trajectory of the architects (Castro, 2022).

Although the trajectory of Pauline Gibling is extensive and includes participation in graphic design projects, editing and musical composition, the work will specifically focus on her collaboration with Schindler in the conception, design, management, and construction of the house on Kings Road, erroneously called the Schindler House. However, it is possible to reconstruct the history of the house from her close collaboration with her ex-husband (Sweeney, 2001). Even during her stay at Taliesin II (Oak Park) in 1919, several years before her establishment in Los Angeles.

These revelations suggest the need for a revision to the narrative about the origin of the house on Kings Road that includes Pauline Gibling and her ideas about community living, politics, and society, which are faithfully represented in the spaces of the house and in the way the meetings and activities that made the house on Kings Road considered a meeting space for the avant-garde of the early twentieth century and that passed into history as an icon of modern architecture (Frampton, 2022).

19.2 Activist from the Beginning

The composer Sophie Pauline Gibling was born in Minneapolis, United States in 1893. She was raised in New York and studied music at Smith College (1915). From a very young age, she developed an interest in and social awareness of community work, which led her to work at the Hull House in Chicago, one of the most important settlement houses of the time, which primarily served European immigrants. She used to work tirelessly and committedly on different initiatives around social and political interest topics (Sweeney, 2001).

She met Schindler in 1918 and a few months later they got married, they were an unconventional couple for the time. The marriage agreed to maintain their economic and sexual independence, parallel to their life together. Schindler started working in Frank Lloyd Wright's studio in 1917, mainly developing projects for his houses in Los Angeles. In 1919, Wright invited the couple to move to Taliesin II, where they settled between Oak Park and Spring Green.

This period of cohabitation near Wright is one of the least politically active periods of Gibling's life, although she maintains her ideology rooted in her way of living and working. Wright called her Schindler's "Bolshevik helper" (Canosa Benitez, 2016).

This period of Gibling's life is reflected in the unpublished manuscript titled Taliesin in 1919, which is also one of the few records about life in Taliesin and Wright's dynamics during that period. In it, the author describes the daily work within the studio, emblematic of visitors like Louis Sullivan. Also, she emphasizes how the architect visited places, always with an approach from the natural landscape. One of the most important records reflected in the text is the progress of important projects like the Imperial Hotel, the arrival of Aline Barnsdall and the project for the Oak Park complex in California, the Hollyhock house, among others. In addition, the text makes clear the relationship that existed between Gibling and Wright, which developed independently of Schindler and persisted even after the Viennese architect established his independent studio in Los Angeles and also broke his relationship with Wright (Castro, 2022).

Although the document does not make it clear that Gibling participated directly as a draftsman in the projects that were developed in Wright's studio, it does prove that she was present on the walks he took with the team around the studio, in important discussions such as the structural decisions of the Imperial Hotel and in the multiple disputes that the architect had with Aline Barnsdall for the Oak Park project. Therefore, at this time Gibling was already part of the studio, although this would never be acknowledged by Wright or Schindler.

In her text, Gibling talks about the striking relationship that the location of Taliesin achieves with the surrounding landscape, showing a sensitivity to architecture, design, and the context that surrounds her. "just as any work of architecture can constitute a philosophical statement, Taliesin speaks and rests, —[in] a harmonious consonance with the earth—" (Gibling, circa 1970: 2).

19.3 First Focuses of Activism in Los Angeles

In 1920, the Schindler couple settled in Los Angeles so that R. M. Schindler could attend to the works of Oak Park and Residence B of the Barnsdall project. For Gibling, this was the opportune moment to resume her political activity, and her husband followed her to some extent, but for Schindler, these encounters were always mostly a means to establish relationships and meetings with potential clients, while for Gibling it meant a genuine boost in her political concerns.

It should be considered that, during the second decade of 1900, the city of Los Angeles was in a moment of boiling, of oil boom that allowed a great economic boost and the vertiginous growth of the city (Banham, 1971). At the same time, the city became the scene of great conflicts between workers' associations, the fight against union leaders and racial conflicts, this was the scenario that would accompany these decades of growth in southern California.

For Gibling, it was an ideal scenario for her political involvement. A few weeks after settling in Los Angeles she had a meeting with Upton Sinclair -one of the most influential left-wing leaders at that time. Through Sinclair, the couple became linked with the Walt Whitman School, a progressive school that had started in 1911 in New York and defended values in education such as equality, respect, sexual education, and the individual development of each student according to their abilities (Parra-Martínez, 2012). Schindler projected for this school a new headquarters in Los Angeles that was never built, but it allows to register the link he had with the association.

In a letter without a known recipient, Gibling expresses her excitement for the activities she is carrying out at the moment and for the connection she has had within the radical movement.

> We are so far and so deeply 'inside' the radical movement these days that we never have a night at home anymore… Meetings of the Worker's Defense League committee, for the Walt Whitman School, large and small conferences, dining in strange places with people who tell us news impossible to obtain except 'word of mouth', conferences; meetings where we only stop long enough to make an announcement before moving on to the next; visits to the printer to review proofs for the school; car trips for a committee composed of a doctor, a lawyer and an alienist, to the hospital to visit a member of the I.W.W. who has been a month in jail awaiting trial, and who has been treated so violently and brutally by the authorities that, in addition to serious bodily injuries, he seems to suffer mentally and is in the psychiatrist's observation room, suspected of madness. Then, on top of all that today… we rushed off to Pasadena… to a meeting at the private residence of a wealthy radical… to listen to Max Eastman… Everyone was there, —and we had a very good chat afterwards… Upton Sinclair introduced me to his wife… Eastman was charming… And we all had a great time… Really a much better time than I have found in Chicago, in general… (Sweeney, 2001, p. 91).

It is evident that while Schindler was immersed in the work of Wright's works, Gibling dedicated herself to cultivating a network of contacts around her political activities, which would later be the basis for the networks that were established in the "garden sessions" at the house on Kings Road. This network of relationships would include her in the field of politics and social struggle, but at the same time would serve as a platform for her husband to meet a good part of the clients that later allowed his definitive establishment in California.

In these first months of establishment in the city, Gibling learns the dynamics around the salons where the meetings of the committees and clandestine organizations she regularly attended were held. This is the first step towards the gestation of her salon in the house on King Road (Sweeney, 2001).

19.4 A Different House, A Community Living

[I am] deeply grateful to you r.m.s... for... this house, which has been so dear to me that in a way it has determined [my] life.

Pauline Glibing to R. M. Schindler. July 9, 1953.

Most of the historiography published today about the house on Kings Road -or Schindler house- directly or indirectly references Pauline Gibling (Castro, 2022). Within Schindler's trajectory, the house on Kings Road is a unique moment and one that was not repeated in later works. While in some elements the influence of Wright can be noticed, this only has to do with the formal. As for the conceptual idea, the distribution and interior spaces, the house is an unprecedented experiment within the architect's trajectory. In his subsequent works, he did not return to propose a scheme of community living or a central courtyard like the one in this house (Crawford, 1997).

This shows the great influence that Gibling had on the initial approach, in the subsequent development of the house and in the activities that took place in it. However, in the historiography of the house, they only mention the composer at the beginning of the project as the generator of the idea. Whereas, as this work tries to prove, Schindler's project is a built manifesto of his wife's ideas about community living and outdoor living.

It is possible to enumerate Gibling's contributions at different times of the house. The first participation was in the idea of living together with another family, the house would be a dwelling for two couples: The Schindlers and the Chaces. Despite the fact that in Gebhard's work on Schindler (1979) it is expressed that Clyde Chase was an old friend of the architect. The truth is that the link with the Chaces was initially established by Gibling, through Marian Da Camara (later Marian Chace), who was a great friend, and they had attended together at Smith College. From there their friendship extended for years, sharing experiences at the Hull House and some teaching assignments during their life in Chicago. At the time of the construction of the house, Clyde Chase was working as an engineer with the architect Irving Gill, deeply admired by Schindler.

In 1921 the two couples started construction with a loan from the bank of $5,000 and another $2,000 contributed by Gibling's parents (Smith et al., 2001). In April 1922, Gibling accepted a teaching job for three months in El Centro to pay for the construction, as the initial budget had been exceeded (Smith et al., 2001). Although the financial contribution is not proof of her authorship, it does demonstrate the commitment she had to the completion of the project.

A second participation is in the spatial distribution of the project. The house was divided into three areas, one for each of the couples and another for guests/tenants. Each area had a study, bedroom, and patio. In the initial approach, the central area was a single shared kitchen. However, years later a second smaller kitchen would be built, to complement one of the other areas. Although this strategy of generating independent zones in a single dwelling was later used by Schindler, it is in this house where it achieved full acceptance and integration among its inhabitants.

Gibling wrote a letter to her mother years before her marriage to Schindler, in which she already outlined some ideas that were realized in the Kings Road house.

> One of my dreams, Mother, is to have, someday, a little joy of a bungalow, on the edge of the woods and mountains near a crowded city, which shall be open just as some people's hearts are open, to friends of all classes and types…. (Gibling in Sweeney, 2001: 87).

One of the most ambitious ideas was that of the central courtyard, in which Gibling imagined an open space for friends "of all classes and types" (Gibling in Sweeney, 2001: 87). This is another contribution of Gibling on the final project, the courtyard as the most important space of the house. In this place, shortly after, the meetings are called *Discussions in the Garden.* Would begin to take shape. Finally, Gibling managed to have her salon in the Schindler's, where she would stand out as an unparalleled hostess.

The house was partially completed in 1923 and both couples moved into a sort of initial camp (Sweneey, 2001). In a short time, the house became a gathering place for musicians, artists, writers, poets, politicians, and architects, who were invited by the couple. It became a sort of institution that accommodated the most authentic ideas, desires, and expressions about the American modern avant-garde (Fig. 19.1).

Some of the most recognized visitors were Upton Sinclair, Scott Nearing, Edward Weston, John Bovingdon and Marta Graham, Igor Stravinsky, John Cage, Galka Scheyer, Max Eastman, Licoln Steffens, Aldous Huxley, Clifford Odets, Anaïs Nin, Theodore Dreiser, Frank Lloyd Wright, Lloyd Wright and the Neutra family (Hines, 2019).

Between 1922 and 1924 different authors refer to the meetings and evenings at the Schindler's house. Gibling continuously writes to his family about the people who visited them, who they were and their contributions in different disciplines. The great legacy of the house for modern architecture extends beyond Schindler's project, it

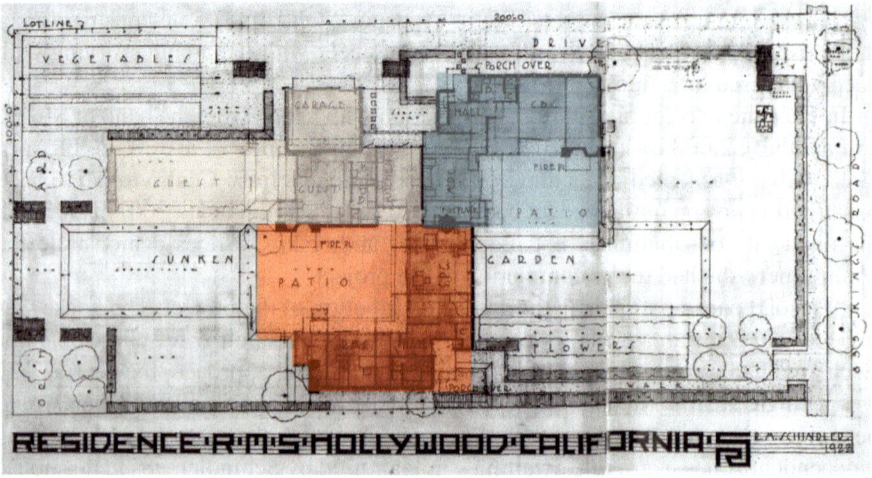

Fig. 19.1 Floor plan of the house on Kings Road with the three differentiated areas. Own elaboration

mainly lies in the intellectual exchanges and the avant-garde personalities that took place in the garden sessions.

> The Schindler's, open house on Sunday evenings attracted the 'arty' intellectuals of post-World War I. These people weren't rebelling against the old; they were disregarding it. They were adventurers reaching out for the new. Hollywood drew them like a magnet.
>
> There was Robert Nichols, a young English poet just come from a year's teaching In Japan. A, a former member of the R.A.F. he was in Hollywood as advisor to William De Mille on the picture, 'Wings Over Europe'. (Hamilton Harris, 1979, 13).

Parallel to the development of the meetings, the problems in the Schindler's marriage became increasingly larger and in 1924 Gibling attempted suicide, in response to Schindler's constant infidelities (Sweneey, 2001). In 1927, Gibling left the house on Kings Road with her son Mark and definitively separated from her husband. She began her independent life in the Halcyron community with a group of artists, intellectuals, and bohemians. Schindler continued living and running his studio from the Kings Road house. In 1930 Gibling returned to the house and settled in the section originally projected for the Chace, where she lived until her death in 1977.

19.5 A Woman Without a Woman's Role

From the already mentioned records, it is possible to observe that Gibling's trajectory was not common to most women in the early decades of the twentieth century. From the beginning, the composer establishes a relationship with her husband that goes beyond the sentimental, she speaks of a collaborative relationship and a leading equity in different aspects of her life, including the professional. But this occurred mainly during the first year of marriage, when they lived in Taliesin and especially in the first three years in Los Angeles.

The equitable relationship of the marriage would change quickly and drastically after Gibling took prominence in her political activism activities. At that time Schindler focuses on the growth of his studio and Gibling on the garden sessions. Also, it is important to note that, in the same way as it happened with Wright, Gibling established friendships independently of her husband, which also lasted after their separation and helped her to establish herself independently in Halcyron, in El Carmel and then back in Los Angeles. This was not common in women of her time and on several occasions, she is judged by the wives of Schindler's colleagues, as is the case with Dione Neutra.

The separation of criteria and objectives in the couple was another breaking point. According to Sweeney, the garden meetings seemed to have a different objective for each of the members of the marriage. Schindler enjoyed and was the life of the party at these meetings, but there was no real activism in him, his focus on Architecture was always his passion, and the meetings allowed him to make contacts and meet potential clients for the survival of his studio. Whereas, in the case of Gibling, one can see a genuine interest in actively advocating for her ideals, promoting and maintaining the garden sessions, as well as documenting in letters and communications the different personalities who visited her home.

The house, as a project and as a completed work, was an enhancer of Gibling's personal way of assuming society, it may be Schindler's project, but it was always Pauline Gibling's home.

> [Schindler] Proposed a dwelling whose material, organizational, and spatial conception expressed the values of a new architecture that was nourished by the essence of California, a special place that finally allowed the lifestyle desired by Pauline to become a reality. It could therefore be said that if the architecture was Rudolph Schindler's, the house was Pauline Gibling's. (Parra-Martínez, 2012, p. 150).

Ultimately, it is possible to define a consolidation of the architectural project beyond its initial purpose. The original idea of the house enhanced its subsequent development and led to the establishment of its legacy. In this sense, the house on Kings Road is important both for its architecture and for the people who met there, the different and constant cultural displays, and the permanence in the memory of those who passed through it. For this reason, the declaration of a shared and collaborative authorship between Schindler and Gibling is considered pertinent.

19.6 Conclusions

The research shows the social, economic, and political context in which Pauline Gibling began her personal trajectory as a political activist, composer, and editor. However, it is possible to see that her marriage to Schindler relegated her to the background and in the same way that the house went down in history as the Schindler House, her legacy around the avant-garde sessions that took place in the garden was also attributed to Schindler.

Pauline Gibling, like other great women who lived in Los Angeles at the beginning of the twentieth century, among whom we can mention Aline Barnsdall, Leah Lovell and later Galka Scheyer were characters whose contributions have been omitted or downplayed, in the history of this moment of boiling and consolidation of the Californian metropolis.

The specific case of Gibling's participation in the house on Kings Road is, moreover, easily verifiable. No previous or subsequent work by Schindler resembles this house, there is no direct inheritance from Wright's work and even less from his master Adolf Loos.

Gibling's participation in the house project, with her imagination, underpins the very particular decisions of this project, which also marked a milestone by proposing a new paradigm about housing and community living. With its *sleeping baskets*, the central courtyard, the single kitchen for the two families, and the work studios inside, the house became a treatise on collective living in the new society that was emerging in early Californian modernity.

It is important to note that this does not nullify or eliminate Schindler's authorship in the project, but rather to make visible a practice system in the development of modern historiography, in which collaborations are discarded. In this specific case, Gibling's participation makes sense of the differences that can be perceived in the architect's later works.

This is a systematic behavior in the construction of modern history, which has made collaborations, especially female figures, invisible (Muxí, 2018). This is not intended to add one more name to a list of personalities, which can be infinite, but to identify and analyze the people who made the development of the modern avant-garde possible. Also, it is possible to verify interdisciplinary collaborations and the value that modern architects gave to the forms of thought that surrounded them.

The history of modern architecture showed the architect of that time as a unique figure and did not reflect the collaboration he received from work teams, cultural and family environment. Nowadays, with a more sensitive view towards collaborative work. The interdisciplinary and multidisciplinary approach makes evident the valuable contribution of these silenced figures, who accompanied the historical architectural processes and their development.

References

Banham, R. (1971–2016). *Los Angeles. The architecture of the four ecologies*. Puente publishers.
Canosa Benítez, S. (2016). Lovell versus Lovell. *REIA European Journal of Research in Architecture, 6*. http://reia.es/REIA6_01.pdf
Castro, A. (2022). The silent gesture. Pauline Gibling Schindler in modern architecture. In *17th Docomomo conference 2022*. Valencia.
Colomina, B. (March 30, 2018). The secret life of modern architecture or we don't need another hero [Conference]. Harvard, GSD. New York.
Crawford, M. (1997). Forgetting and remembering Schindler: The social history of an architectural reputation. In G. Gilli (Ed.) *RM Schindler*, 10 casas. 2G No. 7.
Crawford, M., & Gilli, M (1997). RM Schindler: 10 houses=10 houses. In *2G International Architecture Magazine*. Barcelona: GG.
Frampton, K. (2022). *The other modern movement, architecture 1920–1970*. Yale.
Gebhard, D. (1979). *Rudolph Schindler*. Oikos Tau.
Gibling, P. (circa 1970). Taliesin in 1919 (unpublished document). UC-Santa Barbara Architecture and Design Collections, Schindler Collection and Archives of American Art, McCoy Papers.
Hamilton Harris, H. (1979). Foreword. In E. McCoy (Ed.), *Vienna to Los Angeles: Two Journeys* (pp. 7–15). Art+Architecture Press.
Hines, T. S. (2019). Critic and catalyst: Pauline Gibling Schindler (1893–1977). *Getty Research Journal, 11*(1). https://doi.org/10.1086/702748
Muxí, Z. (2018). *Women, houses and cities*. DPR Barcelona.
Parra-Martínez, J. (2012). *Modern and Militant. California through Harwell Hamilton Harris and Gordon Drake* [Unpublished doctoral thesis]. Polytechnic University of Valencia. https://doi.org/10.4995/Thesis/10251/27713
Smith, E. A., Darling, M., Helfrich, K. G. F., Sweeney, R. L., & Wilson, R. G. (2001). *The architecture of RM Schindler*. Abrams Incorporated Publishers.
Sweeney, R. (2001). Life at Kings road: As it was 1920–1940. In E. A. Smith (Ed.), *The architecture of RM Schindler* (pp. 87–115).

Part V
Processes, Technologies and New Materials: Creation of Possible Futures, Speculative Design, New Aesthetic Universes, New Ecosystemic Relations

Chapter 20
Metaversal Space: Methodology for the Definition of Space in the Metaverse

Daniel Ronderos-López and María Camila Jiménez Frias

Abstract The Metaverse is an area of opportunity for all human activities related to the Internet. It presents a perspective of the future where society's life will unfold (Schumacher, 2022). Understanding architecture as a container of people and activities defined by vertical and horizontal elements, openings and closures, and the metaverse as an environment where the virtual and physical coexist (Tang and Hou in CAADRIA Proceedings, 2022), some fundamental questions arise: what could be the shape of this space? To what extent should these architectures contemplate virtual elements and components of the physical world? To answer these questions, the study addresses the dynamics of information capture and participatory design with people between the ages of 4 and 72. This allows us to observe and compare the surprising skills and spatial sensitivity that children demonstrate concerning the interpretation of distance, angles, and direction (Newcombe and Huttenlocher in Handbook of child psychology, 2007), the transformation of childhood object recognition into spatial abstractions conditioned by formalized representation codes (Newcombe and Huttenlocher in Handbook of child psychology, 2007) and the formation of languages that enable the understanding of the physical world, beyond recognition and coding, which allows the conceptualization behind spatial compositions. The research established that there is a desire to idealize the inhabited space according to what is known at a physical and conceptual level. In this sense, to propose the space in the metaverse, it is necessary to learn from the different ways the spatialities of the real world are understood, including in this reading the diversity offered by different types of users, from childhood to mature adulthood.

Keywords Metaverse · Children · Space · Virtual reality

D. Ronderos-López (✉) · M. C. Jiménez Frias
Universidad de los Andes, Bogotá, Colombia
e-mail: d-ronder@uniandes.edu.co

M. C. Jiménez Frias
e-mail: mc.jimenezf@uniandes.edu.co

M. Vicuña and E. Giorgi (eds.), *Intersections*, Springer Geography,
https://doi.org/10.1007/978-3-031-76402-8_20

20.1 Introduction

This study frames the definition of metaversal space within the term Metaverse coined in the novels of Neal Stephenson and Ernest Cline, "Snow Crash" and "Ready Player One" respectively, where a virtual world is proposed that separates from reality so that people can escape from their daily lives to interact, create, socialize, play and exchange ideas with other users through avatars (Tang & Hou, 2022). This implicitly suggests a confluence between the virtual realm and the real world, which raises a duality in the representation of space in terms of form and function (Tang & Hou, 2022). Under these determinants, the work focused on the development of a methodology aimed at space designers that considers the variables conferred by this dichotomy. Likewise, the flow of the methodology is based on the intersection between design thinking tools and the use of technologies focused on architectural design. This, with the aim of building procedures leading to spatial concepts empathetic with different types of users through simulations that improve the experience in the visualization and understanding of space (Bertol, 1996).

For this, the study first segmented the participants according to the characterizations provided by Hannoun, Jackendoff, Newcombe, and Huttenlocher about the understanding and spatial representation at different stages of life. Second, it used design tools that allowed the understanding and reformulation of the problem (Brown, 2019) and created information capture procedures that respond to each user segment. Third, it defined spatial conception parameters validated from surveys about the virtual experience in spaces built according to the conclusions and findings of the work with the participants. Finally, it built co-creation dynamics using spatial representation media. To prototype the workflow of creating digital scenarios applicable to the Metaverse based on the principles of spatial perception of different user profiles.

20.1.1 Description of the Processes for Capturing Information

The study considered dividing the participants into groups based on the characterization provided by the literature on human spatial perception at different moments in life according to the following descriptions:

Group 1: People who perceive and represent lived space and the space perceived are made up of elements with which they would like to interact directly, understanding them from past physical and direct experiences (Hannoun, 1977). This description is generally associated with childhood.

Group 2: Subjects, generally teenagers and young adults, who have the ability to perceive space despite not having been in direct contact with it (Hannoun, 1977) and demonstrate the ability to integrate coding systems based on metric information and categorized information (Newcombe & Huttenlocher, 2007) to represent it.

Group 3: Individuals, generally adults and mature adults who achieve spatial understanding through the recognition of objects, visual codes, and linguistic components (Jackendoff, 2012), which allows them to attribute meaning to elements that make up the space.

The exercises with each group considered semi-structured interviews (Design Method Toolkit Search results interview, n.d.) using design tools and surveys to establish qualitative criteria. These elements allowed the coding of spatial understanding and representation and enabled subsequent co-creation processes. Thus, the aim is to frame the problem and provide context to understand the people for whom it would be designed (O'Grady & O'Grady, 2017), while simultaneously building a platform for design decisions to create value by building meaningful links with individuals and the research tools used to cultivate a relationship between the user and the project (O'Grady & O'Grady, 2017).

To ensure consistency in the information obtained from the different interactions with the groups of participants, the study centered on the collection of spatial perceptions around a common space for all age groups, asking the question: how do you imagine your ideal home? This decision was made because the domestic space is common for children, young people, adults, and older adults and they have continuously inhabited it, regardless of their origin or life experience.

20.1.2 Information Capture Processes with People Belonging to Group 3

The study contemplated the characterization of space understanding provided by Jackendoff, which indicates that individuals achieve spatial understanding through the recognition of objects, visual codes, and linguistic components. This is because vision enables the construction of physical and conceptual encoded space and language often suggests the meaning of an imagistic component (Jackendoff, 2012). In this sense, a process of capturing qualitative information was carried out that relies on the use of various design tools that allow analyzing and framing the problem that arises from the understanding of space, as well as devising abstract scenarios in search of satisfying needs, and meeting expectations and desires of users. For this, concatenated and sequential dynamics were used to arrive at the capture of parameters that define ideal spatialities based on the question: how do you imagine your ideal home?

The proposed workflow appeals to spatial understanding through the recognition of objects, visual codes, and linguistic components (Jackendoff, 2012), so first the Empathy Map (Design Toolkit | Empathy Map, n.d.) (see Fig. 20.1) is contemplated as a tool that allows determining the perceptions that users have of space, as well as visualizing possible gains and pains. Through this activity, it is discovered what are the relevant themes to be taken into account for the next stage of the exercise. Next, the Experience Map (UX Mapping Methods Compared: A Cheat Sheet, n.d.)

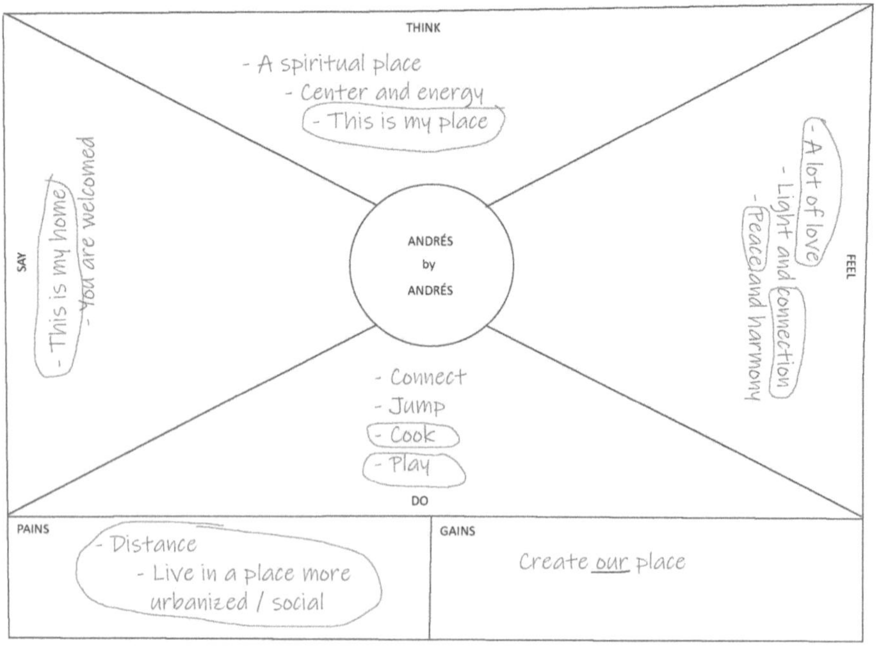

Fig. 20.1 Empathy map (Design Toolkit | Empathy Map, n.d.) user of 55 years

(see Fig. 20.2) shows how, through experiential mapping, the experiences associated with space are understood through the senses. In this dynamic, users must connect the themes stated in the empathy map and the possible sensory perceptions linked to specific spatialities. Thirdly, the User journey or user journey (Kalbach, 2016) (see Fig. 20.3) allows tracking of the activities associated with various spatialities over a period of time. With this tool, it is sought that users deploy the sensory perception connected to spaces over defined atmospheric, chronological, or situational temporalities. Finally, the summary scenario (Patton & Economy, 2014) (see Fig. 20.4) appears as a representation table of the values attributed to space based on what was stated in the first three tools. In this way, users have the possibility of generating a summarized visualization that provides insights about preferences and spatial conceptions.

20.1.3 Information Capture Processes with People Belonging to Group 2

With this group, it was initially sought to use the same methodology used for group 3, however, significant variability was identified in the abilities to express ideas clearly, either through drawings or testimonies. While the subjects of group 3 use

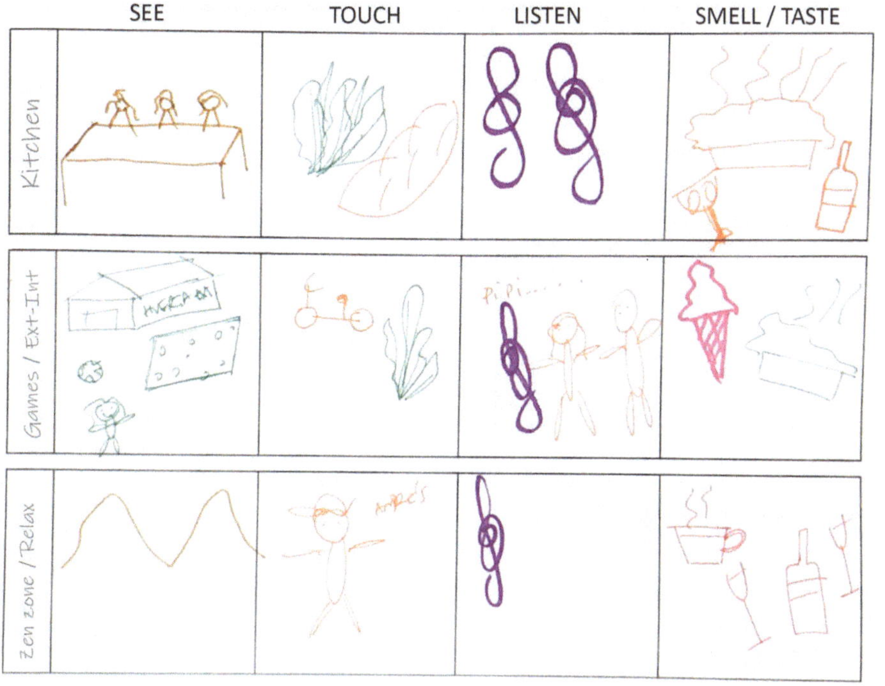

Fig. 20.2 Experience map (UX Mapping Methods Compared: A Cheat Sheet, n.d.) sensory of a 40-year-old subject

schematic drawings and forceful explanations to describe space, users of group 2 seek to make realistic three-dimensional illustrations or detailed architectural plans that do not fully coincide with what is exposed in their testimonies. Consequently, and in order to unify criteria within this sample, the study adopted Hannoun's definitions of the phase where one has the ability to perceive space despite not having been in direct contact with it, where one can distinguish distances and positions of objects in relation to others, and even the notion of geographical space (Hannoun, 1977). Likewise, Newcombe and Huttenlocher's characterization was taken into account, which indicates that subjects show the ability to integrate coding systems based on metric information and categorized information. This implies that the grouping of elements and components is used to understand space through mental subdivisions and represent it in two- and three-dimensional graphic dispositions (Newcombe & Huttenlocher, 2007).

These definitions allowed the use of a mixed information capture process that incorporates the procedures carried out with people from group 3, along with unstructured interviews leading to the elaboration of drawings representative of space from the question: how do you imagine your ideal home? The results of these interactions, (see Figs. 20.5 and 20.6) show that the understanding of space is given from the volumetric representation of objects whose shape is crucial for the configuration

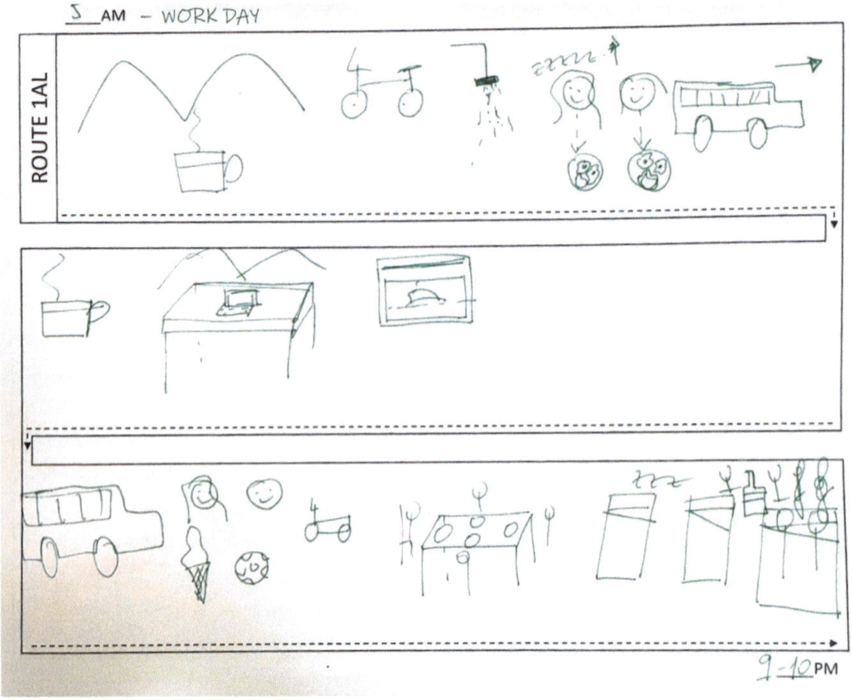

Fig. 20.3 User journey (Kalbach, 2016) of a 40-year-old user

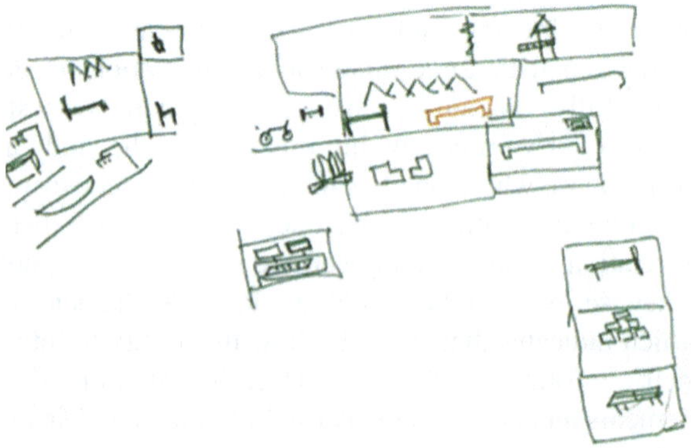

Fig. 20.4 Scenario summary (Patton & Economy, 2014) subject of 55 years old

and reading of the body in space and movement through it (Jackendoff, 2012) and the three-dimensional and two-dimensional coding of space decomposed into parts, establishing a hierarchical order in order to provide structure and narrative to the spatiality that is sought to communicate (Jackendoff, 2012).

Fig. 20.5 Two-dimensional drawing similar to a plan where a 17-year-old person explains the distribution, order, and spatial hierarchy of the same enclosure

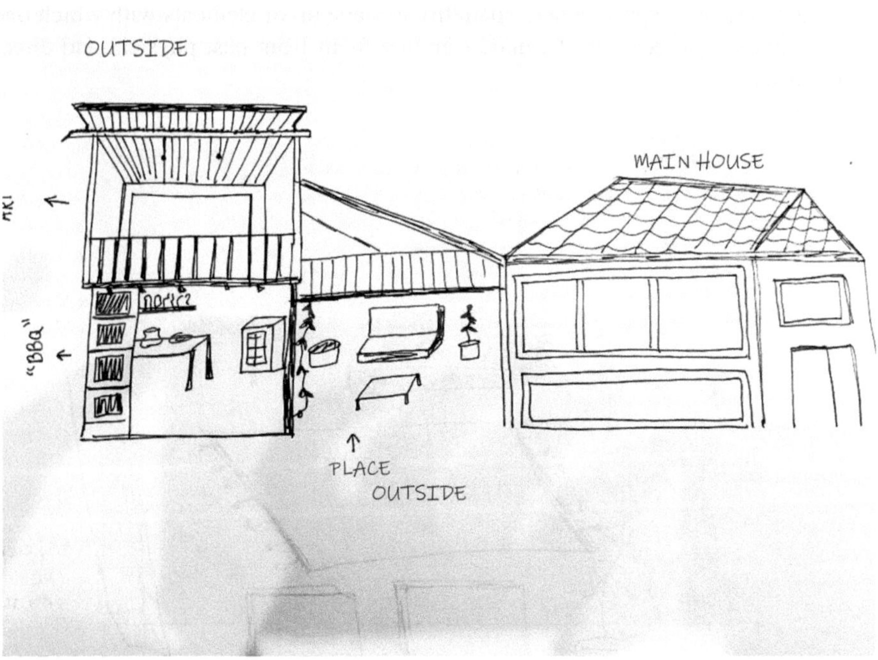

Fig. 20.6 Three-dimensional drawing that seeks to establish the size, position, order, and hierarchy of spaces through a representation that bears similarities to the construction of a perspective with vanishing points

20.1.4 Information Capture Processes with People Belonging to Group 1

For work with this group, the use of design tools applied with group 3 and again implemented with people from group 2 was desisted, because most of the subjects with whom interaction was lacking the skills required to elaborate intelligible schematic drawings and provide testimonies that would allow identifying patterns in the understanding and representation of space.

So, under Hannoun's description of the phases of lived space and perceived space, which indicates that an image of physical space is formed through permanent contact and experience through movement, creating a state of "here" as a direct experience (Hannoun, 1977), unstructured interviews were used that encouraged the making of descriptive drawings of the ideas and reflections that the question raised: how do you imagine your ideal home? Since these spatial representation drawings did not allow a clear reading of the relationship of spaces in horizontal or vertical planes, sequentiality, or proportion of elements. Constituents, models were chosen to be built that will allow visualization of the volume proposed by the participants.

As an example of this, Figs. 20.7 and 20.8 illustrate the exercise carried out which, according to Hannoun, corresponds to spatial representations in the stages of lived space and perceived space, where spatiality is made up of elements with which one would want to interact directly, understanding them from past physical and direct experiences.

Fig. 20.7 Drawing by a 5-year-old child about the elements that should compose his ideal house. *Note* The author's testimony is included: "I would like it to be very large, to have a lot of space to plant vegetables and to plant seaweed to give to the animals that are in the water. That it has a lake to give them algae because they have no food. I wish that all my family's world could fit there and that it is a large pool that everyone could fit in. That it has a nice view and that it has a lot of habitat"

Fig. 20.8 Co-creation model with a 5-year-old child to understand the volume and elements that determine the relationships with the body and movement through space. *Note* In the exercise, it became evident that natural elements are constitutive of space and their representation is faithful to their visualization in reality. The same does not happen with non-natural elements such as the house (which has no depth, it is just a plane) or the pool (which appears as a vertical plane of water), which are defined as simple shapes that do not respond to real physical variables, but appear as instruments to understand the position of the body in space and its movement through it

20.1.5 Mixing Elements for Spatial Formulation

For the formulation of spaces in the metaverse, it is necessary to reach the right mix between the real and the virtual (Tang & Hou, 2022). In this sense, three virtual scenarios were produced from the perceptions and spatial representations described in the information capture process. This medium allowed the participants to confront concrete elements and record their preferences regarding the configuration of space in conceptual, formal, perceptual, and material terms.

Each scenario was experienced by users through virtual tours, produced with real-time rendering applications, and navigated using virtual reality glasses (see Fig. 20.9). The three proposed digital spaces contemplated different elements attributed to the convergence of the perceptions and representations of the three age groups, as indicated by the Venn diagram in Fig. 20.10.

Fig. 20.9 Different users navigating 3D models using virtual reality glasses

Virtual scenario A refers to a visualization of a space totally possible in the real world as illustrated in Fig. 20.11. This approach is generated under the principles exposed by Hannoun and the spatial perceptions and representations about the understanding of lived space and perceived space recorded from groups 1 and 2. The second, as seen in Fig. 20.12, was created using the coding of recognizable elements from a mix between the space decomposed into parts and its linguistic representation. Finally, as shown in Fig. 20.13, the third model, was based on the mix between physical and conceptual coded spatial construction and the spatial relationships declared by elements that allude to past physical and direct experiences.

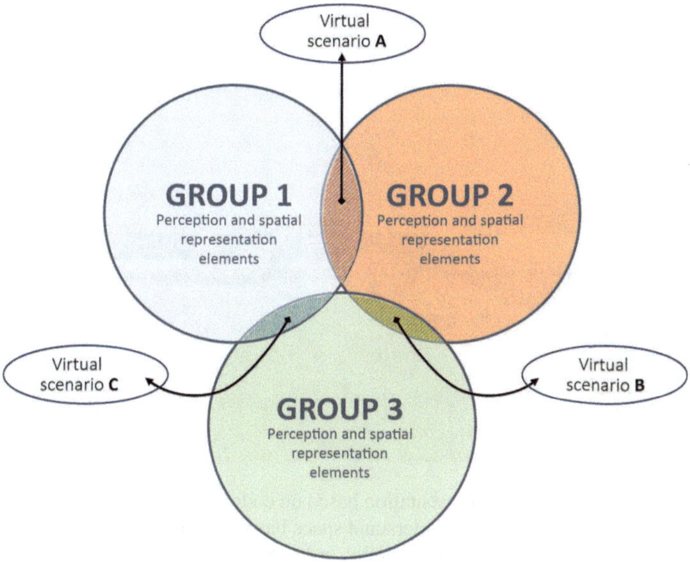

Fig. 20.10 Venn diagram illustrating the convergence of elements of perception and spatial representation of each age group with which the 3D models that participants virtually toured were built

Fig. 20.11 Virtual scenario A: Literal representation of space in a virtual environment. *Note* This modeling seeks to recreate real-world conditions, where the only virtual reading is the relationship with the body and movement through space

Once the scenarios were virtually toured by the users, a series of semi-structured interviews and a survey were conducted that in total gathered the opinions and perceptions of 75 participants. The results revealed that the virtual scenario A was the one most subjects preferred with 63.3%, and the scenario B was the least selected with 6.7%. Likewise, scenario A obtained 100% preference in subjects between the ages

Fig. 20.12 Virtual scenario B: Representation based on codes of representation with recognizable objects. *Note* This modeling seeks to understand space based on a mix of representation codes and literal elements that allow establishing hierarchy, order, scale, proportion, and dimension

Fig. 20.13 Virtual scenario C: Representation based on relationships spaces defined by atmospheric qualities and boundary elements. *Note* This modeling seeks a representation of space at a conceptual level, exalting atmospheric conditions over the material determinants of space

of 4 and 18, 57.7% in users between 15 and 35 years old, and 70% in the group of over 35 years old. Also, the third tour, corresponding to scenario C, was the second most preferred with 30% of the sample and had its greatest acceptance among older people.

Among the reasons that the subjects contributed when choosing their preferences, they argued that tour A had aspects related to fidelity to reality and elements referring to the climate and temperature of the environment mainly. About model C, there was a reference to the sensations it transmitted and the mix between abstract and real elements that defined the sensory parameters in the space. Finally, about scenario B, its lack of preference was due to its spatial nature being unreal and not being familiar with elements with which users could relate.

20.1.6 Considerations for Co-creation from the Learnings of Information Capture and Mixing of Elements for Spatial Formulation

The work done with group 1 provides parameters for formulating spaces where natural elements such as the sun, vegetation, and atmospheric conditions are specific and are represented consistently with reality. However, elements that could determine scale, proportion, position, and dimension could be considered virtual parameters since they do not respond to their physical characterization, but to signs in terms of form and appearance experienced eventually.

The spatial representations made by people in group 2 are based on the representation of codes to express materiality, scale, proportion, and dimension. Less importance is given to the elements that determine atmospheric variables, given that it is assumed that they exist and should not influence spatial characterization. In this sense, the real elements that formulate space are the materiality and specific form of objects, and the virtual elements correspond to the codes that make up a "visual vocabulary" (Jackendoff, 2012) that enables spatial definition.

In the dynamics carried out with people from group 3, natural elements reappear as compositional determinants and recognition of the tension between interior and exterior space, that is, what is outside and what is inside. The representations do not seek to achieve detailed characterizations either formally or materially, and they emphasize the interactions that are expected to be had with objects and other users within the space. It was the only group that referred to spatial qualities in terms of physical comfort.

Similarly, the dynamics of touring digital spaces using virtual reality allow us to outline the preferences of different user groups in terms of the formulation of possible spaces in the metaverse. For this reason, it should be considered that there is not a single formula, but various paths that depend on the segments or profiles with which it seeks to interact. If a metaverse for children is proposed, it must be faithful to reality, since the understanding of space at this age starts from familiarity with objects, and although the spatial representation can be considered as the most abstract, it is actually the one that most seeks the literal visualization of its context.

Both teenagers and young adults seek codes relatable to reality to have their own readings of space and are indifferent to elements that add value in terms of physical comfort. The representation of space goes through achieving realism and mastery of methods and tools of spatial representation; therefore, the visualization of space is mediated by the ability to represent the objects that compose it.

Similarly, it was observed, in users over 35 years old, that the presence of atmospheric elements that define space returns and becomes present in the same way as in early childhood, only that this time there is awareness of the value it adds spatially and in sensory terms, since it allows to anchor the virtual environments with reality and the experience of spaces lived previously. This allows users to maintain a sense of "place in the world" while they seek the desired escape from the every day to imaginaries located in memory that are defined in synthetic terms that abstract what is conceptually relevant.

20.1.7 Co-creation Exercises of Digital Scenarios Applicable to the Design of Space in the Metaverse

The co-creation exercises sought to build mental simulations through the activation of memories, knowledge, and previous physical experiences that would allow future representations of how the world would function (Mitchell, 2019), or in this case, the digital space in the Metaverse. The activities are prototypes of a workflow that consists of asking about a place they would like to *visit* physically, under one of these two conditions: (1) that they have never been there, but have seen it through videos or images, or, (2) that it was an unknown, hypothetical or imaginary space with little or no information about its physical characteristics. With these rules established, images are generated that allow users to visualize the space, and in this way suggest changes and make iterations until reaching a satisfactory scenario. Subsequently, the images are translated into 360 panoramas that finally lead to reflections on their sensory experiences when interacting with them virtually.

20.1.8 Description of the Co-creation Sessions

In the co-creation sessions, semi-structured interviews were conducted about a place to visit under one of the conditions described above. During the discussions, questions were asked about the description of these scenarios in terms of physical and experiential characterization when imagining being in these sites. Using the testimonies, *prompts* or instructions were written to feed a generative artificial intelligence engine (IAGen) for the creation of images that described the provided parameters (see Fig. 20.14). Once the representations were created, it was asked if these met what they had imagined, if edits should be made, or if the instructions needed to be modified to obtain more accurate results.

Once the visualizations were produced according to the desired parameters, they were transformed into 360 panoramas that were experimented with using virtual reality glasses. Finally, and at the same time as the digital representations (see

Fig. 20.14 Images co-created with users from descriptions and instructions given to generative artificial intelligence engines (IAGen)

Fig. 20.15 Participants using virtual reality glasses, touring digital scenarios produced with IAGen and referring to objects in space

Fig. 20.15) of the described space were being traversed, the final part of the interview was conducted where reflections were made on the sensations, emotions, and experiences produced by the co-created scenario.

20.1.9 Co-creation Activities with Participants

Five activities were carried out, two with children and three with adults. The first, with a 5-year-old boy, the second with a 13-year-old girl. In response to the question about the space they would like to visit, the 5-year-old boy said he would like to go to Madagascar, motivated by the animated film of the same name. In the case of the

13-year-old girl, she chose to visit a football stadium, inspired by her love for the sport. Figure 20.16 shows the images produced and edited with IAGen software.

Table 20.1 describes the instructions for each simulation, the edits made after the first result, and the number of iterations, as well as the version chosen by the user.

Based on the navigation and experiences described by the users, it can be concluded that the atmospheric time and the elements that make up the space are attributed to preconceptions rather than visible elements in the simulations. The boy indicated that it is cold in Madagascar because there are penguins, referring to the animated film, because neither were they included in the prompts, nor do they appear in the image. Something similar happens with the girl, who indicates that it is hot in the stadium because "there are many people", although in the instructions to the IAGen, empty stands were specified, and the representation shows it in the same way.

Fig. 20.16 Images produced and edited with IAGen with users aged 5 and 13

Table 20.1 Description of interactions with children to define instructions for the creation of descriptive images of the virtual space

User	Initial instruction	Edits	# Iterations	Chosen iteration
5-year-old boy	Create an image of Madagascar with lions and wild animals, palm trees and other fruit trees, cold weather, a black floor, and a view of mountains	Include zebras, lemurs, and meerkats	3	2
13-year-old girl	Create an image of a football stadium, with a football, with empty stands, with a football shirt at sunset	None	1	1

Three co-creation activities were also carried out with adults of different ages, as shown in the Table 20.2. The first was with a 31-year-old woman who sought to visualize the space of a forest inspired by the planet Pandora from the movie Avatar. The second was a 40-year-old man who chose to *travel* inside the brain, specifically describing the space of a neuron. Finally, a 61-year-old woman preferred to simulate a place in Greece.

The 31-year-old woman described perceiving an environment where she feels "a pleasant cold" attributed to the jacket that the woman in the visualization is wearing, and that the fact that she is in a canoe transmits that there is a sense of direction, where she "knows where she is going", which gives her security (see Fig. 20.17). The 40-year-old man indicated that the space has (see Fig. 20.18) "a very nostalgic and dramatic sensation" and that the surfaces of organic shapes and connected glares agree with what he imagines are the neural spaces. On her part, the 61-year-old woman described that in the visualization of Greece (Fig. 20.19) "it was hot, but the environment was cool due to the sea breeze" and that she knew it was Greece because "there are olives and goats".

In all three cases, the users mentioned words related to their mood based on the space attributing meaning to elements beyond their literal physical characteristics.

20.2 Conclusions

In co-creation with children, it is observed that the representations of space are based on mental simulations constructed from recognizable elements with which they want to interact, along with memories and past knowledge that predict possible scenarios, so much so that they only need physical references with which they have previously had contact, whether in person or through some audiovisual medium to attribute sensations and grant reading to elements of space that are not visualized in digital environments. So, for this type of user, the metaversal space, although literal in its conception and design, can be incomplete and only contain objects relevant to their

Table 20.2 Description of interactions with adults to define instructions for the creation of descriptive images of the virtual space

User	Initial instruction	Edits	# Iterations	Chosen iteration
31-year-old woman	Create an image of a forest at night where the animals and plants glow, like on the planet Pandora, from the movie Avatar. With bodies of water like a waterfall, a lake or a river, with a distant view of snow-capped mountains where I feel safe and calm	Include that the view is in first person navigating in a canoe, on a river that glows, with glowing fish Modify so that the person navigating is a woman	3	3
40-year-old man	Create an image of a very wide place, with soft texture, with many smooth-textured branches of light colors. Where the sky is sparks of cables connected to each other, generating warm and electric backlighting, with white tones and has a nostalgic presence	Modify the ground for a smooth and soft texture Modify the ground for organic shapes, soft texture	6	3
61-year-old woman	Create an image of Greece with many stairs climbing green mountains, where there are villages with white houses among the mountains. The place is hot, you feel peace, tranquility, happiness at sunset	Include the sea as a distant view The view should be from the top of a mountain Include goats Include olive trees	4	4

Fig. 20.17 Digital scenario of a tour through a river in a forest co-created with a 31-year-old woman by writing instructions for an AIGen engine

Fig. 20.18 Digital scenario of the space inside a neuron co-created with a 40-year-old man by writing instructions for an AIGen engine

Fig. 20.19 Digital scenario of Greece co-created with a 61-year-old woman by writing instructions for an AIGen engine

understanding and experimentation. The details and subtleties are not appreciated, since there is an imaginative component that naturally complements the simulation.

In the work with the group of adults, it is observed that the digital simulations must contain indicative and symbolic elements recognizable from past experiences or attributable to previously acquired knowledge. The difference with children lies in that they construct codes with meaning (for example: seeing the sea means that there is a breeze and that the environment is cool) that enables an imaginative component to understand the space from language. Likewise, it is observed that the generation of codes with meaning, is detailed and confers the necessary characteristics to the representation of space. When feedbacking the co-creation activity, these users referred to signs within the scenario to give sense or purpose to the simulation. For example, being in a canoe was indicative of security, since it described a course and one was not lost in a river that crosses a forest. This is important because it shows that indicative elements and subject to interpretation are required to enrich the experience within the simulation.

Finally, the study recommends that for the design of metaversal spatialities, the possible mixtures between the real and the virtual are considered based on the profiling of users according to the phase of perception and spatial understanding described in this article, along with the abilities to perform mental simulations and attribute sense and meaning to the elements that make up digital scenarios. Lastly, it is recommended to contemplate visualization and simulation methods that enable users with enough interactive elements and facilitate the collection and analysis of information on the part of the designers of spaces for the Metaverse.

References

Bertol, D. (1996). *Designing digital space: An architect's guide to virtual reality*. Wiley.

Brown, T. (2019). *Change by design, revised and updated: How design thinking transforms organizations and inspires innovation*. HarperCollins.

Design Method Toolkit Search results interview. (n.d.). https://toolkits.dss.cloud/design/?post_type=method-card&s=Interview

Design Toolkit | Empathy Map. (n.d.). http://design-toolkit.uoc.edu/es/mapa-de-empatia/

Hannoun, H. (1977). *The child conquers the environment: Exploratory activities in primary school*. Kapelusz.

Jackendoff, R. (2012). Language as a source of evidence for theories of spatial representation. *Perception, 41*(9), 1128–1152. https://doi.org/10.1068/p7271

Kalbach, J. (2016). *Mapping experiences: A complete guide to creating value through journeys, blueprints, and diagrams* (1st ed.). O'Reilly Media.

Mitchell, M. (2019). *Artificial intelligence: A guide for thinking humans*. Penguin UK.

Newcombe, N. S., & Huttenlocher, J. (2007). Development of spatial cognition. In *Handbook of child psychology*. https://doi.org/10.1002/9780470147658.chpsy0217

O'Grady, J. V., & O'Grady, K. V. (2017). A designer's research manual (2nd ed.). Updated and expanded: Succeed in design by knowing your clients and understanding what they really need. Rockport.

Patton, J., & Economy, P. (2014). User story mapping: Discover the whole story, build the right product. Van Duuren Media.

Schumacher, P. (2022). The metaverse as opportunity for architecture and society: Design drivers, core competencies. *Architectural Intelligence, 1*(1). https://doi.org/10.1007/s44223-022-00010-z

Tang, S. K., & Hou, J. H. (2022). Designing a framework for metaverse architecture. In *CAADRIA proceedings*. https://doi.org/10.52842/conf.caadria.2022.2.445

UX Mapping Methods Compared: A Cheat Sheet. (n.d.). Nielsen Norman Group. https://www.nngroup.com/articles/ux-mapping-cheat-sheet/

Chapter 21
New Landscape and Emerging Dynamics of Interaction Between Users and Urban Space Produced by Digital Delivery Platforms

Rodrigo Vera

Abstract Globalization and the advancement of information and communication technologies have transformed urban areas, highlighting the emergence of home delivery applications. These applications offer conveniences and job opportunities but also pose challenges such as job insecurity, traffic congestion, and exclusion of marginalized communities. In response to these dynamics, a research methodology based on design is proposed, emphasizing the integration between theory and practice. The research highlights the value of the prototype as a generative and learning tool, demonstrating the feasibility of collecting data through the active participation of citizens. This methodological approach is applied to explore the impact of home delivery applications in urban areas, identifying complexities and emerging dynamics. It focuses especially on the Metropolitan Region, thus contributing to the discipline of urban studies through research through design. This research underscores the importance of considering equity in the city in the context of the collaborative economy and digitalization. Thus, through the prototype of ethnographic instruments, it has been demonstrated that it is feasible to collect data through the active participation of citizens.

Keywords Design · Collaborative economy · Delivery · Digitalization

21.1 Introduction

We are living in an era characterized by the technological revolution of information and communications (ICT), the rise of a form of social organization connected in networks, and the global interdependence of economies and societies (Castells &

The original version of the chapter has been revised. A correction to this chapter can be found at https://doi.org/10.1007/978-3-031-76402-8_43

R. Vera (✉)
School of Design, Pontificia Universidad Católica de Chile, Santiago, Chile
e-mail: rdvera@uc.cl

M. Vicuña and E. Giorgi (eds.), *Intersections*, Springer Geography,
https://doi.org/10.1007/978-3-031-76402-8_21

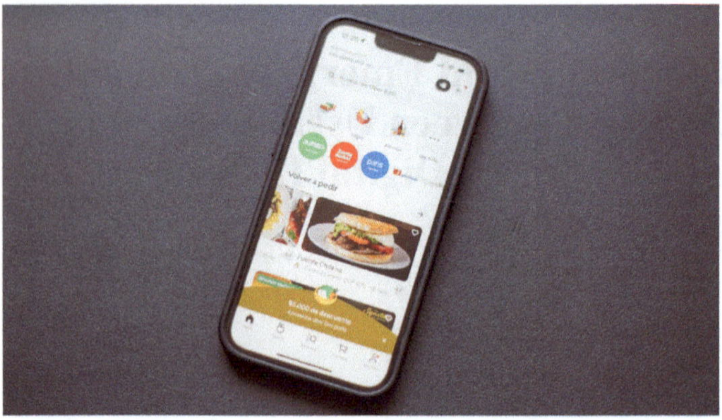

Fig. 21.1 Uber eats delivery application. *Source* Personal archive

Himanen, 2016). The global project of large companies, the emerging capabilities of information and telecommunications technology, and the increase in supranational elements in the functions of the State are jointly forming strategic levels that transcend national borders (Sassen, 2007).

In this context of globalization, delivery applications have proliferated mainly thanks to the so-called collaborative economy (Fig. 21.1). As Murillo (et al. cited by Social Innovation Institute, 2019) mentions, it would be appropriate to perceive the collaborative economy as a new form of neoliberalism through high-value international companies. The collaborative economy is based on the premise that many assets, including vehicles and properties, remain underutilized most of the time (Parker et al., 2016). This approach allows a bidirectional participation of consumers, who can play roles both as providers or receivers of resources, also known as the gig economy. In addition, this distribution modality integrates into what is called the "last mile," which refers to the final stretch in the transport of people and goods.

The massification of these applications, among other facts, has created a phenomenon of "delivery parking" where we can see motorcycles with the backpacks of different delivery companies in some places in the city waiting for the next shipment to go out for delivery (Fig. 21.2). Another phenomenon that can be derived from the use of delivery platforms and the growth of fast food orders is the emergence of ghost kitchens (dark kitchen), a phenomenon that arose with the boom of food delivery applications (Cueto, 2020), and food deserts, places to which deliveries do not reach either due to non-enabled routes or for their own safety. These phenomena, among others, are affecting the quality of life and the experience of city dwellers who use public and private spaces.

21.1.1 Context

The study on fast food consumption in Chile through delivery announced by Rappi in December 2021, revealed that from January to October 2021 the market grew

Fig. 21.2 Image of delivery parking located at Alameda with Santa Rosa (December 2022). *Source* Personal archive

175% compared to the same period in 2020. According to the indicator prepared by the Department of Studies from the National Chamber of Commerce, Services and Tourism in the last quarter of 2021 show that national sales of fast food service recorded a growth of 52.3%.

According to the study conducted by Artool for Rappi (2023), 82% of the Metropolitan Region (RM) of Chile states that they will buy the same or more in 2023. In addition, it is affirmed that Chileans project to increase their purchases in 2023 and that 8 out of 10 users of Generation Z and Millennials claim that they will maintain their purchase level or increase it in 2023. According to the study, the five most popular categories in delivery apps according to respondents are: prepared food, drinks, fresh food, pantry food, and sweets or kiosk products.

If we review data on the growth of delivery, at the beginning of 2021, Statista Digital Market Outlook -which provides data from 90 markets for the digital economy- announced that these home delivery platforms continue to position themselves as a key pillar of "the new normal". In different cities of Chile, delivery services have arrived that carry parcels from one place to another, mainly by motorcycle and bicycle (Peñafiel, 2022), causing an increase in the vehicle fleet of these means. In 2021, the National Association of Motorcycle Importers (ANIM) stated that the marketing of motorcycles in Chile had annually climbed 103% (considered a national historical record) compared to 2020 and 108% compared to 2019 (Orellana, 2022). In its December 2022 report, ANIM reported a total of 54,481 units sold during the period, just 16% below 2021.

From a complex systemic perspective (Sevaldson, 2023), the problem includes various actors who interact in direct and indirect ways (with each other) and who are favored or in some cases harmed by delivery applications and the experience in the city constituting a large complex system (Fig. 21.3). There are the companies that provide the service such as some of the ones already mentioned (Uber Eats, Pedidos Ya, and Rappi among others), there is the provider of the product to be consumed such as restaurants, there is the delivery that makes the shipment and the user/consumer who pays for the service. On the other hand, there are representatives of the public sector such as legislators and territory administrators like mayors who strive to create and apply territorial administration rules and others of various kinds. In addition, within the delivery applications, we also see the designers and developers who create these applications. There are also all those urban studies experts and territorial planners who have observed and studied some impacts of delivery in the city. Finally, there are also the citizens who see their experience in the city affected.

21.1.2 Characterization of Phenomena

The complexity of the challenges facing delivery drivers, and the constant development of the home delivery industry reflects the diversity of dynamics present today. In an ever-evolving urban environment and with the marked growth of home delivery platforms, significant changes, and multifaceted challenges are observed that impact the structure of modern cities.

As Dalla et al. (2022) propose, delivery vehicle drivers face difficulties finding spaces on the sidewalk in urban areas, leading to practices such as cruising and unauthorized parking. The authors also mention that the increase in the fleet of commercial vehicles in cities has not been accompanied by a change in urban infrastructure to support them. In this context, "Cruising" refers to the practice of driving in circles or erratically in search of a parking spot in a congested urban area, a practice that may be common among commercial vehicle drivers making deliveries in urban areas where parking is limited and demand is high (Dalla et al., 2022). Phenomena like this are perhaps seen in large metropolis cities around the world.

According to Reed et al. (2021), the problem with delivery and parking is that delivery drivers spend a significant amount of time looking for street parking, which increases delivery time and reduces delivery efficiency. Studies have found that delivery drivers spend on average 5.8 min and 24 min searching for each parking spot in Seattle and New York City, respectively.

Dark kitchens (Navarrete et al., 2022), also known as ghost kitchens, are a new gastronomic business model that emerged from the demand for platforms to deliver food at home and that increased during the health crisis of the COVID-19 pandemic. According to Navarrete et al. (2022), they are surfaces with separate kitchens that do not have waiters or furniture, and only require kitchen staff, raw materials, stoves, and a site on some home delivery platform. Added to this phenomenon, dark stores (virtual or "dark" stores) are logistical spaces used to store and prepare products for

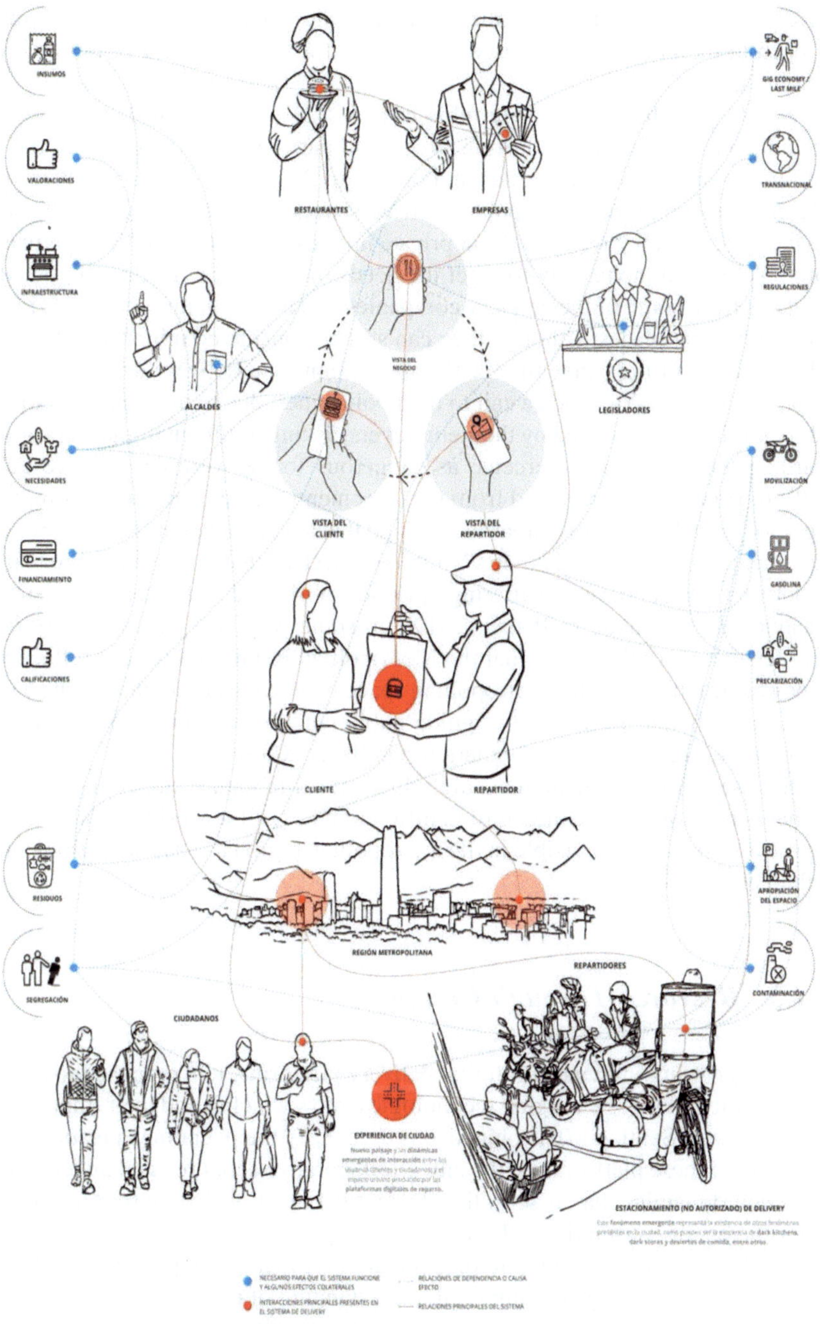

Fig. 21.3 Service Gigamap. *Source* Own elaboration

delivery to customers who buy online. Unlike traditional retail stores, dark stores are not open to the public and are used exclusively for the fulfillment of online orders (Shapiro, 2022).

According to the study by Hurtubia et al. (2021), a group of delivery workers in Santiago de Chile, declared using streets and parks as bathrooms due to the lack of a public one, as well as bottles and alleys. In this sense, they are still far from believing the configuration of public–private spaces that are suitable for the sector is achievable, but they recognize that private initiative in certain Malls has advanced more quickly and inclusively to meet the needs of the sector. To this can be added initiatives from some of the delivery companies together with municipalities to have more formal rest places, although they can still be considered minor initiatives.

In the midst of the pandemic (2019) application food delivery drivers like Uber, Rappi, and Pedidos Ya, made a call to their colleagues, due to the lack of guarantees and social security provided by their employers, asking to fill out digital cartography regarding the places they perceive as dangerous for deliveries (Labbé & Palma, 2019). In general, it is extracted from the aforementioned collaborative cartography (Fig. 21.4) that there are vast areas in Santiago that are devoid of home deliveries, this phenomenon is called "food deserts". It is hypothesized that this is due to the fear that the delivery drivers have of suffering episodes of violence when entering, staying, or leaving them (Labbé & Palma, 2019). In some way, the great disparity in the socio-territorial configuration that the Metropolitan Region shows is a reflection of the power structure over the metropolitan space, where the communes of the eastern cone show a better positioning to guide public and private actions in favor of the public interest of its resident population (Orellana, 2009).

All these territorial phenomena are transnational in nature, appearing according to what is reviewed in the literature in different cities of the world, but also through empirical analysis, most visible in the metropolitan region as has been demonstrated by field observation.

21.1.3 Research Through Design

Given the growing influence of digital delivery platforms, it is proposed to examine the integration of these emerging technologies into social life and their impact on the urban spaces of Santiago. The objective is to identify the phenomena resulting from the impact of these platforms on the city's collaborative economy and understand the interaction dynamics that arise in the changing context of the metropolis, covering various layers of constantly evolving relationships.

This challenge requires an interdisciplinary investigative approach (Dorst, 2018) that can expand the range of empirical understanding of the phenomenon assuming its complexity. It is key to be able to gather information that is collected from contexts that present the so-called 'wicked problems' (Spencer, 2018) -defined as such for their complexity and difficulty to solve- which can have multiple solutions. At the same time, the problem must be addressed from a systemic perspective (Sevaldson,

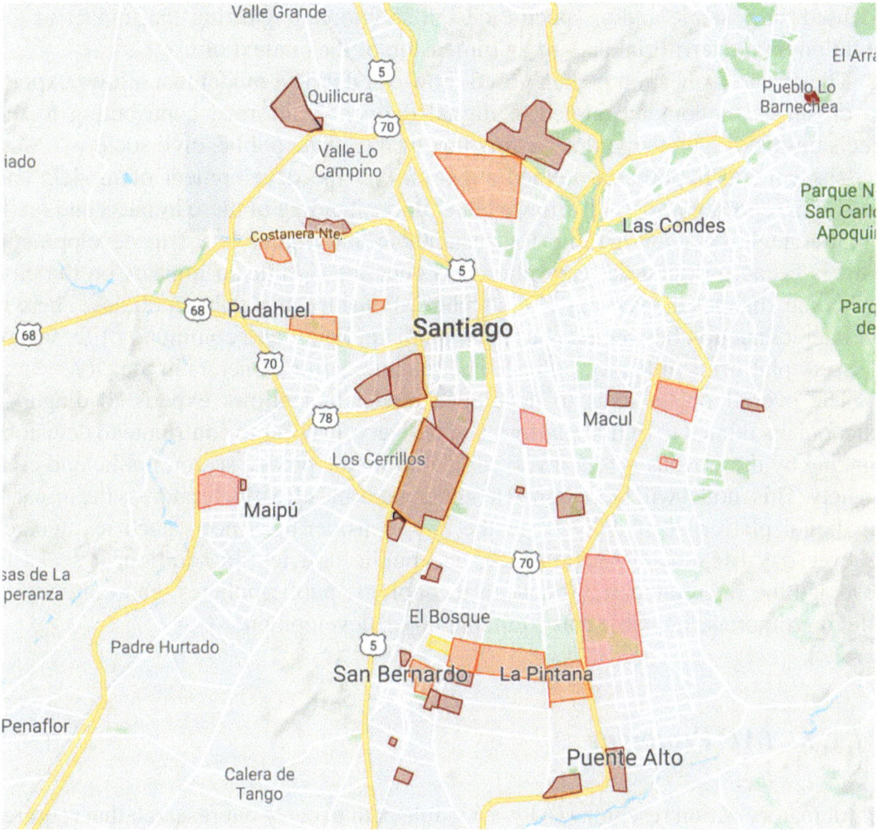

Fig. 21.4 Collaborative map of application workers, 2018. *Source* CIPER

2023) processing and analyzing the multiple variables that affect the problem. An interdisciplinary collaboration (Spencer, 2018) that proposes results that manage to reach preferable, sustainable, and equitable situations. On the other hand, the human, ethnographic, and social dimensions of the phenomenon require the incorporation of a multi-method approach (Muratovski, 2022) focused and applied on the users who inhabit the territory.

Research through design (through or by), is based on the idea that there is no fundamental separation between theory and practice. On the one hand, this approach to research through design is anchored in the project, relying on a particular situation to change the course of things (Findeli & Coste, 2007; cited by Mollenhauer et al, 2020a). From a critique of user-centered design methods, we could say that design has focused particularly on the proximate scale, at hand, that of the user experience (micro-scale) losing focus on other scales in which their designs can impact, such as the meso and macro scale (city). Designers, in the creative perspective, as mentioned by Findeli et al. (2008, cited by Mollenhauer, et al., 2020) use research methods focused on possible and/or specific design solutions, neglecting the impact of the solution on the territorial and urban dimension of the context of use.

The research question is then posed: how can define a model that allows experts to diagnose phenomena related to digital delivery platforms, contributing to the decision-making by the various actors involved (private public, civil society)?

The general hypothesis defined suggests that the development of models and modes of representation will allow a better understanding of these impacts and facilitate strategic decision-making for sustainable and equitable urban development. This is because digital delivery platforms generate significant impacts on the city, including changes in everyday life, social relationships, and public policies. There is an interconnection between technology, public policies, and community life, where delivery platforms influence policies and the interaction of actors in the city.

The overall objective is to develop a model that allows experts to diagnose phenomena related to digital platforms of delivery, in order to contribute to decision-making by the various actors involved, including the private sector, public and civil society. This model will be designed to understand and effectively address the impacts of digital platforms of delivery in the city, considering aspects such as changes in everyday life, social relationships, and public policies. In addition, it seeks to leverage the interconnection between technology, public policies, and community life to promote sustainable and equitable urban development.

21.1.4 IAP Prototype

Participatory action research (IAP) is a viable path to carry out research that requires the development of specific research instruments for complex contexts. Its origin is in social movements and informal processes of pedagogical formation (Freire, 1968; cited by Solís Narváez, 2017) The IAP has been conceptualized as "a process by which members of a group or a vulnerable, oppressed community or in need of change, collect and analyze information, to act on their problems with the purpose of finding solutions and promoting transformations in their community", (Selener, 1997: pp. 17; cited by Balcazar, 2003).

This process implies that the community actively participates in solving their problems (Oliveira, 2015; cited by Ortega-Ibarra et al., 2019). Therefore, the IAP breaks the dichotomy of subject-object and creates an integrated research team, constituted by the researchers who are the facilitators or the agents of change; and the community or the group of individuals where the research is carried out, who are now managers, protagonists and builders of their life project (Bernal, 2010; cited by Ortega-Ibarra et al., 2019).

On the other hand, research through design (Findeli et al., 2008), is based on the presentation of a hypothesis in the form of a solution proposal to a specific identified problem. The main objective of this research consists of the development of this hypothesis through multiple validation studies of the proposals, to the point of validating it as a satisfactory solution (Montalvan et al. 2019).

As a methodological strategy for this study, the following stages are proposed: (i) determination of the relevant variables of each of the instruments raised, (ii) prototyping in a single instrument the relevant variables according to the type of research, (iii) testing of the prototype of the instrument in the specific case and (iv) evaluation of the results of the application of the instrument.

21.1.4.1 Ideation and Prototyping of the Instrument

The purpose of the study is for citizens, who experience the phenomenon to be studied, to be an active part of the research beyond the role of the subject of the study. For this, they must be able to gather data on the emerging phenomena and dynamics that occur in the delivery parking lots.

First, the implementation of a survey was proposed. Mazzini (2015) defines the survey as a research method that involves asking structured questions to a sample of people to collect information about their perceptions, needs, and expectations. According to Manzini, the survey is a valuable tool for designing services because it allows obtaining direct information from users and potential customers, which helps designers better understand their needs and develop more effective solutions.

Second, the implementation of the AEIOU tool was proposed. A good way to address all the dimensions of the service is by conducting an AEIOU check: Activities, Environments, Interactions, Objects, and Users (Figueroa et al., 2017). According to Stickdorn (2012), it is a widely used ethnographic research tool in the field of service design. In this sense, the AEIOU tool is used to collect data on each of these categories and understand how they relate to each other in the context of the service. This can help identify opportunities for improvement in the service design and enhance the user experience.

From the IAP approach and research through service design, the survey and the AEIOU observation guide were integrated into a single instrument. For the particular case for the realization of this prototype, it was considered to use the ArcGIS Survey 123 platform (Fig. 21.5) to build the IAP instrument, which considers the combination of some research instruments such as a survey along with the AEIOU observation guide. It was decided to use ArcGIS Survey 123 (https://survey123.arcgis.com), as it is a complete solution focused on forms to create, share, and analyze surveys. It is used to create forms with exclusion logic, default values, and support in multiple languages. It allows the possibility of collecting data using the web or mobile devices, even disconnected from the Internet, and with the possibility of generating maps with the data.

21.1.4.2 Testing of the Instrument

For the testing of the instrument as a criterion, it was defined to approach the public space of the city without neglecting the intersection with private space. At the same

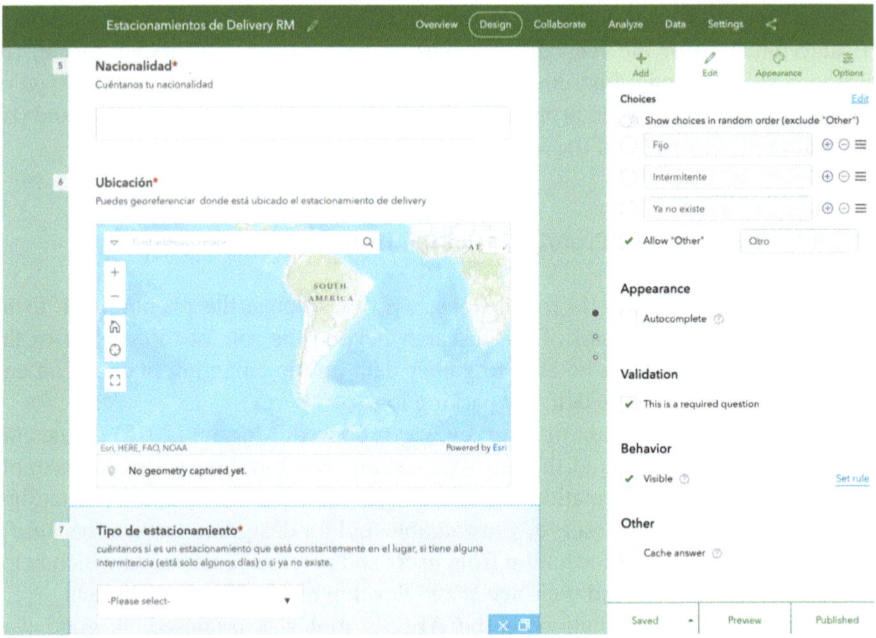

Fig. 21.5 IAP instrument implemented on ArcGIS Survey123 platform. *Source* Own elaboration

time, in the first instance, the closest and most easily accessible users were considered: the final consumers and delivery drivers of the Metropolitan Region.

Evidence of unauthorized parking of delivery motorcycles was sought in unregulated places used as parking lots in the area we call the "Santa Lucía Sur" neighborhood, located precisely at the foot of the hill of the same name and south of Alameda. The area proposed for the study and in this document called "Santa Lucía Sur Sector" is that area of Santiago Centro composed of the neighborhoods Parque Almagro, Matta Norte, Diez de Julio, Santa Isabel, and San Borja. This area is delimited by main avenues such as Manuel Rodríguez, Alameda, Vicuña Mackenna, and Matta. In addition to having high-traffic routes such as Curicó-Tarapacá, Marín-Eleuterio Ramírez, Santa Isabel, and 10 de Julio (Fig. 21.6).

Between December 29, 2022, and February 3, 2023, the IAP instrument was implemented using the ArcGIS Survey123 application (Fig. 21.5) as a tool. To cover the territory considered for the prototype, different routes were made on foot and by car in order to observe in the field aspects of the phenomenon to be studied, delivery parking lots. In addition, 2 routes that were not originally in the planning were added, such as the Barrio Brasil sector and sectors of Providencia and Las Condes in the city of Santiago de Chile.

As a result, 44 (Fig. 21.7) points were recorded on the platform. Of these, 59.09% of the parking lots were categorized as "intermittent" (26) and 38.64% as "fixed" (17). A maximum of 20 people gathered in one place was recorded, with an average

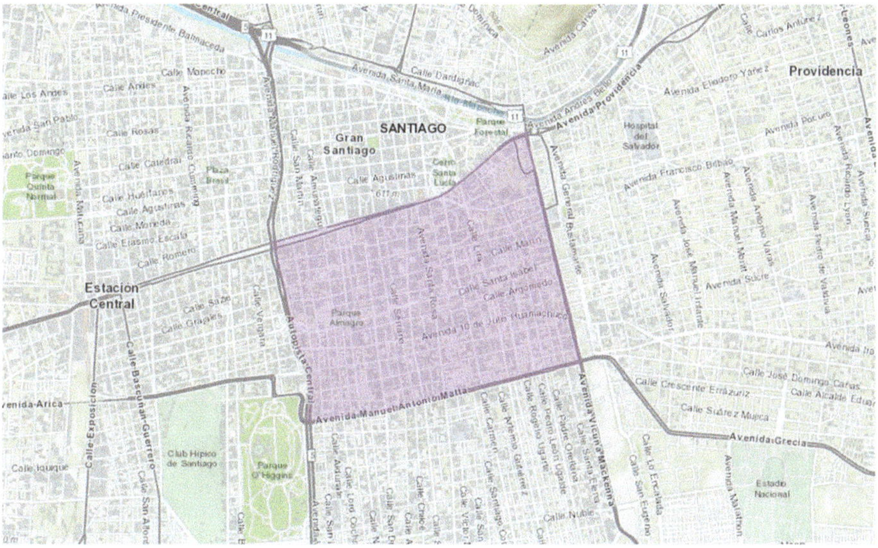

Fig. 21.6 Santa Lucía south sector. Own elaboration

of 5 people gathered per place. In turn, the day with the most records was Thursday, and 50% of the recorded points operated between 20:00 and 24:00 h.

Regarding the delivery users (motorcyclists), the vast majority are identified as males between 25 and 35 years old. Their main observed activity is resting, chatting,

Fig. 21.7 Observed and registered points on the ArcGIS Survey123 platform. *Source* Own elaboration

Fig. 21.8 Visual record of observed and registered points on the ArcGIS Survey123 platform. *Source* Own elaboration

and waiting (probably) before their next deliveries. In most of the places, no external objects are observed at the resting place, except in some cases where cardboard boxes can function as a kind of "seat" to rest the body. In most cases, users use the existing infrastructure, both public and private, to rest (Fig. 21.8).

21.1.5 Results

21.1.5.1 Learnings from the Testing

As a result of the prototype testing, several significant observations have been highlighted:

1. It has been confirmed that it is possible to collect data effectively through the active participation of citizens.
2. Although the use of the instrument requires some dedication of time to load the data, this process does not exceed 10 min, according to researchers from the Institute of Public Affairs (IAP).
3. The instrument adapts to both real-time data collection at the observation site, and remote collection, allowing users to load data after having been at the place of interest.
4. The instrument has become a valuable tool for recording delivery parking phenomena in a georeferenced manner, backed by solid visual evidence.

5. Citizens feel involved in the research by contributing their own experiences of displacement and routes in the city, adding their perspectives and data.

In response to these findings, the following adjustments have been made for its implementation:

- Adjustments have been made in the georeferencing of points, allowing greater proximity in the scale of location. This has significantly improved the accuracy and reliability of the spatial data collected, providing a more detailed and accurate representation of the distribution of the phenomena studied in the urban environment.
- Image fields integrated into the structure of the questions based on the AEIOU framework have been introduced, which has enriched the data collection by providing a substantial visual dimension. This more holistic and multidimensional approach has allowed a more complete capture of the complexity of the observed phenomena.
- A detailed segmentation based on time has been implemented to assess the visibility of the studied phenomena throughout the day in the city. This time segmentation has provided a more dynamic and contextualized perspective of the temporal patterns related to the distribution and occurrence of the observed phenomena.

Opportunities for improvements to be implemented in prototype 2 of the instrument:

1. The possibility of expanding the functionality of the instrument to record and analyze other relevant phenomena, such as "dark kitchens" and "dark stores," whose emerging role in the urban environment poses additional research challenges and opportunities.
2. Although the current focus is on a specific area, it is suggested to consider expanding the observation throughout the entire Metropolitan Region of Chile, and potentially, even at a national level. This geographical expansion would provide a broader view of the patterns and trends at a regional and national level.
3. The possibility of designing a transnational version of the prototype is proposed, which would allow extending the research to other regions of analysis. This expansion could offer a comparative perspective between different urban and cultural contexts.

These opportunities for improvement offer a roadmap for greater sophistication and reach of the instrument, which would provide a solid basis for a deep and holistic understanding of phenomena related to urban mobility and delivery logistics in contemporary urban environments.

21.1.5.2 Preliminary Spatial Analysis

As part of the review of the information collection, it has been decided to enrich the visualization by incorporating layers of information on the map created in ArcGIS (Fig. 21.9). These layers include data on population density per block (INE, 2018), show Territorial Socio Material Index (ISMT), location of shopping centers (Google Maps), service stations (Google Maps), and the identified points of delivery parking.

As a result of the prototype testing, the following was observed in the analyzed sector (Fig. 21.10):

1. The presence of green areas does not seem to have a relevant impact on the selection of locations for parking and the rest of the delivery drivers.
2. It has been found that spaces designated for delivery vehicle parking are concentrated in areas of intense demographic activity, especially in areas of high population density, such as the urban core of Santiago and its surroundings, like Providencia.
3. A trend has been observed for delivery parking to be located in the vicinity of fuel service stations, possibly indicating a logistical convenience strategy that optimizes efficiency in delivery routes and fuel replenishment.
4. Despite the presence of shopping centers, both strip centers and malls, in the studied area, further research is still required to determine to what extent these infrastructures influence the distribution and behavior of parking spaces for delivery drivers. Even so, it is foreseen that two basic uses are fundamental, product withdrawal and bathroom use.

These findings highlight the complexity of the factors that intervene in the spatial distribution of delivery parking, pointing out the need for a holistic understanding of urban patterns and infrastructure planning in the context of mobility and delivery services in metropolitan environments.

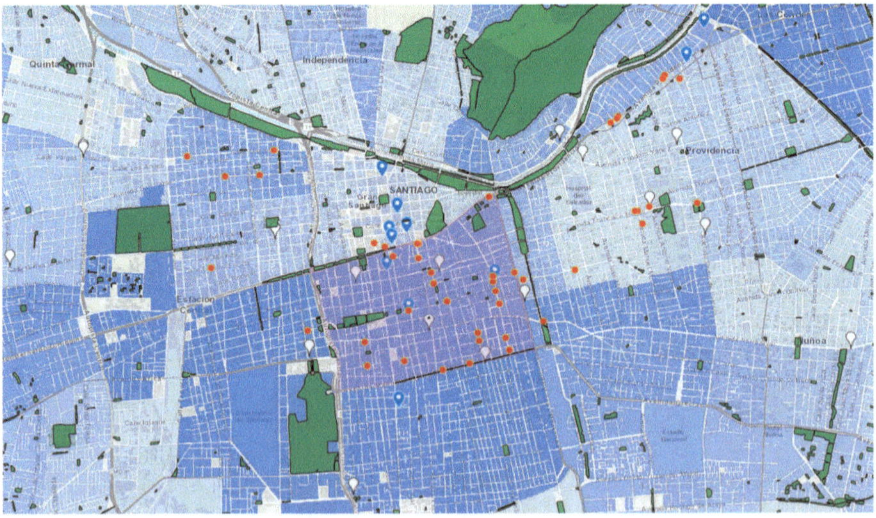

Fig. 21.9 GIS, Prototype of information collection of delivery parking lots "Santa Lucía Sur". Population per block (CENSUS 2017), Shopping centers (Google Maps), Service stations (Google Maps), and raised points of delivery parking lots are shown. *Source* Own elaboration

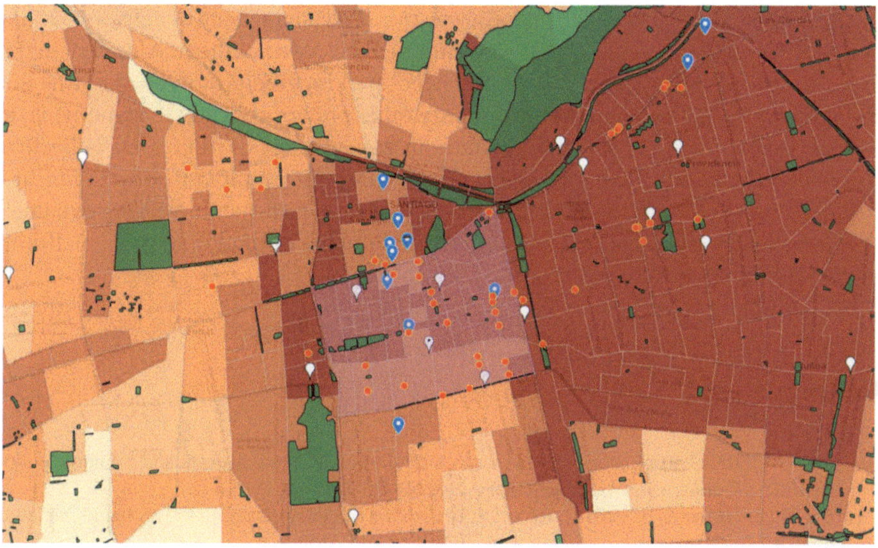

Fig. 21.10 GIS, Prototype of information collection of delivery parking lots "Santa Lucía Sur". Socio Material Territorial Index (ISMT), Metropolitan Region green areas, Shopping centers (Google Maps), Service stations (Google Maps), and raised points of delivery parking lots are shown. *Source* Own elaboration

21.2 Conclusions

The search for evidence through design also becomes a critical design practice, or more precisely forms of criticism that develop through design, and that have become popular in the field of human–computer interaction (Lupetti et al., 2022). From the disciplinary perspective of design, prototypes can also be considered useful as "learning tools" and, consequently, they can exist at any level of resolution, from very coarse to very refined, and can be used at any stage in the design process to explore, evolve and/or communicate ideas (Coughlan et al., 2007).

One of the fundamental values of prototyping lies in its generative nature, which implies that, by engaging in the prototyping process, numerous ideas are generated, some of which can be considered very innovative, while others may not be as prominent. Even the ideas that are not as prominent can serve as a starting point for developing even more outstanding solutions (Zaki, 2009). In this context, two interesting concepts are presented: the prototype as a representation and the prototype as a tool that facilitates learning.

The results of the research show that the instrument designed for the research allowed data to be collected through the participatory action of citizens, and that the instrument can be used both synchronously in person and synchronously remotely, having the ability to collect meaningful information to identify relevant aspects in the city. Citizens feel part of the research contributing from the use of their own routes and paths through the city to add points.

In addition, the instrument became a way to record phenomena of delivery parking in a georeferenced way and with visual empirical evidence of them. Although in this case, it focused on collecting data on delivery parking, the prototype confirms that this methodology can be applied to identify other previously known phenomena, as well as discover new aspects through the subsequent analysis of the data collected.

Finally, the result of this research allowed to raise empirical information in a participatory way about complex problems, phenomena, and new emerging dynamics of interaction in the urban space caused by the impact of digital delivery platforms in the Metropolitan Region.

References

Balcazar, F. (2003). Participatory action research (PAR): Conceptual aspects and implementation difficulties. *Fundamentals in Humanities*, 7–8, 59–77. 7. ISSN 1515-4467.

Castells, M., & Himanen, P. (2016). Reconceptualization of development in the global information age. FCE.

Coughlan, P., Suri, J., & Canales, K. (2007). Prototypes as (Design) tools for behavioral and organizational change. A design-based approach to help organizations change work behaviors. *The Journal of Applied Behavioral Science, 43*, 122–134. https://doi.org/10.1177/0021886306297722

Cueto, J. C. (2020, February 25). What are "ghost kitchens" and how this business is growing in Latin America. *BBC News World*. Retrieved July 5, 2022, from https://www.bbc.com/mundo/noticias-51503421

Dalla Chiara, G., Krutein, K. F., Ranjbari, A., & Goodchild, A. (2022). Providing curb availability information to delivery drivers reduces cruising for parking. *Scientific Reports, 12*(1). https://doi.org/10.1038/s41598-022-23987-z

Dorst, K. (2018, August). Mixing practices to create Transdisciplinary innovation: A design-based approach. *Technology Innovation Management Review, 8*(8), 60–65.

Hurtubia, R., Muñoz, J. C., Tironi, M., Tiznado, I., & Yaksic, M. (2021). Delivery apps and their urban impacts: Challenges and proposals.

Figueroa, B., Mollenhauer, K., Rico, M., Salvatierra, R., & Wuth, P. (2017). *Creating value through service design*. Pontifical Catholic University of Chile.

Findeli, A., Brouillet, D., Martin, S., & Moineau, C. (2008). Research through design and transdisciplinarity: A tentative contribution to the methodology of design research the construction of memory, emotional and perceptive judgments: body and metacognition View project Memory compensation View project. https://www.researchgate.net/publication/235700599

Institute of Social Innovation. (2019, October 15). The dark side of the collaborative economy. Do Better (by Esade). Retrieved May 13, 2022, from https://dobetter.esade.edu/es/economia-colaborativa-lado-oscuro

Instituto Nacional de Estadísticas [INE]. (2018). Resultados CENSO 2017. Retrieved June 13, 2023, from http://resultados.censo2017.cl/

Labbé, G., & Palma, P. (2022). Estudio revela que envíos de comida a domicilio crecieron 175% en 2021 y tiendas de conveniencia expanden negocio. El Mostrador. Retrieved April 25, 2022, from https://www.elmostrador.cl/mercados/2022/02/27/estudio-revela-que-envios-de-comida-a-domicilio-crecieron-175-en-2021-y-tiendas-de-conveniencia-expanden-negocio/

de Lange, M., Merx, S., & Verhoeff, N. (2019). Urban interfaces: Between object, concept, and cultural practice.

Li, C., Mirosa, M., & Bremer, P. (2020). Review of online food delivery platforms and their impacts on sustainability. *Sustainability, 12*(14), 5528. https://doi.org/10.3390/su12145528

Lupetti, M. L. (2022). Understanding critical design tensions with progression through historical practices. *Diid — Disegno Industriale Industrial Design, 77*, 12. https://doi.org/10.30682/diid772022g
Mazzini, E. (2015). *Design, when everybody designs: An Introduction to design for social innovation.* MIT Press.
Montalván, J., Morales, C.S., Hopkins, A., Ascue, R., & Atijo, E. (2019). Research guide in art and design. Vice-rectorate of Research. Research Management Directorate. PUCP. https://cdn02.pucp.education/investigacion/2016/06/12214732/guia-de-investigacion-en-diseno.pdf
Mollenhauer, K., Moraga, E., & Bernasconi, R. (2020a). Instrumental model for complex projects. Poliedro Value Proposition Academic. Urban Territorial Logbook, 30. https://doi.org/10.15446/bitacora
Mollenhauer, K., Moraga, E., & Bernasconi, R. (2020b). Modelo instrumental para proyectos complejos. Poliedro Propuesta de Valor Académico. Bitácora Urbano Territorial, 30. https://doi.org/10.15446/bitacora
Muratovski, G. (2022). *Research for designers: A guide to methods and practice* (2nd edn).
Navarrete, M. del C., García Muñoz Aparicio, C., & Sánchez Rosado, O. B. (2022). Nuevo modelo de negocios gastronómico. Las Dark Kitchen. Ventajas y desafíos. *Vinculatégica EFAN, 8*(6), 52–60. https://doi.org/10.29105/vtga8.6-222
Orellana, V. (2022, February 16). The reasons behind the "boom" of motorcycle sales in Chile and the most demanded models Source: Emol.com. Retrieved April 19, 2022, from https://www.emol.com/noticias/Economia/2022/02/16/1046952/aumento-venta-motos-urbano-chile.html
Orellana, A. (2009). La gobernabilidad metropolitana de Santiago: la dispar relación de poder de los municipios. *EURE (Santiago), 35*(104), 101-120. https://doi.org/10.4067/S0250-71612009000100005
Ortega-Ibarra, Ilse & Ortega Ibarra, Edú & Jiménez, Andrea. (2019). La investigación de acción participativa: ¿una alternativa de investigación o una estrategia de solución?. Universita Ciencia. 20. 61. https://doi.org/10.5281/zenodo.7011291
Parker, G., Van Alstyne, M. W., & Choudary, S. P. (2016). *Platform revolution: How networked markets are transforming the economy-and how to make them work for you.* W. W. Norton & Company.
Parker, S. (2020). From pillar to post: Culture, representation and difference in the urban world. In *Urban theory and the urban experience* (pp. 162–182). https://doi.org/10.4324/9780203122402-19
Peñafiel, C. (2022). Delivery de alimentos, un mercado en ascenso. BioBioChile. Retrieved March 31, 2022, from https://www.biobiochile.cl/noticias/opinion/tu-voz/2022/03/23/delivery-de-alimentos-un-mercado-enascenso.shtml
Reed, S., Campbell, A. M., & Thomas, B. W. (2021). Does parking matter? The impact of search time for parking on last-mile delivery optimization. http://arxiv.org/abs/2107.06788
Sassen, S. (2007). *Global cities, the recovery of place and social practices. A sociology of globalization* (1st edn., pp. 125–164). Katz Publishers.
Shapiro, A. (2022). Platform urbanism in a pandemic: Dark stores, ghost kitchens, and the logistical-urban frontier. *Journal of Consumer Culture.* https://doi.org/10.1177/14695405211069983
Sevaldson, B. (2023). Designing complexity the methodology and practice of systems oriented design. Common Ground Research Networks
Spencer, J. (2018). Designing for dynamic challenges: Key attributes of designers for leading interdisciplinary research and projects. *Diseña, 13*, 84–109. https://doi.org/10.7764/disena.13.84-109
Solís Narváez, N. (2017). Towards multidisciplinary and interdisciplinary research: Reflection on the application of participatory methodologies and participatory action research in the social sciences. *Humanism and Social Change Journal. 74.* https://doi.org/10.5377/hcs.v0i5.4996
Stickdorn, M. (2012). *This is service design thinking: Basics, tools, cases* (1st Edición). Wiley.

Sznel, M. (2020, May 5). The time for environment-centered design has come—UX collective. Medium. Retrieved June 28, 2022, from https://uxdesign.cc/the-time-for-environment-centered-design-has-come-770123c8cc61

Zaki, T. (2009) *Prototyping—A practitioner's guide*. Rosenfeld Media.

Chapter 22
Interdisciplinary Implementation of the National Rural Development Policy in Local and Regional Strategic Planning Instruments in Chile

Katherine Mollenhauer, Daniel Moreno, and Arturo Orellana

Abstract In Chile, the development and planning of rural territory has historically lagged behind urban planning. The National Rural Development Policy (PNDR) has highlighted these gaps as well as introduced a new approach to rurality. However, there is evidence that this policy has had a low level of implementation in strategic planning instruments at the local level (community development plan, PLADECO) and regional level (regional development strategy, ERD) (Orellana et al., *North Great Geography Journal*, 31–49, 2020). Understanding this problem, the question arises of how to address the implementation of this national policy in local and regional planning instruments. In this context, the Office of Agricultural Studies and Policies (ODEPA) asked a group of researchers from the Governance and Territorial Planning Core (NuGOT) of the Catholic University, to develop an innovative methodological strategy that would integrate traditional methods of planning and urban and territorial management with ethnographic methods applied to the design of public services (Mollenhauer et al., II Interdisciplinary congress of research in architecture, design, city and territory (pp. 205–223), 2017). The presentation discusses how the components, methods, techniques and instruments that make up the toolkits (practical guides) become relevant tools for municipal officials, regional governments and consulting firms involved in the formulation of PLADECO's and ERD's to engage in public policy (Molokwane and Lukamba, The 3rd annual international conference

K. Mollenhauer (✉)
School of Design, Research Center on Governance and Territorial Planning (NuGOT), Pontificia Universidad Católica de Chile, Santiago, Chile
e-mail: kamollenhauer@uc.cl

D. Moreno
Research Center on Governance and Territorial Planning (NuGOT), Pontificia Universidad Católica de Chile, Santiago, Chile
e-mail: dlmoreno@uc.cl

A. Orellana
Institute of Urban and Territorial Studies, Research Center on Governance and Territorial Planning (NuGOT), Pontificia Universidad Católica de Chile, Santiago, Chile
e-mail: amorella@uc.cl

M. Vicuña and E. Giorgi (eds.), *Intersections*, Springer Geography,
https://doi.org/10.1007/978-3-031-76402-8_22

on public administration and development alternatives (pp. 192–199), 2018) through the application of the criteria of the PNDR despite it being an indicative policy. It also discusses the interdisciplinary, participatory and co-creative methodological process (Sanders and Simons, A social vision for value co-creation in design, 2009) from the design of services and its contribution to urban planning (Miettinen and Sarantou, Arts-based methods for decolonising participatory research (pp. 123–145), 2021) for the development of both toolkits and as a strategy to reduce uncertainty in the process of innovation in public services (LIP, 2017).

Keywords Planning · Interdiscipline · Service design

22.1 Introduction

Chile has 346 communes of which 78 are mixed and 185 are rural. In this country, the development and planning of rural territory have historically lagged behind urban planning. The National Rural Development Policy (PNDR) has highlighted these gaps as well as introduced a new approach to rurality. However, there is evidence that this policy has had a low level of implementation in strategic planning instruments at the local level (community development plan, PLADECO) and regional level (regional development strategy, ERD) (Orellana et al., 2020) due to its indicative and non-normative nature.

Due to the scale and complexity of the problem, the integration of multidisciplinary approaches is necessary that can, from an interdisciplinary R&D process, create instruments that allow the incorporation of the PNDR approach in the PLADECO and ERD processes, contributing to the improvement of the planning of the territories.

In response to this, the NUGOT research group of the Pontifical Catholic University of Chile (PUC), during the second semester of 2019 carried out an R&D project to support municipal officials and regional governments in the task of implementing this policy in the preparation of PLADECO and ERD, through the creation of a methodological kit.

From a conceptual and methodological perspective, the research team maintains that this type of input should be co-created with the target user group. This means moving towards a direct involvement of those affected by the problem and the proposal of solutions from their perspective, contributing in various aspects to reduce uncertainty during planning processes as well as to improve the appropriation and adoption of the designed solution. For this, methods and techniques from different disciplines such as urban planning, economics, agronomy, geography and the design of public services.

From the project, the question arises: How can we incorporate the PNDR approach into the PLADECO and ERD processes, contributing to the improvement of territorial planning? The answer proposes the following professional hypothesis: The design of user-centered tools that can address the macro, meso and micro scale of the incorporation of the PNDR enables the actors of the public–private ecosystem in the incorporation into the formulation processes of the PLADECO and ERD contributing to the improvement of territorial planning.

However, from academia, thinking about disciplinary convergence, another question and corresponding hypothesis is also raised: What are the challenges faced by the interdisciplinary team when incorporating the PNDR as an indicative instrument in local and regional planning? As a possible answer, it is suggested that the incorporation of a new perspective on the rural proposed by the PNDR as an indicative instrument, requires an interdisciplinary methodological approach that addresses from the design of public services, the territorial complexity at the local and regional scales. In this sense, Design, from a multiple approach, understands complexity as the relationships and interactions between various elements (Sevaldson, 2022; Sevaldson & Jones, 2019), considering those human and/or non-human from a territorial point of view (Moreno et al., 2023).

With both approaches and from this interdisciplinary methodological approach, it seeks to contribute to the solution of complex problems that arise in the planning of territories. It should be highlighted that it is an exploratory process as it aims to solve a specific requirement as well as move towards a proposal for comprehensive and interdisciplinary analysis. However, the results obtained are generalizable to other situations of similar characteristics.

22.2 Theoretical Framework

The concept of territory is multidimensional and has approaches from various disciplinary fields and has become an important and widely used category for the analysis of various processes. It is a concept that encompasses political and social characteristics (Castillo, 2020). Therefore, it constitutes a theoretical concept and at the same time an empirical object whose analysis is adjusted according to the disciplinary and interdisciplinary point of view. It is a concept that exists materially and therefore has representation, although its limits or determination depend on social spatialization and the relationships that sustain it. Likewise, politically and economically it is used as a tool to demarcate power and exchange (Llanos-Hernández, 2010).

The delimited territory refers to the limited social space, occupied and used by different social groups due to the implementation of their territoriality or the field of power over a space exercised by the dominant institutions (Capel, 2016). However, the determination of the predominance and configuration of the territories remains a cultural construction where social practices occur with different interests, with different perceptions, valuations and territorial attitudes (Nates Cruz, 2011). In contrast with this, today the unitary, centralist and sectorial attempts to order and plan the territory for the challenges of the future are unthinkable from the political discourse and, moreover, lack the foundation from the technical point of view (Arenas & Orellana, 2019). The Chilean territory, characterized by a rich and geographically diverse landscape, presents significant imbalances in terms of its sociodemographic configuration and in terms of its social, economic and environmental development, like much of the Latin American countries (ECLAC, 2017).

Any territorial management and planning system requires that normative and indicative instruments support a multi-scale integration and articulation with national policies and their guidelines (Orellana et al., 2020), as happens with the PNDR. The understanding of the ecosystem in the design of a territorial planning instrument and its integration with national policies requires the identification of key actors and their participation. This is necessary and urgent to promote deliberative democracy, giving voice to all types of actors and collectives, and facilitating dialogue among them (Pinazo-Dallenbach & Tutistar-Rosero, 2020). The main objective is to solve problems and conflicts strategically, attending to the demands of each one, as well as understanding the interpretations of national policies and how they should adapt and/or integrate into the planning of the territory at regional and local scales. However, this analysis is exploratory insofar as it seeks to solve a problem with directly involved actors, whose process is also part of the product.

In this context, municipal and regional government officials, as well as consultants participating in these processes, are the actors responsible for implementing the PNDR in the construction of planning instruments for mixed and rural communes. This implies that it is urgent to incorporate the needs, desires, and expectations of all users who inhabit the territory, promoting their active participation and recognizing the particularities of each one. The above becomes more urgent in Chilean rural territories, which have historically been left behind in terms of planning in contrast to urban areas and their urban regulatory plans.

For Kees Dorst (2018), the application of Design and its methodologies in complex contexts, on the one hand, enable the design team to identify the critical elements of the problem and to reduce the uncertainty about the adoption of the solution by the users. On the other hand, they enable the members of the multidisciplinary team to define an interdisciplinary solution and, in the case of the participation of stakeholders in the process, it allows progress towards a transdisciplinary solution that responds to the complexity present in the context. For Sevaldson and Jones (2019) and Aguirre (2020), Systemic Design and System-Oriented Design are an open praxeological field (Cross, 1999) and of continuous development where the thought and practice of Systems and Design converge. This facilitates the convergence of a multiplicity of perspectives and disciplines which often become tensions, but if these are managed from Design with co-creation methods, they become spaces for creativity, collaboration, and problem-solving of complex problems by all stakeholders (Moreno et al., 2023). In this sense, Service Design is a discipline that makes services more useful, usable, desirable for its users, and efficient and effective for organizations and is widely used in other countries for the development and implementation of public policies. It is an interdisciplinary approach that combines different methods and techniques, constituting a means for addressing complex problems, and being able to recognize five principles: user-centered, co-creative, sequenced, evidenced, and holistic (Stickdorn et al. 2019). These allow users to actively participate in the definition of their needs and in the design of their solutions, reducing uncertainty

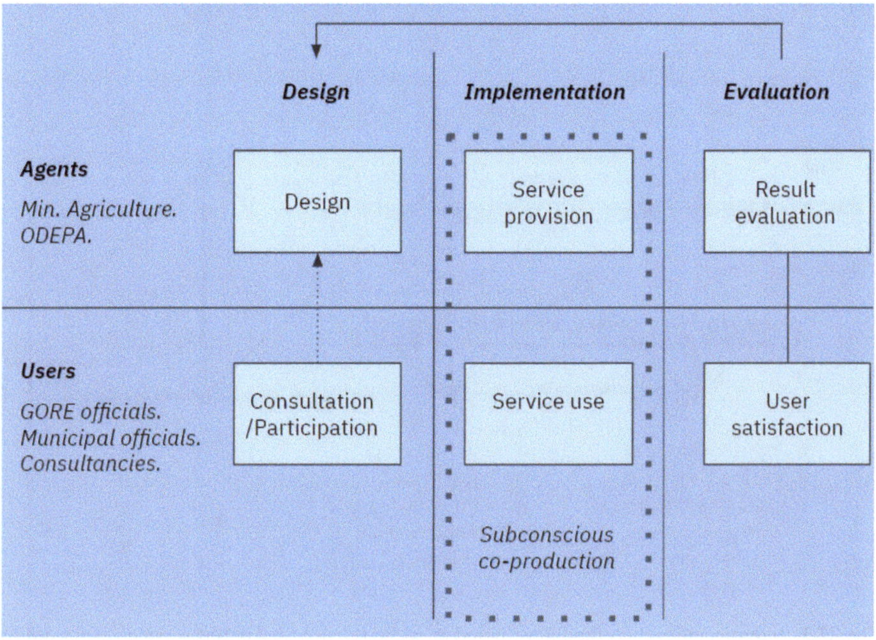

Fig. 22.1 Traditional service provision model and the actors of the PNDR ecosystem linked to the project. *Source* Based on Public Innovation Laboratory (2020)

when adopting and implementing them (LIPUC, 2017; Miettinen & Sarantou, 2019; Sanders & Simons, 2009; Soto et al., 2021).

The authors highlight the importance of focusing on the user through the use of co-creative design methodologies, always maintaining a holistic view of the process, allowing participants to incorporate capabilities that allow them to have more complex perspectives of the problem, as well as being more empathetic and tolerant with greater knowledge of potential errors and iterations. The above allows a horizontal exchange between the actors, generating tacit knowledge about the problem, new ideas, and own solutions, minimizing the risks associated with design, and facilitating the implementation of the solutions proposed by the users themselves (OECD, 2017 in LIPUC, 2017) (Fig. 22.1).

22.3 Methodology

For the design of the interdisciplinary methodological strategy, a multimethod investigative approach was used (Muratovski, 2022) integrating methods and techniques from the social sciences and public service design (LIP, 2017; Mollenhauer et al., 2017).

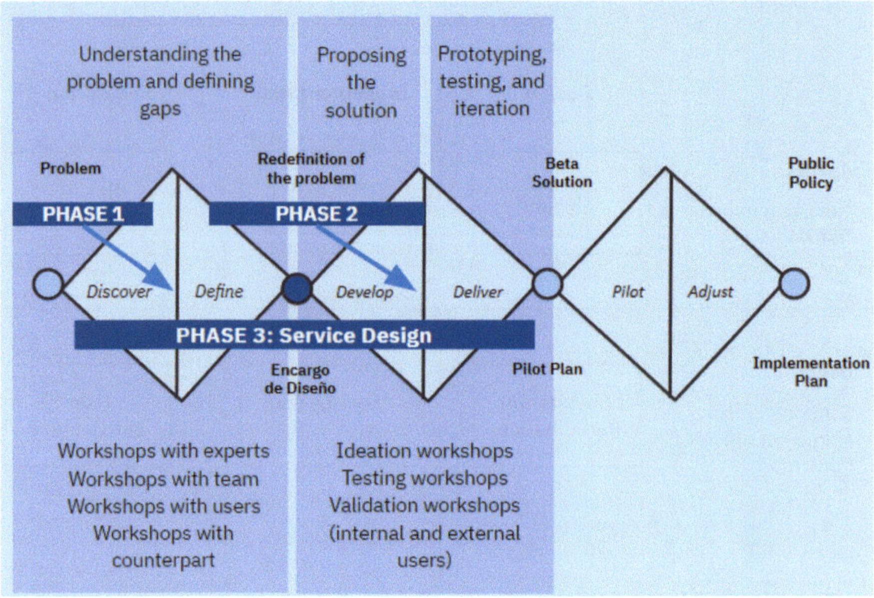

Fig. 22.2 Interdisciplinary methodology based on social sciences and public service design. *Source* Based on LIPUC (2017)

According to the Participation Management Ladder in Design (Salvatierra, 2020), co-creation, as a key method of Public Service Design, aims to involve users in the three highest levels of participation during the process. This implies moving from "Information", "Opinion" and "Evaluation" to "Collaboration" where the research team works together with the users, to "Co-creation" where the parties participate at the same level in decision-making and to "Self-management", where decision-making is shared with the user.

In each of the 3 phases, co-creation workshops were designed and carried out with the different actors of the ecosystem, seeking to generate a high sense of user ownership of the project in practice for a smooth implementation of its results (LIP, 2017) (Fig. 22.2).

22.3.1 Phase 1: Systematization of Regional and Communal Experiences

In the first phase, a certain number of regional and communal experiences were systematized around the incorporation of the rural development perspective in their planning, management, and development instruments. 30 Communal Development Plans (PLADECO), 15 Regional Development Strategies (ERD), and 6 Regional Territorial Planning Plans (PROT) were reviewed.

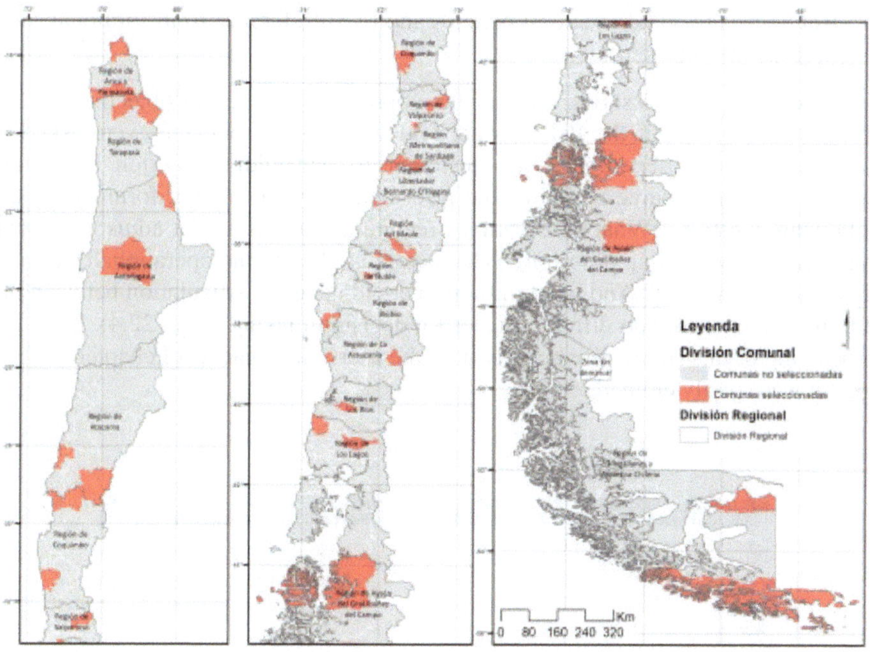

Fig. 22.3 Communes selected for analysis. *Source* Orellana-Ossandón et al. (2020)

The results show the scarce proximity that the current PLADECOs of rural communes have with the contents of the PNDR. Mainly from the point of view of their linkage with other normative and indicative planning instruments, sectoral policies, and their articulation with institutional actors. The variables considered are presented in phase 2, distributed in three groups (Orellana et al., 2020) (Fig. 22.3).

22.3.2 Phase 2: Distributed Content Analysis

In the second phase, a methodology of analysis distributed in three blocks was applied:

1. Group A Content Analysis: (a.1) Empirical basis, (a.2) Definition of the concept of rural development, (a.3) Objective on rural development, (a.4) Governance for rural development,
2. Group B Predominant focus on areas of the PNDR: (b.1) Social Welfare, (b.2) Economic Opportunities, (b.3) Environmental Sustainability, (b.4) Culture and Identity, and
3. Group C Cross References: (c.1) Reference to other IPT (Regional Development Strategy, Intercommunal Regulatory Plans, Communal Regulatory Plans), (c.2)

References to ad hoc sectoral policies and (c.3) Reference to institutional roles in rural development).

In co-creation workshops, each of the approaches, gaps, and objectives set by the guiding criteria of the PNDR were reviewed, and participants were asked to make comments, observations, and adjustments to each guiding criterion. Key actors and actions for resource targeting and prioritization were sought, formulating the following questions: What observations, recommendations and/or adjustments can we make to the guiding criterion? What actors are key to the operationalization of this guiding criterion? And what initiatives could generate articulation between the different actors and their different scales for decision-making? (Fig. 22.4). From the processing of this co-creative activity with the users, the contents of chapter 1 of the toolkit were partly defined.

Fig. 22.4 Model of the instrument for the content analysis of the distributed analysis methodology of the second phase of the interdisciplinary methodology, applied to a certain guiding criterion. *Source* PUC (2019)

22.3.3 Phase 3: Ideation, Prototyping and Testing of the Solution Within the Framework of Public Service Design

In the third phase, the R&D process of the solution was carried out based on the results of the first and second phases of the methodology using an approach from the design of public services focusing on the development of co-creation processes between the research team and the users of the solution(s). This phase of R&D of the solution considered four stages: (i) Problem definition, (ii) Development of the conceptual proposal, (iii) Prototyping, testing and iteration of the solution, and (iv) Design of the beta solution of the toolkit. The details of the objectives, types of activities and participants are detailed below (Fig. 22.5).

Stage	Objectives	Types of Activities	Related Actors
1. Problem Definition	Define the guiding criteria that determine the rural perspective based on the paradigm proposed by the PNDR	Workshop with experts	- Professional consultants - DESE technical team - ODEPA team - Communal and regional professionals
	Systematize regional and communal experiences around incorporating the rural development perspective in their instruments	Internal workshop with technical team 1	
		Communal workshops	
	Identify the factors that facilitate or obstruct the incorporation of PNDR guidelines in the instruments	Regional workshops	
		Internal workshop with technical team 2	
2. Development of the Conceptual Proposal	Develop the contents of the regional and communal guides, considering a design suitable for the final user	Counterpart coordination meeting	- DESE technical team - ODEPA team - ODEPA advisors
3. Prototyping, Testing, and Iteration	Reflect the developed contents in didactic material for easy transfer / communication / dissemination / distribution (toolkit) for application and adoption by its users	Internal validation workshop	- Communal and regional professionals - Professional consultants - DESE technical team
		Internal workshop with technical team	
4. Beta Solution Toolkit Design		Presentation of the Methodological Kit	- DESE technical team - ODEPA technical team and advisors - Local teams
		Capacity transfer workshop	

Fig. 22.5 Methodological strategy of the R&D project. *Source* Own elaboration based on LIP (2017)

Before starting stage 1, a stage 0 or preparation stage was carried out, preparing the internal project team and the users through internal training processes of the research team for a multimethod methodological alignment and the installation of innovation capabilities in the design of public services centered on users. Then, the results of the first, second and third stages were integrated, through the development of internal co-creation workshops facilitated by the research team with the users of the solution(s) (Figs. 22.6, 22.7 and 22.8).

Stage 2 of phase 3: In stage 2, and based on the results, the research team carried out a synthesis process of the requirements to consider in the co-creation process with the users for each of the components of the methodological kit, according to the following detail (Table 22.1).

Stage 3 of phase 3: In stage 3, co-creation techniques were applied again with the users in the prototyping and testing of each prototype regarding the contribution of the kit to the incorporation of the PNDR guidelines in the planning instruments at regional and communal scales, respectively. Based on the results, adjustments

Fig. 22.6 Workshops with municipal officials from the commune of Melipeuco. *Source* PUC (2019). Stage 1 of phase 3: in stage 1, the identification of the factors that favor and/or hinder the incorporation of the guidelines of the PNDR in the planning instruments at regional and communal scales was carried out. For this, workshops were held with municipal officials and regional governments who are the users of the solution

Fig. 22.7 Workshops with officials from the regional government of Coquimbo. *Source* PUC (2019)

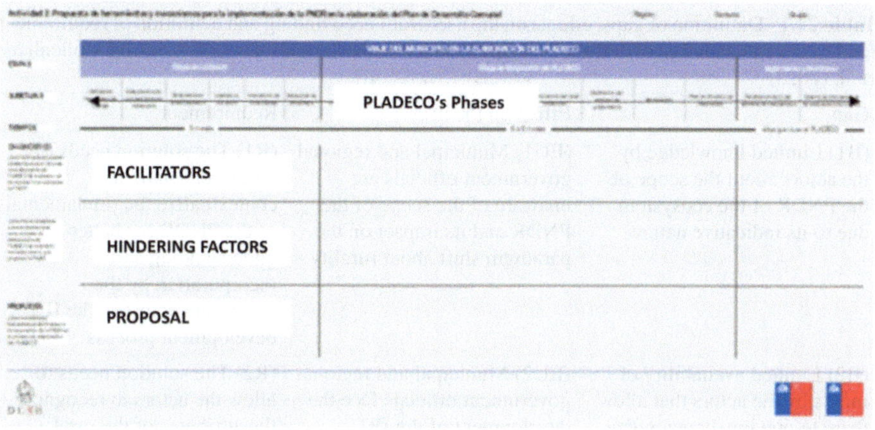

Fig. 22.8 User journey instrument of the PLADECO in the third phase of the interdisciplinary methodology. *Source* PUC (2019)

were made for the delivery of the two practical guides or toolkits to be used in the development of PLADECO and ERD, respectively (Fig. 22.9).

22.4 Results

The results of this R&D project are two practical guides or toolkits available at https://www.masvidarural.gob.cl/guia-pndr-para-comunas/ and https://www.masvidarural.gob.cl/guia-pndr-para-regiones/.

These practical guides constitute relevant tools for municipal officials, regional governments, and consulting firms involved in the formulation of PLADECO's and ERD's to engage in public policy (Molokwane & Lukamba, 2018) by applying the criteria of the PNDR despite it being an indicative policy.

Based on the identification of needs and gaps of the potential users, the requirements that needed to be solved by the solution were defined in a co-creative way with the users. In this way, chapter 1 was developed with a theoretical orientation and provides a conceptual framework for the PNDR. Chapter 2 proposes a work methodology to be carried out prior to the formulation of the planning instrument. And chapter 3, provides four types of information: reminders, recommendations, tools, and alerts, which guide the process of formulating the PLADECO or ERD as appropriate. In addition, annexes were incorporated with a set of guiding and complementary tools (Table 22.2).

In December 2020, at the start of the ERD development process for Lagos 2030, chapter 2 of the practical guide for ERD was applied with regional government officials to integrate the view on rurality in the Los Lagos Region. Subsequently,

Table 22.1 Definition of gaps, effects on the user from need finding and definition of requirements for the conceptual design of the practical guides based on the logical matrix of design by Mollenhauer et al. (2023)

Gap	Effect on the user	Requirement
(B1) Limited knowledge by the actors about the scope of the PNDR of the ecosystem due to its indicative nature	(EU1) Municipal and regional government officials are unaware of the scope of the PNDR and its impact on the paradigm shift about rurality	(R1) The solution needs to describe, detail, and contextualize the fundamentals of the PNDR for better understanding and incorporation by the ecosystem actors in the IPT development process
(B2) Limited availability of inputs by the actors that allow them to previously recognize the values of the territory with a focus on the rural before the IPT development process	(EU2) Municipal and regional government officials face the development of the IPT without a rigorous systematic and consensus process about the rural characteristics of the territory	(R2) The solution needs to allow the actors to recognize the attributes of the rural territory prior to the planning instrument development
(B3) Limited resources that guide the responsible actors of the incorporation of the PNDR in the IPT development process in the stages and sub-stages of preparation, development, and follow-up	(EU3) Municipal and regional government officials carry out the IPT development process without clear guidance on how to incorporate the PNDR during the process	(R3) The solution needs to specify how to incorporate the PNDR at each stage and sub-stage of the IPT development
(B4) Limited resources that support the responsible actors of the incorporation of the PNDR in the IPT development process in the stages and sub-stages of preparation, development, and follow-up	(EU4) Municipal and regional government officials face the IPT development without having tools that support them in the process	(R4) The solution needs to provide a set of tools for better operationalization of the PNDR at each stage and sub-stage of the IPT development by the teams

Fig. 22.9 Testing of the toolkit components with users from the regional government of Santiago. *Source* PUC (2019)

Table 22.2 Definition of the contents of the practical guides from the definition of requirements based on the logical design matrix of Mollenhauer et al. (2023)

Requirements	Toolkit chapter	Contents:
R1	Chapter 1: framework of contents of the national policy of rural development	Conceptual contents that allow users to understand the foundations and conceptual framework of the PNDR for its incorporation in the process of elaboration of the IPT
		Objectives of the PNDR
		Areas of the PNDR
		New rural outlook
		Definition of rural territory
		Principles of the PNDR
		Governance of the PNDR
R2	Chapter 2: analysis of the communal/ regional rural context (MTR)	Work methodology to be carried out prior to the formulation of the planning instrument that values the rural characteristics of the territory so that the teams can prepare before starting the process of developing the IPT by recognizing economic activities, natural and cultural attractions (religious, historical, etc.) and risks of the territory. Contents
		Foundations
		Preliminary communal analysis: rural territorial map
		Key questions
R3	Chapter 3: PLADECO/ ERD development process as appropriate	Methodology that guides the incorporation of the PNDR in the process of developing the PLADECO or ERD as appropriate. Provides information of four types: (i) reminders, (ii) recommendations, (iii) tools and (iv) alerts. Specifies who the information is for: (i) technical team and (ii) consulting team
		Contents
		Preparation stage and sub-stages
		Development stage and sub-stages
		Monitoring stage and sub-stages
R4	Annexes	Set of tools that allow the team of professionals to delve into the IP development process before, during and after

the same workshop was held with the inhabitants of the 30 communes in the region (Fig. 22.10).

22.5 Conclusions

The incorporation of Service Design, as a methodology focused on the user, allowed to articulate the team and enable it for an interdisciplinary methodological approach. Similarly, the incorporation of co-creation as a main technique turns out to be a contribution to traditional planning methodologies, as it recognizes the diversity of

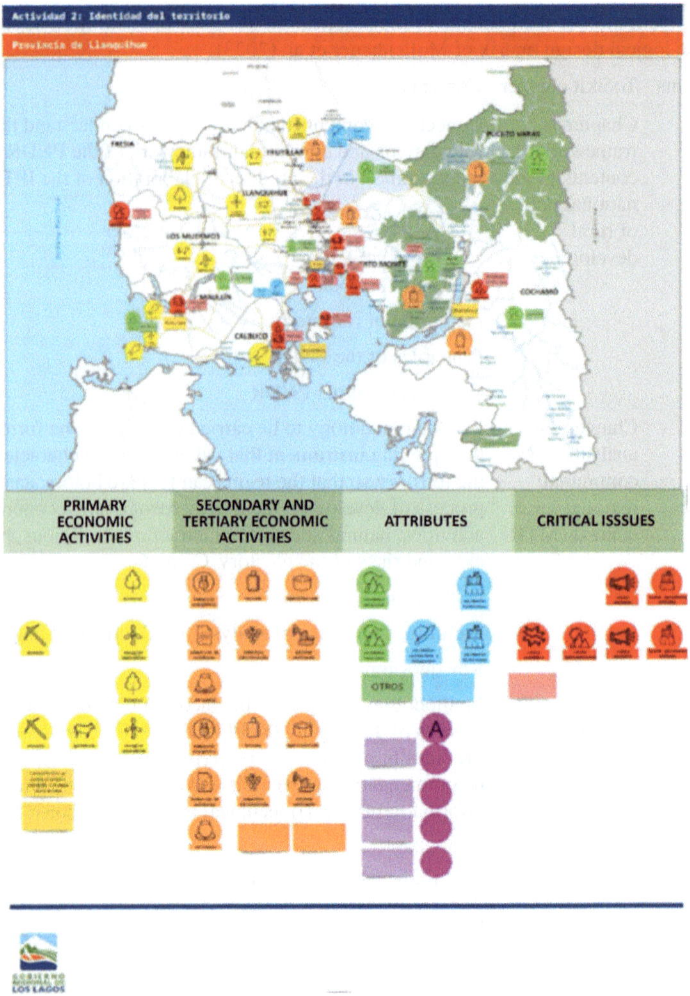

Fig. 22.10 Application of the practical guide to integrate the PNDR in regional development strategy (ERD). *Source* PUC (2021)

tacit knowledge of each actor in the territory and transforms it into explicit knowledge to be incorporated into the process.

Regarding the process and the results of R&D we can conclude that in the creation of solutions that are oriented to the users of public services, the interdisciplinary methodological approach is key to integrate knowledge coming from the study of secondary sources (phases 1 and 2)—very typical of social sciences—with empirical information coming from primary sources (stage 1 of phase 3). This integration facilitates the empirical knowledge that the users themselves—who are ultimately the true experts on what they need as a solution—be systematized and incorporated as components of the solution.

In the same vein, the application of various methods and techniques for the systematization of results through two practical guides or toolkits allows the dissemination of the necessary information for the incorporation of the PNDR in the preparation of the IPT, the replicability and/or scalability of the PNDR, recognizing the diversity of each territory. At the same time, it ensures that the principles of the policy are appropriately implemented, as the needs and expectations of each type of actor in the ecosystem are recognized and resolved from a multi-level approach.

It is worth noting that the academic team carried out a process of "research through design" or "*research through design*" (Findeli et al., 2008), systematizing the results and building an interdisciplinary R&D methodology. The integration of multidisciplinary approaches such as economics, agronomy, geography and service design, among others, allows the creation of interdisciplinary knowledge and generates a differentiated academic value proposition based on the ability to address highly complex problems (Page, 2009) such as those related to territorial planning.

The construction of this interdisciplinary R&D methodology allows the team to install new elements within the planning processes. Among them, the creation of innovative solutions by reducing the uncertainty regarding the appropriation and adoption of the solution designed by the users (Public Service Design (LIP, 2017). Also, to seek a transformation of the state of affairs, allowing the situation of the users to improve without simplifying the complexity of the problem (System-oriented Design, Sevaldson, 2022). Finally, that, as far as possible, they scale, replicate and adapt the designed solutions to other territories and institutions, through the systematization of the practice carried out (Single Case Design, Ottenberger, 2000).

References

Aguirre, M. (2020). *Transforming public organizations into co-designing cultures: A study of capacity-building programs as learning ecosystems* [Ph.D. thesis, Oslo School of Architecture and Design].

Arenas, F., & Orellana, A. (2019). Contributions from territorial planning to the political and administrative decentralization of the country. In Von Baer, H., & Bravo, N. (Eds.), *Collaborative territorial development; decentralizing, power, competences and resources* (pp. 153–160). Ediciones Universidad de la Frontera.

Blomkamp, E. (2018). The promise of co-design for public policy. *Australian Journal of Public Administration, 77*(4), 729–743. https://doi.org/10.1111/1467-8500.12310

Capel, H. (2016). Social sciences and the study of territory. Biblio3W. *Bibliographic Review of Geography and Social Sciences, 21*. https://doi.org/10.1344/b3w.0.2016.26319

Castillo, G. (2020). The territory as a sociopolitical appropriation of space. Between deterritorialization and multiterritoriality. *Geographical Research*, (103), e60127. https://doi.org/10.14350/rig.60127

Cross, N. (1999). Design research: A disciplined conversation. *Design Issues, 15*(2), 5–10. https://doi.org/10.2307/1511837

Dorst, K. (2018) Mixing practices to create transdisciplinary innovation: A design-based approach. *Technology Innovation Management Review*, *8*(8), 60–65. https://doi.org/10.22215/timreview/1179

ECLAC. (2017). Overview of territorial development in Latin America and the Caribbean. In *Global development agendas and multilevel planning*. Santiago de Chile.

Findeli, A., Brouillet, D., Martin, S., Moineau, C. Tarrago, R. (2008). Research through design and transdisciplinarity: a tentative contribution to the methodology of design research. *Focused current design research projects and methods. Berna: Swiss Design Network*, pp. 67–91. https://complete-works-alainfindeli. ch/publications/Xt-tmhAAAB4AUtAJ

Government of Chile. (2020). National rural development policy. In *User co-production in public services* (Vol. 1). Public Innovation Laboratory, Santiago, Chile. https://www.lipuc.cl/publicaciones/coproduccion

Llanos-Hernández, L. (2010). The concept of territory and research in social sciences. *Agriculture, Society and Development*, *7*(3), 207–220. Retrieved December 29, 2023, from http://www.scielo.org.mx/scielo.php?script=sci_arttext&pid=S1870-54722010000300001&lng=es&tlng=es

Laboratorio de Innovación Pública. (2017). La co-producción del usuario en los servicios públicos. *Documento de Trabajo No. 1.* [en línea]. https://www.lipuc.cl/publicaciones/coproduccion

Miettinen, S., & Sarantou, M. (2019). *Managing complexity and creating innovation through design*. Routledge.

Mollenhauer, K., Figueria, B., Tello, C., & Wuth, P. (2017). Use of the prototype in the co-creative design of public services: FONDART case. In F. Encinas, A. Wechsler, W. Bustamante, & F. Díaz (Eds.), *Intersections 2016. II Interdisciplinary congress of research in architecture, design, city and territory* (pp. 205–223). ARQ Editions. ISBN: 9789569571480.

Mollenhauer, K., Rio, C., Rodriguez, J., Bianchi, D., & Riveros, S. (2023). Logical design matrix (LDM): How to operationalize systemic projectual design thinking in complex contexts. In W. Karwowski, T. Ahram, M. Milicevic, D. Etinger, & K. Zubrinic (Eds.), *Human systems engineering and design (IHSED 2023): Future trends and applications. AHFE (2023) international conference* (Vol. 112). AHFE Open Access, AHFE International, USA. https://doi.org/10.54941/ahfe1004135

Molokwane, T., & Lukamba, M. (2018). Citizen involvement in the formulation of public policy. In *The 3rd annual international conference on public administration and development alternatives* (pp. 192–199). https://www.researchgate.net/publication/328828793

Moreno, D., Mollenhauer, K., & Orellana, A. (2023). Empowering through design: Regional development strategy of Los Lagos as an intersectional case. In *Diid industrial design industrial design* (No. 1 2023, p. 332). ISBN: 979-12-5477-329-1. https://www.diid.it/diid/index.php/diid/article/view/t2-moreno-mollenhauer-orellana/t2-moreno-mollenhauer-orellana

Muratovski. (2022). *Research for designers: A guide to methods and practice* (2nd ed.). SAGE Publication Inc.

Nates Cruz, B. (2011). Theoretical and ethnographic supports on concepts of territory. *Coherence Journal, 8*(14), 209–229.

Orellana, A., Arenas, F., & Moreno, D. (2020). Territorial ordering in Chile: New scenario for regional governance. *North Great Geography Journal*, *77*, 31–49. https://doi.org/10.4067/S0718-34022020000300031

Orellana, A., Mollenhauer, K. Rey, C., Altamirano, P., Moreno, D., Irizarri, D., & Icaza, Mª del C. (2021a). Practical guide to integrate the national rural development policy into the regional development strategy. Office of Agricultural Studies and Policies (ODEPA) of the Ministry of Agriculture. https://www.odepa.gob.cl/dpto-desarrollo-rural/herramientas/ruralidad-en-chile/guia-pndr-comunas

Orellana, A., Mollenhauer, K. Rey, C., Altamirano, P., Moreno, D., Irizarri, D., & Icaza, Mª del C. (2021b). Practical guide to integrate the national rural development policy into the plan of "Communal Development". Office of Agricultural Studies and Policies (ODEPA) of the Ministry of Agriculture. https://www.odepa.gob.cl/dpto-desarrollo-rural/herramientas/ruralidad-en-chile/guia-pndr-regiones

Orellana-Ossandón, A., Moreno-Alba, D., Irizarri-Otárola, D., & Mollenhauer-Gajardo, K. (2020). Analysis of the perspective of integration of the national policy of rural development. Communal development plans in Chile. *Urbano Magazine, 23*(42), 66–79. https://doi.org/10.22320/07183607.2020.23.42.06

Ottenberger, A. (2000). *El estudio de casos en la investigación social.* Universidad Tecnológica Metropolitana.

Page, S. E. (2009). *Understanding complexity: Course Guidebook.* Chantilly, Virginia: Teaching Company.

Pinazo-Dallenbach, P., & Tutistar-Rosero, D. (2020). Evaluation of planning and territorial management processes: A methodological contribution from a review of approaches. *Geography Notebooks of the University of Valencia. 29.* https://doi.org/10.7203/CGUV.104.17304

Pontifical Catholic University of Chile, Directorate of Extension and External Services. (2019). *Consultancy for the development of a methodological kit that allows incorporating the perspective and guidelines in the instruments of planning, management and rural development at the regional and communal level.*

Pontifical Catholic University of Chile, Directorate of Extension and External Services. (2021). Update of the regional development strategy, Los Lagos Region BIP Code No. 40016120-0.

Salvatierra, R. (2020). Managing strategic participation through design principles: A model for value co-creation in service-based organizations. In W. Karwowski et al. (Eds.), *Human systems engineering and design III. IHSED 2020.* AISC 1269 (pp. 1–8). https://doi.org/10.1007/978-3-030-58282-1_12

Sanders, L., & Simons, G. (2009). *A social vision for value co-creation in design.* Open-Source Business Resource. http://TIMREVIEW.CA/ARTICLE/310

Sevaldson, B. (2019). What is systemic design? Practices beyond analyses and modelling. In *RSD8 Symposium, Chicago* (Working Paper). http://openresearch.ocadu.ca/id/eprint/3233/

Sevaldson, B. (2022). *Designing complexity the methodology and practice of systems oriented design.* Common Ground Research Networks.

Sevaldson, B., & Jones, P. (2019). An interdiscipline emerges: Pathways to systemic design. *She Ji: The Journal of Design, Economics, and Innovation, 5*(2), 75–84. https://doi.org/10.1016/j.sheji.2019.05.002

Soto, M., Mollenhauer, K., Sarantou, M., & Miettinen, S. (2021). Building a community through service design and responsiveness to emotions. In T. Seppala, M. Sarantou, & S. Miettinen (Eds.), *Arts-based methods for decolonising participatory research,* (pp. 123–145). Routledge. https://doi.org/10.4324/9781003053408

Stickdorn, M., Lawrence, A., Hormess, M., & Schneider, J. (2019). *This is service design doing applying service design thinking in the real world.* O'Reilly.

Chapter 23
New Biobased Materials: Exploratory Analysis of the Actors Developing New Biobased Materials in Chile

Andrea Wechsler Pizarro, Yesenia Briones Castro, and Pablo Domínguez González

Abstract This manuscript presents an analysis of the state of the art of several actors who are currently working on the development of new materials in Chile in pursuit of a circular economy. In Chile, some units of developers have been gradually researching new materials aiming to add value to the organic solid waste generated by small and medium national industries through their application in new materials and products. These communities have the potential to generate and drive knowledge not only in the product field, but also manufacturing industries can manufacture new materials, reintegrate by-products as new resources and establish new productive processes. To carry out this study, the main activities, the types of projects and ways of reaching the community of these actors were collected and analyzed through an exhaustive bibliographic review, surveys and interviews with the leaders of these spaces. With the abovementioned, it was possible to study and systematize the information collected, managing to establish the profile of these units. The technological resources they have, means of dissemination and their connection with the environment were also identified. As a result, a certain disconnection is perceived between these units, a high concentration in the Metropolitan Region and very heterogeneous dissemination of results, with a rather massive and inefficient strategy in linking actors. This analysis addresses an opportunity and a challenge for Chile. Challenge, because these new materials lack homologation, and their mechanical performance is slightly lower than traditional materials. Opportunity, since the productive capacity of the Chilean fruit, forestry and aquaculture industry at a global level, the quantity and quality of the waste resulting from industrial processes is high and they are feasible to be used as inputs for new materialities.

A. W. Pizarro (✉) · Y. B. Castro · P. D. González
Facultad de Arquitectura y Urbanismo, Universidad de Chile, Santiago, Chile
e-mail: andreawechsler@uchilefau.cl

Y. B. Castro
e-mail: yessy.bc@uchile.cl

P. D. González
e-mail: pdominguez@uchile.cl

M. Vicuña and E. Giorgi (eds.), *Intersections*, Springer Geography,
https://doi.org/10.1007/978-3-031-76402-8_23

Keywords Biobased materials in Chile · Material design · Waste valorization · Industrial ecology · Circular economy

23.1 Introduction

This manuscript presents an analysis of the relevant characteristics of the various working groups that to date work on the valorization of waste from national industries. These units seek to valorize waste, applying them to new biobased materials and in new products.

Much is said that garbage is a design error, and partially it is. The development and manufacturing phases of products are the ones that have the greatest impact on the environment. These impacts are linked to the decision-making about materials, design and operation (Cai et al., 2019; Fu et al., 2020). From this, the United Nations Environmental Program (UNEP) developed a general policy framework called "Sustainable Consumption and Production" (SCP) (UNEP, 1996, 2007; UNESCO 1978), whose approaches are ecodesign and industrial ecology. Ecodesign arises to decrease the environmental impacts of a product throughout its life cycle (Aguayo et al., 2011). Industrial ecology, on the other hand, studies the flow of energy and materials throughout the life cycle of a product, with a particular focus on the impacts that industrial activities have on the environment (Garner & Keoleian, 1995). On the other hand, the circular economy, a term coined recently, is a model of production and consumption that seeks to extend the life cycle of products as much as possible through strategies such as sharing, reusing, repairing and recycling, reducing waste to a minimum (European Parliament, 2023).

In Chile, there are initiatives to move towards a circular economy, such as the implementation of the Extended Producer Responsibility Law (REP), Clean Production Agreements (APL), CORFO actions, with its innovation line and emblematic municipal ordinances, such as the prohibition of plastic bags, for example. However, except for the REP law, these parallel initiatives have not managed to incorporate a systemic view, so it would be necessary to ensure greater coherence and linkage between the different initiatives.

According to studies commissioned by Rubik & Gobierno de Chile (2019), collaborative work with innovation in which managers of development initiatives of materialities with a design perspective are appreciated has not yet been detected. Some of the identified gaps are the lack of material suppliers, lack of infrastructure and lack of innovation in materials and processes (MMA Chile, 2021; Ramos et al., 2021; Rubik & MMA Chile, 2019). Based on these identified gaps, it is proposed to identify how the groups that work generating new materials for a circular economy link with society and industry.

In Chile and the world there are various communities that promote and develop research and development projects that link innovation, technologies and design in materials (Araújo & Franco, 2021). In fact, for some decades now, actors who have worked on the conversion of waste to feed new industrial processes have appeared.

These groups, sponsored by universities, range from laboratories that arise from research projects and that respond to local needs, to productive collaborative communities, such as the "*fablabs*" or independent entrepreneurs (Donoso & Wechsler, 2020; Meroni, 2007; Schwab, 2016).

Although Chile presents itself as a "natural laboratory" thanks to its diversity of ecosystems, with immense potential to take advantage of its resources from the local (mining, water, wood, horticultural, aquaculture, among others) through different solutions according to the location, it is feasible and necessary to seize the opportunity to develop sustainable materials from the waste generated by the industry (Rubik & MMA Chile, 2019). Therefore, if there is industrial and technical capacity to carry out these developments, the question arises, which is the main reason for the gap of lack of material suppliers, infrastructure and innovation of materials and processes not yet bridged?

One hypothesis is that these new materials, to achieve their industrial production and insertion in the market, should be compatible with current production processes, with an equal or lower price and better quality than traditional materials (Cores Irago, 2018). However, maintaining a productive level and competitive performance is very complex to achieve, since a large number of these new materials lack homologation and usually present a mechanical performance lower than traditional materials. These new materials, despite their potentialities, must be functional, reproducible and attractive to current society. To bridge this gap, the development of interdisciplinary research that considers actors from waste-generating and managing companies, professionals from the creative area to local communities, scientists and mass production industries becomes vital.

Other the questions which arises, relates with the current situation of the actors who develop new materials for a circular economy and how do these actors insert and link with the national reality?

To answer these questions, this study seeks to identify and categorize the different actors and working systems of the groups that develop biobased materials in Chile from a circular economy perspective.

23.1.1 Methods: Identification of Groups Generating and Developing Biobased Materials in Chile

To carry out this study, a survey was conducted, consulting in a first instance secondary sources. These included a literature review, starting with the systematic digital search (websites, social networks and digital media) of the different groups working on biobased materials up to the review of reports, theses and academic articles. Terms focused on the design and development of biobased materials, laboratories, exploration, innovation and interdiscipline were searched. Among the search criteria are that they are local units, that they have at least one biobased material developed at laboratory scale and that they work from the valorization of waste.

In a second instance, primary sources were approached, these were contacted, and a survey was sent to more than 30 actors who develop biobased materials. Simultaneously, interviews and visits were conducted with 16 actors identified in the Metropolitan Region, and online meetings were generated with actors from other regions. They were asked about their specific geographical location, the type of dissemination they use to present their results to different communities (academic, professional, industrial or others), the level and type of technological equipment they have, the type of projects they carry out, their source of funding and the type of results they obtain. Another variables to analyze were the essential concepts that distinguish these centers according to their own perceptions and approaches. Among these concepts are innovation, technology, resources, impact, sustainability, territory, academia, avant-garde, productive sector, scaling, industrial level, among others.

23.2 Results and Discussion: Types of Institutions Active in the Development of Biobased Materials

The first stage of the study resulted in the construction of a database with secondary sources of 30 actors from different spaces listed. These actors were characterized and mapped. From these, primary information was obtained from 20 units that met the requirement of developers of biobased materials in Chile. These dimensions were defined depending on the declared and/or identified profile by the researchers of this study. In this way, the number of institutions and characteristics analyzed were normalized to be able to compare them on equal terms according to each type of actor. These were organized into the following five dimensions:

- Teaching, research and development,
- Independent researchers and entrepreneurs,
- FabLabs,
- Biofabrication research centers and
- Research and development centers.

They were organized in this way based on the types of projects they carried out, on their infrastructure, on their source of funding and the type of results they obtained. Given the low reach and visibility of these spaces, it was complex to contact them and obtain further information referred to the addressed actors. There were spaces that did not respond, remaining outside the analysis.

The first result that draws attention and is corroborated with other studies in the line is that the highest percentage of actors (70%) are located in the Metropolitan Region, as well as the highest percentage belongs to independent and private units (MMA Chile, 2021; Ramos et al., 2021; Rubik & MMA Chile, 2019). The exception is seen in the FabLabs and in the biofabrication research centers, which are located 50% in regions and 50% in the Metropolitan Region, this is shown in Table 23.1.

Table 23.1 Analysis of the location of actors developing biobased materials in Chile by group

	National presence of identified groups %	
	RM	Other regions
Teaching, research and development space	90	10
Independent researchers and ventures	80	20
FabLabs	50	50
Biofabrication research centers	50	50
Research and development centers	70	30
Average	70	30

Figure 23.1 shows a summary of the spaces identified in this study, the type of dissemination they develop and the type of infrastructure they have. Broadly speaking, very varied profiles can be seen among these different groups, particularly in their way of disseminating results. Most of these groups have a website, even though it is noted that independent ventures (85%) and teaching spaces (67%) do not fully have this resource.

Fig. 23.1 Summary of the 5 types of spaces identified (own elaboration)

There is quite a lot of dissemination on social networks, however, the most used is Instagram. The use of *Facebook*, on the other hand, is varied. *Twitter*, is more used by FabLabs, and finally *YouTube*. Three out of 5 groups of units claim to use *YouTube* actively. If the social networks used to disseminate from the units are compared broadly, they can be associated, according to the CADEM (2022) survey *Instagram* is the platform most used and generates the most trust among users between 13 and 23 years old (Generation Z), followed by the generation between 24 and 40 years old (Millennials), the generation between 41 and 55 years old (Generation X) and between 56 and 71 years old (Baby Boomers). *Facebook*, on the other hand, is used by a slightly more mature age group, led by Generation X, followed by Millennials, Boomers and finally the youngest generation (Generation Z). On the other hand, *YouTube* presents a younger audience, led by Millennials, followed by Generation X and finally Boomers.

In terms of research, all the actors represented here declared to carry out interdisciplinary academic research in different ranges, led by biofabrication research centers and research and development centers. However, academic publications are low in terms of those who declare to carry out research. On the other hand, there is an active dissemination based on talks and workshops.

While many of these units have developed formulas or recipes, only the FabLabs (70%) have open-source formulations to the community, the rest of the groups do not have open-access formulations.

In terms of technological infrastructure, it is also varied, although all groups declare to have infrastructure, it is led by private biofabrication research centers and research and development centers. Regarding 3D printing, only the FabLabs declared 70% to work with 3D technologies, followed by the research and development centers.

The 5 dimensions in which the participating actors of this study are presented and detailed below.

23.2.1 Teaching, Research and Development Spaces

They are laboratories inserted in an academic-university environment. They are dedicated to developing projects resulting from academic activities. They are commonly directed by an academic who coordinates these spaces, in addition to having an active participation of upper-level students and postgraduates. Although they have the support of the university that hosts them, several depend on external grants for their operation and acquisition of equipment and supplies. These spaces have a greater presence and have in common the work with industrial waste and post-consumer waste. Among the types of research carried out they are rigorous in terms of the performative characterization of the developed materials (Fig. 23.2). The identified concepts that govern them are innovation, impact, academia, society, culture, transdisciplinary, sustainability, valorization, transfer and technology.

Fig. 23.2 Summary diagram of main characteristics of spaces of teaching, research and development (own elaboration)

Although in Fig. 23.2 it is seen that about 75% of these groups carry out research, this is not reflected in academic publications, rather their results are disseminated through digital resources (50%), social networks, particularly Instagram and Facebook and talks or workshops to the community (75%), as a more dynamic way of spreading. The material formulations developed here are not public domain (10%) and they have intermediate technological infrastructure (55%).

Some of the units identified here are the Laboratory of Adhesives and Composite Materials (LAMC), the Center for Biomaterials and Nanotechnology (CBN, 2016) and the Prodima-Lab (2022), all from the University of Bío-Bío (Faculty of Engineering University of Bío-Bío, 2018) (Fig. 23.3), along with the Biobased Materials Laboratory, BioLab FAU from the University of Chile (2019) (Fig. 23.4) and the LincUDP (2013).

23.2.2 Independent Researchers and Ventures

In this category, the spaces have mostly emerged under the university teaching umbrella and are self-sustaining. They work with biobased materials from waste or organic elements and their profile is exploratory (Fig. 23.5). It was identified that

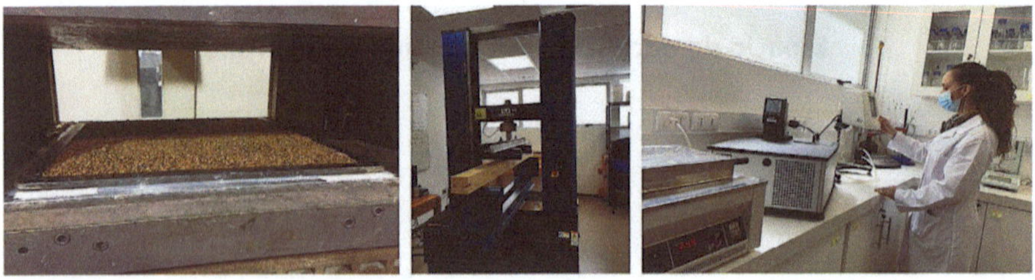

Fig. 23.3 Prodima-Lab workspace (2022)

Fig. 23.4 Projects carried out by industrial designers at the BIOLAB FAU of the University of Chile (2019)

the concepts that govern them are local, territory, valorization, resources, production and collaboration.

In terms of dissemination media used, there are websites and social networks, particularly *Instagram*. They also declare to carry out research (40%) and 50% recognize dissemination through talks and workshops. They have very low technological infrastructure (10%).

Some of these ventures are that of Carolina Pacheco, designer with an exploration of calcareous materials (Pacheco, 2019) (Fig. 23.6); Savia Circular (2020) (Fig. 23.7) and Margarita Thalep (2019; Desintegrame, 2018) (Fig. 23.8). Both designers explore hydrocolloid materials. Valnux® (2019), on the other hand, presents an antibacterial material made from walnut shell waste and a thermoplastic material. It should be mentioned that none of these ventures are currently actively producing and selling their products, even though they started as very promising ventures.

Fig. 23.5 Summary diagram of the group "Researchers and independent ventures" (own elaboration)

Fig. 23.6 Some projects carried out by Pacheco are Calcáreo (2019), Lugae (2020), and Estratos, an exploratory project in collaboration with Cosmosapiens (Local Encounter, 2021)

23.2.3 FabLabs

FabLabs were born as digital fabrication workshops capable of producing objects on a local scale using global information. Their particularity lies in their strong connection with society rather than with industry. FabLabs proposes that production

Fig. 23.7 Material exploration carried out by Savia Circular (2020)

Fig. 23.8 Independent projects Desintegrame and Kanten by Thalep (2019)

should be developed locally, considering the recycling of materials and the solution to local problems (Diez, 2018). Almost all FabLabs in Chile carry out workshops to transfer knowledge to the community, besides implementing technology transfer projects for companies, entrepreneurs, public and private institutions, among others (Valenzuela Zubiaur, 2022) (Fig. 23.9).

In terms of media used for dissemination of results, websites and social networks such as *Facebook* (100%), *Instagram* (100%), *Twitter* and *YouTube* are used to a lesser extent (75%). They also claim to generate research at an intermediate percentage (50%) and disseminate it through talks and workshops. It is striking that, unlike the previous groups, they develop less research than dissemination through talks and workshops. Very few claim to generate academic publications (15%). Among their results are developed material formulas, many of which are open source. FabLabs do not declare a high technological infrastructure (40%), however, they do declare a higher percentage (68%) of 3D printing use. The concepts that govern them are local, innovation, society, transdisciplinary, sustainability and technology.

Among the national FabLabs we found the Design Lab of Adolfo Ibáñez University (2022), the FabLab of the University of Chile (2022) (Fig. 23.10), the Austral Fablab of the Austral University (2021), ProteinLab (2017) of the Metropolitan Technological University (UTEM, 2021) (Fig. 23.11), the former Ecotechnological FabLab, now EcoFabLab (2024) (Fig. 23.12) and Sinestesia.cc (2022) (Fig. 23.13), the latter are independent.

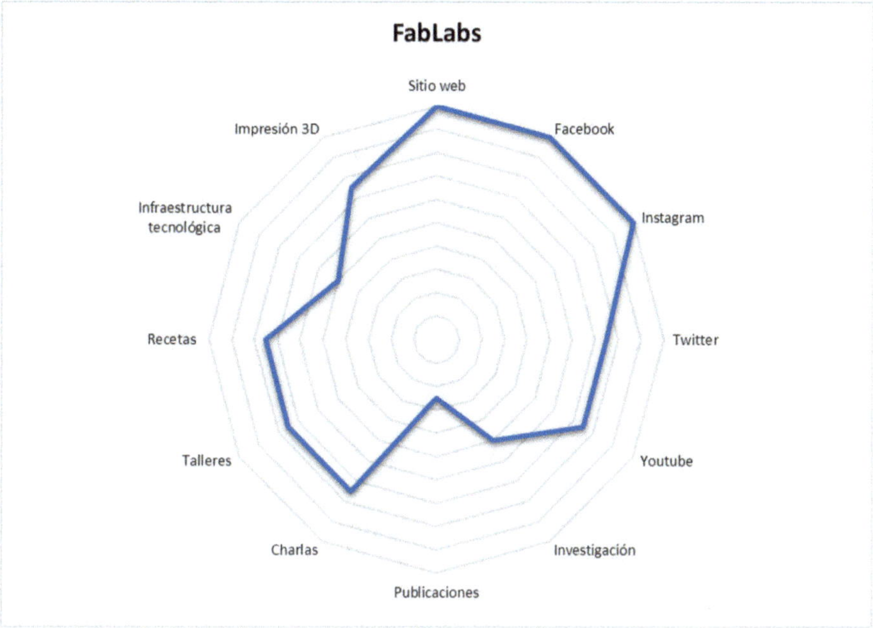

Fig. 23.9 Summary diagram of the group "FabLabs" (own elaboration)

Fig. 23.10 Projects carried out by FabLabs of the University of Chile (2022)

23.2.4 *Biofabrication Research Centers*

They are transdisciplinary spaces that explore local materialities. They have a wide exploration of new materials and develop knowledge for medium and small companies, intensely promoting local research and experimentation. This is reflected in a distribution of 50% in Santiago and another 50% in regions (Table 23.1).

The materials developed here are characterized by being low-cost technologies and open source. For example, biomaterials that arise from the cultivation of organisms (GIY—*Grow it Yourself*) or generate new formulations (CIY—*Cook it Yourself*).

In terms of dissemination media used, they have websites and social networks such as *Facebook, Instagram* and to a lesser extent *YouTube.* All declare to develop

Fig. 23.11 Projects carried out by ProteinLab (2017) from UTEM

Fig. 23.12 Liquid wood projects carried out by EcoFabLab (ExFabLab-Olmué) (2024)

Fig. 23.13 Laboratory and projects carried out by Sinestesia.cc and residents of the space (2022)

research and disseminate it to a lesser extent through academic publications, talks, and workshops (50%). Among their results they declare to have developed material formulations (50%), with a high technological infrastructure; however, they do not declare the use of 3D printing (Fig. 23.14). The concepts that govern them are local production, academia, territory, citizenship, culture, transdisciplinary, sustainability, fair trade.

Some examples of these are the Valdivia LABVA Laboratory (2023) (Fig. 23.15), the UC Bio-manufacturing Laboratory (2019) of the Catholic University of Chile (Fig. 23.16), who have experience and development in materials that are grown.

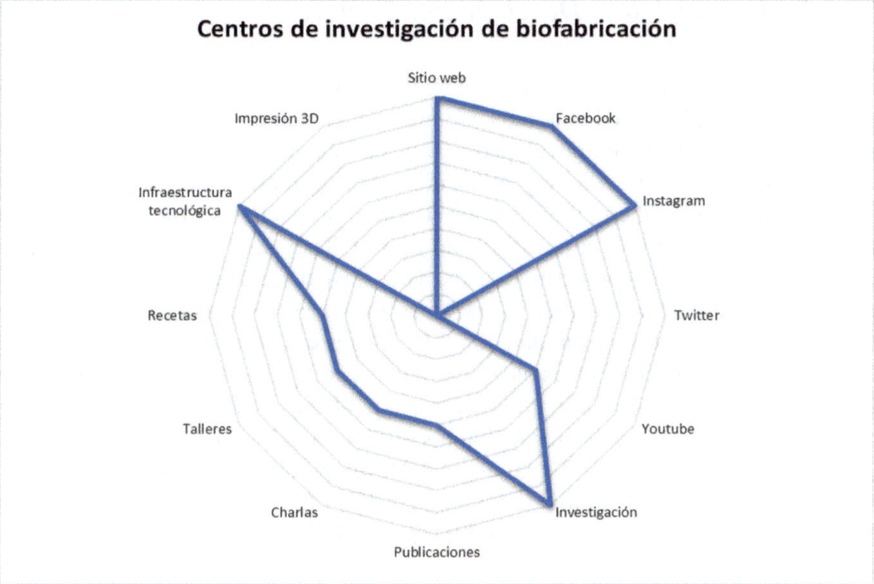

Fig. 23.14 Summary diagram of the group "Biofabrication research centers" (own elaboration)

Fig. 23.15 Projects and material exploration carried out by LABVA (2023)

Fig. 23.16 Projects and material exploration carried out by Bio-fabrication Laboratory UC (2019)

23.2.5 Research and Development Centers

These are spaces that have high technological infrastructure for work, from thermoplastic polymers, biobased materials and material characterization. In these centers, materials and products are developed in a closed manner through consultancies to companies, in addition to training. What distinguishes these centers are the essential concepts that govern them, these are avant-garde, productive sector, scaling, industrialization, sustainability, innovation and impact. They are financed both from the private sector and with competitive funds. They have dissemination of their research through institutional web pages in 100%, as for social networks 68% declare that they use *Facebook, Instagram* and *YouTube*. The use of *Facebook* shows that they target an audience over 40 years old. 70% of these units declare to carry out research, however, no institution in this group actively publishes academically, instead they disseminate their results also through talks (70%) (Fig. 23.17).

Some examples are the Technological Development Unit (UDT) (1996), the Research Center in Technologies for Society (C+) (2020) from the University of Development (Briones & Cepeda, 2020) (Fig. 23.18) and Leitat Chile (2021).

Fig. 23.17 Summary diagram of the group "Research and development centers" (own elaboration)

Fig. 23.18 Projects carried out by the research center in technologies for society (C+, 2020)

23.3 Conclusion

The systematic analysis and presentation of the various working groups that to date carry out material development in Chile, show their main characteristics in terms of dissemination, research and technological infrastructure.

It can be observed that despite the variety of the origins of the institutions identified here, they share the main constitutive characteristics of each one. However, when investigating the works developed in parallel and unlinked between each institution, there is an opportunity to work collaboratively. In addition, there is a need for linkage according to the capabilities of each of these groups, in order to be able to technically evaluate and enhance the materials explored in terms of their characteristics and creation of design products with a sustainable focus. This need is reflected in the low rate of academic publications and technology transfer. This could be solved with the insertion of more researchers trained in the subject of study, promoting the development of innovation from materiality to design products. It is noted that the way of disseminating results is quite heterogeneous, using rather mass media and although they provide information, this does not manage to link the key actors with whom they should work, who are industries that generate waste, product manufacturers, customers and end users. The profile of dissemination is aimed at a young audience and the information provided is rather general, being what social networks allow to deliver, although dynamic, they do not allow to delve into the contents.

In this chapter it can be observed that the capacity both technical and practical, although it exists, lacks to enhance associativity and adequate dissemination to be able to reach the insertion of products and new biobased materials in waste in the market, promoting a productive symbiosis in view of a circular economy.

Acknowledgements The authors thank the collaboration of the various research groups studied and the National Fondart Project 531807 "Research and development of an open source 3D prototyping system for biobased materials".

References

Aguayo González, F., Peralta Álvarez, M. E., Lama Ruiz, J. R., & Soltero Sánchez, V. M. (2011). Ecodesign. In *Sustainable engineering from cradle to cradle (C2C)*. RC Books.

Araújo, R., & Franco, M. (2021). The use of collaboration networks in search of eco-innovation: A systematic literature review. *Journal of Cleaner Production, 314*, 127975.

Biofab UC. (2019). UC Biofabrication Laboratory. Retrieved July 22, 2022, from http://biofab.cl/category/investigacion/

BioLab FAU. (2019). In A. Wechsler, & Y. Briones (Eds.), *Biobased materials laboratory* (p. 28). Faculty of Architecture and Urbanism, University of Chile. https://drive.google.com/file/d/1de7RatJaumhXjXPP7kfowbqZons-YZ74/view

Briones, Y., & Cepeda, A. (2022). Development and exploratory design of a biobased textile in tea, coffee and yerba mate. *Base Design and Innovation, 7*(7), 107–122. https://doi.org/10.52611/bdi.num7.2022.807

C+. (2020). *Research center in technologies for society. Innovation engineering*. Faculty of Engineering University of Development. Retrieved July 12, 2022, from https://ingenieria.udd.cl/ii/centroid/

CADEM. (2022). *The upcoming Chile: Media and social networks*. https://cadem.cl/wp-content/uploads/2022/02/Chile-que-Viene-Enero-2022-Medios-y-Redes-Sociales.pdf

Cai, W., Lai, K. H., Liu, C., Wei, F., Ma, M., Jia, S., Jiang, Z., & Lv, L. (2019). Promoting sustainability of manufacturing industry through the lean energy-saving and emission-reduction strategy. *Science of the Total Environment*, *665*, 23–32. https://doi.org/10.1016/j.scitotenv.2019.02.069

CBN UBB. (2016). Biomaterials and Nanotechnology Center University of Bio-Bio. CBN UBB. Retrieved July 22, 2022, from http://cbn.ubiobio.cl/

Cores Irago, I. (2018). Biobased materials. Design. Valencia, Escolad'Art I Superior de Disseny de Valencia. Product Designer: 84.

DesignLab UAI. (2022). School of Design of the Adolfo Ibáñez University. FabLab. Design Lab UAI. Retrieved July 12, 2022, from https://designlab.uai.cl/

Desintegrame. (2018). Desintegra.me-biomaterial. Instagram. Retrieved July 22, 2022, from https://www.instagram.com/desintegra.me/

Diez, T. (2018). Fab City. *The mass distribution of (almost) everything*. https://issuu.com/iaac/docs/fabcitymassdistribution

Donoso, S., & Wechsler, A. (2020). *The biobased materials and the Latin American developmentalist paradigm: Perspectives from Industrial Design* (p. 114). Notebooks of the Center for Design and Communication Studies.

European Parliament. (2023). Circular economy: Definition, importance and benefits. *News*. 20151201STO05603.

EcoFabLab. (2024). Ecofablab Ecotecnológico. https://www.ecofablab.cl/

Fab Lab Austral. (2021). Welcome | FAB LAB AUSTRAL. The world's southernmost digital fabrication laboratory. Retrieved July 22, 2022, from https://www.fablabaustral.org/

Fab Lab U. de Chile. (2022). *Completed projects*. FABLAB University of Chile. Retrieved July 22, 2022, from http://www.fablab.uchile.cl/proyectos/

Faculty of Engineering University of Bío-Bío. (2018). *Laboratories and centers*. Faculty of Engineering UBB. Retrieved July 22, 2022, from http://fi.ubiobio.cl/laboratorios-y-centros/

Fu, Y., Wang, L., & Li, L. (2020). Automatic conceptual design scheme generation and decision-making considering green demand. *Procedia Manufacturing*, *43*, 407–414. https://doi.org/10.1016/j.promfg.2020.02.194

Garner, A., & Keoleian, G. A. (1995). *Industrial ecology: An introduction*. National Pollution Prevention Center for Higher Education, University of Michigan.

LABVA. (2023). *We are LABVA [collaborations] [social network post Instagram]*. Instagram. Retrieved July 22, 2022, from https://www.instagram.com/somoslabva/

Leitat Chile. (2021). *Leitat Chile: Private and non-profit technological center*. Retrieved July 25, 2022, from https://www.leitat.cl/

LINCudp. (2013). *Innovation and creativity laboratory*. Diego Portales University. LINC UDP. Retrieved July 22, 2022, from https://linc.udp.cl/

Local Encounter. (2021). Local Encounter 2021—Emerging talent: Caro Pacheco. Local Encounter 2021. Retrieved July 22, 2022, from https://encuentrolocal.cl/expo-talento-emergente/caro-pacheco/

Lugae. (2020). Lugae. *Algae-based biomaterial*. Retrieved July 22, 2022, from https://www.lugae.cl/

Meroni, A. (2007). *Creative communities. People inventing sustainable ways of living*. Polidesign.

MMA Chile. (2021). Roadmap for a Circular Chile by 2040. https://faolex.fao.org/docs/pdf/chi211134.pdf

Pacheco, C. (2019). *Calcareous*. Caro Pacheco. Retrieved July 25, 2022, from https://www.caropacheco.work/calcareo

Prodima-Lab. (2022). Laboratory of engineering products based on wood and adhesives from the Department of Civil and Environmental Engineering of the University of Bío-Bío (DICA). Retrieved July 22, 2022, from http://www.dica.ubiobio.cl/historia-mision-vision/laboratorios/prodima-lab

ProteinLab UTEM. (2017). Prospective and Technological Innovation Program—Community Engagement. Metropolitan Technological University. Retrieved July 22, 2022, from https://vtte.utem.cl/prospectiva-e-innovacion-tecnologica/

Ramos, M., Pérez, J., Dumbay, C., Robinson, J., Florenzano, A., & Salvo, R. (2021). *Sector diagnosis, clean production agreement: "Transition towards a Circular Economy"*. https://accionempresas.cl/content/uploads/informe-de-diagnostico-sectorial-apl-transicion-hacia-la-economia-circular-accion-empresas

Rubik, & MMA, Chile (2019). *Circular economy in Chile final report*. https://www.paiscircular.cl/wp-content/uploads/2019/08/Econom°a-Circular-_-Informe-1.pdf

Rubik, & Gobierno de Chile, M. (2019). Economía Circular En Chile Informe Final. https://www.paiscircular.cl/wpcontent/uploads/2019/08/Econom°a-Circular-_-Informe-1.pdf

Savia Circular. (2020). Circular Sap. *Biodesign solutions*. Instagram. Retrieved July 22, 2022, from https://www.instagram.com/saviacircular/

Schwab, K. (2016). *The fourth industrial revolution*. Debate.

Sinestesia.cc. (2022). Creative Industries | Synesthesia | Santiago. Retrieved July 25, 2022, from https://www.sinestesia.cc

Talep, M. (2019). *Material exploration/biomaterials/craftsmanship*. Margarita Talep. Retrieved July 22, 2022, from https://margaritatalep.com/

UDT (1996). *Technological development unit. Science, technology and innovation in bioeconomy*. University of Concepción. Retrieved July 23, 2022, from https://www.udt.cl

UNEP. (1996). *Life cycle assessment: What it is and how to do it*. United Nations Environment Programme, Industry and Environment.

UNEP. (2007). *Buildings and climate change—Summary for decision-makers* (p. 62). United Nations Environment Programme Sustainable Building and Climate Initiative, Paris.

United Nations Educational, Scientific and Cultural Organization (UNESCO)—United Nations Environment Programme (UNEP). (1978). *Intergovernmental conference on environmental education, Tbilisi, Georgia, 1977* (Final report, ed/md/49, Paris, pp. 11–12).

UTEM. (2021). ProteinLab UTEM wins a FONDART 2021 project in the design line to develop biomaterials applied to textile design. Linkage with the environment—Metropolitan Technological University. Retrieved July 22, 2025, from https://vtte.utem.cl/2021/03/26/proteinlab-utem-se-adjudica-un-proyecto-fondart-2021-en-la-linea-diseno-para-desarrollar-biomateriales-aplicados-al-diseno-textil/

Valenzuela Zubiaur, M. (2022). *Contribution of the Chilean Fab Labs Network to the local innovation ecosystem to achieve sustainable development* [Doctoral dissertation, Universitat Politècnica de València].

Valnux. (2019). Instagram. Retrieved July 22, 2022, from https://www.instagram.com/valnux/

Chapter 24
Multidisciplinary Approach for the Development of Materials from the Byproduct of the Opening of the Cashew Nut in Vichada

Camilo Ayala-Garcia, Carolina Pérez Rodríguez, Clara Forero Lesmes, Jessica Rodríguez Briceño, Santiago De Francisco Vela, Leonardo Parra Agudelo, Tatiana Carolina Cruz Perea, Juan Sebastian Porras Cardenas, Alejandro Marañón León, Camilo Hernandez Acevedo, Oscar Álvarez Solano, and Alicia Porras Holguín

Abstract The Vichada region in eastern Colombia is known for its cashew production. Its pseudofruit, oil, and nut are agro-industrial products of great value in international markets. It is estimated that by 2025, cashew production in the region will reach approximately 100,000 cultivated hectares, which means a significant flow

C. Ayala-Garcia (✉) · C. Pérez Rodríguez · C. Forero Lesmes · J. Rodríguez Briceño · S. De Francisco Vela · L. Parra Agudelo
Department of Design, Universidad de los Andes, Bogotá, Colombia
e-mail: c.ayala954@uniandes.edu.co

C. Pérez Rodríguez
e-mail: c.perez1816@uniandes.edu.co

C. Forero Lesmes
e-mail: cl.forero2722@uniandes.edu.co

J. Rodríguez Briceño
e-mail: jm.rodriguez17@uniandes.edu.co

S. De Francisco Vela
e-mail: s.defrancisco@uniandes.edu.co

L. Parra Agudelo
e-mail: leonardo.parra@uniandes.edu.co

T. C. Cruz Perea · J. S. Porras Cardenas · A. Marañón León
Department of Mechanical Engineering, Universidad de los Andes, Bogotá, Colombia
e-mail: tc.cruz10@uniandes.edu.co

J. S. Porras Cardenas
e-mail: js.porras@uniandes.edu.co

A. Marañón León
e-mail: emaranon@uniandes.edu.co

M. Vicuña and E. Giorgi (eds.), *Intersections*, Springer Geography,
https://doi.org/10.1007/978-3-031-76402-8_24

of the different products and byproducts of this crop to national and international markets. Using material-driven design methodologies, possibilities for creating material alternatives from the cashew shell, a byproduct of the nut opening, are explored. This allows for proposing new production and consumption scenarios, in which, on the one hand, the substitution of synthetic polymers within the productive chain. On the other hand, it involves the participation of the community in the development process, encouraging them to actively participate in research processes, contributing their knowledge and skills. This article reports a multidisciplinary approach to the development of materials from the agro-industrial byproducts of the cashew nut.

Keywords DIY-Materials · Circular design · Multidisciplinary development · Cashew · Vichada · Colombia

24.1 Sustainable Materials Driven by Design

In the department of Vichada, there are particular climatic and geographical conditions that favor the cultivation of a variety of foods. Located in eastern Colombia, it is known for the production of forestry, soy, rice, corn. In particular, since the 1980s it has made a significant incursion into the cultivation and production of different varieties of cashew, a fruit whose components are agro-industrial products of great value in international markets (FAO, 2020). The cashew apple (pseudofruit) is used to make products such as juices, wines, and gums (Akinwale, 2000). The nut has a high nutritional value of edible almond, protected by a shell that contains oil (CNSL, for its acronym in English: Cashew Nutshell Liquid) rich in phenols such as cardol and cardanol.

The cashew nut shell (CNS, for its acronym in English: Cashew Nutshell) represents approximately 75% of the nut, and is currently considered as a waste of production. This is why it is usually returned to the soil or used as an input for combustion during the nut cooking stage in the transformation process, which can lead to an environmental problem if not handled properly (Prakash et al., 2018). At the same time, the increase in global demand, which rises by more than 7% annually, has expanded cashew crops in the territory. It is estimated that by the year 2025, cashew production in the region will reach about 100,000 cultivated hectares, representing

C. Hernandez Acevedo
Department of Mechanical Engineering, Escuela Colombiana de Ingeniería Julio Garavito, Bogotá, Colombia
e-mail: camilo.hernandez@escuelaing.edu.co

O. Álvarez Solano · A. Porras Holguín
Department of Chemical and Food Engineering, Universidad de los Andes, Bogotá, Colombia
e-mail: oalvarez@uniandes.edu.co

A. Porras Holguín
e-mail: n-porras@uniandes.edu.co

a significant flow of the different products and byproducts of cashews in national and international markets (Cardona, 2017). As a result, a notable increase in the volume of underutilized waste is expected in a region like Vichada, where problems are currently beginning to emerge regarding the management of these byproducts.

For the purpose of taking advantage of this waste, the design discipline, traditionally distant from the technical processes of material transformation, recognizes materials and their management as critical points for sustainable production and consumption (Vezzoli, 2018, p. 25). The material, which was previously considered a step in the design process, now becomes the focus of the project (Karana et al., 2020; Rognoli et al., 2015). Therefore, material-driven design becomes an essential tool for sustainability and the vision of an effective ecological transition.

Material design, different from the technical development of materials, can envision alternatives for waste transformation, as in this case, the shell of the cashew nut, through speculative scenarios. Instead of focusing on the development of a material to improve performance or a specific market need, innovative solutions for contemporary problems are sought.

Several contributions have emerged in the past 10 years in the transformation of material resources through a creative process driven by design in the international arena (Ayala-García, 2023). Some material development alternatives come from understanding the capabilities to grow or harvest a material thanks to agricultural techniques (Zhou et al., 2021), fermentation methods (Cohen et al., 2020) or the manipulation of organic waste. Some other contributions have developed a creative use of parts (hair, skin, bones, etc.) of animals (Studio Swine, 2023) or through the collaboration of animals or microorganisms such as bacteria (Lee, 2023).

Many of these materials developed in recent years do not come exclusively from scientific laboratories, as was the standard, but take advantage of the possibilities of interdisciplinary work typical of design, which when combined with other experts in multiple fields such as Biology, Engineering, or Agriculture (D'Olivo & Karana, 2021) achieve, through a multidisciplinary approach, alternatives to counteract the use and abuse of materials and product consumption in the current linear path of our industrialized society. The multidisciplinary and, in some cases, multicultural approach to material development is becoming the new standard (MaDe, 2021). The different methods available today allow us to imagine alternative and sustainable futures. The conscious use of resources, minimizing energy consumption, along with new access to technologies through open source (Gershenfeld, 2012), technological democratization (Bettiol & Micelli, 2014; Tanenbaum et al., 2013) and the rediscovery of ancient artisanal practices (Sennett, 2008), provide the necessary tools to propose different industrializations. (Anderson, 2012). One of the drivers to achieve circularity in industry and agroindustry comes from understanding the value of all the different resources involved in the process (Manzini & Vezzoli, 2003). This aforementioned approach has guided different multidisciplinary teams led by Design, including ours, to discover potentials for elaboration and transformation of new material by taking advantage of the waste product of the cashew nut shell or CNS.

The development of this research is focused on a collaborative co-design methodological approach where the active participation of multiple actors in the design process including the community around cashew cultivation, together with engineering and design teams generates a constant reflection in the action of the project (Schön, 1983). This methodology seeks: (1) The inclusion of users and actors who can contribute with their knowledge and differential approaches. (2) The multidisciplinary interaction of teams from various branches of knowledge that feed the development of solutions, products or services resulting in accurate proposals to the needs of those involved and the particular conditions of the challenges presented.

Co-design is a method to cover a 360° panorama for the design of products, services or experiences by involving different stakeholders in the creative process, seeking to improve the quality and relevance of the solutions. It incorporates the user from the beginning of the project instead of the traditional design process, which includes them only in the verification or presentation of results phases. This collaborative approach involves the use of participatory techniques (Buur and Matthews, 2008). Through the generation of joint design workshops (users, direct and indirect actors, researchers, experts among others), multidisciplinary brainstorming sessions, rapid prototyping at low and high resolution and reflection during the process, it is possible not only the application of technical knowledge, but also provides the opportunity to adapt and adjust design decisions during each of the phases of the process. The purpose of this research methodology is to understand the particularities of the context and to foster open communication and collaboration among the participants to achieve more effective solutions, user-centered and in line with local reality.

Within the framework of this project, agro-industrial by-products derived from the transformation of the cashew nut have been the basis for exploring possible opportunities with the aim of generating new developments that could potentially benefit the cashew chain itself. For this, the use of CNS in the development of a composite material with a view to functional packaging developed through a process that includes:

- Derivatives in the territory: tours within the region to understand consumption dynamics and resource availability in chain.
- Understanding of decisive factors: access to resources, costs of acquiring inputs and the use of high environmental impact materials in the chain, to identify relevant findings for the development of the material proposal.
- Spaces for ideation and creation: by the design team through experimentation and DIY practices for the development of material samples.
- Iteration with the engineering team: for the evaluation, definition of parameters and proposition of the model of replicability and feasibility of material developments.
- Co-creation models with the community: to build from and for the vision of the region through the transfer of knowledge, in addition to the validation of the different stages of material development.

The above with the aim of establishing a mechanism that contributes to strengthening the production chain, increasing the competitiveness of the region, establishing a differentiating factor and therefore promoting the economic growth of the sector. Similarly, it is proposed to provide inputs for the consolidation of a circular economy model (Sillanpää & Ncibi, 2019; Ellen MacArthur Foundation, 2022) that encourages participation, self-production and the well-being of the community.

As part of this strengthening strategy, the developments and partial results corresponding to the process of material design advanced with an interdisciplinary team, in which Design, Chemical and Food Engineering and Mechanical Engineering converge. A group established under the premise of ensuring an integral vision of the material development model. For the material development, it was decided to hybridize the methodology *Material Driven Design—MDD* (Karana et al., 2015), whose purpose is to facilitate design processes driven by materials, with the Design-Build-Test-Learn methodology (Wheelwright & Clark, 1992), whose purpose is to perform iterations and experiments in controlled environments to obtain improvements in the properties and characteristics of a material. The first methodology proposes four main stages:

- Understanding the material, which includes the analysis of the current situation of the resource and the relationship of it with the physical properties of the matter.
- Understanding sensory knowledge, that users have about the material (Cashew shell), as well as the experience derived from praxis with it.
- The creation of a vision of material experience, where after the analysis and understanding of the material, both positive and negative impacts of the matter are manifested and the intention from the design field to create the "new materials" is visualized.
- The manifestation of patterns of material experience, which stems from a process of ideation and creation of qualities and properties based on context and opportunities; and the design of material or product concepts, where the transition from conceptual intent to material production, its evaluation and the final development of the product takes place.

The second, typical of engineering, comes to enhance each one of the stages described above with processes of parallel experimentation and fine-tuning of the different material proposals through controlled laboratory tests, evaluation of results and preliminary characterization of properties and behaviors of the proposed material.

We observe with this methodological hybridization, that, in most cases, the design team chooses to start its material creation process from biologically based and biocompatible sources available in the local environment, generating innovative material alternatives that combine different elements present in the region, which can function as a binder, stabilizer or reinforcement within the proposed material. For their part, engineers take these bases or "material concepts" and transform them with their technical knowledge into viable alternatives for a possible packaging application. The most interesting thing about this fusion is, as explained in detail throughout the chapter, the interaction of the multidisciplinary team with the actors of the cashew chain in the region. Constantly reviewing the progress with those who interact daily with the nut and its waste, becomes the vehicle for innovation towards the chain itself. These interactions, through co-creation workshops, enrich the material development process and at the same time, allow it to remain in and for the context of its development. These practices promote environmental sustainability and social innovations by adding value to an agricultural production chain, which needs to reinvent itself to move towards the future.

24.2 Interdisciplinary Development of the Material from the Cashew Shell Byproduct

The research presented in this chapter focuses on the development of new sustainable materials through the use of the cashew shell byproduct. The initial phase for the development of the material proposal starts from the approach to the territory and the cashew production process; it is based on an approach of recognition of possible material sources, identification of production processes, understanding of the raw material, and the environmental conditions that may influence the behavior of the materials. It is fundamental for the success of the project, to recognize the situated knowledge present on the area. Therefore, the entire multidisciplinary team travels to the Vichada region (Fig. 24.1). Actor mapping workshops are created with the community (Gopal & Clarke, 2015) together with social cartography visualizations (Moore & Garzón, 2010). This allows both the community to understand what is wanted within the framework of the project, as well as the engineering teams, to recognize the value of gathering information from people, above the physico-chemical formulations of a material. Once the information obtained in the region is compiled and analyzed, a second phase of experimentation begins, which through DIY-Materials development approaches (Rognoli et al. 2015; Rognoli and Ayala-Garcia, 2021) or free experimentation through tinkering (Parisi et al., 2017) promotes the production of a series of quickly executed material samples, with a wide variability of results and a technical diversity that opens a range of broad-spectrum possibilities.

The resulting material samples are evaluated and parameterized, with the purpose of selecting the best results, and subsequently, to proceed with the replication of the same under controlled laboratory conditions in conjunction with the engineering teams. From these material replicas, the properties and material qualities are established in order to determine the opportunities, requirements and processes for the in situ serial development of the same.

Once the first "potential materials" were obtained, they were presented to the community of the Vichada region through a material sample format. For this purpose, the presentation language is designed and workshops are held to socialize the different material samples. The workshops begin with an expressive-sensory evaluation (Rognoli and Levi, 2005), along with the users. This workshop not only allows the transfer of knowledge based on the community's previous experiences with the materials and processes in the cashew chain, but rather amplifies the perception and validates the findings for the creation of the material concept and the development

Fig. 24.1 Multidisciplinary visit of the research team to the territory of Vichada. Photo by Carolina Pérez

of the product. For their part, the research team nourishes the transfer from sharing the technical understanding of the material and the visualization of future phases of the process developed in the creation of the sample. Additionally, the acceptance of the materials is verified, blurring any prejudices that may be held about it, and the future scenario of these proposals is proposed through the co-creation of possible application alternatives by the community itself.

24.2.1 Material Impulse Through DIY Practice

Following the Do-it-yourself Materials approach as a starting point for the development of the samples, the group of designers makes a broad and experimental approach to the subproduct of the shell. Starting from visits to the territory and the visualization of the diversity and abundance of resources available in the region, the first material proposals are formed using free experimentation as a creative vehicle that allows mixing of various sectors of knowledge. The use of various production techniques, the combination of raw materials and the hacking of processes from other disciplines were the resources used to obtain diverse and innovative results.

24.2.2 Iteration of Design and Engineering Disciplines

The creative process developed by the designers proposes the leveraging of their results through the iteration between design and engineering. This iterative factor allows to strengthen the very practice of material development through the characterization of the samples obtained in the previous experimentation phase by the engineering team, who applies their methodologies to standardize and guarantee the replicability of the sample.

Based on the principle of physical integrity and the incorporation of own and available materials in the region, a pre-selection of the samples to be replicated in laboratory conditions (drying temperature, mixing speed, concentrations) was made by the engineering team. Starting from an ideation session between the teams, the bases of the formulation were established to obtain a standardized response from the possible materials, which would guarantee replicability in the territory, and thus contribute to the strengthening of the cashew chain and drive the community towards a possible circular economy model for part of this chain.

The results of the phase of new formulations for the "potential materials" were obtained through processes of component integration in an internal mixer at 180 °C and 60 rpm for 7 min each. Subsequently, they were manufactured in the form of films using the compression molding method at 180 °C and 110 bar. These new polymeric compounds integrated polylactic acid (PLA) as a matrix to guarantee their elasticity and mechanical resistance.

24.2.3 Expressive-Sensory Evaluation with the Community

In the search for the formation of an integral model driven by the material, a series of workshops with the community were proposed for the socialization of the replicated material samples. Using an expressive-sensory evaluation tool, is possible to grasp from users, the perception of the different materials samples, and understanding the possibilities and the limitations for the integration of the materials in the cashew productive chain and outside of it. This approach will be scaled in the region with other material iterations to obtain a broad and validated material vision that allows to improve the materials in the process of development. This cycle of iterations is repeatable as many times as required to achieve the established objectives.

24.3 Results

The DIY-Material experiments developed by the design team (Fig. 24.2) have a manifest variability of shape, sensation, stability and composition that allowed an understanding of the opportunities (potential as filler for ceramic or polymeric compounds, the high oil content and the antifungal properties of the same, etc.) as well as limitations (loss of structure, the growth of biological agents, complexity related to the particle size of the material) of the cashew shell as an input for the configuration material. Low-tech processes (fragmentation of shells with common tools, use of molds available in the market, use of easily accessible inputs) and the rapid approach allowed the development of several iterations and the combination of various material sources were key to achieve proper samples for validation with the rest of the team. Through this multidisciplinary experimentation, rich and diverse material samples were obtained, allowing to project in later phases, multiple possibilities for product creation with actors and users, expanding the portfolio of possible applications for the future use of the waste byproduct along the cashew production chain.

It is key to highlight the importance of developing a format for step-by-step consignment, which allows for relevant information about the process such as: Elements and compounds used, quantities, times, processes and external factors. This allows for control of the processes and provides a structure to establish feasibility principles with the engineering teams.

From the result, DIY-Material samples were obtained with potential physical (hardness, resistance, molding capacity) and qualitative (textural variations, color and finishes) (Fig. 24.3) to be replicated under laboratory conditions (Fig. 24.4).

Fig. 24.2 DIY-Material experiments with cashew shells

Fig. 24.3 DIY-Material samples chosen for further development under laboratory conditions

Fig. 24.4 Process of replicating DIY-Material samples under controlled laboratory conditions

The process of multidisciplinary co-creation between the teams generated a space to cross specific information about the manufacture of materials in the laboratory, with the qualitative and experiential information product of the DIY material experiments.

In this development, a space was generated for ideation, where the technical knowledge of the engineering team about manufacturing processes was complemented with the qualitative experience obtained from the DIY material experiments carried out by the design team, achieving 16 samples replicated under controlled laboratory conditions (Fig. 24.5) maintaining their physical and sensory properties for being taken to user interaction tests. It is important to note that within the ideation sessions between the teams, factors such as: availability of resources in the territory, the possibilities of manufacturing these materials in the region, the possible environmental impacts and the viability of them to contribute to the productive chain were analyzed. The selected and replicated samples were used as input for the co-creation workshop with the community.

Through the expressive-sensory evaluation workshop carried out with the community (Fig. 24.6), the qualities of each of the material samples were recognized, and their opportunities and considerations were identified, generating associations to products and to existing materials. Examples of applications that are part of the cashew production chain, as well as others, that are part of their daily life within the territory were highlighted by the users. With the sensory scales (Fig. 24.7), a qualitative comparison was obtained between the samples allowing to establish differences in perception linked to the experiences of the different users. Likewise, the sensory scale works as a tool that characterizes the material beyond its mechanical properties.

The sensations, meanings and emotions that the samples evoked in the users as a result of the performative interaction were recorded, identifying the negative and positive judgments that were awarded to them, obtaining a description of the sensory character of each of the evaluated materials.

Similarly, in the socialization session by the research team and under the knowledge transfer model, a base of material verification criteria that are replicable and comparable in future workshops and the following phases of material development was obtained. A roadmap was established for the execution processes between the research team and the community, ensuring viability within the cashew chain and in other possible scenarios.

By understanding the territory where the research project is carried out, it is possible to understand the importance of proposing developments of this type hand in hand with the actors of the cashew production chain. Their particularities impose the need to arrive at material proposals that must be capable of meeting specific requirements to arrive at a solution that contributes to the consolidation of a local and circular model that encourages community participation and the integration of various allied actors. It then becomes, as a result of the research, the consolidation of innovative proposals for the optimization of processes within and outside the cashew chain and the formation of a model that capitalizes on the acquired knowledge contributing to the productivity of the region and the well-being of the community.

Fig. 24.5 Material samples replicated under controlled laboratory conditions

Fig. 24.6 Expressive-sensory evaluation of the material samples by the community of Vichada. Photo by Carolina Pérez

Placing the material in the context of the territory becomes a priority in this search, taking advantage of the synergy between the disciplines involved to consolidate a material design process characterized by: the innovation offered by DIY experimentation; functionality backed by material replicas made in laboratory conditions; the constant validation of processes carried out during fieldwork and workshops in the region; the exploitation of community knowledge that arises from sensory evaluations; and the two-way knowledge transfer that enhances material development.

Fig. 24.7 Development of sensory scale matrix with the community. Photo by Carolina Pérez.

24.4 The Material Vision

The creation of materials under the premise of this research is enriched by the visions of each discipline. From engineering, the manufacture of the material and the implementation of manufacturing methods that standardize DIY experiments are obtained, which facilitates the replicability of the material for an application in the region. This, coupled with the qualities of the materials defined from the design, where the user and their direct interaction with the samples are proposed as the central axis, makes a co-creation of the material possible. In turn, a recursive and agile process is presented that allows for a greater number of material proposals with fewer resources. Iterations with the community allow the future material to be enriched sensorially in future applications. The use of the sensory scales tool allows materials to evolve from 3 levels: affective, interpretive, and performative. (Giaccardi & Karana, 2015), in conjunction with the properties and characterizations that are being fine-tuned during the project. The interaction between the research team, the material, and the users (Cashew community of Vichada) promotes participatory discussion as an effective method of knowledge transfer around material results, allowing us to understand how to innovate with meaning, with local relevance, and above all with environmental awareness.

Acknowledgements The authors thank the Ministry of Science, Technology and Innovation and the OCAD of ACTeI who carried out the feasibility, prioritization and approval of this research with resources from the General System of Royalties—SGR in Call No. 6 of the Project "Use of Agroindustrial by-products in the production of cashew in the department of Vichada"—BPIN 2020000100571. Likewise, thanks are given to the government and community of the department in general for their interest and participation in the activities carried out to date.

References

Akinwale, T. O. (2000). Cashew apple juice: Its use in fortifying the nutritional quality of some tropical fruits. *European Food Research and Technology, 211*(3), 205–207.

Anderson, C. (2012) *Makers: The new industrial revolution.* Crown Business.

Ayala-García, C. (2023) *The materials generation: The emerging experience of DIY-materials.*

Bettiol, M., & Micelli, S. (2014). The hidden side of design: The relevance of craftsmanship. *Design Issues, 30*(1), 7–18.

Buur, J., & Matthews, B. (2008). Participatory innovation. *International Journal of Innovation Management, 12*(3), 255–273. https://doi.org/10.1142/S1363919608001996

Cardona, A. O. (2017). *Cashew cultivation is one of the most potential export crops in Colombia.* Agronegocios. https://www.agronegocios.co/agricultura/maranon-es-uno-de-los-cultivos-con-mayor-proyeccion-2623049

Cohen, N., Sicher, E., & Yavuz, S. U. (2020). Designing with microbial cellulose to feed new biological cycles. *International Journal of Food Design, 4,* 155–171.

D'olivo, P., & Karana, E. (2021). Materials framing: A case study of biodesign companies' web communications. *She Ji: The Journal of Design, Economics, and Innovation, 7*(3).

Ellen MacArthur Foundation. (2022). *What is a circular economy?* Ellen MacArthur Foundation. https://ellenmacarthurfoundation.org/topics/circular-economy-introduction/overview

Food and Agriculture Organization of the United Nations—FAO. (2020). http://www.fao.org/faostat/en/#rankings/countries_by_commodity

Gershenfeld, N. (2012). How to make almost anything. The digital fabrication revolution. Foreign Affairs (November/December).

Giaccardi, E., & Karana, E. (2015). Foundations of materials experience: An approach for HCI. In *CHI'15 proceedings of the 33rd annual ACM conference on human factors in computing systems* (pp. 2447–2456).

Gopal, S., & Clarke, T. (2015). *System mapping: A guide to developing actor maps.* FSG. http://www.nonprofitjourney.org/uploads/8/4/4/9/8449980/final_system_mapping_guide.pdf

Karana, E., Barati, B., Rognoli, V., & Zeeuw van der Laan, A. (2015). Material driven design (MDD): A method to design for material experiences. *International Journal of Design, 9*(2), 35–54.

Karana, E., Rognoli, V., & Jacob Dazarola, R. (2020). The function of design in the development of new materials: Interview with Elvin Karana. *Diseña,* (17), 46–55. https://doi.org/10.7764/disena.17.46-55

Lee, S. (2023) *Microbes are "the Factories of the Future".* Suzanne Lee, Dezeen. Retrieved September 6, 2023, from https://www.dezeen.com/2014/02/12/movie-biocouture-microbes-clothing-wearable-futures/

MaDe. Llorach, P. (2021) *Circular design and circular material design in: MaDe material designers.* http://materialdesigners.org/book

Manzini, E., & Vezzoli, C. (2003). *A strategic design approach to develop sustainable product service systems: Examples taken from the 'environmentally friendly innovation' Italian prize.*

Moore, E., & Garzón, C. (2010). *Social cartography: The art of using maps to build community power.* http://www.friendsofrpe.org/files/Moore.Garzon.17-2.pdf

Parisi, S., Rognoli, V., & Sonneveld, M. (2017). Material tinkering. An inspirational approach for experiential learning and envisioning in product design education. *The Design Journal*, *20*(1), S1167–S1184.

Prakash, A., Vadivel, V., Banu, S. F., Nithyanand, P., Lalitha, C., & Brindha, P. (2018). Evaluation of antioxidant and antimicrobial properties of solvent extracts of agro-food by-products (cashew nutshell, coconut shell and groundnut hull). *Agriculture and Natural Resources, 52*(5), 451–459.

Rognoli, V., & Ayala-Garcia, C. (2021). Defining the DIY-materials approach. In *Materials experience 2*. Butterworth-Heinemann.

Rognoli, V., Bianchini, M., Maffei, S., & Karana, E. (2015). DIY materials. *Materials and Design, 86*(2015), 692–702.

Rognoli, V., & Levi, M. (2005). *Materials for design: Expressivity and sensoriality [Materiali per il design: Espressività e sensorialità]*. Polipress.

Schön, D. A. (1983). *The reflective practitioner: How professionals think in action*. Basic Books.

Sennett, R. (2008). *The craftsman*. Yale University Press.

Sillanpää, M., & Ncibi, C. (2019). Getting hold of the circular economy concept. In The circular economy: Case studies about the transition from the linear economy (pp. 1–35). Elsevier. https://doi.org/10.1016/B978-0-12-815267-6.00001-6

Studio Swine. (n.d.). *Hair Highway*. Retrieved September 06, 2023, from https://studioswine.com/work/hair-highway/

Tanenbaum, J. G., Williams, A. M., Desjardins, A., & Tanenbaum, K. (2013). Democratizing technology: Pleasure, utility and expressiveness in DIY and maker practice. In *Proceedings of the SIGCHI conference on human factors in computing systems, Paris, France*.

Vezzoli, C. (2018). *Design for environmental sustainability: Life cycle design of products* (2nd ed.). Springer.

Wheelwright, S. C., & Clark, K. B. (1992). *Revolutionizing product development: Quantum leaps in speed, efficiency and quality*. The Free Press.

Zhou, J., Barati, B., Wu, J., Scherer, D., & Karana, E. (2021). Digital biofabrication to realize the potentials of plant roots for product design. *Bio-Design and Manufacturing, 4*, 111–122.

Part VI
Territories, Intermediate Cities and People in Movement

25.1 Challenges of the National Urban Parks Policy

Parks constitute important public spaces in cities, which provide key functions and services for urban quality of life, and, can become potential attractions for local populations, as well as, a contribution to local economy. Parks provide well-being to the population; are spaces for recreation, meeting, and enable the performance of sports and cultural activities. Parks and green areas contribute to the improvement of neighborhood's and enhance urban quality of life; on the contrary, their deterioration implies a detriment to neighborhood's life, an impact on the built environment, and an increased perception of insecurity, among others. The National Urban Parks' Policy (MINVU, 2021, hereinafter PNPU) recognizes that one of the main challenges of its implementation is management and maintenance. While, on one hand, it highlights the role that urban parks and the system of green areas play being "*an essential part of urban public goods, fulfilling multiple functions and providing social, environmental and economic benefits to the cities and territories where these are located and to the people who inhabit them*" (MINVU, 2021: 2); on the other hand, it identifies a series of shortcomings that threaten their sustainability, particularly once constructed, requiring greater governance for better and more integrated management. In this sense, considering that municipalities are usually the institution that performs the functions of maintenance and management of communal green areas, the definition of a typology and scale of park, according to their different geographical, demographic, socioeconomic, and, administrative and financial capacities, allows to define tools adapted to their needs to accomplish these purposes.

At the same time, the Urban Parks' Policy recognizes, among its foundations, the need for "*greater social integration in cities, and greater equity in the distribution of existing public goods*" (MINVU, 2021: 2). As achieving this goal is vital: the research identified the need to democratize knowledge and provide guidelines for good management of urban parks adapted to diverse social contexts.

The research characterizes the different scales of parks and green areas of the communes of the Region of La Araucanía, recognizing various realities of availability and management that are interweaved with unequal socioeconomic conditions as shown in Fig. 25.1 where it is observed that the communes with the least provision of green areas are also those that present the highest income poverty indicators (Fig. 25.1).

Considering these conditions, from a perspective of territorial equity and decentralization, the role that cities and intermediate communes play in their potential to attract and retain their inhabitants is not only local but also regional. Therefore, the quality of life they can offer over the regional capitals and the appropriate maintenance and development of their urban amenities are a challenge.

The generation of communal investment strategies in terms of urban parks, is usually defined based on the availability of land and the existing gap in the surface of green areas, as defined by the "Methodology for the Construction of Public Spaces' Projects"[1] (MDSF) and the "Methodology for the Construction of Communal and

[1] NIP—Multisectoral Sector—Urban Development Public Spaces Construction Projects.

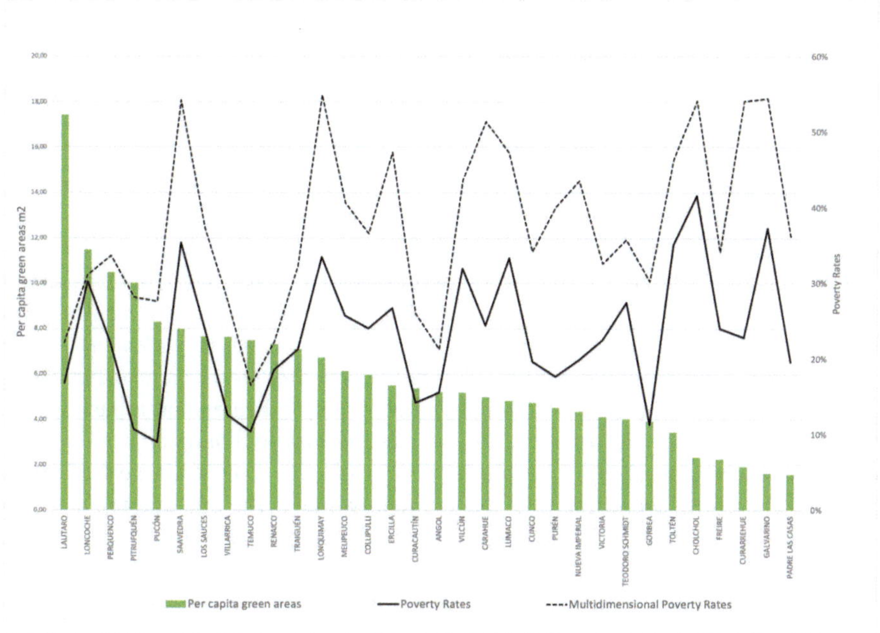

Fig. 25.1 Per capita green areas at the municipal level and poverty rates in La Araucanía region. *Source* Own elaboration based on SINIM and CASEN (2022)

Intercommunal Parks Projects"[2] (MDSF) established by the agency in charge of the assessment of regional public investment. The methodologies above will consider only the conditions and capacities of municipal management through a specific management plan project for the park or the new or existing green area improved; without considering the global management capacities of the municipality in terms of green area maintenance.

In this sense, to consider conditions that avoid oversizing and find the "ideal" scale of feasibility for the investment; and, for the maintenance of parks and green areas, this research seeks to define typologies that reflect the diversity of communal conditions, and, proposes recommendations for the scale of investment's initiatives.

The mechanism used to evaluate the communes of La Araucanía is presented below, considering the indicators chosen to understand the communal conditions in relation to: the investment needs in green areas, and their pertinency considering the municipal maintenance's capacities to ensure the sustainability of the parks. The exercise was also replicated in the O'Higgins region, which allowed to test the methodology, considering that the groupings by types of commune, although tendential, are also specific to each region.

[2] NIP—Multisectoral Sector—Urban Development Projects for the Construction of Communal and Intercommunal Parks.

25.2 Green Areas and Urban Parks in Chilean Regulations

The National Urban Parks' Policy proposes to municipalities and to the regional ministerial secretariats of housing and urban planning the construction of new urban parks in consolidated cities where these must comply with the urban standards, established in the General Ordinance of Urbanism and Construction and in the Communal or Intercommunal Regulatory Plans, and also, with administration standards subject to the organic of different institutions.

The General Ordinance of Urbanism and Construction (OGUC) defines a Green Area is a "*surface of land preferably destined to recreation or pedestrian circulation, generally made up of vegetable species and other complementary elements*" and a Public Green Area as a "*national good of public use that meets the characteristics of green area*". And, this same normative document, indicates that a park will be understood as a "*free space of public use, wooded eventually, equipped with facilities for leisure, recreation, sports and cultural practices and others*" (Article 1.1.2 of the OGUC). Therefore, according to this document, which is the one that regulates the planning and construction in the urban areas of the country, the difference between a green area and a park would be the provision of installations or equipment.

For this reason, the National Urban Parks' Policy defines urban parks "a*s the free spaces of public use, located within the urban limits, with a minimum surface of two hectares and a predominant vegetal coverage, which can be eventually equipped with facilities for the development of recreation, sport, cultural or others activities, and, that provide environmental and social benefits in the territory where are located*". Considering the above, the variable that would allow to classify a green area as a communal or intercommunal urban park, is the size of the area to be intervened, specifying, later that over 1 hectare of the initiative can be considered a park, thinking that a public square does not exceed the hectare, on average.

At the same time, in the title referring to urban planning and its instruments, the OGUC in articles 2.1.30 and 2.1.31 makes a difference between green areas that are national goods for public use (BNUP) and those that can be owned by a private individual, a municipality or another public entity. Article 2.1.31 also specifies, that in green areas, the following facilities are always understood to be admitted as complementary and compatible destinations: scientific, worship and culture, sports and recreation.

Therefore, urban parks correspond to existing areas that were defined as such in the foundation of the city, or, were built in its subsequent expansion, or, that are projected in planning; and therefore correspond to areas, recorded as such, in land planning instruments. Parks and green areas can be defined by both intercommunal regulatory plans and communal regulatory plans. Within the communal regulatory guidelines of the plan, the characteristics of tree planting, vegetation, external lighting, sidewalks and median strips of public spaces can be defined (Article 2.1.10 bis). In turn, as the regulatory plans can define areas of nature protection (coastal, maritime, lake or river edges, national parks, national reserves and natural monuments), risk areas and non-buildable zones, these may, according to the corresponding studies, accommodate parks and green areas, in some cases, floodable.

The investment initiatives for new communal or intercommunal parks are uploaded to the National Investment System (SNI), which is administered by the Ministry of Social Development and Family (MDSF), according to the methodology of the "Multisectoral Sector—Urban Development"; and, for its evaluation, in addition to the common matters to green area projects, a Management Model is required.

In the information requirements for the application of investment initiatives (RIS) of the Ministry of Social Development and Family, indicate that the Management Model will be understood as a document that analyzes the sustainability of the initiative, maximizing the benefits that the park will deliver to the population. The Management Model must identify the relevant actors and possible financing routes for the operation, maintenance and administration of the park to encourage the efficient use of resources. The Management Model must consider the maintenance criteria, the operation and administration plan of the park, as well as the estimation of associated costs and revenues.

Urban park's projects that pass into the National Investment System are evaluated according to cost-efficiency analysis considering investment, operation, maintenance and administration costs. However, it is noted that the components of the Urban Park project are not classified in sectors according to the same rules, instructions and procedures of the public investment process and may involve different operating models (for example, cafeterias, event halls or others) that are susceptible to being concessionable, must be evaluated on the merit of their own costs and benefits to justify their inclusion in the project. Likewise, it is indicated that in general, it is expected that these components are self-sustainable.

For the administration of public urban parks, two options have been defined: parks can be managed by the Regional Housing and Urban Planning Secretariats or by the Municipalities.

Consequently, the proposed methodology for investment recommendations in terms of green areas and public spaces is based on the need to identify the maintenance, operation and administration potential capacity of the municipalities, according to the requirements previously exposed and the sectoral, regional and communal investment challenges posed by the National Urban Parks' Policy.

25.3 Methodology

According to the bibliography related to the maintenance costs of parks; and, therefore, with their sustainability standard (Fundación mi Parque, 2013; Reyes-Päcke et al., 2014), the variables involved in the maintenance of municipal parks and green areas are: (i) the maintenance value per m^2, (ii) the municipal expenditure allocated to green areas with maintenance, both per m^2 and per inhabitant; and, (iii) the surface of green areas and parks both total and prorated per inhabitant. Likewise, in the document "Challenges and proposals for the administration of parks and squares in Chile" (Center for Public Policies UC, 2018) it is pointed out that one of the gaps to achieve the adequate maintenance of parks is the inequality and lack of permanent resources that municipalities have for their maintenance.

In this sense, to understand the feasibility of municipal administration for green areas and communal parks, the current proposal considers to characterize two aspects: (i) the potential maintenance and operation expenditure of green areas that is associated with the surface area; and, (ii) the availability of resources.

To characterize the budgetary availability for the maintenance of green areas, the following financial indicators available in the National System of Municipal Information (SINIM) were considered: (i) the Permanent Own Income (IPP) of the municipalities, (ii) the Income from the Common Municipal Fund (FCM); and, (iii) other income, which corresponds to income from the casino's law, mining patents, among others. As can be seen in Fig. 25.2 there is a correlation between the municipalities that have a higher proportion of permanent own income and other income and the percentage of expenditure in municipal green areas; while, the municipalities that have greater dependence on the Common Municipal Fund,[3] also have less availability for expenditure on maintenance of green areas (Fig. 25.2).

To have a first approximation of the social, economic and territorial conditions in each commune, a compilation, review and processing of the communal background of the region of La Araucanía[4] was carried out. When comparing the green areas per capita with the communal poverty rates, it is understood that there is an inverse proportion between the surface of green areas per inhabitant and the poverty rates, especially when reviewing the contrast with the multidimensional poverty.[5] In this sense, in those communes in which there is a higher proportion of inhabitants below the poverty line, there are also fewer public green areas per inhabitant, therefore the public communal space largely replicates the social conditions of its inhabitants (Fig. 25.1).

In addition to the above, demographic information was collected from the communes and their geographical location was analyzed in relation to the principal axis of the country which corresponds to Route 5, a highway that runs Chile from north to south, marking conditions of accessibility and centrality.

In total, the indicators presented in Table 25.1 are considered for the characterization and classification by types of communes in relation to their potential for maintenance, operation and administration of public green areas and urban parks.

Using population, geographical position, and income indicators, the communes of the region are grouped, differentiating their capacities (Table 25.2), which is illustrated in Fig. 25.3.

---

[3] The Municipal Common Fund constitutes the main source of financing for Chilean municipalities, it corresponds to a mechanism of solidarity redistribution of own income among the municipalities of the country.

[4] Using open data obtained on digital platforms from SINIM, INE, CASEN, SII and SIIT BCN.

[5] CASEN 2017 (Poverty rate by income and multidimensional).

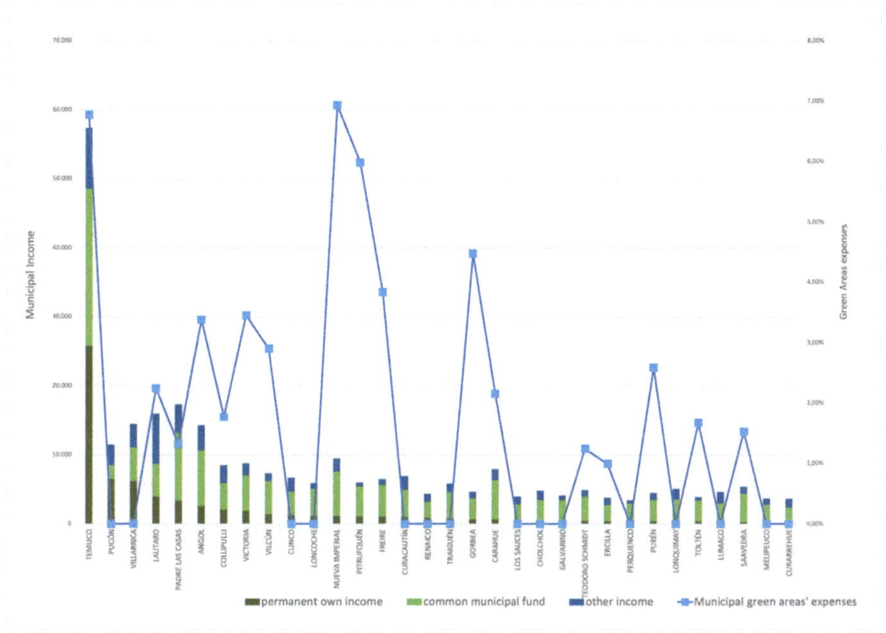

Fig. 25.2 Municipal income by origin (in thousands of pesos) and proportion of municipal expenditure allocated to the maintenance of green areas. Own elaboration from SINIM data (community garden maintenance services, code BSMJSC, and municipal expenses or total accrued expenditure, code IADM11). Note that some municipalities declare maintenance cost $0, despite declaring to have municipal maintained green areas

Table 25.1 Indicators used for the characterization by types of communes

Indicator	Source
Total population projection 2021	BCN/SIIT/communal reports 2021
Urban/rural population	BCN/SIIT/communal reports 2021
Income poverty rate	BCN/SIIT/communal reports 2021
Multidimensional poverty rate	BCN/SIIT/communal reports 2021
Percentage of permanent own income (IPP)	SINIM code IADM40
Percentage of common municipal fund (FCM) dependence	SINIM code IADM75
Municipal income: total income received (IT)	SINIM code IADM01
Other income = IT − FCM − IPP	Own elaboration based on SINIM
Total green areas surface with maintenance	SINIM code MAVM
Parks surface	SINIM code MMPQC
Surface of green areas (not park) with maintenance	Own elaboration based on SINIM
Square meters of green areas per inhabitant	Own elaboration based on SINIM

Note BCN: National Congress Library; SIIT: Integrated Territorial Information System; SINIM: National Municipal Information System, http://datos.sinim.gov.cl/

Table 25.2 Communal groupings of the La Araucanía region. Own elaboration

Group	Main characteristics
1	Urban communes and conurbations
2	Mixed communes, touristic, with high IPP and low dependence on FCM
3	Mixed communes, with medium IPP and high dependence on FCM
4	Small communes, rural, with high poverty rates and dependence on FCM

Note IPP municipalities (Permanent Own Revenues), FCM (Common Municipal Fund)
Source Own elaboration

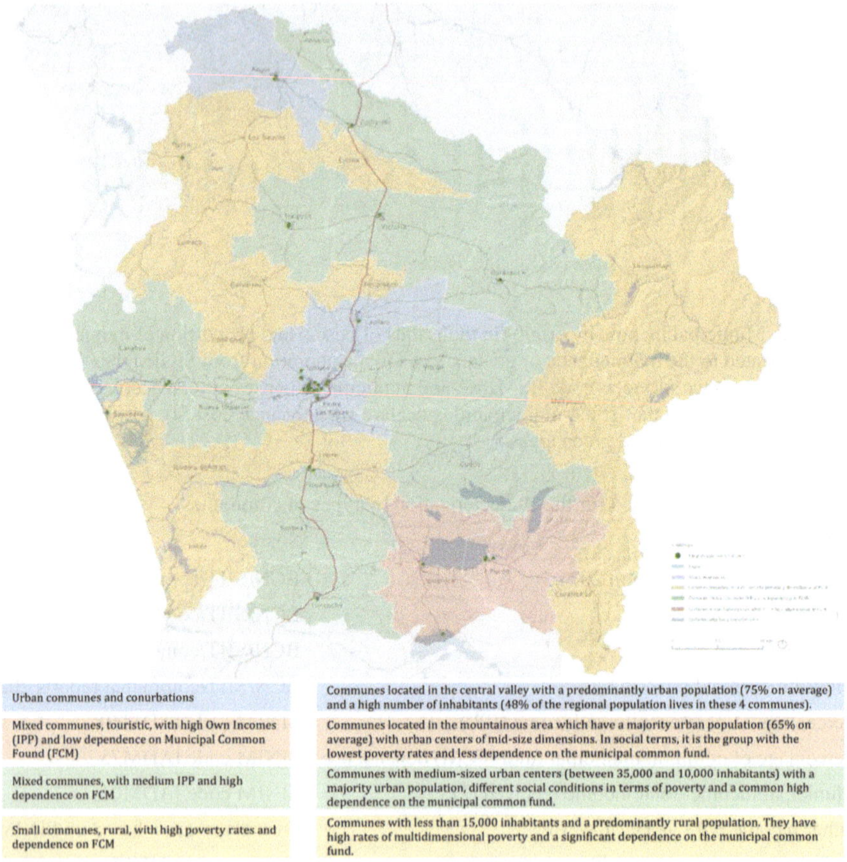

Fig. 25.3 Characterization and grouping of the municipalities of La Araucanía region. Own elaboration

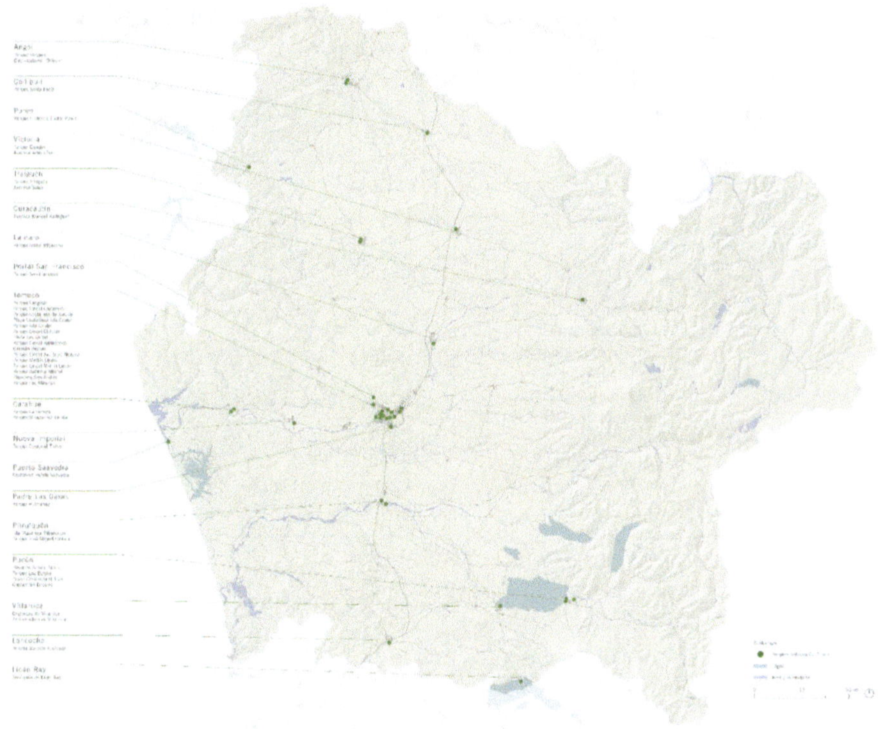

Fig. 25.4 Registry of communal parks and green areas. *Source* Own elaboration

Subsequently, a census of parks and regional green areas was carried out from georeferenced databases[6] (Fig. 25.4). This mapping was complemented with an analysis by commune, considering the communal capitals and other populated centers, and was recorded in a communal datasheet like the one presented in Fig. 25.5. Cartographies of urban centers and conurbations were made identifying parks and other urban green areas. In this way, it was possible to understand the dimensions of each public space of the green areas' type, its relation with the urban grid and topography, as well as to identify differences in shapes and typologies. Thus, urban green areas are recognized to be related to water courses and hills, others to be central avenues, also the use of urban edges, neighborhood squares and main squares, among others.

Considering the location and concentration of larger green areas and existing parks, a classification of parks and urban green areas was proposed considering their dimensions, shapes and capacity for uses (Table 25.3).

In the final stage, considering the group of communes and the higher-level green area per commune, this information was cross-referenced with the current availability of green areas in each territory. Thus, the current availability of green areas

[6] Obtained on digital platforms from IDE, MINVU and INE.

Comunas mixtas y turísticas		VILLARRICA
Recomendación	Parque mediano (Villarrica) + AAVV menores	
Población	59.554	
Urbana	36.480	65,8 %
Rural	18.998	34,2%
IPP	$ 6.156.394 (M)	
IPP/hab.	$ 103 (M)	
Dep. FCM	44,3 %	
Superficie AAVV	454.585 m2	
AAVV/hab.	7,6 m2	
Mantenidas/hab.	4,6 m2	
AAVV menores/hab.	2,9 m2	
Aporte Parques	62 %	
Ciudad 1	Villarrica	
	31.780 habitantes	
	Costanera de Villarrica (T5)	
Pueblo 1	Lican Ray	
	2.642 habitantes	
	Península de Licán Ray (T2)	
Pueblo 2	Ñancul	
	1.620 habitantes	
	Sin Parque	

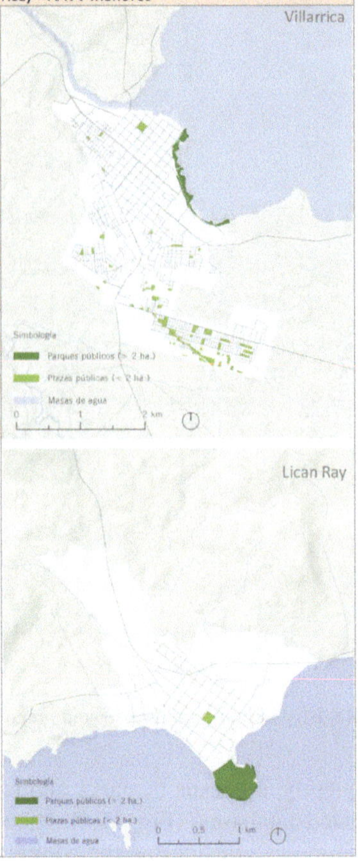

Fig. 25.5 Example of commune file. *Source* Own elaboration

per inhabitant was compared with and without considering the park's surfaces, to finally determine, according to the differentiated management capacities of each commune (assimilated to the IPP and the dependence on the FCM), what would be the investment recommendations in terms of parks and green areas that are proposed to be incorporated for each municipality.

Following the criteria defined by the studied indicators and the established classifications, investment recommendations are determined according to the types of parks required by groups of communes. These recommendations are based on the communal need for new green areas that are in line with the communal reality (Table 25.4). The National Council for Urban Development has defined as a standard for urban green areas: (1) Green area surface (squares and parks) per inhabitant: from 10 m^2 per inhabitant; (2) Distance from a block to its nearest square: up to 400 m; (3) Distance from a block to its nearest park: up to 3,000 m.

Table 25.3 Classification of parks and urban green areas. Own elaboration

Type	Classification	Dimensions	Description	Communes	Reference
T1	Very large park	More than 18 Ha	Park of larger dimension that is configured as a public space of reference at a regional level and involves intercommunal uses. It has different types of spaces with different associated uses. It can host massive activities, regional meetings and fairs. It positions itself as a relevant ecological area for the city	Lautaro Temuco	Isla Cautín Park, Temuco
T2	Large park	Between 18 and 8 Ha	Large park located in mid-sized or larger cities and positions itself as a public space for communal use. In cases of touristic communes, it is a space that can be associated with touristic activities and local differentiation. It can host medium-sized activities, typical fairs, local and cultural meetings	Villarrica Pitrufquén Loncoche Pucón	Municipal Island Pitrufquén
T3	Mid-sized park	Between 8 and 4 Ha	Park of medium-sized dimensions that positions itself as a public space and green area at a local scale. In mid-sized cities with lower population density, it can be recognized as the main green area. It can host smaller-scale activities such as thematic fairs, local and cultural gatherings	Collipulli, Temuco Angol Nueva Imperial Traiguén Padre Las Casas	Thiers Community Park, Nueva Imperial

(continued)

Table 25.3 (continued)

Type	Classification	Dimensions	Description	Communes	Reference
T4	Small park	Between 4 and 2 Ha	Park of smaller dimensions that positions itself as a local or community public space. In medium-sized and small cities, it can be recognized as the main green area. It can host sports and recreational spaces, sporadic fairs, traditional and community activities	Angol Temuco Carahue Purén Collipulli Victoria Pucón	Historic Park Fort Purén
T5	Water-front	Various	Parks of various dimensions that are distinguished by bordering a body of water. It serves as a public transition space between the water and the city. It can host recreational spaces, traditional and thematic fairs, local and cultural gatherings of smaller scale. It can position itself as a differentiating public space that makes the city recognizable for tourists	Saavedra Villarrica Pucón	Puerto Saavedra Waterfront
T6	Linear park	Various	Long and narrow parks that connect two points of the city, in this sense, its main quality is connecting. It can have infrastructure for pedestrian routes and bike lanes. In addition, it can be associated with a watercourse, functioning as an ecological corridor and transition space between the river and the city	Temuco , Victoria, Traiguén Pitrufquén, Curacautín, Cunco	Arturo Prat Avenue, Victoria

(continued)

Table 25.3 (continued)

Type	Classification	Dimensions	Description	Communes	Reference
T7	Mid-sized green area	Less than 2 Ha	Squares, central medians and smaller public green areas are public spaces for local and neighborhood's use. Sometimes they can be spaces of communal identity due to their historical traditions, such as the main squares. When they have larger dimensions, they can host fairs, recreational spaces, and small-scale cultural gatherings	All	José Donat Square, Freire

Table 25.4 Investment recommendations by groups of communes. Own elaboration

No.	Description criterion	Recommendation
1	For mixed or urban communes, with parks that exceed 40% of the total green areas, and smaller green areas less than 4 m^2 per inhabitant. In cases of communes that meet these criteria, but one of their cities does not have a park according to its population dimensions (T3 for communes over 30,000 inhabitants, T4 for communes with a population less than 30,000 inhabitants), it is proposed to add to that recommendation a park according to these dimensions	Mid-sized areas (and park where applicable)
2	For mixed or urban communes, with parks that exceed 40% of the total green areas and smaller green areas greater than 4 m^2 per inhabitant	Improvements in existing green areas
3	For mixed or urban communes, without urban parks or with parks that are less than 40% of the total green areas, and smaller green areas less than 4 m^2 per inhabitant	Mid-sized green areas and urban park (according to the corresponding size)
4	For mixed or urban communes, without urban parks or with parks that are less than 40% of the total green areas, and green areas less than greater than 4 m^2 per inhabitant	Urban park (according to the corresponding size)
5	For rural communes without urban parks	Small urban park
6	For rural communes with urban parks, and green areas of less than 4 m^2 per inhabitant	Mid-sized green areas
7	For rural communes with urban parks, and green areas greater than 4 m^2 per inhabitant	Improvements in existing green areas

Additionally, considering a certain proportionality between the availability of parks and smaller green areas, it was estimated as an ideal that the surface of parks is not less than 40% of the total green areas, and that the green areas (not park) are not less than 4 m^2 per inhabitant, considering that although this measure is far from the "ideal" standard, it is more likely to be implemented in the medium term.

25.4 Results

In the region of La Araucanía, there are communes with relevant socioeconomic differences that influence the needs and challenges of managing their urban parks. The smaller municipalities that have fewer resources have a larger gap in the availability of green areas while having less permanent income for their management, which translates into less municipal spending on green areas. Considering the communes with less green area surface per inhabitant, it is important to mention that almost all are small communes without urban parks.

Regarding the recommendations (Table 25.5), in the case of mixed and urban municipalities with low permanent own income that have a large gap in the availability of total green areas, the recommendations are aimed at improving the availability of urban parks and smaller green areas (neighborhood green areas, which are not parks), the former focused on uses expanded to the city and the municipality, while the latter focused on meeting and recreation spaces at the neighborhood scale.

Additionally, there are mixed and urban municipalities with better economic conditions that have a good availability of urban parks, directing the recommendation to improve the availability of smaller green areas considering the criterion of access and proximity of the housing and the future consolidation of a green infrastructure network.

Finally, municipalities with greater resources are identified that have good indices in the availability of green areas, so the recommendation is oriented towards the realization of improvements associated with better maintenance, connection and accessibility of these spaces with respect to their urban context and surrounding rural areas, designing and strengthening the incipient or consolidated green infrastructure network.

In several cases of small and rural municipalities, the incorporation of a small park type, according to the classification made, is recommended, in order to integrate new uses to their urban green areas that could not be developed in smaller green areas.

The sequence of analysis is presented below. Its replicability is proposed as a mechanism that can be applied by regions and local governments when defining, sizing and prioritizing investments in green areas and urban parks (Table 25.6).

As for prioritization, the main criterion can be considered the largest gap in terms of availability of m^2 green areas per inhabitant.

Table 25.5 Results of application of recommendations in La Araucanía Region. Own elaboration

Commune	Classification	Population	% Rural	IPP/per capita	FCM dependency (%)	% of parks in the total green areas surface (%)	M2 parks/ inhabitants	M2 green areas/ inhabitants	Higher existing green area	Recommendation
Temuco	1	304.871	6.8	85	46.9	52	3.9	3.6	T1: Very large park	Small park (2d city) + mid-sized green areas
Padre Las Casas	1	83.009	40.3	40	74.9	37	0.6	1.0	T3: Mid-sized park	Mid-sized park + mid-sized green areas
Angol	1	56.289	8.2	45	75.9	32	1.7	3.5	T3: Mid-sized park	Mid-sized park + mid-sized green areas
Lautaro	1	41.065	28.1	94	55.4	69	12.0	5.4	T1: Very large park	Improvements
Villarrica	2	59.554	34.2	103	44.3	62	4.7	2.9	T2: Large park	Mid-sized park + mid-sized green areas
Pucón	2	30.052	35.7	215	23.8	71	5.9	2.4	T2: Large park	Mid-sized green areas
Victoria	3	35.512	27.3	52	73.4	42	1.7	2.4	T4: Small park	Mid-sized green areas
Nueva Imperial	3	33.898	41.9	32	85.6	40	1.7	2.6	T3: Mid-sized park	Mid-sized green areas

(continued)

Table 25.5 (continued)

Commune	Classification	Population	% Rural	IPP/per capita	FCM dependency (%)	% of parks in the total green areas surface (%)	M2 parks/ inhabitants	M2 green areas/ inhabitants	Higher existing green area	Recommendation
Vilcún	3	31.258	45.2	44	77.8	0	0.0	5.2	T7: Mid-sized green area	Small park (city 1) + small park (city 2)
Collipulli	3	26.271	25.4	76	65.7	66	4.0	2.0	T3: Mid-sized park	Mid-sized green areas
Pitrufquén	3	26.245	33.4	40	80.2	77	7.8	2.3	T2: Large park	Mid-sized green areas
Carahue	3	25.505	46.0	24	90.1	41	2.1	2.9	T4: Small park	Mid-sized green areas
Freire	3	25.472	68.0	40	81.7	0	0.0	2.2	T7: Mid-sized green area	Small park + mid-sized green areas
Loncoche	3	24.766	29.7	45	77.9	56	6.5	5.0	T2: Large park	Improvements
Traiguén	3	19.315	23.8	42	82.5	60	4.2	2.9	T3: Mid-sized park	Mid-sized green areas
Curacautín	3	18.196	26.3	52	80.5	24	1.3	4.1	T6: Linear park	Small park
Cunco	3	18.065	49.5	69	73.2	24	1.1	3.6	T6: Linear park	Small park + mid-sized green areas

(continued)

Table 25.5 (continued)

Commune	Classification	Population	% Rural	IPP/per capita	FCM dependency (%)	% of parks in the total green areas surface (%)	M2 parks/ inhabitants	M2 green areas/ inhabitants	Higher existing green area	Recommendation
Gorbea	3	15.137	29.5	41	82.9	0	0.0	3.9	T7: Mid-sized green area	Small park + mid-sized green areas
Purén	3	12.167	36.0	30	89.6	51	2.3	2.2	T4: Small park	Mid-sized green areas
Renaico	3	10.884	18.5	79	72.9	0	0.0	7.3	T7: Mid-sized green area	Small park
Teodoro Schmidt	4	15.784	64.1	26	89.7	0	0.0	4.0	T7: Mid-sized green area	Small park
Saavedra	4	12.764	78.1	20	94.1	73	5.8	2.2	T5: Waterfront	Mid-sized green areas
Galvarino	4	12.623	65.4	33	87.7	0	0.0	1.6	T7: Mid-sized green area	Small park
Cholchol	4	12.427	62.5	37	86.4	0	0.0	2.3	T7: Mid-sized green area	Small park
Lonquimay	4	11.071	63.1	32	90.0	0	0.0	6.7	T7: Mid-sized green area	Small park

(continued)

Table 25.5 (continued)

Commune	Classification	Population	% Rural	IPP/per capita	FCM dependency (%)	% of parks in the total green areas surface (%)	M2 parks/ inhabitants	M2 green areas/ inhabitants	Higher existing green area	Recommendation
Toltén	4	10.027	60.2	34	90.1	0	0.0	3.4	T7: Mid-sized green area	Small park
Lumaco	4	10.015	57.8	31	89.6	0	0.0	4.8	T7: Mid-sized green area	Small park
Ercilla	4	8.446	52.9	46	85.7	0	0.0	5.5	T7: Mid-sized green area	Small park
Curarrehue	4	7.826	69.6	28	90.5	0	0.0	1.9	T7: Mid-sized green area	Small park
Los Sauces	4	7.510	39.2	65	82.6	0	0.0	7.7	T7: Mid-sized green area	Small park
Perquenco	4	7.241	48.4	52	87.1	0	0.0	10.5	T7: Mid-sized green area	Small park
Melipeuco	4	6.283	54.3	38	91.4	0	0.0	6.1	T7: Mid-sized green area	Small park

Table 25.6 Sequence of analysis for the classification of parks and urban green areas. Own elaboration

No.	Stage	Activities
1	Review of municipal background related to green area indicators	Compilation of SINIM, INE and MINVU information and database creation
		Analysis and cross-referencing of information by municipality
		Compilation of results
2	Inventory of parks and green public spaces and areas	Compilation of georeferenced databases: INE and MINVU
		Processing of information in QGIS, separating by type of Green Areas (park or smaller)
		Creation of synthetic cartographies for the region and urban centers
3	Classification of parks and green public spaces and areas	Study of the dimensions and shapes of the existing parks in the region
		Segmentation of parks and green public spaces and areas into typologies
		Description of typologies and analysis of referential cases at the regional level
4	Creating groups of communes to make recommendations for management and sustainability for parks and green public spaces and areas	Testing of groups according to criteria defined by communal backgrounds and their green areas
		Definition of final criteria for their grouping
		Analysis of group characteristics, description and adjustments
5	Recommendations for park typologies by groups of communes	Definition of criteria to consider for the making of recommendations
		Creation of criteria or rules of communal indicators according to recommendation
		Adjustments of criteria and drafting of results

25.5 Conclusions

Considering the need to establish recommendations that can be adapted to the diversity of conditions that exist at the regional level, the mechanism used seeks to focus investment recommendations on the availability of green areas and management possibilities considering both the expectations of activation of public space pursued by the National Policy of Urban Parks as well as the municipal capacities for operation and maintenance.

In addition, the study of the specificities of the communes and the opportunity for methodological replicability allows sustained planning for progress over time, adapting the recommendations based on these advances and changes in conditions, in an updated horizon to execute a strategy in terms of availability and management of urban parks.

The same capacity for methodological replicability contemplates the opportunity to extend its application to other regions of the country, using the criteria for grouping and defining recommendations. In this sense, it was identified that the criteria for grouping by typologies of communes is the one that pose the greatest methodological challenge, considering that there are usually regional characteristics that determine them.

However, while each region, commune, locality or city has its own characteristics in which green areas, network of public spaces and the park or parks play a structuring role at different scales, understanding the sequence of actions to implement for ensuring their conservation over time as well as their effective activation for the community's benefit is a challenge for local governments.

In particular, in the case of Chile, the gap in terms of human and financial capacities of small municipalities, hinders the possibility of implementing effective strategies in the different matters proposed by the PNPU. That is why, despite the fact that local actors clearly identify the needs for green areas and parks, which differ depending on the existing provision, the vocation of the territories and their demographic and geographical realities, the possibilities for administration, maintenance and operation are scarce. In this sense, the methodology sought to accompany the region and the municipalities of Araucanía in the implementation of instruments aimed at improving the maintenance and management of urban parks, in the sense of what is proposed by the National Policy of Urban Parks, proposing recommendations that, without being exclusive, cover a wide range of the situations identified in the case studies.

Acknowledgements This research was developed within the framework of the project "Sustainability and Management Model of Urban Parks: Pilot Implementation in Isla Cautín Urban Park" funded by Public Goods for Competitiveness, Region of La Araucanía, Innova Chile, CORFO (Code 21BPCR-191127) developed by the School of Architecture of the Pontifical Catholic University and the UC Local Development Center (Clients: SEREMI MINVU Araucanía and Municipality of Temuco) by Moreno, O., Arizaga, X., Rojo, S., Tapia, J., Salazar, G.

References

CASEN. (2022). National Socioeconomic Characterization Survey. https://observatorio.ministeriodesarrollosocial.gob.cl/encuesta-casen

MDSF. (n.d.). Ministry of social development and family. *NIP—Standards, instructions and procedures for the public investment process.*

Mi Parque Foundation. (2013). *Maintenance cost of m^2 of green areas in the municipalities of Greater Santiago.* https://www.miparque.cl/wp-content/uploads/2017/03/informe-estudio.pdf

MINVU. (1992). *Decree 47 sets general ordinance of urban planning and constructions.* https://bcn.cl/3ggd1

MINVU. (2021). Decreto 17/2021 Apruebab Política Nacional de Parques Urbanos (PNPU). Ministerio de Vivienda y Urbanismo. http://bcn.cl/2q062DE

MINVU. (2022). Ministry of Housing and Urban Development Center of Studies, Green áreas Database. https://ide.minvu.cl/datasets/MINVU::areas-verdes-minvu-ine/explore

National Institute of Statistics. (2020). *Quality indicators of urban parks and squares in Chile—Results report.* Retrieved March 06, 2023, from https://geoarchivos.ine.cl/Files/Calidad_PlPq/INDICADORIN%20DE%20CALIDAD%20%C3%81REAS%20VERDES.pdf

INE, National Institute of Statistics. (2022) *Urban development indicators and standards system.* Retrieved March 06, 2023, from https://storymaps.arcgis.com/stories/b801e9968b894f60ae7e3de0f7ed139f

Reyes-Päcke, S., De La Barrera, F., Dobbs, C., Spotorno, A., & Pavez, C. (2014). *Maintenance costs of urban green areas in Chile.* UC Public Policy Center.

UC Public Policy Center. (2018). *Challenges and proposals for the administration of parks and squares in Chile.* https://politicaspublicas.uc.cl/publicacion/desafios-y-propuestas-para-la-administracion-de-parques-y-plazas-en-chile/

Chapter 26
The Scales of Vulnerability. Mobility and Accessibility of the Vulnerable Active Population in Santiago de Chile

Juan Correa, Felipe Ladrón de Guevara González, and Giovanni Vecchio

Abstract Social exclusion associated with urban mobility is a multi-scalar phenomenon. On the one hand, the possibility of accessing essential activities such as work depends on the availability of job opportunities (at the metropolitan level) as well as the availability of modal alternatives (at the neighborhood-community level); on the other hand, socioeconomic vulnerability can determine specific requirements for modes of transport—public/private, which may be less accessible in vulnerable neighborhoods. Considering these elements, the work proposes to determine the conditions of accessibility (urban and neighborhood) for workers to the public transport system in the Metropolitan Area of Santiago (MAS), focusing on 40% most vulnerable households. The analysis considers urban-scale accessibility, considering average travel times in public transport, and accessibility to public transport at the neighborhood scale, considering access points, frequencies and relative times. The results show that workers from more vulnerable households have long travel times (30.1% travel more than 60 min to the city's business center) and often live in areas with very low or low accessibility to public transport (53%). The results show that it is necessary to consider different scales to address the problem of accessibility, which affects the vulnerable active population of the city, considering that the relationship between accessibility and vulnerability allows to highlight priority areas for

J. Correa (✉)
Center for Space Production, Universidad de las Américas, Providencia, Chile
e-mail: jrcorrea@uc.cl

F. L. de Guevara González
School of Architecture, Pontifical Catholic University of Chile, Santiago, Chile

School of Architecture, Universidad Andrés Bello, Santiago, Chile

F. L. de Guevara González
e-mail: felipeldgg@uc.cl

G. Vecchio
Institute of Urban and Territorial Studies, Pontificia Universidad Católica de Chile, Santiago, Chile
e-mail: giovanni.vecchio@uc.cl

M. Vicuña and E. Giorgi (eds.), *Intersections*, Springer Geography,
https://doi.org/10.1007/978-3-031-76402-8_26

considering average times of travel by public transport and accessibility to public transport at neighborhood scale, estimating access points, frequencies, and relative times. We propose to address a fundamental dimension of social exclusion related to mobility, but we do so considering different spatial scales and focusing on vulnerable population groups.

26.2 Materials and Methods

The present research used an inductive approach of an exploratory nature, focused on the use of quantitative analysis techniques, through the use of geographic information systems (GIS), to determine the conditions of accessibility, both at urban and neighborhood scale, to the public transport system for workers belonging to the 40% most vulnerable households in the Metropolitan area of Santiago (Fig. 26.1).

The study area corresponds to the city of Santiago, the capital of Chile and the largest urban conurbation in the country, covering 34 municipalities in a consolidated urban area of 112,783 hectares, where 6,059,079 inhabitants reside, according to figures from the 2017 Population and Housing Census (INE, 2017).

Regarding the methodological proposal, first, the total number of workers belonging to the 40% of most vulnerable households in the city was determined, which was done using the methodology of the Territorial Socio Material Index (ISMT) of the UC Cities Observatory (2022), which estimates the socioeconomic level of households from the analysis of the microdata from the 2017 census (INE, 2017), using the variables of the head of household's level of education, quality of the housing according to its materiality in walls, floor and roof, level of overcrowding (relationship between number of people and bedrooms) and cohabitation situation (existence of 2 or more households in the same dwelling).

It is from these 4 variables that the final weighted score of each household was estimated, thanks to which it is possible to determine to which percentile of the population each household belongs, and therefore, determine if they belong to the 40% of households with the highest vulnerability in the study area. Subsequently, the total number of people over 18 years old and under 65 years old was estimated for each household, which was treated as the target population, corresponding to the potential workers who move around the city. All the previous analyses were worked at the census zone scale, which corresponds to a cluster of blocks, which does not exceed 2000 dwellings and could be considered similar to a neighborhood.

Secondly, city-scale accessibility to the public transport system was determined, by estimating the average travel times in the transport system (buses and urban trains) from all over the city to the area known as "Estación Tobalaba", considered as the *Central Business District* (CBD) for practical purposes of this research, based on the results of the 2012 Origin–Destination survey (SECTRA, 2012) that point to this sector as the main attractor of work trips, and also for its identification as one of the main commercial subcenters in terms of commercial trips and built surface according to Truffello and Hidalgo (2015). The estimation of travel times was made

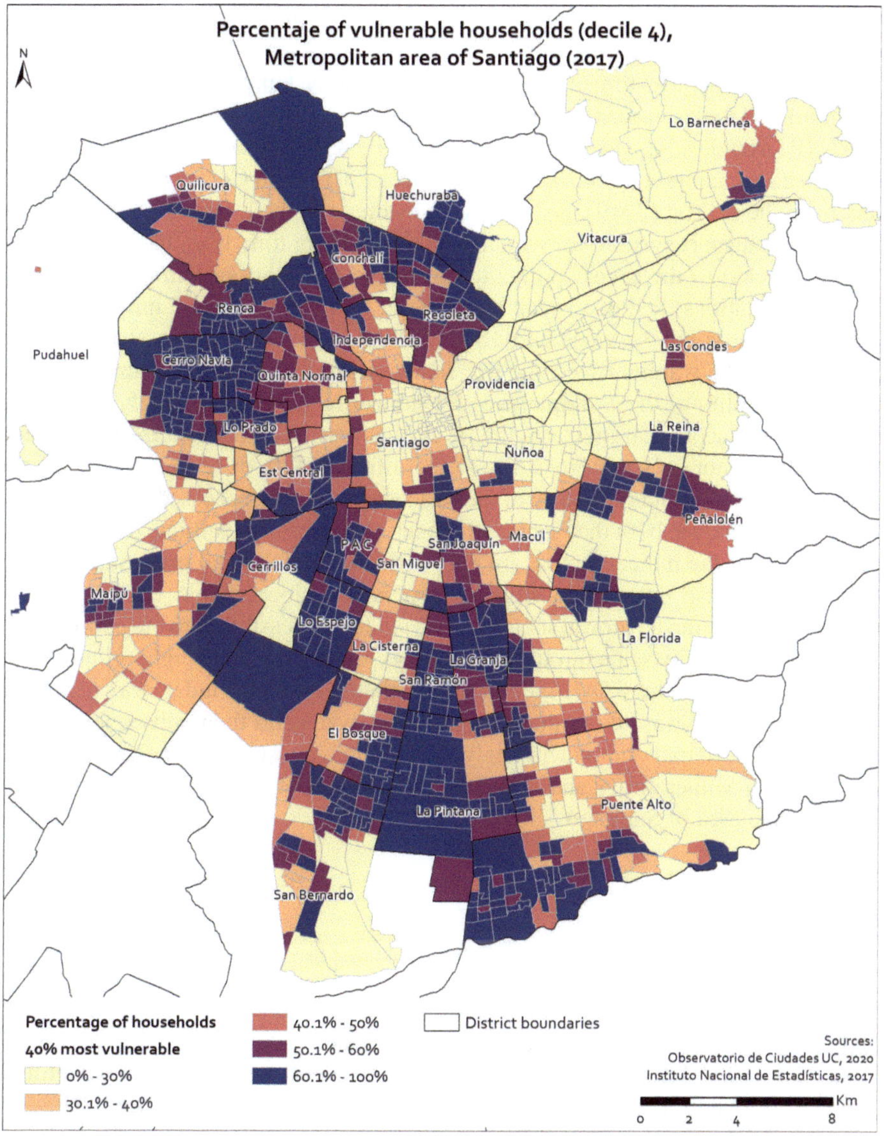

Fig. 26.1 Percentage of households belonging to the 40% most vulnerable, Metropolitan area of Santiago, 2017

from the module *Network Analyst* of the GIS software *ArcGis* 10.7, by assigning the average speeds of routes reported by the Metropolitan Public Transport Directorate and Metro de Santiago (2019) to the city's road arcs and simulating the movement on a working day during peak morning hours (between 6 and 8 in the morning), the travel time estimates were made at the census block scale, to later estimate the average by census zone, to thus homologate the territorial scale of the social vulnerability index through the ISMT.

Thirdly, in order to estimate the accessibility to the system of public transport at the neighborhood scale, an adaptation of the methodology *Public Transport Accessibility Levels* (PTAL) developed by *Transport for London* (2010) and adapted Previously for the Chilean context by Shirahiga and Correa (2015) and Correa (2016) based on data from the DTPM (2019), generating an estimate of the levels of connectivity to the public transport system, for which both the walking time from the origin points (census blocks) to the access points to the public transport system (bus stops and urban train stations), using a speed of 60 m per min, considering 5 min of walking (300 m) towards bus services and 15 min (900 m) towards metro stations. Also estimated are the waiting times, the frequency of services (buses or trains per hour) and the reliability of these frequencies for the entire system, ultimately generating a correlational score that indicates the level of accessibility to the transport system, which reflects not only the proximity to the infrastructure or services, but also the quantity, frequency and reliability of the services that can be accessed. This indicator was also estimated at both the block and census zone scale, to facilitate its operability with the rest of the indicators that are part of the methodological proposal.

It is from these three dimensions of analysis: socioeconomic vulnerability of workers based on the territorial socio-material index, accessibility to the public transport system at the city scale through travel times to the CBD and neighborhood-scale accessibility, using the PTAL methodology, that a synthetic index was generated from the normalization of the three variables analyzed and their subsequent reduction through principal component analysis (PCA), to construct an indicator capable of measuring vulnerability and accessibility at multiple scales for workers in the Metropolitan area of Santiago.

26.3 Results

From the analysis of the microdata from the population and housing census 2017 (INE, 2017) and the application of the territorial socio-material index (OCUC, 2022) a total of 1,750,804 households in the Metropolitan area of Santiago were identified, in which 2,997,848 workers live. Of these more than 1.7 million households, 664,815 belong to the 40% most vulnerable in the region, in which live 1,175,272 workers.

Figure 26.2 shows the concentration and spatial distribution of the most vulnerable households in the city, reflecting a territorial pattern that accounts for the inequality and segregation that has historically characterized the city of Santiago (Agostini et al., 2016; Garretón et al., 2022; Sabatini et al., 2001), where the vast majority of these households are concentrated in the northern, western and southern outskirts of the city, as an effect of the housing policies developed between the years 1980–2000, where the priority was purely the provision of housing, above architectural quality and especially urban planning, generating large territories of socioeconomic homogeneity and social exclusion (Ducci, 1997; Hidalgo, 2007; Rodríguez and Sugranyes,

2004), which must face various socio-spatial problems, such as precisely the lack of accessibility and connectivity to the various goods, services and opportunities offered by the city, where the public transport system plays a fundamental role in this gap (Rodríguez and Sugranyes, 2004).

Regarding the levels of urban-scale accessibility to the public transport system, it was determined that the average travel time on public transport during peak morning

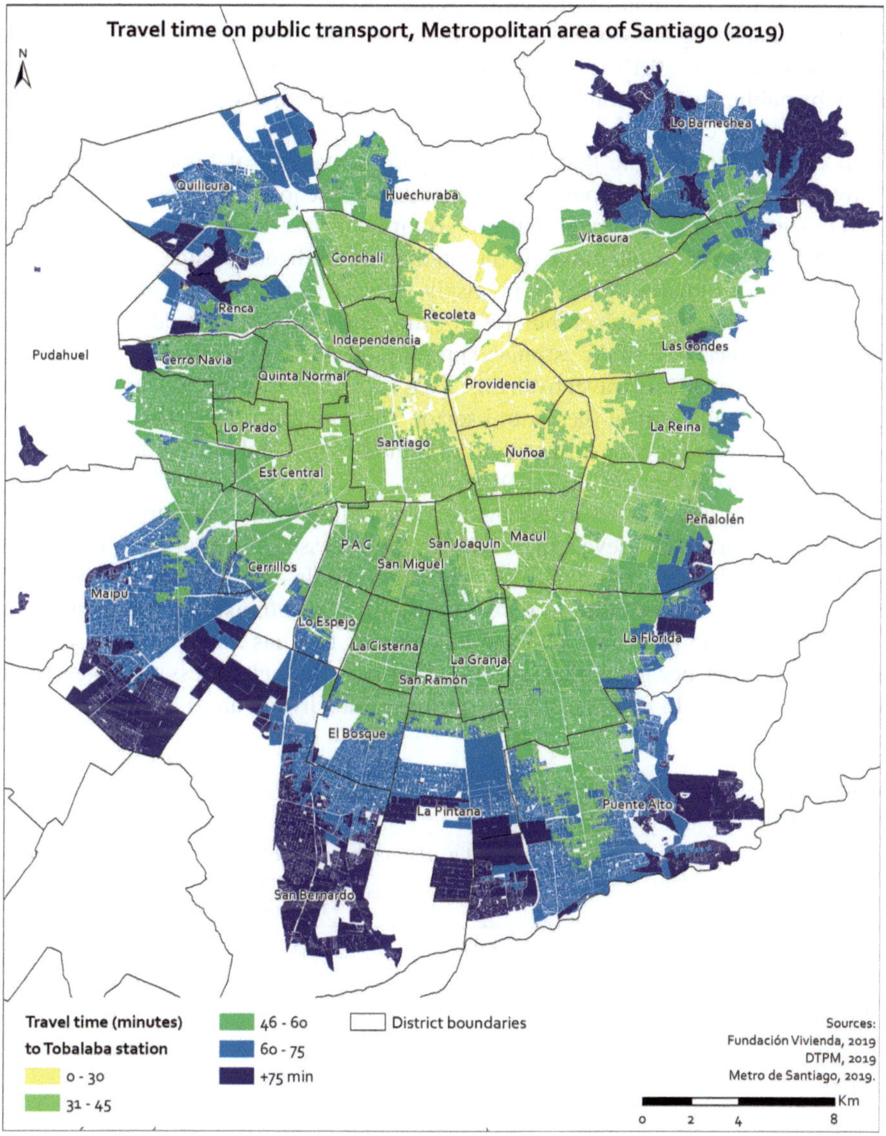

Fig. 26.2 Travel time on public transport, Metropolitan area of Santiago, 2019

working hours to the Tobalaba sector fluctuates between 0 and 120 min, with an average of 37.6 min and a median of 41 min. In spatial terms, a clear centrifugal pattern is observed, but with a bias towards the western and southern areas of the city, where large areas of the city with high travel times, exceeding 75 min, are observed.

From the above, it was estimated that 25.8% of households live more than 45 min away by public transport from the CBD in Tobalaba, while only 14.6% of the city's households live less than 15 min away (Table 26.1).

When analyzing the workers belonging to the 40% most vulnerable households, it is estimated that 30.1% of vulnerable workers (353,757) live more than 45 min away by public transport, and only 4.8% (56,413 workers) live less than 15 min from the city's nerve center.

It is important to note that, if we compare the results of Figs. 26.2 and 26.3, corresponding to the distribution of vulnerable households and travel times to the CBD, it is precisely observed that the sectors with the longest travel times from the north, west, and south outskirts (communes of Quilicura, Maipú, San Bernardo, La Pintana, and Puente Alto) are precisely the sectors with the highest population belonging to the most vulnerable 40% of the city, where households belonging to this group exceed 50% of the estimated proportion. While in much part of the sectors that are less than 30 min travel from the CBD, vulnerability levels tend to be less than 30%, since a large part of the central sector, and especially, the eastern area of the city, is the place where the vast majority of middle-high and high strata households are concentrated (Agostini et al., 2016).

On the other hand, when analyzing the accessibility to the transport system itself, Fig. 26.3 illustrates the levels of neighborhood accessibility from the PTAL index (TFL, 2010), showing that 51.3% of AMS households live in areas classified with very low or low accessibility to public transport, which translates into living more than 300 m from the nearest stop and having only two bus routes with frequencies less than 6 buses per hour (that is, routes that pass through access points every 10 or more minutes). In contrast, 15.8% of households live in areas classified as high

Table 26.1 Total households and workers according to the level of vulnerability and travel time to the city's CBD

Travel time (min)	Total household (%)	Total worker (%)	40% of vulnerable households (%)	40% vulnerable workers (%)
0–15	14.6	12.9	5.1	4.8
15–30	29.6	29.6	27.2	27.4
30–45	30.0	31.0	37.5	37.7
45–60	18.7	19.4	23.5	23.4
60–75	6.0	6.1	6.0	5.8
+75	1.0	1.0	0.8	0.8
Total general	100.0	100.0	100.0	100.0

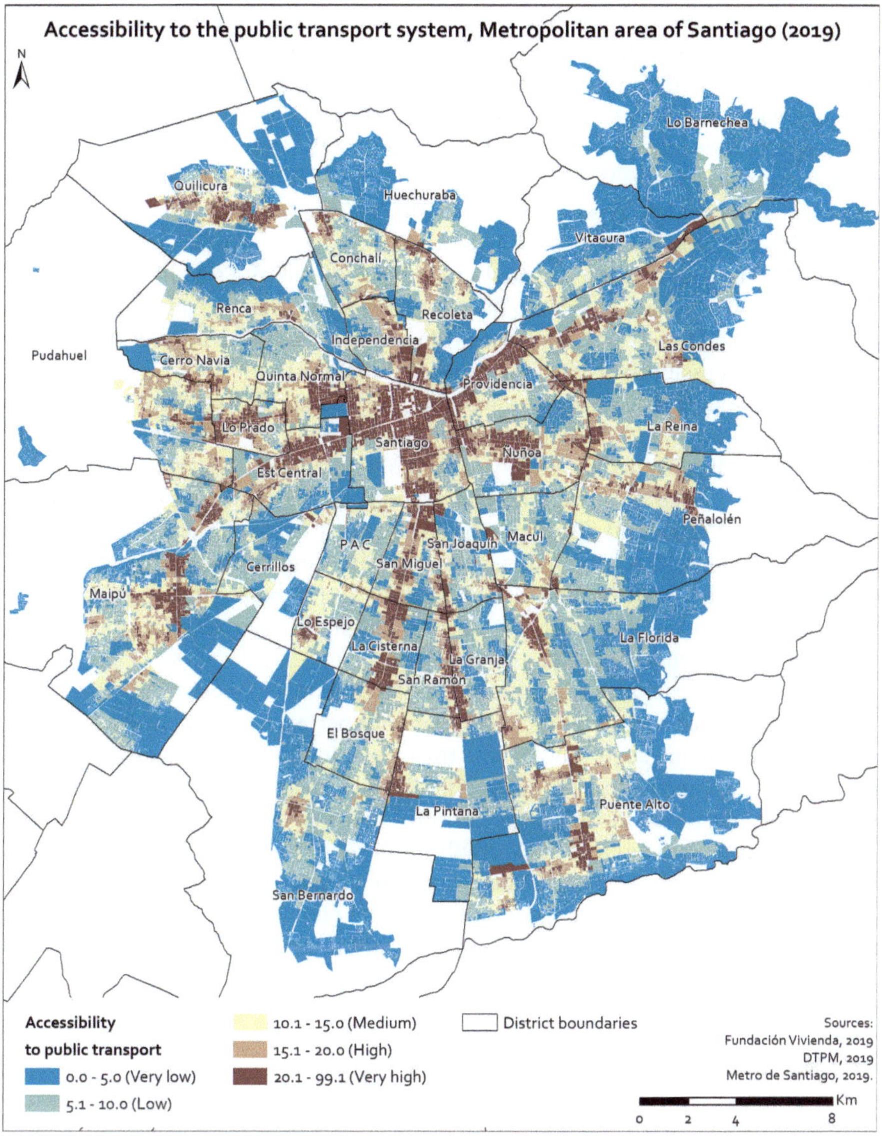

Fig. 26.3 Accessibility to the public transport system, Metropolitan area of Santiago, 2019

accessibility, characterized by having access to the urban train system and having more than 15 bus routes available in their environment.

Similar figures are obtained when looking at the situation of the most vulnerable households in the city, where it is observed that 53% of the workers live in areas with very low or low accessibility to the transport system at the neighborhood scale and that only 7.2% of these vulnerable workers reside in areas that have a high level of neighborhood accessibility to the public transport system (Table 26.2).

Finally, by integrating the three dimensions studied, through a principal component analysis, it is found that, by observing the structure and behavior of the data, it is possible to reduce the three variables into one, within which the relative weights of each component are 37.16% for city-scale accessibility (travel time), 33.05% for neighborhood-scale accessibility, and 29.79% for the level of vulnerability of households and workers in these census areas.

From the above, it was estimated that 43.5% of the city's vulnerable workers live in critical areas with respect to the public transport system, followed by 46% in average areas and finally only 10.5% of vulnerable workers live in areas of very high or very high accessibility to the system itself (Fig. 26.4).

In conclusion, it is estimated that of the most vulnerable workers in the AMS, 30.1% travel more than 1 h to their workplaces and 53% live in neighborhoods with precarious conditions of accessibility to the public transport system.

On the other hand, the study determines that 43.5% of workers live in critical areas in terms of urban and neighborhood accessibility to the public transport system. This aspect conditions access to transport in specific neighborhoods, such as peripheral sectors, which suffer from less accessibility to the public transport system, or are further away from access points, such as bus stops, metro stations or points of combination and modal transfer.

This study, at the same time, shows that the longest travel times are made by people living in vulnerable sectors of the city, mainly in peripheral municipalities, and far from the high-income cone, where these trips are reduced in time, identifying factors of exclusion and inequality in the city, in this aspect, in relation to municipalities with higher and lower incomes, accessibility to transport and the optimization of times in relation to the journey to workplaces.

On the other hand, the aspects mentioned above can condition access to specific job opportunities, which are concentrated in certain municipalities of the AMS, and in this way the search area is restricted, limiting the radius of closer options, since the time factor can lead to discarding the furthest options.

Table 26.2 Total households and workers according to vulnerability level and level of accessibility to the transport system

Accessibility level	Total household (%)	Total workers (%)	40% of vulnerable household (%)	40% vulnerable workers (%)
Very low	15.9	16.5	12.9	13.0
Low	35.1	35.2	40.1	40.0
Average	24.7	24.9	27.4	27.6
High	11.4	11.4	12.2	12.2
Very high	12.9	12.1	7.5	7.2
Total general	100.0	100.0	100.0	100.0

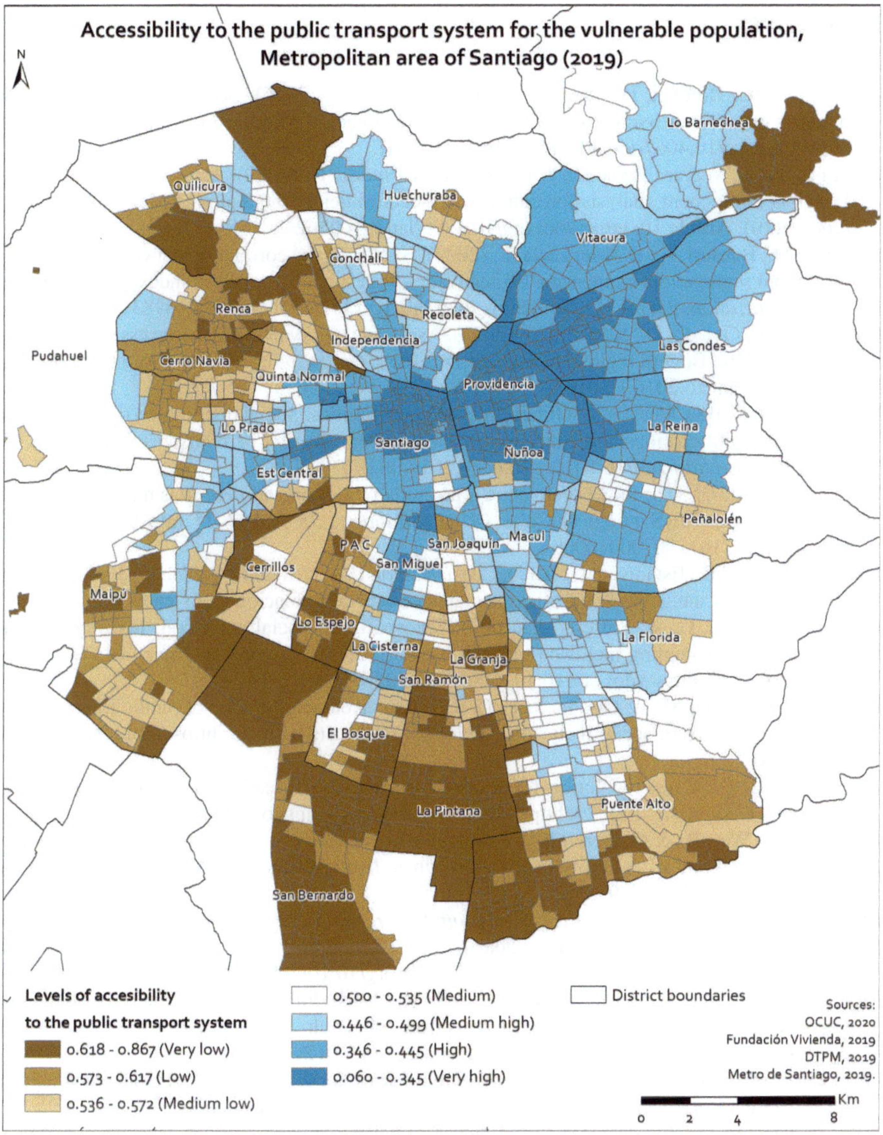

Fig. 26.4 Accessibility and vulnerability index to the public transport system, Metropolitan area of Santiago 2019

Finally, the need for qualitative studies on accessibility experience is evident, in relation to conditions of public space, barriers and opportunities in the "home-stop" journey to understand, specifically, the experience of people who use public transport, to complement, with the scope of this dimension, the study presented.

References

Agostini, C. A., Hojman, D., Román, A., & Valenzuela, L. (2016). Segregación residencial de ingresos en el Gran Santiago, 1992–2002: Una estimación robusta. *Eure, 42*(127), 159–184. https://doi.org/10.4067/S0250-71612016000300007

Correa-Parra, J., Vergara-Perucich, J. F., & Aguirre-Nuñez, C. (2020). Towards a walkable city: Principal component analysis for defining sub-centralities in the Santiago Metropolitan Area. *Land, 9*(10), Article 10. https://doi.org/10.3390/land9100362

Garretón, M., Basauri, A., & Valenzuela, L. (2022). Exploring the correlation between city size and residential segregation: Comparing Chilean cities with spatially unbiased indexes. *Environment and Urbanization, 32*(2), 569–588. https://doi.org/10.1177/0956247820918983

Guzman, L. A., Oviedo, D., & Rivera, C. (2017). Assessing equity in transport accessibility to work and study: The Bogotá region. *Journal of Transport Geography, 58*, 236–246. https://doi.org/10.1016/j.jtrangeo.2016.12.016

Hernandez, D. (2018). Uneven mobilities, uneven opportunities: Social distribution of public transport accessibility to jobs and education in Montevideo. *Journal of Transport Geography, 67*, 119–125. https://doi.org/10.1016/j.jtrangeo.2017.08.017

Hidalgo, R. (2007). ¿Se acabó el suelo en la gran ciudad?: Las nuevas periferias metropolitanas de la vivienda social en Santiago de Chile. *EURE (Santiago), 33*(98), 57–75. https://dx.doi.org/10.4067/S0250-71612007000100004

Instituto Nacional De Estadísticas. (2017). *Censo Nacional de Vivienda y Población 2017*. Availabe at: https://www.ine.gob.cl/estadisticas/sociales/censos-de-poblacion-y-vivienda

Kenyon, S., Lyons, G., & Rafferty, J. (2002). Transport and social exclusion: Investigating the possibility of promoting inclusion through virtual mobility. *Journal of Transport Geography, 10*(3), 207–219. https://doi.org/10.1016/S0966-6923(02)00012-1

Lättman, K., Friman, M., & Olsson, L. E. (2016). Perceived accessibility of public transport as a potential indicator of social inclusion. *Social Inclusion, 4*(3), 36–45. https://doi.org/10.17645/si.v4i3.481

Lucas, K. (2019). A new evolution for transport-related social exclusion research? *Journal of Transport Geography*, 102529. https://doi.org/10.1016/j.jtrangeo.2019.102529

Lucas, K., van Wee, B., & Maat, K. (2016). A method to evaluate equitable accessibility: Combining ethical theories and accessibility-based approaches. *Transportation, 43*(3), 473–490. https://doi.org/10.1007/s11116-015-9585-2

Martens, K. (2017). *Transport justice: Designing fair transportation systems*. Routledge.

Martens, K., Singer, M. E., & Cohen-Zada, A. L. (2022). Equity in accessibility. *Journal of the American Planning Association*, 1–16. https://doi.org/10.1080/01944363.2021.2016476

Moreno-Monroy, A. I., Lovelace, R., & Ramos, F. R. (2018). Public transport and school location impacts on educational inequalities: Insights from São Paulo. *Journal of Transport Geography, 67*, 110–118. https://doi.org/10.1016/j.jtrangeo.2017.08.012

Observatorio de Ciudades UC. (2022). *Metodología Índice Socio Material Territorial*. Available at: https://observatoriodeciudades.com/portfolio/ismt-paquete-de-r-ismtchile/

Pucci, P., Vecchio, G., Bocchimuzzi, L., & Lanza, G. (2019). Inequalities in job-related accessibility: Testing an evaluative approach and its policy relevance in Buenos Aires. *Applied Geography, 107*, 1–11. https://doi.org/10.1016/j.apgeog.2019.04.002

Rodríguez, A., & Sugranyes, A. (2004). El problema de vivienda de los "con techo". *EURE (Santiago), 30*(91), 53–65. http://dx.doi.org/10.4067/S0250-71612004009100004

Sabatini, F., Cáceres, G., & Cerda, J. (2001). Segregación residencial en las principales ciudades chilenas: Tendencias de las tres últimas décadas y posibles cursos de acción. *EURE (Santiago), 27*(82), 21–42. https://dx.doi.org/10.4067/S0250-71612001008200002

Secretaría de Planificación de Transporte. (2012). *Encuesta Origen Destino Santiago 2012*. Availabe at: https://www.sectra.gob.cl/encuestas_movilidad/encuestas_movilidad.htm

Tiznado-Aitken, I., Lucas, K., Muñoz, J. C., & Hurtubia, R. (2020). Understanding accessibility through public transport users' experiences: A mixed methods approach. *Journal of Transport Geography, 88*, 102857. https://doi.org/10.1016/j.jtrangeo.2020.102857

Tiznado-Aitken, I., Muñoz, J. C., & Hurtubia, R. (2018). The role of accessibility to public transport and quality of walking environment on urban equity: The case of Santiago de Chile. *Transportation Research Record, 2672*(35), 129–138. https://doi.org/10.1177/0361198118782036

Tiznado-Aitken, I., Vecchio, G., Guzman, L. A., Arellana, J., Humberto, M., Vasconcellos, E., & Muñoz, J. C. (2023). Unequal periurban mobility: Travel patterns, modal choices and urban core dependence in Latin America. *Habitat International, 133*, 102752. https://doi.org/10.1016/j.habitatint.2023.102752

Transport for London (TFL). (2010). *Measuring public transport accessibility levels*, London, United Kingdom.

Truffello, R., & Hidalgo, R. (2015). Policentrismo en el Área Metropolitana de Santiago de Chile: reestructuraci6n comercial, movilidad y tipificaci6n de subcentros. *EURE (Santiago), 41*(122), 49–73. https://dx.doi.org/10.4067/S0250-71612015000100003

Van Wee, B. (2016). Accessible accessibility research challenges. *Journal of Transport Geography, 51*, 9–16. https://doi.org/10.1016/j.jtrangeo.2015.10.018

Van Wee, B., & de Jong, T. (2023). Differences in levels of accessibility: The importance of spatial scale when measuring distributions of the accessibility of health and emergency services. *Journal of Transport Geography, 106*, 103511. https://doi.org/10.1016/j.jtrangeo.2022.103511

Vecchio, G., & Martens, K. (2021). Accessibility and the capabilities approach: A review of the literature and proposal for conceptual advancements. *Transport Reviews, 41*(6), 833–854. https://doi.org/10.1080/01441647.2021.1931551

Vecchio, G., Tiznado-Aitken, I., & Hurtubia, R. (2020). Transport and equity in Latin America: A critical review of socially oriented accessibility assessments. *Transport Reviews, 40*(3), 354–381. https://doi.org/10.1080/01441647.2020.1711828

Chapter 27
The Construction of a Segregated Metropolis: Urban Segregation and Natural Preexistences in the Metropolitan Area of Concepción

Daniela Villouta, Ricardo Trufello, Margarita Greene, and Andrea Fernández

Abstract Urban segregation has been the center of discussion in various theories of urban sociology (Ruiz-Tagle, 2016), gaining great importance in recent years. The problem has been mainly addressed from residential location and socioeconomic stratification, although there are other factors that can complement its understanding and representation. Among them, the level of segregation according to socioeconomic level and its representation in space (Trufello et al., North Grande Geography Journal 160:81–131, 2022) play a key role in the construction of an area in the process of metropolitanization. This article seeks to observe the way in which segregation has been constructed and evolved in the Metropolitan Area of Concepción (AMC) from its representation in space. Using a quantitative methodology, first, homogeneous areas are constructed regarding territorial segregation attributes in the AMC from the geolocation of an entropy index and socioeconomic variables. Second, typologies of urban environments are constructed considering the natural preexistences in the territory. These data are processed through spatial statistics from the regionalization method and its cartographic representation. The results contribute to understanding socio-spatial segregation from its representation in space at the metropolitan level.

D. Villouta (✉)
Department of Civil Engineering, Universidad Católica de la Santísima Concepción, Concepción, Chile
e-mail: dvillouta@ucsc.cl

R. Trufello · M. Greene
Center for Sustainable Urban Development (CEDEUS), Pontificia Universidad Católica de Chile, Santiago, Chile
e-mail: rtruffel@uc.cl

M. Greene
e-mail: mgreenez@uc.cl

A. Fernández
Faculty of Architecture, Urbanism and Geography, Universidad de Concepción, Concepción, Chile
e-mail: afernandezc@udec.cl

M. Vicuña and E. Giorgi (eds.), *Intersections*, Springer Geography,
https://doi.org/10.1007/978-3-031-76402-8_27

With this, it is expected to minimize the error of representative samples for the generation of the surveys that will be carried out after this study and for the selection of metropolitan-level case studies.

Keywords Segregation · Cartographic representation · Regionalization · Natural elements

27.1 Introduction

In recent studies of urban segregation, numerous criticisms have been made regarding the type of measurement necessary to address segregation from a mainly economic and sociological approach (Ruiz-Tagle & López, 2014). Much of the literature deals with segregation in cities in a traditional way, understood as "the degree to which two groups live separately from each other, in different parts of the urban environment" (Massey & Denton, 1988, p. 282). Thus, segregation has been mainly addressed from residential location and socioeconomic stratification, but there are other factors that can complement its understanding and representation.

Part of the criticism has focused on measurement and representation strategies; this is due, in part, to the importation of methodologies that are incomplete for understanding urban segregation from the particularities of the territories (Ruiz-Tagle & López, 2014). Thus, it seems necessary to delve into mechanisms that allow for the complementation of existing studies and that manage to account for the particularities of the segregation process in the territory through its representation and measurement in space. This is particularly important in cases where natural pre-existences can play a key role in the structuring of the city.

On the other hand, segregation studies have mainly been based on traditional statistics, from the randomness and independence of observations. In this way, the geographical space has been downplayed in the representation and measurement of processes such as segregation, which could be an important tool to minimize the error of representative samples in the territory (Truffello et al., 2022). Faced with the need for a complementary approach, quantitative geography becomes important in the generation of methodologies that allow for the effective representation and measurement of a certain urban process.

Taking into account the above, this article highlights the importance of measuring and representing urban segregation from the construction of homogeneous areas regarding territorial attributes of segregation and natural pre-existences in the Metropolitan Area of Concepción (AMC). With this, it is expected to minimize the error in the generation of representative samples for surveys that will be conducted after this study and for the selection of case studies at the metropolitan level. To observe segregation, a quantitative geographical methodology based on the geolocation of the cross between the Theil entropy index (Parisi et al., 2011) and socioeconomic variables of the AMC is proposed. To identify natural pre-existences, a geographical survey of the main elements in the territory was carried out. The data were processed using spatial statistics based on the regionalization method (Duque et al., 2007, 2012; Folch & Spielman, 2014; Truffello et al., 2022) and its cartographic representation. The results contribute to a better understanding of socio-spatial segregation from its representation in space at the metropolitan scale.

27.1.1 Urban Segregation and Its Representation in a Complex Territory

Defining urban social segregation, in a field of broad theoretical and methodological disagreements, is complex. The influence of American empirical literature on the analysis of segregation has historically dichotomized the phenomenon as an urban problem of race, with clear conceptual and methodological biases. Ruiz-Tagle (2016) recognizes four lines that explain the formation of segregation, (i) changes in housing location, (ii) subdivision and densification of modern cities and their subsequent separation between rich and poor (iii) diversification of cultural traits and (iv) racism.

From a local approach, segregation is understood as the "degree of spatial proximity or territorial agglomeration of families belonging to the same social group, whether this is defined in ethnic, age, religious preference or socioeconomic terms" (Sabatini et al., 2001, p. 27). Thus, traditional definitions address segregation in a socioeconomic way and how it is distributed in space, leaving aside the role of the geographical environment in the formation of the phenomenon. In the case of Chile, the social housing policy, which significantly reduced the quantitative housing deficit from the 90s, in practice created and reinforced urban problems such as socioeconomic segregation, by focusing on quantitative aspects, without paying attention to the location of the houses and their access to minimum urban public goods. In this line, the subsequent strategies implemented to overcome segregation have been strongly centered on a territorial dimension. This conception, however, remains limited, as it emphasizes only the mix in places of residence, without achieving the mix and interaction between social groups.

Existing studies on urban segregation and its expression in space mainly address its consequences in terms of general quality of life (Sabatini et al., 2010); access to education (Sanhueza & Larrañaga, 2008) and discrimination (Sabatini & Wormald, 2013). Most of these studies address the periphery of the city or more recently the expansion areas (Rasse et al., 2020). However, spatial segregation at the metropolitan level has not been addressed, except for a few exceptions, generating biases when categorizing and understanding metropolitan complexity.

In the case of the AMC, the recent study by Medina (2020) shows a distribution of socio-spatial segregation in six types of neighborhoods at the metropolitan level. The results position neighborhoods integrated with positive qualities for the population in the eastern sector of San Pedro de la Paz, downtown Concepción, Pedro de Valdivia, Nonguén access, and a significant part of the access to the municipality of Talcahuano. As segregated neighborhoods with a concentration of poverty, the study identifies the western sector of San Pedro de la Paz, Hualpén, and the high zone of Talcahuano. In this way, while it is possible to address the phenomenon of segregation at the metropolitan level from a quantitative approach, this approach generalizes segregated groups without distinguishing their socioeconomic level in an integrated manner.

27.1.2 Natural Preexistences in the Metropolitan Area of Concepción (AMC)

The AMC is part of the Biobío Region, located at the southern limit of the central zone of Chile, and corresponds to a conurbation developed in the extension of a coastal axis of 60 km, centered at the mouth of the Biobío River. The system of cities is positioned over the coastal spaces between the Coastal Mountain Range and the coastline of the province of Concepción (Hernández, 1983). The Metropolitan Area comprises ten municipalities: Concepción, Coronel, Chiguayante, Lota, Penco, San Pedro de la Paz, Talcahuano, Tomé, Hualpén, and Hualqui, which cover a total surface area of 248.55 km^2 with a population of 1,080,000 inhabitants. Thus, the AMC is the second conurbation in the country after the Metropolitan Area of Santiago, (AMS) (Fig. 27.1).

The territory occupied by the AMC is characterized by various geographical elements: the coastal edge, with the bays of Concepción, San Vicente and Coronel; the Biobío and Andalién rivers; the coastal mountain complex called the Coastal or Nahuelbuta mountain range; and the low sites composed of wetlands and lagoons. These elements have been influential in the processes of settlement, configuration, and growth of its urban cores (Salinas & Pérez, 2011). Functionally, the AMC settlements have been organized in a polycentric model in the municipalities of Concepción and Talcahuano; the rest of the municipalities have not been able to develop and consolidate their own subcenters and their inhabitants continue to depend on the center of Concepción to meet their needs. Paradoxically, the deeper the process of metropolization, the greater the concentration of specialized services in the traditional center of the regional capital, increasing the dependence of the peripheral municipalities of the urban system, characterized by a dispersed and centrifugal occupation in the territory (Baeriswyl, 2009; Napandesky, 2020; Pérez & Salinas, 2009). Added to this, the expansion process derived from the neoliberal market model and weak territorial planning has increased new forms of growth towards natural areas transforming and significantly modifying existing ecosystems. Studies indicate a reduction in the surface area of wetlands by close to 40% since 1975 and from 2001 it decreased by 700 hectares in just three years (Smith & Romero, 2009).

The literature suggests that the growth of a city should be guided by the identity form that its essential components and natural elements represent (Rivas, 2009), the shape of the topography, the presence of elevations or depressions, the waterfalls in the valleys, the dominant heights, and be linked with the discourse on the existing urban identity. Mumford (1966) and McHarg (1969) analyze the birth of the city

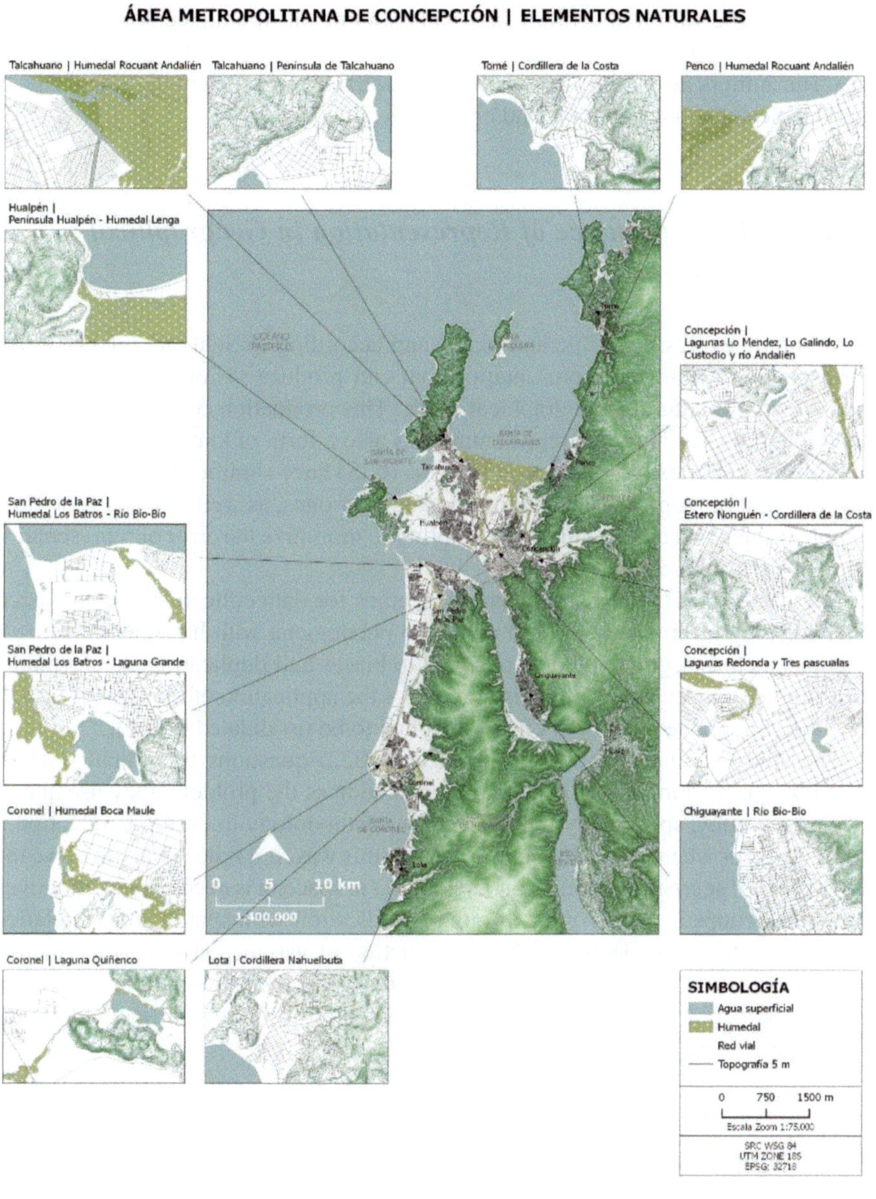

Fig. 27.1 Location of natural elements in the metropolitan area of Concepción

considering the values of ecology in urban planning, attributing ecological associations to the complex forms of the soil and its transformation in the formation of the city. The authors also argue that the modification and replacement of these forms by marketable substitutes disrupts and endangers cities.

27.1.3 The Importance of Representation in Geographical Space

Cartography expresses a time and place, and according to Schlögel (2017), a map corresponds to a mode of presentation that can produce visibility and an overall view of what is naturally geographical space. This production contains an order and direction; thus, cartographic representation is also a form of knowledge production. In this line, geographic information systems (GIS) have significantly influenced the development of maps, virtual and traditional paper ones, as well as when generating sample frames for surveys, due to their ability to minimize the error of representative samples (Truffello et al., 2022).

The use of traditional random sample frames for data collection does not accurately display information when it comes to phenomena with high spatial autocorrelation, such as segregation, and can lead to biases and duplication of information (Griffith, 2005). A clear example is the case of the application of the Census in 2012 in Chile, where due to its high error rate it had to be invalidated (9.6%).

The study conducted by Truffello et al. (2022) questions traditional statistics based on the randomness of observations. To address the problem, they use quantitative geography, specifically regionalization methodology to ensure a reduction of sampling error when implementing surveys. In this way, if it is necessary to delve into the process of segregation and its understanding in space, it is necessary to effectively know the groups in the territory through stratification with spatial autocorrelation, generation of homogeneous zones and reduction of variance. Through the method of regionalization with local heuristic optimization for the creation of homogeneous areas, a reduction of up to 20% in error (compared to traditional methodologies) (Truffello et al., 2022) is achieved.

27.2 Methodology

To construct homogeneous groups regarding territorial attributes of segregation in the AMC, the synthetic indicator of sociodemographic stratification, the Territorial Socio-Material Index (ISMT), which summarizes the socio-material level based on four census variables: education level of the head of household; level of overcrowding of the household; household overcrowding and materiality of the dwelling (OCUC, 2017) was used as a base.

Based on the ISMT (2017), the Theil Entropy Indicator was constructed using the Geo-Segregation Analyzer software. Entropy or diversity index (Apparicio et al., 2014) is used to detect completely homogeneous spatial units (where individuals of a single group reside, H2 = 0) or spatial units with maximum diversity (all population groups are of equal size, H2 = 1), where

$$H2 = \sum_{i=1}^{n} \left[\left(P_{ij}/P_j\right)\ln\left(P_{ij}/P_j\right)\right]/\ln n$$

- n number of Groups
- P_j sum of the population of the groups in n for the spatial unit j
- P_{ij} population of group i in spatial unit j
- t_i Total population of unit i
- T Total population in the total area
- x_i Population of group X in spatial unit i
- X Population of group X in the total area.

To spatially stratify the data, the regionalization method was used, which has been validated in various research for its ability to reduce sampling error (Duque et al., 2012; Folch & Spielman, 2014; Truffello et al., 2022). Specifically, Grouping Analysis (GA) was used, which considers multivariable clustering algorithms, using the K-Means method with a heuristic component, although not for local optimization (Truffello et al., 2022). In addition to the Entropy Indicator, the socioeconomic level was integrated for the creation of four *clusters,* subsequently, the two smaller *clusters* (2 and 3) are combined for a better interpretation of the data.

Finally, for the survey of natural pre-existences in the AMC, the database of the Center for Sustainable Urban Development (CEDEUS) on urban wetlands, bodies of water (lagoons, rivers) and topography (hills) was used. The data were processed using Qgis 2018 and integrated with the results of homogeneous groups obtained previously.

27.3 Results

27.3.1 Composition of a Segregated Metropolis: Territorial Socio-material Indicator and Theil's Entropy Level

The results indicate the existence of three predominant homogeneous areas in the AMC: (i) *Cluster* 1, which corresponds to socially integrated groups of middle income; (ii) *Cluster* 2, which corresponds to low-income groups socially segregated, and (iii) *Cluster* 3, corresponds to high-income groups socially segregated. *Cluster* 1

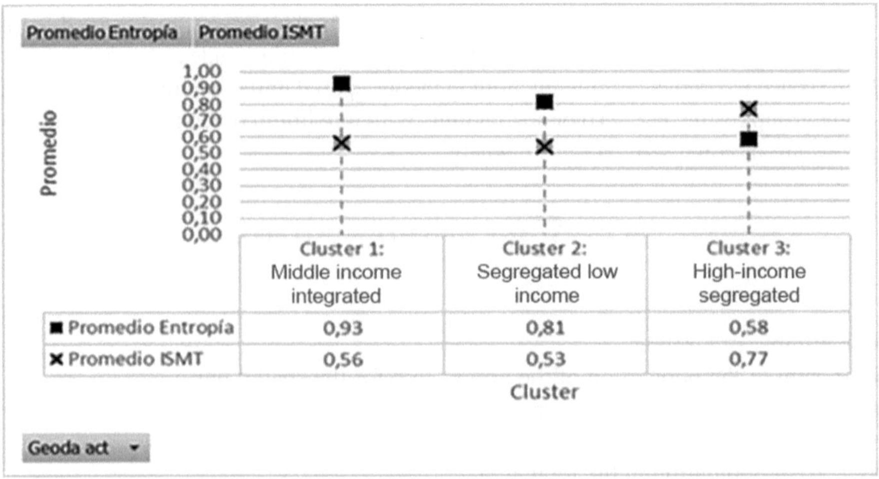

Fig. 27.2 Composition of predominant typologies in the AMC based on the average of ISMT and average entropy

mainly contains middle-income households with 30% of D stratum[1] and 25% of the C3 stratum. *Cluster* 2 has a similar composition to the integrated groups of middle income (*Cluster* 1), but with a higher composition of E stratum households. Finally, *Cluster* 3 is composed of more than 80% of high-income groups with 60% ABC1, followed by 20% of C2 households. In contrast, low-income D and E households represent less than 5%.

Regarding the composition of ISMT and entropy level for each typology, Fig. 27.2 shows few differences in ISMT between the first two groups; on the contrary, the entropy level shows more significant differences between them. Specifically, the typology of integrated households presents the highest entropy values (0.93). On the contrary, the typology of high-income segregated households presents the lowest entropy values (0.58), as it is composed of households of the same type.

27.3.2 Predominant Typologies in the Metropolitan Area of Concepción: Social Segregation and Natural Pre-existences

The urban growth of the AMC has been conditioned from its beginnings by the predominant geography of the territory defined by various natural elements. Urban

[1] Households are classified into seven groups: AB (upper class), C1a (affluent middle class), C1b (emerging middle class), C2 (typical middle class), C3 (lower middle class), D (vulnerable middle class) and E (poor). The ABC1 group would be the sum of the individuals classified in the AB, C1a and C1b groups (AimChile, 2019).

occupations are mostly located on the surface of the valley formed between the coastal mountain range (Nahuelbuta) to the east and the coastal edge to the west. The mountain edge guides and at the same time limits urban growth marked by the foot of the mountain, island hills (Cerro Caracol, Cerro Chepe, Cerro La Pólvora, Cerro Amarillo, Cerro Macera, Cerro Lo Galindo, among others) and the peninsulas of Hualpén and Talcahuano. The water system composed of the Biobío and Andalién rivers—both of which traverse a large part of the territory—represent natural barriers that have formally and spatially guided urban growth and defined the administrative boundaries of the AMC. Likewise, the low-lying areas mainly formed by wetlands (Rocuant-Andalién, Paicaví-Tucapel, Vasco da Gama, Price, Los Batros, Lenga) and urban lagoons (Grande, Chica, Redonda, Lo Custodio, Lo Méndez, Lo Galindo), have been reached by urban growth, and today form an important part of the AMC's urban landscape.

Natural elements, around wetlands, foothills and hills with steep slopes, have been highly modified for residential land use. Among them, both social housing projects driven by the state, (Candelaria in San Pedro de la Paz, Higueras sector in Talcahuano) and real estate projects for high-income families, have contributed to social segregation and have been located on lands with environmental restrictions and limitations for urbanization, such as the high risk of flooding, waterlogging, landslides and tsunamis. The extension of new urbanizations on lands not suitable for urbanization, such as coastal wetlands and hill sectors with steep slopes, also implies higher costs in their enablement (Pérez & Salinas, 2009).

The results show a trend of occupation of the coastal edge and edge of the Biobío River by segregated low-income groups and an occupation of the high zones by segregated high-income groups, while the low zones (flat) are associated with occupation in a mosaic mode, both of socially integrated groups and segregated low-income groups, but without the presence of segregated high-income groups. In particular, the segregated low-income groups are located in coastal areas of the metropolis (for example, the Boca Sur neighborhood in San Pedro de la Paz, historic center of Coronel and Lota) with the exception of the surface of the traditional center of Concepción, residential and industrial sector of Hualpén. However, in the peripheral communes of the AMC's northern zone, corresponding to Tomé, Penco and Talcahuano, they are in the high zones (Fig. 27.3).

On the other hand, it was found that the type of urbanization of the segregated high-income groups is associated with the natural element, especially those linked to bodies of water: San Pedro del Valle—Los Batros Wetland; Brisas del Sol—Rocuant Wetland; Andalue—Hill and Grande and Chica Lagoon. They correspond to closed urbanizations (cul de sac type), which privatize the natural element taking advantage of its landscape conditions, have high vegetation coverage, a low-density residential use and low rates of impermeabilization. This translates into environments of high environmental quality, but inaccessible to those who do not belong to this group, evidencing the inequity of accessible public spaces for the communal population.

Finally, the integrated groups of middle incomes are mainly located in high and low areas of the AMC, which are generally located in the central areas of each commune. These areas coincide with modern urbanizations developed between 1950 and 1960

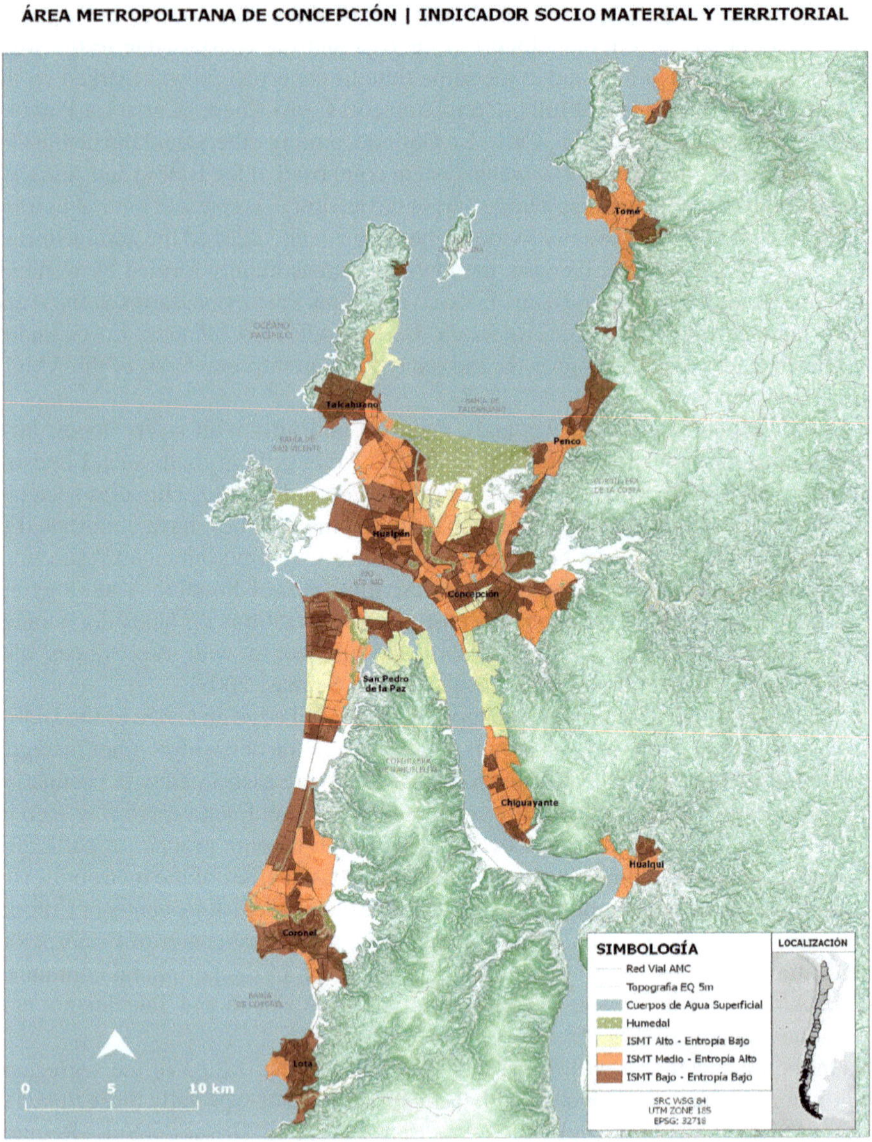

Fig. 27.3 Homogeneous areas in the AMC according to social segregation and socio-economic level (own elaboration based on ISMT, graphic by Yabel Arévalo)

in the AMC, which integrate elements of the landscape in their urban development. They correspond to the sectors of Higueras (Villa Presidente Ríos) in Talcahuano; A. Alarcón del Canto sector in Hualpén; El Golf and Lorenzo Arenas sectors in Concepción; all are residential areas built to provide a solution to worker housing derived from the industry, which are configured through roads, passages, patios and gardens, contemplating public and collective social meeting spaces (Fuentes, 2012; Fuentes & Pérez-Bustamante, 2007).

From a communal approach, a double-entry matrix was used to observe the configuration of the AMC communes in relation to the predominant typology (segregation and socioeconomic level), type of composition (mixed, dual) and natural configuration (according to elements of natural configurants). Table 27.1 shows that the most integrated communes correspond to Tomé (74% integrated households) and Chiguayante (65% integrated households); both are located between the coastal edge and water systems and are protected by hills. However, as Chiguayante contains all three typologies, it is considered of mixed type. On the contrary, the most segregated communes (no high-income segregated communes are recorded) correspond to Lota (90% segregated households), Hualpén (65% segregated households) and San Pedro de la Paz (61% segregated households). These last two are configured by complex and diverse natural systems (Fig. 27.4).

27.4 Conclusions

This article demonstrates the importance of cartographic representation from the construction of typologies or *clusters* regarding territorial attributes of segregation and natural preexistences in the Metropolitan Area of Concepción (AMC). Also, it values the role of quantitative geography in reducing error when trying to generate representative samples as part of the sampling framework of surveys that will be carried out after this study and for the selection of critical case studies at the metropolitan level.

From a metropolitan perspective, this research provides a macro reading of the municipal composition using Census data (2017) seeking to improve the understanding of the phenomenon of segregation from the natural particularities of the territory. From a communal scale analysis and taking into account the data analyzed, it is possible to conclude that neither the condition of the peripheral commune nor the configuration between the coastal border and hill, as is the case of Lota and Tomé, condition the level of segregation that their households present. Likewise, patterns of occupation by the level of social segregation and the socioeconomic level of households are observed, where socially segregated groups of low income are located in coastal areas, socially segregated groups of high income occupy high areas and finally, the flat areas (between the coastal border and high areas) are occupied by integrated groups and to a lesser extent segregated low-income households.

The implemented methodology allows observing particularities of the case of the AMC, however, it is necessary to advance in the understanding of the distribution

Table 27.1 Communal composition matrix in the AMC

AMC communes	Typology: segregation and rate of predominant SES	Type of composition	Natural configuration
Chiguayante	Middle-income integrated (65.7%)	Mixed (typology 1, 2 and 3)	Between river edge and hill
Concepción	Low-income segregated (46.7%)	Mixed (typology 1, 2 and 3)	Between river edge and hills, lagoons and urban wetlands
Coronel	Low-income segregated (59.5%)	Dual (typology 1 and 2)	Between hill and coast, urban wetlands
Hualpén	Low-income segregated (65.7%)	Dual (typology 1 and 2)	Between river edge and urban wetlands and peninsula
Hualqui	Low-income segregated (54.0%)	Dual (typology 1 and 2)	Between river edge and hill
Lota	Low-income segregated (90.2%)	Dual (typology 1 and 2)	Between hill and coast
Penco	Segregated low-income (55.2%)	Dual (typology 1 and 2)	Between hill and coast
San Pedro de la Paz	Segregated low-income (61.3%)	Mixed (typology 1, 2 and 3)	Between hill and coast, lagoons and urban wetlands
Talcahuano	Segregated low-income (50.7%)	Mixed (typology 1, 2 and 3)	Between double coast and urban wetlands
Tomé	Integrated middle-income (74.3%)	Dual (typology 1 and 2)	Between hill and coast

of the segregation of the groups in the territory from a representative breakdown of a smaller scale. For this, this project will continue with the characterization of homogeneous areas using 50% of the most representative census zones of each *cluster*. In this way, it will be possible to advance in the selection of case studies concrete for the implementation of surveys with low sampling error.

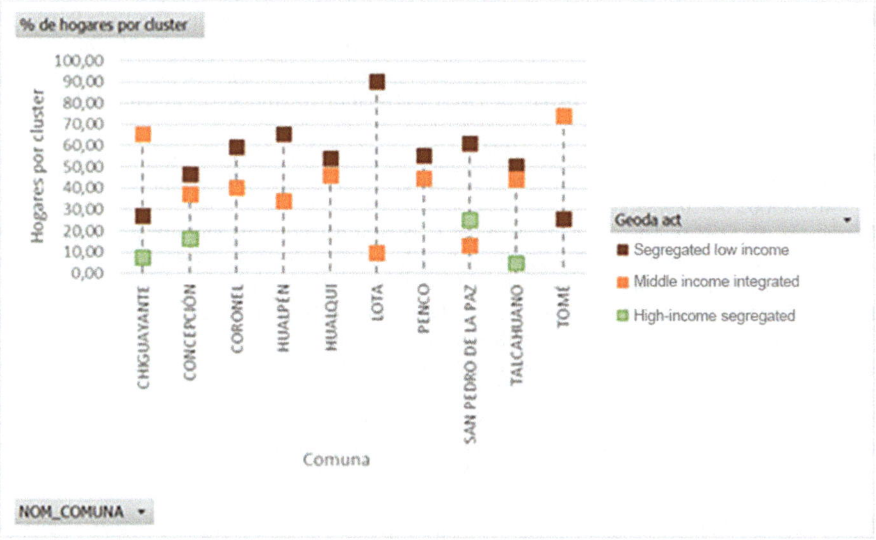

Fig. 27.4 Municipal composition according to typologies

Acknowledgements To the Center for Sustainable Development, CEDEUS, ANID/FONDAP Project 1523A0004 and the funding from ANID 2020 National Doctorate 21202161.

References

Aim. Association of Market and Public Opinion Researchers. (2019). Busting myths of the GSE: The imaginary in the geographic distribution of the ABC1a sector. https://aimchile.cl/derribando-mitos-de-los-gse-el-imaginario-en-ladistribucion-geografica-del-sector-abc1a/

Apparicio, P., Martori, J., & Fournier, E. (2014). Geo-segregation analyzer: A tool for the analysis of residential segregation. In *International conference on regional science*. University of Zaragoza. https://reunionesdeestudiosregionales.org/conferences/

Baeriswyl, S. (2009). Urban Growth of Greater Concepción, Patterns and Trends of a Metropolization. In R. Hidalgo, C. De Mattos, & F. Arenas (Ed.), *Chile: From urban to metropolitan country*. Geo Libros

Duque, J., Anselin, L., & Rey, S. (2012). The max-p-regions problem. *Journal of Regional Science, 52*, 397–419. https://doi.org/10.1111/j.1467-9787.2011.00743.x

Duque, J., Ramos, R., & Suriñach, J. (2007). Supervised regionalization methods: A survey. *International Regional Science Review, 30*, 195–220. https://doi.org/10.1177/0160017607301605

Folch, D. C., & Spielman, S. E. (2014). Identifying regions based on flexible user-defined constraints. *International Journal of Geographical Information Science, 28*, 164–184. https://doi.org/10.1080/13658816.2013.848986

Fuentes, P. (2012). Housing complexes in the metropolitan area of Concepción institutionalization of the modern city, 1939–1973. *Revista, 180*(29), 54–59. http://www.revista180.udp.cl/index.php/revista180/article/view/105

Fuentes, P., & Pérez-Bustamante, L. (2007). Worker housing in Concepción: The Villa Presidente Ríos first modern city in Chile. In M. J. Castillo, & R. Hidalgo (Eds.), *1906/2006: One hundred years of housing policy in Chile* (pp. 85–106). UNAB https://bibliotecadigital.uchile.cl/discovery/fulldisplay?vid=56UDC_INST:56UDC_INST&tab=Everything&docid=alma991002511989703936&context=L

Griffith, D. A. (2005). Effective geographic sample size in the presence of spatial autocorrelation. *Annals of the Association of American Geographers, 95*, 740–760. https://doi.org/10.1111/j.1467-8306.2005.00484.x

Hernández, H. (1983). The great Concepción: Historical development and urban structure. First part: Genesis and evolution: From military foundations to industrial conurbation. *Geographical Research, 30*, 47–70. https://doi.org/10.5354/0719-5370.1983.27676

ISMT, Índice Socio Material Territorial. (2022). Observatorio de Ciudades UC. https://ideocucocuc.hub.arcgis.com/maps/c83a1ea2c31b4850b65a481b21e4919f

Massey, D., & Denton, N. (1988). The dimensions of residential segregation. *Social Forces, 67*, 281–315. https://doi.org/10.1093/sf/67.2.281

McHarg, I. L. (1969). *Design with nature.* Natural History Press.

Medina, D. (2020). *Segregación socioespacial y tipologías de barrio para 6 comunas del área metropolitana de Concepción.* Tesis de Magíster, Universidad de Concepción. https://repositorio.udec.cl/handle/11594/5713

Mumford, L. (1966). *The city in history: Its origins, transformations and perspectives.* Pepitas de Calabaza.

Napadensky, A. (2020). An old center, for new uses? Spatial distribution of knowledge intensive services [KIS] in the Metropolitan Area of Concepción [MAC], Chile. *Architecture, City and Environment, 15*. https://doi.org/10.5821/ace.15.44.9210

OCUC. (2017). Territorial socio-material indicator. *Cities Observatory, Santiago.* https://ismtchile.mra-portal.dev/home

Parisi, D., Lichter, D. T., & Taquino, M. C. (2011). Multi-scale residential segregation: Black exceptionalism and America's changing color line. *Social Forces, 89*, 829–852. https://doi.org/10.1353/sof.2011.0013

Pérez, L., & Salinas, E. (2009). City fragments. Forms of growth of the Metropolitan Concepción. In R. Hidalgo, C. De Mattos, & F. Arenas (Eds.), *Chile: From the urban country to the Metropolitan country.* GeoLibros. https://www.academia.edu/11106129/Chile_del_pa%C3%ADs_urbano_al_pa%C3%ADs_metropolitano

Rasse, A., Robles, M. S., Cáceres Quiero, G., Trebilcock, M. P., & Sabatini, F. (2020). Segregations: Inhabiting the popular periphery in Santiago, Concepción and Talca. *Urban Territorial Logbook, 31*(1), 223–235. https://doi.org/10.15446/bitacora.v31n1.86855

Rivas, J. L. (2009). The most transparent journey. *The vision of Córdoba, Málaga and Granada from their city street* [Doctoral thesis, University of Granada]. https://digibug.ugr.es/handle/10481/19760

Ruiz-Tagle, J., & López, E. (2014). The study of residential segregation in Santiago de Chile: Critical review of some methodological and conceptual problems. *EURE, 40*, 25–48. https://doi.org/10.4067/S0250-71612014000100002

Ruiz-Tagle, J. (2016). Segregation and integration in urban sociology: Review of approaches and critical approaches for public policy. *INVI Journal, 31*(87), 9–57. https://doi.org/10.4067/S0718-83582016000200001

Sabatini, F., Cáceres, G., & Cerda, J. (2001). Residential segregation in the main Chilean cities: Trends of the last three decades and possible courses of action. *EURE, 27*, 21–42. https://doi.org/10.4067/S0250-71612001008200002

Sabatini, F., Wormald G., Sierralta, C., & Peters, P. (2010). Residential segregation in Santiago: Trends 1992–2002 and effects linked to its geographical scale. In F. Sabatini, R. Salcedo, G. Wormald, & R. Cáceres (Eds.), *Trends of segregation in the main Chilean cities: Census analysis 1982–2002.* PUC and INE.

Sabatini, F., & Wormald, G. (2013). Segregation of social housing: reduction of opportunities, loss of cohesion. In F. Sabatini, G. Wormald, & A. Rasse (Eds.), *Segregation of social housing: Eight sets in Santiago, Concepción and Talca* (pp. 12–31). UC Urban Studies Collection. https://www.researchgate.net/publication/301485647_Segregacion_de_la_vivienda_social_reduccion_de_oportunidades_perdida_de_cohesion
Sanhueza, C., & Larragaña, O. (2008). The consequences of residential segregation for the poorest. *Economic Observatory*, *19*, 1–8. https://doi.org/10.11565/oe.vi19.309
Salinas, E., & Pérez, L. (2011). Recent urban processes in the metropolitan area of Concepción: Morphological transformations and occupancy typologies. *Revista de Geografía Norte Grande*, *49*, 79–97. https://www.scielo.cl/scielo.php?script=sci_arttext&pid=S0718-34022011000200006
Schlögel, K. (2017). *In space we read time (on history of civilization and geopolitics)*. Siruela.
Smith, P., & Romero, H. (2009). Effects of urban growth of the metropolitan area of Concepción on the wetlands of Rocuant-Andalién, Los Batros and Lenga. *Revista de Geografía, Norte Grande, 43*, 81–93. https://doi.org/10.4067/S0718-34022009000200005
Truffello, R., Flores, M., Garreton, M., & Ruz, G. (2022). The importance of geographical space to minimize the error of representative samples. *North Grande Geography Journal, 160*, 81–131. https://doi.org/10.4067/S0718-34022022000100137

and international markets. The demand for cashews has a global growth of 7% per year, while the rate of production only reaches 6%. Various production and transformation processes have been established in the region thanks to community experimentation and iteration, which has attracted the attention of the public and private sector who see in it opportunities to boost the economic and social development of the region. The proposal of a project to take advantage of cashew by-products and the development of a prototype opening machine that encourages the participation of small and medium producers in the sector is an example of this. The execution of this project was carried out under a participatory approach, to recognize existing dynamics associated with cultivation practices, production and transformation. We implemented three workshops that allowed: (1) to visualize relevant actors, (2) to identify technological and human capacities throughout the entire cashew chain, and (3) to co-create models of opening machine proposals. In this chapter, we present the results of the workshops, which evidence the extensive knowledge of the territory by the local community and allow to build a more detailed picture of the factors that will later influence the dynamics of the use of the machine. At the same time, they allow to recognize both the most important actors as well as the less visible ones, thus facilitating the execution of a much more inclusive technological transfer by recognizing the need to propose solutions that contribute to the development of design and technology led circular model, which contemplates the differences present in the daily life of the different members of the community, associated with the particularities of the territory.

Keywords Participatory design · Circular design · Interdisciplinary · Cashew · Colombia

28.1 Introduction

To the east of Colombia is Vichada, the second largest department in the country -after the Amazon-, which with a surface area of 105,947 km^2 is one of the least populated, with just over 107,000 inhabitants. Its territory spans from the Eastern Plains of the foothills of the Eastern Cordillera, to the Orinoco River basin (Fig. 28.1) that surrounds the border with Venezuela. Its capital is the city of Puerto Carreño with a population close to 21,000 inhabitants (National Administrative Department of Statistics [DANE], 2023), which together with the municipalities of Cumaribo, Santa Rosalía and La Primavera make up the department. Among its economic activities stand out agriculture, fishing, livestock and mining.

O. Á. Solano · A. P. Holguín
Department of Chemical and Food Engineering, Universidad de los Andes, Bogotá, Colombia
e-mail: oalvarez@uniandes.edu.co

A. P. Holguín
e-mail: n-porras@uniandes.edu.co

Fig. 28.1 Aerial view of the Vichada department. Photo by Jessica Rodriguez

The indigenous legacy is of high importance to its culture, as around 60% of the people belong to this population group. In the language of the Sikuani, the main indigenous people living in this territory, vichada means "where the savannah becomes jungle" (National System of Cultural Information [SINIC], n.d.), highlighting a characteristic factor of the region: its natural contrast. This condition, added to the internal armed conflict that marked part of the history of the country, and the low previous social investment in science, technology and innovation projects, has generated low industrial capacities in this department (CONPES 3582, 2009), causing it to be considerably disconnected from the rest of the national territory in a land of men for men without land, (Rutas del conflicto, n.d.), limiting the development of infrastructure (Fig. 28.2), technology and trade (Salazar, 2016). Thanks to its natural wealth, the region offers a variety of foods, including cashews, whose derivatives are used for various food and industrial products. The cashew nuts from the Corpoica Mapiria Ao1, Corpoica Yopare Ao2 and Corpoica Yucao Ao3 clones (Fig. 28.3) developed at the regional research center, have outstanding characteristics due to their jumbo size and high yield under minimal resource conditions (Fig. 28.4) (Clímaco et al., 2016). This makes them highly attractive in different national and international markets, where demand has increased by 7% annually, while local production only reaches 6% (Rodríguez-Polanco & Orduz-Rodríguez, 2022).

In the process of obtaining the nut, associated with the productive chain of the cashew, by-products such as the shell (Fig. 28.5), consisting of oil and natural fibers are generated. With the appropriate transformation processes, these residues can be used to develop innovative products such as packaging and additives. They can also leverage processes related to the generation, accumulation, socialization and application of technological knowledge to improve production, post-harvest, transformation, and diversification of products, as pointed out by the Strategic Plan for Science, Technology and Agricultural Innovation—PECTIA (2017–2027).

Fig. 28.2 Residential street on the outskirts of Puerto Carreño. Photo by Jessica Rodríguez

Fig. 28.3 Cashew fruit in harvest season. Photo by Jessica Rodríguez

Therefore, it is necessary to recognize that part of the success in the execution of activities that contribute to regional economic growth depends largely on the autonomy with which they can be carried out by local communities. Not only because of the volume of agro-industrial production expected to be developed (Mateus et al., 2022), but because it is an opportunity to reduce the productive lag of the region (Departmental Government of Vichada, 2020).

It is also essential to take into account the local knowledge—empirical or formal—of the inhabitants, as they have accumulated deep knowledge through their life experiences associated with the territory (Pardo et al., 2021). In this way, we respond to Donna Haraway's (1991) call for the production of situated knowledge, where

Fig. 28.4 Cashew almond in its final state of transformation. Photo by Jessica Rodríguez

Fig. 28.5 Shell of the cashew nut after being opened. Photo by Jessica Rodríguez

the team designs and implements solutions appropriate to the particularities of the territory.

Thus, in response to the interest of the Colombian state in promoting research and development projects in different regions through budget allocation, through the 2019 call 6 of the Ministry of Science, Technology and Innovation—MinCiencias we have advanced an interdisciplinary project that seeks to strengthen the productivity of the Vichada department, developing a cashew nut opening machine to facilitate the sustainable use of: (1) agro-industrial by-products in the production of cashew and

(2) residues from the process, to expand the range of resources available in the territory. To achieve this, we implement processes that address research in laboratories, participatory design, knowledge transfer and capacity development.

This project is funded by the General System of Royalties of Colombia—SGR through the Science, Technology and Innovation Fund—FCTel It is led by the University of the Andes in alliance with the Colombian School of Engineering Julio Garavito, the Government of Vichada, and the National Learning Service—SENA.

28.2 Methodology

Our interdisciplinary approach is based on design and engineering tools. Participatory Design (Halskov & Hansen, 2015) is an approach that recognizes people associated with the situation of study as direct participants in the design process. This opens the door for particiapatory innovation (Buur and Matthews, 2008). It is based on 5 fundamental aspects, understood in this project as: (1) the inhabitants of the territory are fundamental for research and development as they are experts on the subject, (2) people impacted by a decision—such as the development of the machine for their productive activities—should have the opportunity to influence it, (3) the territory and the community are the starting point of research, (4) the tools implemented are means for the community to gain influence over design decisions, (5) the product of participation is the design of alternatives that improve the quality of life of the community.

Based on this, the project contemplates the planning, development and execution of 6 workshops that seek to directly understand the needs and characteristics of the territory and incorporate the community as part of the ideation and validation, with the aim of designing meaningful solutions for Vichada. In this chapter, we report the progress of 3 workshops carried out to date.

Similarly, in Engineering, the Systematic Approach of Pahl and Beitz (2007) is used, which recognizes a four-stage process: (1) task definition, (2) design conceptualization, (3) design tangibility, and (4) design details. Within each stage, activities to be carried out are listed, which once completed are reviewed to decide whether they are repeated, or the next phase continues. This process is carried out with the premise of generating the smallest number of loops possible, which is why the inputs obtained through the participatory design approach are of great use, as they guide this exploration from the beginning.

In this section, the workshops are presented linearly, although the analysis of each one was done independently and staggered, so the approach of each workshop is informed by learnings from the preceding workshops.

28.2.1 Community Workshop: Social Cartography

In the first workshop, we used social cartography (Navarro-Sanint, 2013) to identify relevant actors and their relationships, classify labor and social dynamics, and make visible the spaces important for the community and the project. 74 people from the region participated, during 6 h of joint work.

It was developed in three moments: (1) welcome and introduction, (2) identification of key actors, and (3) exploration of the cashew production chain in Vichada. In the first one, we welcomed the participants, socialized the project's objective, and introduced the research team. We continued with an icebreaker activity about the meaning of cashews for each participant. Some socialized their reflection and subsequently, each participant introduced themselves.

In the second, we inquired about the main actors, the productive centers, the socialization points, the supply of inputs, transportation means, and water sources. We asked to discuss in groups these points and locate them geographically on a map of the territory drawn by themselves (Fig. 28.6). Later, we delved into spaces of collective interest, other productive activities, and other actors. Each team socialized their cartographies to all those present.

In the third, we explored the cashew production chain based on the previous information. For this, we asked about the typologies of crops in the department and their relationship with the cashew harvest, and we identified the active participants in the

Fig. 28.6 Participants of the social cartography workshop mapping routes within the Vichada department. Photo by Clara Forero

production chain. To conclude, we invited the attendees to express their appreciation for the workshop discussions.

28.2.2 Technological Capabilities Workshop: Mapping the Cashew Productive Chain

For the second workshop, we mapped and specified the technological capabilities related to the chain, based on the AEIOU tool: Activities, Environments, Interactions, Objects, and Users (Hanington & Martin, 2019). We invited the attendees to pose challenges, opportunities, and solutions to the perceived problems around this, and we identified various practices in planting, harvesting, production, transformation, storage, and marketing processes. 50 people participated for 6 h of joint work.

We welcomed, presented the objectives, and continued with general instructions for the session. Later, we conducted an icebreaker activity to socialize learnings from a trip related to cashews that some participants made to Ghana, Africa. Afterward, we asked people with different roles within the production chain to introduce themselves and socialize their perspectives on this panorama in Vichada. Then we defined groups that brought together different actors from the chain.

We started the application of the AEIOU in two parts: in the first, we delivered an input (Fig. 28.7) with which we inquired about the general components of cashew production, including preparation, production, and transformation. This with the intention of bringing up knowledge about these processes that would become inputs for the second part. Recognizing the possibility that the workshop attendees did not know in depth the entirety of the cashew production chain processes, it also aligned the understanding of the members to facilitate dialogue among themselves.

For the second, we proposed a rotating participation by tables, where each one corresponded to an AEOU letter (the letter I is discarded due to a lack of relevant elements for the workshop). We arranged a resource (Fig. 28.8) with its definitions and a blank sheet with the corresponding letter. As the participants went through them, we asked them to discuss and write elements of the production chain only associated with the letter assigned to the table. They focused on A: understanding the activities, E: identifying the spaces and places, O: specifying the objects, and U: mapping the users. From their responses, we obtained a compilation of elements that we later analyzed to identify their relevance according to their frequency of appearance.

Finally, we moved on to ideation to establish challenges and opportunities associated with the above. For this, the participants chose one of the AEIOU parameters, proposed solution alternatives to problems identified in group conversations, and socialized their process (Fig. 28.9) emphasizing technological or territorial strengths and limitations of the chain.

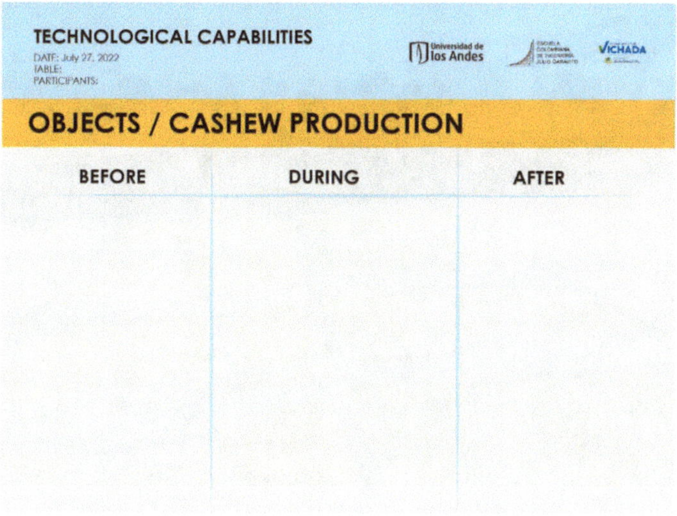

Fig. 28.7 Format for stating objects (letter O), differentiated by their use before, during and after the cashew production chain. Self-made input

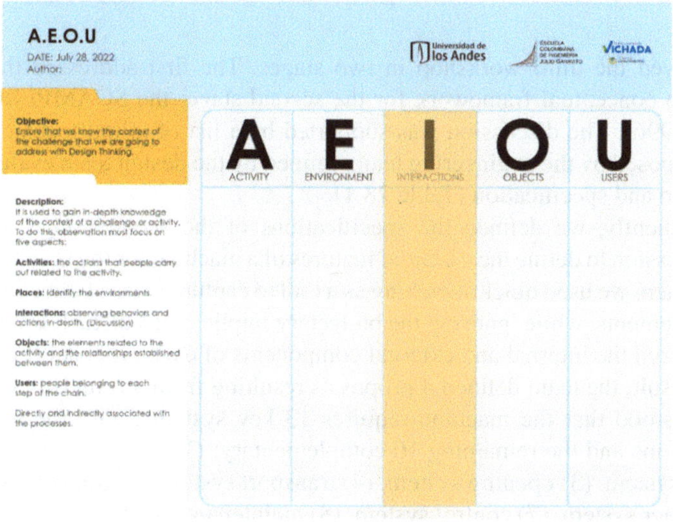

Fig. 28.8 Definitions of the AEIOU acronym present at each table. Self-made input

Fig. 28.9 Participants of the technological capabilities workshop preparing for the socialization of their discussions and conclusions. Photo by Carolina Pérez

28.2.3 SCAMPER Workshop: Model Construction

We deployed the third workshop in two stages. The first addressed the preparation of the conceptual framework for the second stage: the SCAMPER workshop (Eberle, 1996). The discussion was supported by a list of initial technical requirements proposed by the engineering team, refined by the design team evaluating their importance and specification (Table 28.1).

Subsequently, we defined the specifications of the second stage, through an ideation session to define the essential features of a machine for the transformation of the nut. There we used quick sketching as a tool to capture ideas about configurations and key elements, while ignoring the budgetary implications of each solution. Then, we socialized the internal and external components of each proposal (Fig. 28.10).

As a result, the team defined 4 proposals resulting from 4 rounds of exploration. We understood that the machine requires 13 key systems, with the first 3 being basic systems and the remaining 10 complementary: (1) energy source system, (2) structure system, (3) opening system, (4) transport system, (5) transmission system, (6) container system, (7) control system, (8) maintenance system, (9) safety system, (10) selection system, (11) separation system, (12) waste system, and (13) cleaning system.

These systems were materialized concepts by the community in the workshop of the second stage, through their representation with three-dimensional chips (Fig. 28.11) added to the SCAMPER tool (Eberle, 1996), with which we facilitated ideation based on relational conditions of Substitution, Combination, Adaptation, Modification, Proposition, Elimination, and Reorganization. The meeting brought

Table 28.1 Initial specifications for the cashew nut opening machine. Own elaboration

Item	Parameter	Type
Geometry	Nut dimensions	Requirement
	Machine dimensions	Restriction
Forces	Opening	Requirement
	Maximum load	Function
	Separation	Requirement
Energy	Electrical	Requirement
	Mechanical	Desire
Materials	Corrosion resistance	Requirement
	Harmlessness	Requirement
	Resistance to ambient temperature	Requirement
Control	On and off switch	Function
	Emergency stop	Function
Safety and ergonomics	Auditory safety (physical and chemical)	Requirement
	Ocular safety (physical and chemical)	Requirement
	Muscular safety (physical and chemical)	Requirement
Production	Opening performance	Requirement
	Separation performance	Requirement
	Operating hours	Requirement
	Operating intervals	Requirement
Quality control	% complete almonds	Requirement
	% open nuts	Requirement
	Contamination	Restriction
Assembly	Installation at SENA (Vichada)	Restriction
Transportation	Transfer from Bogotá to Vichada	Restriction
Operation	Trained personnel	Restriction
Maintenance	Trained personnel	Restriction
	Time interval	Method
	Ease of cleaning	Requirement
Construction	Supplier	Requirement
Cost	Low cost for the region	Desire
Schedule	Delivery June 2024	Restriction
Instruments and accessories	For cleaning	Requirement
	For maintenance	Requirement
	For operation	Requirement

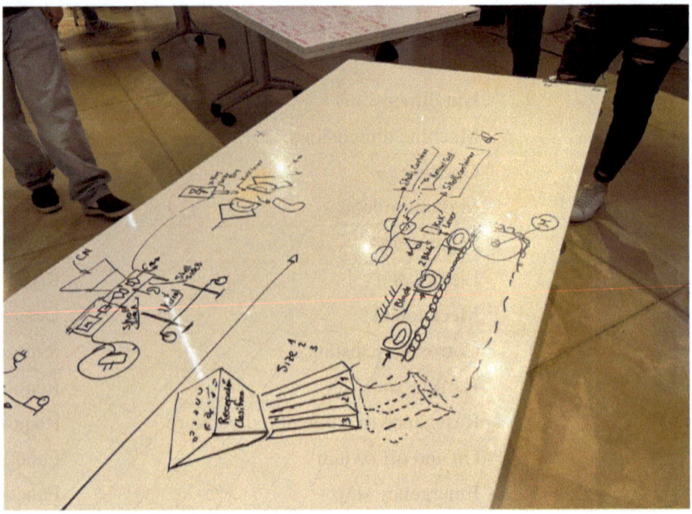

Fig. 28.10 Example of a solution proposal built in the ideation workshop between mechanical engineering and design teams. Photo by Clara Forero

together 3 groups at a time, for 2 h and 30 min, adding up to a total of 28 people distributed in 9 work teams.

We divided this workshop into three moments: in the first, we contextualized and did an icebreaker activity where each participant chose a representative chip, socialized their reasons and explained their motivations for attending. In the second we explained the rules and emphasized the safety of the space for the exploration

Fig. 28.11 Participants of the workshop exploring their representation of the basic systems using the chips. Photo by Jessica Rodriguez

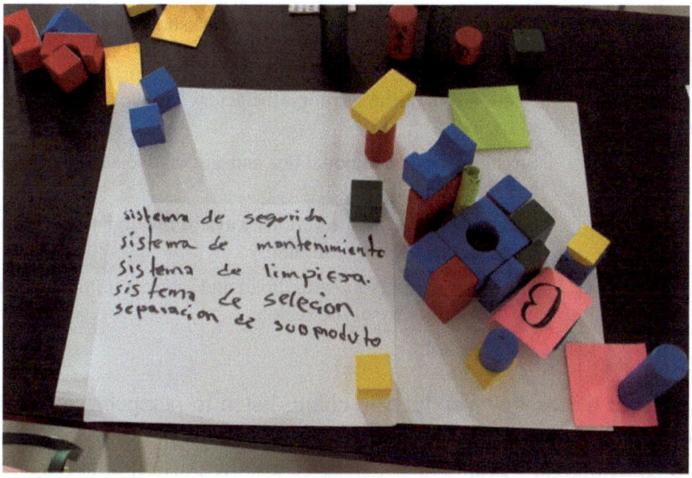

Fig. 28.12 Example of a machine proposal made by one of the participating teams. Photo by Jessica Rodriguez

of ideas and collective construction. We explained the dynamics of SCAMPER and the chips to build three-dimensional visualizations of the proposals and facilitate feedback.

The third moment consisted of developing the SCAMPER to explore different configurations of nut-transforming machines by the participants. Initially, the construction of models was limited to the use of 6 chips to position the 3 base systems. When each team consolidated a consensual configuration, we socialized the proposals. Then, they were asked to add 5 of the 10 complementary systems, without limitation of chips (Fig. 28.12), name their machine proposal and indicate its human scale. We socialized the models discussing the possible scenarios and dynamics of use of each one, taking into account the particularities of the territory.

28.3 Results

In this section, we present the results obtained from the three workshops explained above, discriminated by workshop.

28.3.1 Community Workshop Results

Being the first point of contact with the community, this workshop allowed us to make contact and introduce the project. We were able to meet people involved in

Table 28.2 Summary of results obtained in the Community workshop

Workshop No. 1	Results
Community workshop: social cartography	Recognition of the territory, the actors and the dynamics transversal to society
	Strengthening of regional ties and associative actions based on trust and mutual understanding
	Characterization of the various ways of life and activities associated with the vichadense community
	Recognition of actors with direct and indirect influence on the productive reality of the region

different roles in the cashew production chain, listen to perspectives on the factors that influence this, and make a recognition of the territory (Table 28.2).

Starting from the cartographies constructed by the participants (Fig. 28.13), the team was able to recognize the geographical characteristics of the region, such as rivers and water sources, its municipalities and population concentrations, productive centers, food crops, livestock and poultry points, tourist centers, the airport, ports and docks, access roads, the residences of the participants, and the distances between these and their workplaces. The flow of people, and the connection of activities and actors in the territory, especially associated with inputs for economic activities and everyday life interactions, was also evident.

Additionally, we obtained a general perspective of the activities that make up the life of the community, such as work schedules, traditions or leisure activities around the rivers. The impact of the lack of infrastructure was also observed, by visualizing adaptations developed in response to the intermittency of the electricity service, and the absence of a mass public transport system, within and between municipalities. Finally, actors with relevance in regional productivity, associated with food crops including cashews, were identified.

This space served as a first meeting for many people associated with the chain who did not know each other, where the workshop conversations helped the participants to recognize each other as possible allies for future productive activities.

28.3.2 Results of the Technological Capabilities Workshop

The second workshop allowed us to characterize the dynamics and activities present along the cashew production chain, enabling the understanding of local capabilities, their needs and possible solutions to current problems (Table 28.3).

From the compilation of elements obtained, we identified the most important components within the chain. In activities (A) (Fig. 28.14), the most important are the cultural tasks, recognized by the community as fertilization, fumigation, irrigation, hole digging, plowing, and liming. The inspection of the selected plant to ensure the adherence of the graft used for cultivation is also considered an intrinsic activity to

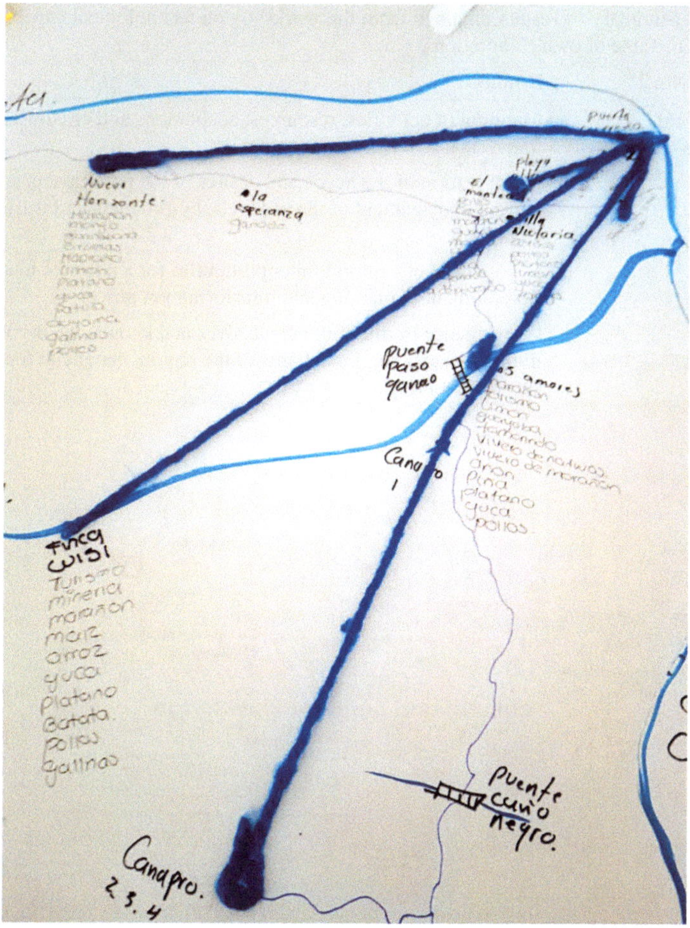

Fig. 28.13 Example of a social cartography that shows actors, resources, and connections of the activities in the territory. Photo from personal archive

these. Transportation, the selection of the land and the identification of water sources emerge as transversal and conditioning factors for all the activities.

The most relevant spaces (E) (Fig. 28.15) are the areas for liquid, solid, or organic waste. Having access routes and cleaning spaces that allow for proper handling of elements are conditioning and transversal factors of the spaces related to the chain because the processes are usually carried out in different locations. Related to this, among the objects (O) (Fig. 28.16) trucks or vans and packaging stand out, as the safety associated with the storage and transport of supplies appear as the conditioning factors.

Regarding the users (U) (Fig. 28.17), we identified *Marallanos Vichada* and *Florez Rojas* as the original ventures of the region (SMEs) with the longest history, related to the production and marketing of cashew, followed by *CANAPRO* and *Cashews of the*

Table 28.3 Summary of results obtained from the workshop on technological capabilities of the cashew chain. Table of own elaboration

Workshop No. 2	Results
Technological capabilities workshop: mapping in the cashew production chain	Mapping of activities, resources, adaptations and circumstances transversal to the cashew chain
	Identification of the needs and desires of the participants around the current management of the products, by-products and waste of the chain
	Recognition of exploration opportunities for a possible material development in planting and transformation processes
	Convergence of different perspectives and knowledge derived from multiple actors and inhabitants of the region, beneficial for the subsequent ideation processes

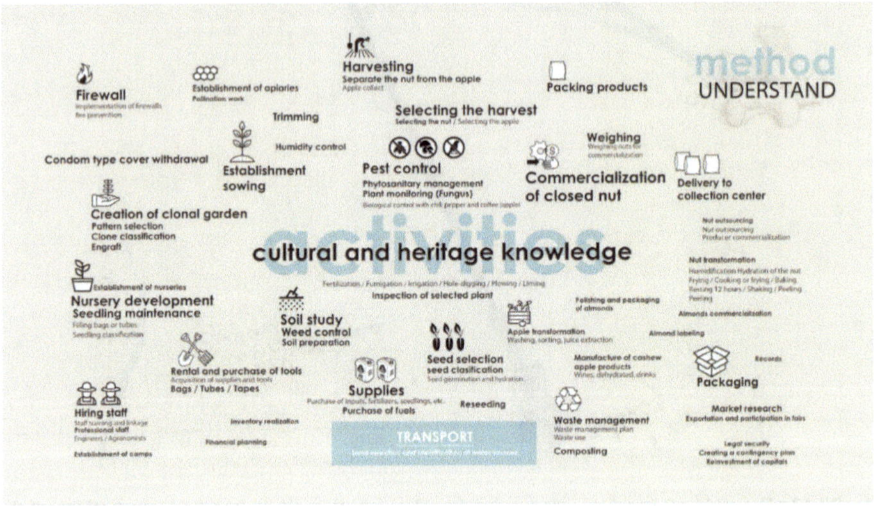

Fig. 28.14 Consolidation of the activities (A) associated with the cashew production chain. Map of own elaboration

Orinoquia (large companies). We also recognized government institutions, financial entities, technical and scientific centers, producers, and transformers. Participants perceive a complete interconnection between all the actors influencing the chain, regardless of their role within it.

We observed that the knowledge about cashew production is broad and diverse, largely composed of empirical knowledge of the inhabitants, which has led to the establishment of concepts, processes, practices, and development of products suitable for the particularities of the region. For example, the *despitonado* (Fig. 28.18) refers to the separation of the nut and the cashew apple by means of a string, with which pressure is applied around their union until the cut is achieved. It is carried out from

Fig. 28.15 Consolidation of the spaces (E) associated with the cashew production chain. Map of own elaboration

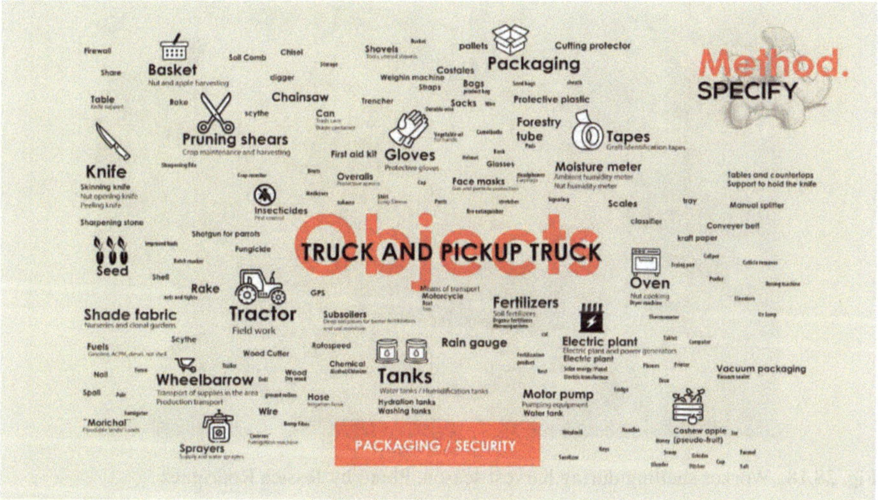

Fig. 28.16 Consolidation of the objects (O) associated with the cashew production chain. Map of own elaboration

8:00 a.m. under a tree of the crop, given the intensity of the sun in the region, after having collected between 5:00 and 7:00 a.m., a time with a lower temperature.

On the other hand, only some attendees were in phases of cashew production greater than 4 years, while the rest were in the initial planting phase. The analysis of the data shows an increasing trend in cashew planting, including large-scale plans in

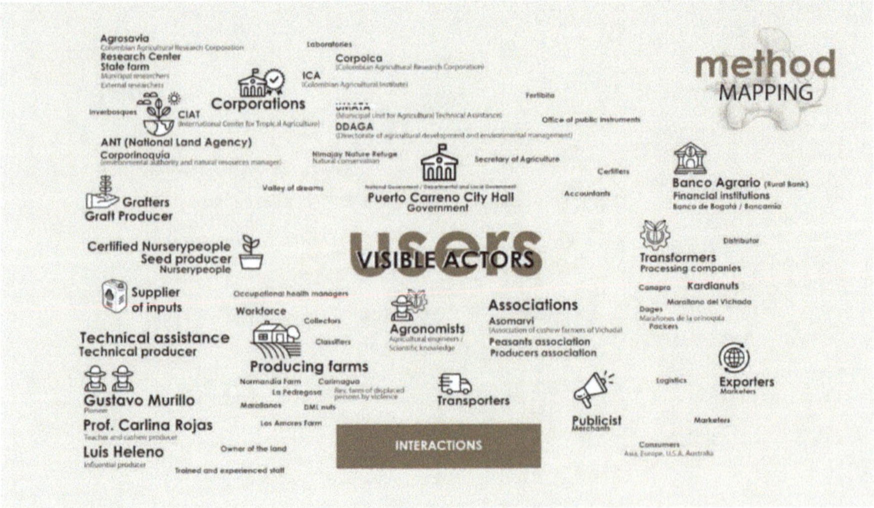

Fig. 28.17 Consolidation of the users (U) associated with the cashew production chain. Map of own elaboration

Fig. 28.18 Worker shelling during harvest season. Photo by Jessica Rodríguez

various areas of the department. Few attendees carry out the transformation processes of the nut, and a percentage of these incorporate manual processes. Most of the producers market the closed nut or must outsource their opening and peeling process of the almond. Finally, we found that few actors have the capacity to market the almond (final product).

Table 28.4 Summary of results obtained in the SCAMPER workshop

Workshop No. 3	Results
SCAMPER workshop: model construction	9 possible configurations of a machine for opening and separating cashews, derived from the situated knowledge of both the community and the research team
	Validation of the relevance of the 10 complementary systems to be considered in the construction of the machine prototypes
	Construction of guiding criteria for the definition of components, requirements, and use scenarios of the machine, according to the particularities of the region

28.3.3 SCAMPER Workshop Results

From the third workshop, three-dimensional models were obtained on possible machine configurations for opening and separating the cashew nut, and the definition of essential systems and requirements for its operation in the territory (Table 28.4).

Given the needs and expectations of the participants, the 9 models proposed by the community (Fig. 28.19) demonstrate the relevance of the systems initially proposed by the team. After analyzing the data, we developed cards (Fig. 28.20) that collect the detailed configuration characteristics for each model built, with the intention of synthesizing their approach. From the sum of this, we identified general requirements also collected in cards (Fig. 28.21). Both inputs aim to contribute to the final design of the machine.

The most relevant proposals were: (1) the power source uses solar energy, given the fluctuations that occur in the electricity service of the region; (2) the structure corresponds to a medium-scale machine, enabling its installation and use by small and medium producers, according to the spaces they have at their disposal; (3) the nut is opened by cutting or hitting by fall; (4) transport is understood as the internal transfer of the nut, and the ease of moving the machine without affecting its stability; (5) the selection system, supported by containers, addresses from the entry and exit of the nut, to the separation of waste that can be used as organic resources in other productive processes; (6) the cleaning process is simple, requires few inputs and is periodic (not daily); (7) the well-being of the operator and the safety are priorities of safety and control.

28.4 Discussion

The methodological approach was established under a participatory model that allows the development of the project, taking into account the particularities of the territory and its cashew production chain. The workshops highlighted the importance of approaching the community, as they become spaces where its members gain a voice

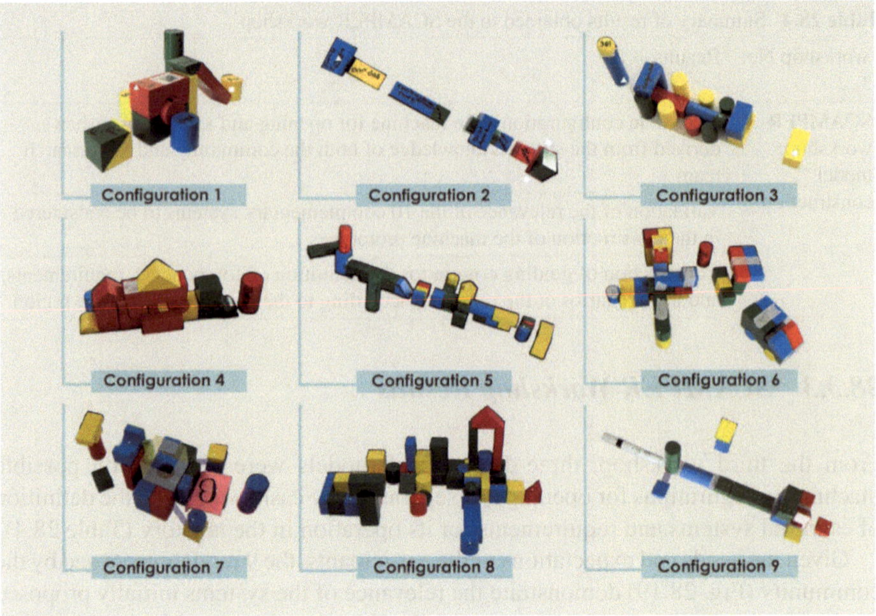

Fig. 28.19 The 9 machine configurations conceived by the teams in the workshop. Own elaboration

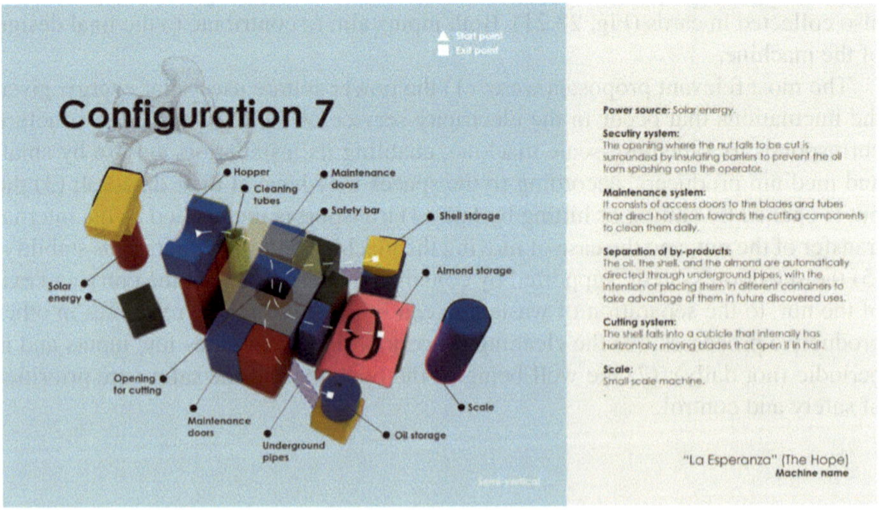

Fig. 28.20 Card of configuration number 7 named "La Esperanza". Own elaboration

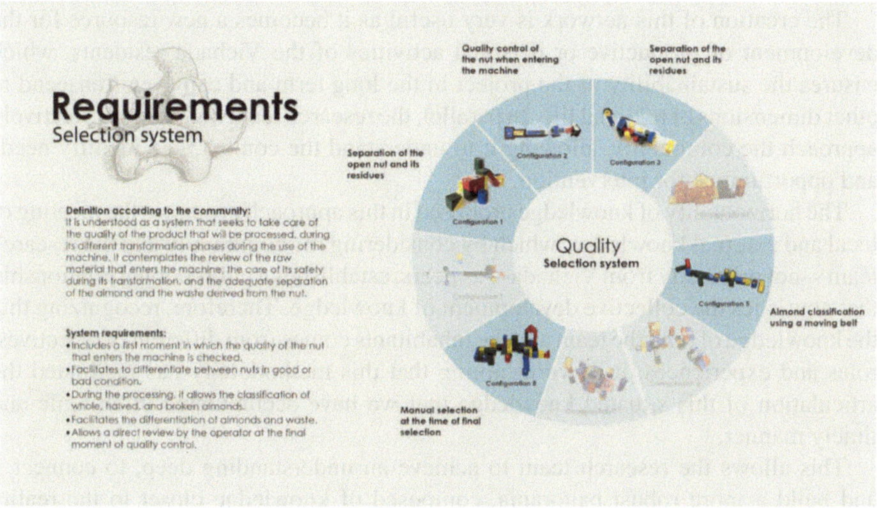

Fig. 28.21 Requirement card for the selection system from the 5 configurations that incorporate it. Own elaboration

through the recognition of their experiences and knowledge. Being a territory considerably disconnected from the rest of the country, these meetings become a source of information that reveals new data about living conditions and agro-industrial productivity, essential for the execution of a science, technology and innovation project like this.

This methodology promotes the construction of design requirements agreed between academia, community and government, which by giving the inhabitants the opportunity to participate in the process, favors their agency in favor of regional development. This is relevant given that if done in another way—one closer to the traditional process of machine development—the interdisciplinary team would not have the possibility to learn in depth from the knowledge in situ of the people. Nor could it devise with them, being a valuable opportunity that, guided by design and engineering, increases the creative capacity of the team in the search for suitable solutions.

Additionally, the execution of workshops in a segmented way, linked to the research phases, helps the community gain power over the decisions associated with each of the stages of the process. Likewise, it blurs the hierarchies among members of the same community, facilitating the creation of a proactive work environment together, in which the contribution of all participants is valuable regardless of their role within the production chain.

Likewise, considering that many of them did not know each other—a situation exacerbated by the great distances and infrastructure limitations—each workshop contributes to the strengthening of the social fabric by promoting the recognition of the inhabitants among themselves, as part of the same community.

The creation of this network is very useful as it becomes a new resource for the development of productive or personal activities of the Vichada residents, which ensures the sustainability of the project in the long term and can even transcend to other dimensions of regional life. In parallel, the research team manages to effectively approach the community, allowing it to understand the context, and identify needs and opportunities for intervention.

The horizontality of knowledge proposed in this approach promotes the meeting of local and external knowledge, which by considering the community and the research team—not originally from Vichada—as peers, establishes a bidirectional relationship that nourishes the collective development of knowledge. Therefore, recognizing that the knowledge of both the team and the inhabitants comes from different perspectives, roles and experiences, it is worth noting that this methodology has facilitated the articulation of this situated knowledge that we have accumulated in an agile and timely manner.

This allows the research team to achieve an understanding deep, to connect it and build a more robust panorama, composed of knowledge closer to the reality of the region and its processes. Therefore, it directly contributes to decision-making, guiding them towards relevant results in the face of different production and transformation scenarios that may arise in the daily life of Vichada.

In summary, the implementation of this methodology has facilitated the production of situated knowledge that can be used as a reliable input to guide high-impact decision-making in the region, by intervening in its cashew production chain. It permeates the tangible and intangible borders existing within the territory—and with the rest of the country—, representing significant benefits for the process of territorial research.

This stands out when dealing with a government-scale project that, through conversation with the community, enhanced by academia through the participatory design approach, supports bottom-up cooperation to guide top-down decision-making processes, that improve the quality of life in Vichada.

These learnings can be useful for territorial studies in other Latin American regions, where there is a need to advance projects conditioned by similar regional factors, since approaching the community allows understanding of the territory and situate knowledge while strengthening the agency of the inhabitants and the governance of the region.

28.5 Conclusions

To contribute to the strengthening of the productivity of the Vichada department, we seek to expand the available resources by developing a cashew nut opening machine that contributes to the use of its agro-industrial by-products. To achieve this, we hybridized design and engineering methodologies, based on the principles of participatory design to involve the community as part of the process. We proposed the development and execution of 6 workshops to directly understand the needs and characteristics of the territory, and design meaningful solutions for the region. In this chapter, we report the results of 3 workshops executed to date.

We genuinely understood and learned from the community, allowing us to propose and evaluate solution possibilities that respond to the particular dynamics of Vichada. We also fostered the construction of important bonds between community members, which will continue to strengthen as the workshops are developed. This is essential to contribute to the construction of an installed capacity that, at the end of the project, ensures the sustainability over time of the results achieved.

By triggering horizontal conversations between the knowledge of the research team and the community, we built a deep panorama to strengthen the relevance of development decisions, expanding our capacity for innovation and transcendence in areas of the project that go beyond the machine as a product.

In summary, the machine begins to be recognized as a leading element in the interactions associated with the cashew production chain. With the joint construction between knowledge of the community, industry, state, and academia, we seek a real, sustainable change for the department and its distinguishing characteristics.

Acknowledgements The authors thank the Ministry of Science, Technology and Innovation and the OCAD of ACTeI who carried out the feasibility, prioritization and approval of this research with resources from the General System of Royalties—SGR in the Call No. 6 of the Project "Use of Agro-industrial by-products in the production of cashew in the Department of Vichada—BPIN 2020000100571. Likewise, thanks are given to the governorship and community of the department in general for their interest and participation in the activities carried out to date.

References

Buur, J., & Matthews, B. (2008). Participatory innovation. *International Journal of Innovation Management*, 255–273. https://doi.org/10.1142/s1363919608001996

Clímaco, J., Guevara, E., Navas, A., & Arango, L. (2016). Corpoica Mapiria Ao1, Corpoica Yopare Ao2, Corpoica Yucao Ao3: Cashew clones for the flat highlands of the Colombian Orinoquía. Colombian corporation for agricultural research–Agrosavia. https://repository.agrosavia.co/handle/20.500.12324/11535

Departmental Government of Vichada. (2020). *Departmental development plan "Work for all Vichada" 2020–2023*. https://vichada.micolombiadigital.gov.co/sites/vichada/content/files/000331/16528_plan-de-desarrollo-aprobado-ordenanza.pdf

Departamento Nacional de Planeación. (2009). Documento CONPES 3582: Política nacional de competitividad y productividad. Retrieved from https://colaboracion.dnp.gov.co/CDT/Conpes/Económicos/3582.pdf

Eberle, B. (1996). *Scamper: Games for imagination development*. Prufrock Press Inc.

Halskov, K., & Hansen, N. (2015). *The diversity of participatory design research practice at PDC 2002–2012*. Science Direct. https://www.sciencedirect.com/science/article/pii/S1071581914001220#t0040

Hanington, B., & Martin, B. (2019). *Universal methods of design expanded and revised: 125 Ways to research complex problems, develop innovative ideas, and design effective solutions*. Rockport Publishers.

Haraway, D. (1991). *Science, cyborgs and women: The reinvention of nature*. Cátedra Editions.

Mateus, D., Fernández, L. A. P., Montenegro, C. J., González, C. F., Cano, F., Orduz-Rodríguez, J. O., Forero, O. I., Álvarez, S. J., Rodríguez, J. F., Pisco, Y. C., Rincón, A., & Aguilera, A. (2022). *Cashew: An approach to the productive, social, environmental and agro-industrial context in the highlands of Vichada.* https://repository.agrosavia.co/bitstream/handle/20.500.12324/37279/Ver_Documento_37279.pdf?sequence=6&isAllowed=y

National Administrative Department of Statistics—DANE. (2023). *Capital information.* https://www.dane.gov.co/files/noticias/29032023-Informacion-Capital.pdf

National System of Cultural Information—SINIC. (n.d.). *Colombia cultural—Vichada.* https://www.sinic.gov.co/SINIC/ColombiaCultural/ColCulturalBusca.aspx?AREID=3&SECID=8&IdDep=99&COLTEM=216

Navarro-Sanint, M. (2013). Social cartography for social innovation: A design approach. In *Proceedings of the XVII conference of the Iberoamerican society of digital graphics—SIGraDi: Knowledge-based design.* https://doi.org/10.5151/despro-sigradi2013-0017

Orduz-Rodríguez, J., & Rodríguez-Polanco, E. (2022). *The cashew (Anacardium occidentale L.) a crop with productive potential: Technological development and perspectives in Colombia.* https://www.scielo.sa.cr/pdf/am/v33n2/2215-3608-am-33-02-00024.pdf

Pahl, G., & Beitz, W. (2007). *Engineering design: A systematic approach.* Springer. https://doi.org/10.1007/978-1-84628-319-2

Pardo, N., Espinosa, M., González-Arango, C., Cabrera, M., Salazar, S., Archila, S., Palacios, N., Prieto, D., Camacho, R., & Parra-Agudelo, L. (2021). *Worlding resilience in the Doña Juana Volcano-Páramo, Northern Andes (Colombia): A transdisciplinary view.* https://doi.org/10.1007/s11069-021-04662-4

Rutas del Conflicto. (n.d.). Vichada: Tierra de hombres sin tierra. Retrieved June 17, 2024, from https://rutasdelconflicto.com/especiales/vichada/tierra_hombres_sin_tierra.html

Salazar, G. S. (2016). El conflicto armado y su incidencia en la configuración territorial: Reflexiones desde la ciencia geográfica para la consolidación de un periodo de pos negociación del conflicto armado en Colombia. Bitácora Urbano Territorial, 26(2), 45–57. https://doi.org/10.15446/bitacora.v26n2.57605

Part VII
Critical Methodology, Project and Didactics of Research in Architecture, Design, City and Territory

Chapter 29
Methodologies for Co-constructing Narratives Aimed at Understanding Resilience and Adaptation Processes of Urban Situated Nature

Alfo Puppo Stuardo

Abstract The phenomenon of unregulated urbanization is the central axis of social and ecological transformations to coexist with ecosystems worldwide. In Latin America, there is a challenge to understand these impacts from everyday life, as there are various initiatives and resistances when interaction with nature has been threatened. This document proposes Narrative Productions as a novel research technique to identify sociocultural valuations towards urban nature and thus know the thoughts and actions that are strained by the impacts of urbanization in a municipality located in southern Chile. The chapter indicates the theoretical aspects of the research project to situate social and ecological problems of urbanization in everyday life; describes the origin and procedure of the Narrative Productions technique and presents the main results to understand the sociocultural valuations of nature situated in the municipality of Chiguayante, which in this case express the loss of native forests due to business logics that promote the growth of the city under particular interests and the need of the inhabitants to recover ancestral wisdom related to food supply.

Keywords Sociocultural valuations · Narrative productions · Situated urban political ecology · Situated urban nature

---

A. P. Stuardo (✉)
Facultad de Arquitectura y Urbanismo, Universidad de Chile, Santiago, Chile
e-mail: alfo.puppo@ug.uchile.cl

M. Vicuña and E. Giorgi (eds.), *Intersections*, Springer Geography,
https://doi.org/10.1007/978-3-031-76402-8_29

29.1 Introduction

29.1.1 Sociocultural Valuations: Thoughts and Attitudes When Our Interaction with Nature Is Threatened

The phenomenon of unregulated urbanization is the central axis of social and ecological transformations of ecosystems worldwide (Forman, 2014; Heynen, 2006). In Latin America, this scenario is promoted by a neoliberal urban model (Harvey, 1996; Vásquez et al., 2017), which is expressed in the (de)regulated planning of the territory to meet the interests of land use value (Tutor, 2021). This scenario implies that everyday life in the city is marked by differentiated experiences of privilege and exclusion, including our possibilities to have a link with nature (Quimbayo & Vásquez, 2016).

In this scenario, there are insurgent and small-scale strategies to interact with nature on the Latin American continent, such as the creation of community gardens, urban orchards, the appropriation of green spaces (Inostroza et al., 2020) even through daily rituals aimed at protecting and bringing good luck (Castro, 2021) (Fig. 29.1). From a scientific and ecological perspective, these actions derive from a series of concerns expressed through language, which has been conceptualized as sociocultural valuations towards ecosystems (VSC) because they are knowledge systems and worldviews aimed at establishing a relationship between human beings and nature, or human and more than human worlds (Chan et al., 2016; Kenter et al., 2019; Muraca, 2016).

The VSC represents those conflicts and resistances that emerge when there are threats to interact with nature (Kenter et al., 2016; Muraca, 2016). In Latin American cities, there is a lack of scientific research dedicated to understanding this concept even when these thoughts and actions allow us to understand experiences linked to the territory and possible solutions to address these dynamics (Chan et al., 2016). The study of the VSC for the urban ecosystems of the world has been predominantly studied, however, they rely on the interpretation of scientific tradition to report them without guaranteeing the protagonism of those threatened by these social and ecological transformations (Kenter et al., 2019). The purpose of this chapter is to present a novel research technique to understand the sociocultural valuations between inhabitants and nature in contexts of accelerated urbanization, through collaborative work that allows the protagonists who experience these transformations to write their own history and the resistance actions they articulate to face these challenges.

29.1.2 Objective

The following chapter is related to ongoing research that aims to understand the sociocultural, institutional, and biophysical evaluation of urban nature from the point of view of everyday life, in cities located in the south of the Chilean Mediterranean.

Community orchard initiated in 2019 during the Estallido Social. City of Valparaíso, Chile.

Use of Ruda (Ruda graveolens) branches. Plant used for protection. Downtown Lima, Perú in 2018.

"Pacas" self-managed initiative. Compost system in 2023 in Parkway. City of Bogotá, Colombia

Sale herbs for health in 2023. Cali City Market, Colombia

Fig. 29.1 Photographs of social interaction with urban nature in Latin American cities (author Alfonsina Puppo)

This chapter exposes the contribution of the Narrative Productions research technique to understand the first dimension of the indicated research, associated with identifying and understanding the sociocultural valuations of urban nature from the perspective of inhabitants in the city of Chiguayante in southern Chile.

In this way, the organization of the chapter is as follows: (i) the theoretical aspects of the research project are presented; (ii) the origin and procedure to apply the Narrative Productions technique; and finally, iii. the study area is described, and the main results are exposed to understand the sociocultural valuations of nature expressed by a community that is part of this research project.

2 Situated Urban Nature: Theorizing the Particularities of the Territory

In the Chilean context, the increase in impermeable coverings as part of urban infrastructure has fragmented urban ecosystems (De la Barrera et al., 2015; Hernández-Moreno & Reyes-Paecke, 2018), promoting unequal socio-ecological landscapes through protected ecosystems in sectors with high added value, to the detriment of vulnerable and popular sectors (Banzhaf et al., 2018; Vásquez et al., 2017). From a Latin American perspective, these problems impact the lives of urban inhabitants with reactions and expressions that exceed the explanatory capacity of urban theory developed by schools of thought located in the global north, because these conceptual frameworks were not created to provide solutions to these particular contexts (Lawhon et al., 2014; Roy, 2009). For these reasons, the development of the subdiscipline Situated Urban Political Ecology has begun (Quimbayo & Vásquez, 2016), an approach that seeks to decentralize those imaginaries that are rooted in stereotypes introduced by hemispheres alien to their context, a matter that is possible from the recognition of the material and particular history of the various human and more than human worlds that are linked to places and their ways of life (Chakrabarty, 2000).

These antecedents seek to make visible the multiple struggles for urban nature, broadening the debate on how to value it, and transcending the traditional recognition that characterizes studies on these problems (Muraca, 2016; Quimbayo, 2019). From an axiological and situated point of view, there are a series of thoughts, intentions, and attitudes to establish a direct interaction with nature, which surpass the influence of modern thought based on the dialectical imaginary that culturally separates us from it (Loncón, 2022) and that are susceptible to theorizing relevant frameworks that analyze the network of relationships that is created in these contexts (Lawhon et al., 2014). In this way, the interaction towards urban ecosystems on the continent implies understanding the multiple visions that express their relationship with the territory identifying the pluralism of values linked to ancestral traditions that remain in force (Chan et al., 2016; Kenter et al., 2019), but also of the mestizo views that have been gestated with the influence of the hegemonic knowledge of modernity (Rivera Cusicanqui, 2018).

29.2 Methodology

29.2.1 Co-constructing Narratives to Understand the Relationship with Urban Nature from Its Protagonists

The research technique Narrative Productions (NP) is presented below, aimed at recovering experiences of oppression from the procedure of telling and writing stories. This technique comes from feminist epistemologies and critical social constructivism (Galaz & Álvarez, 2020). It is characterized by developing a collaborative work with the subjects immersed in the research problem and its main objective is to theorize experiences so that the written product can be cited as a new production of scientific knowledge (Balasch & Montenegro, 2003). It mainly differs from interviews, focus groups and other techniques associated with the interpretation of semantics because: (i) it does not involve an analytical coding phase to report the data obtained; and because (ii) the written result is conditioned to the constant validation of its interlocutors to make visible and denounce such experiences of oppression.

Stages for the application of the technique

The NP are developed through work workshops with the participants of the study consecutively. There are 4 main stages to develop the technique (Fig. 29.2):

i. Step 1. Project presentation: The needs and positions that the participants will take when producing the dialogue sessions (negotiation) are considered, with the aim of understanding the dimensions and components of the story to be produced.
ii. Step 2. Meetings: In the case of working with territorial problems, the strategy of establishing a question that guides all the agreed dimensions of analysis is considered. This question is reiterated to open each session and organize the dialogue. The criterion of proposing a single question throughout the entire narrative and writing process is established for two reasons: (i) so that the researcher has the least possible intervention in the dialogic process, trying to be a mediator of the dialogues; and (ii) to carry out an iteration process that allows guiding the conversations without dispersion, but at the same time flexibly. In this way, the consideration of new information that allows enriching the understanding of the territorial processes linked to the narratives is sought.
iii. Step 3. Textualization: a phase that begins with the drafting of the first corpus of dialogues. This writing emerges from the first session, transcribing the emerging voices. In this way, the document prepared by the researcher is taken to the second session to be read by its interlocutors with the purpose of analyzing, enriching and editing the text. The number of sessions to carry out this process is subject to the needs of its participants, who decide when to finish the theoretical corpus (Galaz & Álvarez, 2020; Galaz & Rubilar, 2019). The process

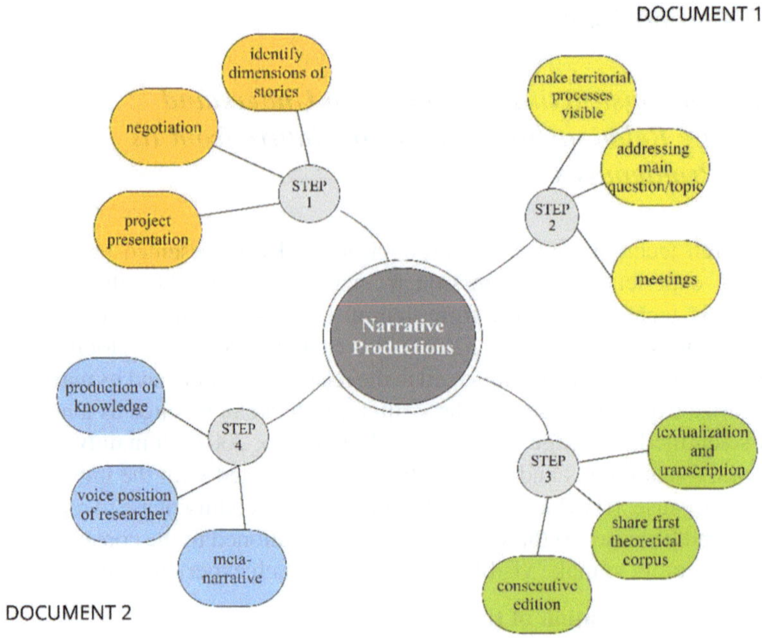

Fig. 29.2 Steps and procedure of narrative productions technique. *Source* The author (2023)

concludes with the signature and validation of the participants involved in the final document, a procedure valid as an ethical consent document.

iv. Step 4. Metanarrative: is the last phase of the technique, whose objective is to position the voice of the researcher (Balasch & Montenegro, 2003). This is a document written by the researcher, independent of the texts created by its protagonists to establish the analysis of the created narratives or establish a own writing that centers its interest in the personal intentions to carry out the investigative process (Galaz & Rubilar, 2019). In this case, the dimensions linked to the sociocultural valuations of urban nature in the context of accelerated urbanization are interpreted.

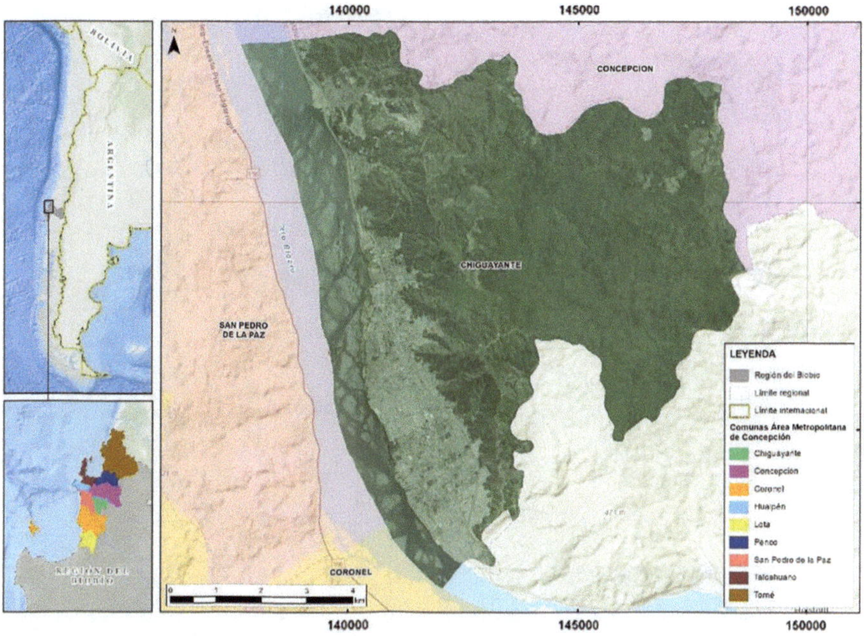

Fig. 29.3 Chiguayante map, located in the Metropolitan Area of Concepción, Biobío Region, Chile. *Source* The author (2023)

29.3 Study Area: Chiguayante, Bedroom City in Metropolitan Area of Concepción

The research is located in the commune of Chiguayante (Fig. 29.3), an administrative area founded in 1996 in the Bío Bío Region in the south of Chile. This municipality belongs to the Metropolitan Area of Concepción[1] and is considered a bedroom city due to the rapid urbanization of its agricultural lands (Campos Petron, 2019) (Fig. 29.2). It is located in an urban ecosystem characterized by its biodiverse landscape in bodies of water, hills, and forests, and its name derives from Mapuzungun[2] as *Chiguay Antu,* which means "sun among the fog" (Espinoza, 2018). This name probably owes to the location of this settlement, in the middle of an elongated valley where the Biobío River is on one side, and on the other, the Foot of the Manquivámiva Hill.

To identify the study case, it is established as a selection criterion than the green space recives maintenance from its residents. Recinto Caupolicán Square (Fig. 29.4) is located in Altos de Chiguayante sector and belongs to a housing complex granted

---

[1] The Metropolitan Area of Concepción is located in the Bíobío Region, center-south of the country (36.82° S, 73.05° W). It is composed of 11 municipalities, among them Chiguayante (Fig. 29.3), with an area of 71.5 km^2 and an approximate population of 91,805 inhabitants (BCN, 2023).

[2] Language belonging to the Wallmapu territory where the ancestral Mapuche peoples live.

Fig. 29.4 Location of Recinto Caupolicán Square, Chiguayante. *Source* The author (2023)

to the working class of the Caupolicán Textiles industry in the 80s'. At that time the Square was a bare slab without vegetation, so its inhabitants decided to plant trees and organize to maintain the square.

The textile industry is no longer operational and part of its infrastructure is used by the municipal administration, which has boosted the economic valuation of the sector. This situation concerns a group of inhabitants who belong to the Recinto Caupolicán Neighborhood Association No. 12, who are interested in the research project and decided to construct a narrative aimed at making visible their concerns and attitudes to interact with the urban nature that surrounds them.

29.4 Results

29.4.1 Values of Territorial Memory and Ancestral Wisdom

The group of neighbors belonging to the Recinto Caupolicán Neighborhood Association No. 12, work in 6 workshop sessions during the spring of 2022 on the writing titled "Between scent of Boldos and Poets of Recinto Caupolicán". The complete text is available in a written and audiovisual format to make visible the constructed narratives.[3] The components of the story that were agreed (step 1 of technique) were 3: i. narrate stories linked to ecological environment around the neighborhood; ii. describe neighbordhood transformations over the last three decades; and iii. explain neighbors strategies to confront these transformations.

As its protagonists indicate,

> The rural life that characterized Chiguayante along with its industries, are the memories we seek to rescue. We live in what was once called Caupolicán Enclosure, a place surrounded by nature. There were forests of Litre—which we carefully did not climb to avoid their allergies—, Quillay, Acacias and Boldos, which refreshed the air with their intense aroma. We had Palqui, Maquis, Apple trees, Raulí and also Oaks of different ages, like the Hualle and the Pellín. (…) This is our sector; this is where we belong. (…) That was the lung we had. (Neighbors Recinto Caupolicán No. 12, p. 1)

The beginning of the narrative mentions the transition from everyday life linked to peasant tradition, through the disappearance of the native tree forests that surrounded the housing sector. In addition, the names of the species with which a direct interaction was established for recreation and breathing are indicated, identifying in the case of the Oak (Nothofagus *obliqua*) a different name according to its age. This difference has a direct relationship with the Mapuzugún language used by the ancestral Mapuche communities where the young Oak is distinguished by the name of Hualle, and when it ages in Pellín.

As indicated at the beginning of the story, the purpose is to recover the memory linked to the sector where the inhabitants of Recinto Caupolicán live. From this purpose, the narrative adds

> We are proud of our hill, which we remember with an abundance of native trees before it was filled with Pines. That already tells us who the owners of the hill are. However, we do not forget the various species that the Manquimávida houses: its Copihues, Digüeñes, Nalcas and Hazelnuts. Species with which we fed ourselves (Neighbors Recinto Caupolicán No. 12, p. 1).

For Recinto Caupolicán No.12, neighbors Manquimávida Hill is an element associated with their territorial identity, which they express with pride. The relationship with this hill is formed by their coexistence and by their valuation associated with the provision of food. However, the narratives about the hill are also in tension, expressing the disturbances exerted by the business sector towards wildlife, which

---

[3] https://www.ecosistemasurbanos.cl/post/historias-situadas-para-conocer-el-valor-de-la-naturaleza-urbana-en-dos-comunas-del-gran-concepci%C3%B3n.

has implied the disappearance of various species of native trees. In this way, the inhabitants indicate that the main role of the forestry companies in the sector has been the privatization of the hill for the plantation of pine monocultures (Pinus *radiata*).

The disappearance of species from the native forest is expressed in this narrative through a critical perspective in relation to the process of urbanization of the commune, indicating:

These forests began to disappear 30 years ago, through the purchase of sites by private companies, with the aim of building new homes. Little by little the trees began to be cut down and little by little SERVIU also arrived urbanizing and walling the sector. (…) They began to cut down the trees at night, because we opposed resistance (…) Thus began to disappear all the green area that was here (Neighbors Recinto Caupolicán No. 12, p. 1).

The impact of urbanization is not only perceived in the surroundings through the Manquimávida hill but also on a neighborhood scale through the described real estate interests, indicating their actors and mode of operation to carry out housing projects. In this way, the participation of real estate companies in conjunction with the public service that executes the political guidelines of the Ministry of Housing and Urbanism (SERVIU) to carry out these initiatives is highlighted. The replacement of trees and forests by impermeable coverings is a situation that the inhabitants reject with resistance actions, therefore public and private actors coordinate to execute their interests at specific times and thus avoid conflicts with the population.

On the other hand, the writing constructed by inhabitants of Recinto Caupolicán No. 12 seeks to delve into their concerns as witnesses to human intervention towards the main biophysical elements that characterize the city's ecosystem.

> We thought Chiguayante wouldn't grow so much. In its beginnings, the commune had 15,000 inhabitants, today we have already surpassed 90,000. We have seen how the river has been paved, and we already have news about the urbanization of Cerro Manquimávida with new homes. (Neighbors Recinto Caupolicán No. 12, p. 2)

This fragment indicates how the inhabitants have observed the accelerated growth of the city in its 28 years of history throughout their lives. They mention the constructions on the edge of the Bíobío River and how the urbanization of Cerro Manquimávida is projected, narratives that describe situations of a past, a present, and a future connected by the process of urbanization. Faced with this scenario, the inhabitants of Recinto Caupolicán maintain:

> We already know that this country belongs to 12 or 15 families. They are cutting down the trees of the territory to place their interests, whether they are housing, forestry, or for crops that are in fashion. We recognize their *modus operandi*: buying land to plant pines, because they grow quickly. Buying sites to urbanize and build more homes, to earn more money for their pockets. They are the ones responsible for changing the cycles of the earth, and that is why we must organize to face the changes of our nature. (Neighbors Recinto Caupolicán No. 12, p. 12)

The narrative generally associates the relationship between the territory, power, and business sectors associated with the country's elite group. In addition, the inhabitants describe these relationships through a dynamic process that operates through the logic of capital accumulation to promote the increase of interest. They maintain that these relationships are responsible for the social and ecological transformations of Chiguayante as a territory, which in their words they describe as their own place through the phrase "our nature". This sense of belonging allows them to position themselves from the neighborhood organization sphere and express their motivations to face these problems.

The writing "Between scent of Boldos and Poets of Recinto Caupolicán" adds a final part, indicating the alternatives they seek as an organization in this scenario,

> Our previous generations taught us to cultivate the land, to know how to feed ourselves from nature. But that has been lost. With the arrival of supermarkets, that fruit and vegetable we buy, we feel it full of chemicals and without its own flavors. That's why we have wanted to plant our food, recover those knowledges that have been forgotten over time, through activities that are developed around a community garden (Neighbors of Caupolicán Precinct No. 12, p. 13).

The main strategy of this organized group is to recover and enhance the knowledge that was developed in Chiguayante prior to the social and ecological transformations that have stimulated the urban consolidation of the territory. The processes of urbanization have promoted the arrival of new forms and services of supply, changing the logic of food provision through practices of flower, mushroom, and fruit collection in Cerro Manquimávida, or through the agricultural fields of the sector that have currently been transformed into new housing complexes. The new interaction with food through supermarkets is problematized in the narrative questioning the origin of the products and their treatment with fertilizers. In this way, the inhabitants make this writing visible to their community, with specific purposes,

> All people who want to seek another way of relating to nature, with food, and who are willing to work with their community are invited. We exchange plants, knowledge, and we work with our hands on the earth. (…) With this story, we want to transmit how we perceive the changes in our commune, and let them know from our experiences, what we do to stop the environmental crisis that affects us all (Neighbors Recinto Caupolicán No. 12, p. 14).

The final section of this narrative seeks to be a space of openness towards the community to establish solutions to the problems of the environmental crisis perceived in the commune of Chiguayante, through the written work of this story and the development of workshops on urban gardens for supply and recovery of ancestral knowledge (Fig. 29.5). In this way, the dialogue is established that has allowed them to exchange stories associated with the territory and its drastic transformations.

29.5 Conclusion

The following chapter is focused on understanding the sociocultural valuations of nature in contexts of accelerated urbanization according to the expression of inhabitants who are impacted by these contexts in their daily lives. In this way, this chapter exposes the results associated with the social dimension of ongoing research,

Route to identify species planted by the community in the Plazoleta Recinto Caupolicán since the 1980s.

Textualization stage of the Narrative Productions research technique during the year 2022.

Urban orchards workshop for the reclaiming of ancestral knowledge in the Plazoleta Recinto Caupolicán in the spring of the year 2022.

Narrative validation entitled "Entre aromas de Boldos y Poetas del Recinto Caupolicán" created in 2022 by the group of inhabitants of the Junta de Vecinos Recinto Caupolicán n°12.

Fig. 29.5 Photographs of the work process during the co-production of the narrative created by the inhabitants of Recinto Caupolicán (author Alfo Puppo)

proposing a novel research technique to recognize the axiological system of human thoughts and attitudes towards nature when it is threatened.

To identify and understand the sociocultural dimension of the sociocultural valuations of urban nature, this proposal works with the Narrative Productions technique, with the aim of promoting the expression and visibility of writings elaborated from the experience and voice of its protagonists. In this case, the exposed writing expresses the loss of wildlife from the surrounding native forests due to the development of the real estate and forestry sector. In addition, the inhabitants indicate that these transformations are the consequences of the impact of particular economic interests to preserve the privileges of the elites linked to Chilean business.

The social and cultural valuations of the inhabitants of Recinto Caupolicán express an impact on their relationship with Cerro Manquimávida and the forms of appropriation they maintain with the territory for recreation and food supply. The interaction towards nature is expressed with the relationship of native trees of the territory, especially the Oak species is highlighted by association with the Mapuzugún language to express its life stage. In this way, the inhabitants resist the construction processes of new housing complexes and establish organization mechanisms to recover the knowledge linked to food production that characterized the history of Chiguayante before its rapid urbanization began.

29.6 Discussion

The possibility of re-thinking how to theorize the relationship with urban nature in contexts located in Latin America implies understanding that we live in combined territories; expressions intertwined by the inherited knowledge of the territory and the neoliberal urban model that predominates in the planning of cities around the world. In this case, the city project oriented towards its growth and its own expectations of the sociocultural legacy of Recinto Caupolicán mix through dynamic processes that transform the social and ecological relationship of its inhabitants. These combinations are expressed in attitudes linked to power, where the coordination of private companies and state actors impose projects to favor the growth of the city without the consent of its inhabitants who react with various strategies to not lose their bond with nature.

Chiguayante is in a transition process oriented towards urban growth in an imposing way before the historical dynamics linked to agricultural daily life, based on direct interaction with native forests for recreation and food search. These values are expressed through a mestizo language between Spanish and Mapuzugún, Mapuche language, to indicate that the Oak, a sacred tree for this ancestral community, has disappeared due to the dynamics of urbanization, altering the system of knowledge related to nature. The exposed narrative recognizes this imposition through specific practices that establish the control and dominance of the interests of the business sector, the main promoters of the neoliberal urbanization process. Therefore, everyday life in these contexts becomes a place of dispute and resistance.

The novel research technique of Narrative Productions applied to territorial contexts allows us to recognize those problems and prefigurative initiatives (the indissoluble link between thought and action) that the inhabitants of the Caupolicán Enclosure establish to face the impacts of the growth of the city they inhabit. In this way, they propose as a self-managed solution the recovery of knowledge linked to the land through gardens, which as they indicate, critically responds to the arrival of new supply services for the city that promote a lifestyle that disconnects them from those foods that have historically been linked to the territory. Therefore, applying this research technique can also be an opportunity to link particular experiences with the urban ecosystem in multiscale terms, where the perception and understanding of the body connects with the territorial dynamics through food; and the neighborhood from a microscale connects with the Manquimávida Hill through its valuation and resistance to the loss of native forests.

The relationship with urban nature in contexts of accelerated urbanization is an opportunity to understand that there is a historical imposition that is replicated through the neoliberal model but that impacts territories in a situated way depending on their social and ecological valuation. There is a multiple and differentiated set of thoughts and attitudes to interact directly with ecosystems that have been scarcely observed and researched because the scientific tradition observes the relationship between human beings and urban nature from generic conceptual frameworks that primarily identify the instrumental benefits of nature to people without delving into the ancestral and mestizo interaction that has been forged through the processes of urbanization in Latin American contexts.

Understanding the concept of sociocultural valuations through collaborative research methodologies such as Narrative Productions, allows us to raise new epistemic categories by decentralizing the problems and solutions that traditional urban theory develops. Therefore, deepening the understanding of situated natures can be a possibility to identify how social and ecological discomfort and well-being are perceived in the accelerated processes of urbanization and thus promote concrete and relevant solutions to their context.

Acknowledgements This document has been made possible thanks to the efforts of the INTERSECTIONS Congress, the National Doctorate Scholarship 21200462 and the support of the FONDECYT Project 1231859; research funded by the Chilean National Research Agency ANID.

References

Balasch, M., & Montenegro, M. (2003). A methodological proposal from the epistemology of situated knowledge: Narrative productions. *Encounters in Social Psychology*, *1*(3), 44–48. https://www.academia.edu/762651/Una_propuesta_metodol%C3%B3gica_desde_la_epistemolog%C3%ADa_de_los_conocimientos_situados_Las_producciones_narrativas

Banzhaf, E., Reyes-Paecke S., & De la Barrera, F. (2018). What really matters in green infrastructure for the urban quality of life? Santiago de Chile as a showcase city. In S. Kabisch et al. (Eds.), *Urban transformations. Future cities* (Vol. 10, pp 281–300). Springer, Cham. https://doi.org/10.1007/978-3-319-59324-1

Campos Petron, C. (2019). Crisis in latency: Landscape configuration as urban-natural transition: reforestation project and containment against risks of landslides and floods in Slopes and Ravines in the Chigüayante sector [Thesis to opt for the title of Architect and Master in Landscape Architecture, Pontifical Catholic University of Chile]. https://doi.org/10.7764/tesisUC/ARQ/26871

Castro, G. (2021). Introductory manual of plants for tranquillity (1st ed.). Fanzinoteca Yerbatera.

Chakrabarty, D. (2000). *Provincializing Europe: Postcolonial thought and historical difference* (p. 31). Princeton University Press.

Chan, K., Balvanera, P., Benessaiah, K., Chapman, M., Diaz, S., Gomez-Baggethun, E., & Luck, G. (2016). Opinion: Why protect nature? Rethinking values and the environment. *Proceeding of the National Academy of Sciences of the United States of America, 113*(6), 1462–1465. https://doi.org/10.1073/pnas.1525002113

De la Barrera, F., Bachmann, A., & Tironi, E. (2015) The research of ecosystem services in Chile: A systematic review. *Geographical Research, 50*, 3–18. https://ultimadecada.uchile.cl/index.php/IG/article/download/41171/42708/

Espinoza, L. (2018) The gold route in the old frontier of Bío Bío. *Editions of the Historical Archive of Concepción*. MMXVIII. https://www.archivohistoricoconcepcion.cl/wp-content/uploads/2022/01/la_ruta_del_oro.pdf

Forman, R. (2014). *Urban ecology*. Cambridge University Press.

Galaz, C., & Rubilar, M. (2019). Professional experiences in psychosocial intervention: the narrative exercise as a methodology of reflexivity and epistemological vigilance. *Latin American Journal of Methodology of the Social Sciences (Relmecs), 9*(1). http://sedici.unlp.edu.ar/handle/10915/78048

Galaz, C., & Álvarez, C. (2020). Feminist Qualitative Analysis: a critical proposal from narrative productions. In K. Duarte (Ed.), *Separate to build. Qualitative analysis of information, Social Editions*. University of Chile.

Harvey, D. (1996). *Justice, nature and geography of difference*. Dream Traffickers. https://www.traficantes.net/sites/default/files/pdfs/PC18_Harvey_web.pdf

Hernández-Moreno, A., & Reyes-Paecke, S. (2018). The effects of urban expansion on green infrastructure along an extended latitudinal gradient (23°S–45°S) in Chile over the last thirty years. *Land Use Policy, 79*, 725–733. https://doi.org/10.1016/j.landusepol.2018.09.008

Heynen, N. (2006). Green urban political ecologies: Toward a better understanding of inner-city environmental change. *Environmental and Planning, 48*, 499. https://doi.org/10.1068/a37365

Inostroza, L., Garay, H., & Andrade, G. (2020). *Ecosystem services in Latin America. Opportunities for sustainable urban development, climate action and biodiversity management*. Center for Sustainable Development Goals for Latin America and the Caribbean CODS (No. 4). https://cods.uniandes.edu.co/wpcontent/uploads/2020/01/Documento_CODS_ecosistemicos_enero20-1.pdf

Kenter, J., Bryce, R., Christie, M., Cooper, N., Hockley, N., Irvine, K. N., Fazey, I., O'Brien, L., Orchard-Webb, J., Ravenscroft, N., Raymond, C. M., Reed, M. S., Tett, P., & Watson, V. (2016). Shared values and deliberative valuation: Future directions. *Ecosystem Services, 21*, 358–371. https://doi.org/10.1016/j.ecoser.2016.10.006

Kenter, J., Raymond, C., van Riper, C., Azzopardi, E., Brear, M., Calcagni, F., Christie, I., Christie, M., Fordham, A., Gould, R., Ives, C. D., Hejnowicz, A., Gunton, R., Horcea- Milcu, A., Kendal, D., Kronenberg, J., Massenberg, J., O'Connor, S., Ravenscroft, N., ... Thankappan, S. (2019). Loving the mess: Navigating diversity and conflict in social values for sustainability. *Sustainability Science, Special Feature: Theoretical Traditions of Social Values for Sustainability, 14*, 1439–1461. https://doi.org/10.1007/s11625-019-00726-4

Lawhon, M., Ernstson, H., & Silver, J. (2014). Provincializing urban political ecology: Towards a situated UPE through African urbanism. *Antipode, 46*, 497–516. https://doi.org/10.1111/anti.12051

Loncón, E. (2022). Azmapu. Contributions from Mapuche Philosophy for the care of the Lof and Mother Earth, Ariel, Chile.

Muraca, B. (2016). Relational values: A Whiteheadian alternative for environmental philosophy and global environmental justice. *Balkan Journal of Philosophy, 8*, 19–38. https://doi.org/10.5840/bjp2016813

Neighbors of Recinto Caupolicán. (2022). Between scent *of Boldos and Poets of Recinto Caupolicán*. https://www.ecosistemasurbanos.cl/post/historias-situadas-para-conocer-el-valor-de-la-naturaleza-urbana-en-dos-comunas-del-gran-concepci%C3%B3n

Quimbayo, G., & Vásquez, F. (2016). Towards a political ecology of urbanization in Latin America. *Political Ecology*. Notebooks for international debate. July, 4. E-ISSN: 2604-6091. https://dialnet.unirioja.es/descarga/articulo/5605786.pdf

Quimbayo, G. (2019). Territory, sustainability, and beyond: Latin American urbanization through a political ecology. *Nature and Space*. https://doi.org/10.1177/2514848619887933

Rivera Cusicanqui, S. (2018). *A ch'ixi world is possible: Essays from a present in crisis*. Tinta de Limón.

Roy, A. (2009). The 21st-century metropolis: New geographies of theory. *Regional Studies, 43*(6), 819–830. https://doi.org/10.1080/00343400701809665

Tutor, A. (2021). Neoliberal(urban)ism. Socioterritorial transformations and popular struggles in Chile, Spain and Mexico. *Bitácora Urbano Territorial, 31*(3), 275–277. https://doi.org/10.15446/bitacora.v31n3.95470

Vásquez, A., Lukas, M. S., Mayorga, M. (2017) Urban environmental (in)justice in Latin America: The case of Chile. In R. Holifield, J. Chakraborty, & G. Walker (Eds.), *The Routledge handbook of environmental justice*. Routledge.

Chapter 30
Photoethnography as a Research Methodology for Manifestations in Public Space in Times of Revolt

Pablo Hermansen Ulibarri and Roberto Fernández Droguett

Abstract Considering that the public space of large cities has been reconfigured by political demonstrations that not only use it as a stage but also intervene and potentially transform and resignify it, it is a relevant theoretical-methodological challenge today to perfect existing research methods to approach and better understand these phenomena, particularly in current times when various social revolts occur in which the public space is the fundamental substrate of them. Based on previous photoethnography work on political demonstrations, in this paper, we propose to present some foundations of this methodology and its application to the field of demonstrations in public space with their effects on the city and the territory. For this, we will explain the techniques and problems associated with the production of photoethnographic data, and then present the procedures for analysis and presentation of results. First, we will distinguish between the representative dimension of photography and its performative dimension, and how this distinction invites us to assume and problematize both the authorial and contextual conditions of photography and its ethical–political consequences. Then, we will propose that photoethnographic fieldwork involves a situated deployment of the researcher, who must seek different photographs in terms of planes, distances, actors, actions, and contexts. Finally, we will argue that the presentation of results should seek to articulate the photographs with the ethnographic text in order to preserve and highlight the constitutive importance of the image for phenomena such as the political occupation of public space.

Keywords Public space · Political demonstrations · Qualitative research methodologies · Photoethnography

P. H. Ulibarri (✉)
School of Design, Pontificia Universidad Católica de Chile, Santiago, Chile
e-mail: phermans@uc.cl

R. F. Droguett
Department of Psychology, Universidad de Chile, Santiago, Chile

M. Vicuña and E. Giorgi (eds.), *Intersections*, Springer Geography,
https://doi.org/10.1007/978-3-031-76402-8_30

30.1 Introduction

Although revolts and social outbreaks have had the street as their main stage of development, during the twenty-first century the appropriation of public space for political purposes has been reconfigured. Those who demonstrate situate themselves in the public space, transforming it while also changing with it. This interaction with the places where they demonstrate develops based on elements such as urban infrastructure; the analog and digital ways of recording, signifying, inhabiting, and transiting them; or the strategies for visibility and dissemination of the occupation, intervention, and eventual transformation of these spaces. As Butler (2017) and Castells (2015) point out, the processes of social revolt of the cycle that begins with the demonstrations of 2011 in different countries of the world are played out both in what happens in the streets and squares and in their recording and dissemination through the media—traditional and electronic—and social networks. In short, they occur in a digitally augmented way (Hermansen & Chilet, 2010). As various authors (Butler, 2016, 2017; Didi-Huberman, 2016; Fernández & Hermansen, 2009; Hermansen & Fernández, 2020; Voirol, 2005) argue, social struggles are always struggles for visibility, both of those who demonstrate and their demands and claims.

In Chile, during the social revolt that began on October 18, 2019, citizens made massive and intensive use of the streets and squares in different cities of the country, developing a multiplicity of spatial practices and interventions in the public space in which they inscribe through slogans and images the citizens' claims. Campos and Bernasconi (2021) conceptualize this phenomenon as a graphic outbreak.

> Graphic inscription is one of these spatial practices. Through the production of graffiti, tags, and murals, among other forms, individuals and collectives materially inscribe their presence and social existence, as well as their demands, interests, and visions of reality, their ways of relating to memory and history, to the present and the future (Campos Medina & Bernasconi Ramírez, 2021, p. 113).

As we see in Image 30.1, the statue of General Baquedano, the main monument of the environment that also gives its name to the Plaza, is intervened with graphic inscriptions and objects, constituting a stage for, in this case, the sound intervention of a group of bronzes that interpret songs related to the outbreak.

Under this logic in which visibility plays a central role, photography has become a fundamental tool for recording, dissemination and analysis of social protest. Even though its availability has been present since the nineteenth century and there are notable photographic records of social revolts, with the digitization of photography and the existence of social networks, today anyone can not only take photographs but also publish and share them on social networks (Hermansen & Chilet, 2010), whose reach, in principle, seems to have no type of limitation, except for possible censorship by the authorities. In fact, the most important transformation that the image undergoes with its digitization is not its dematerialization, but the weakening of its representative function in favor of its performative power, which multiplies its records, the dissemination of its meanings and visual practices in the social body (Martínez, 2019).

Image 30.1 Monument to general Baquedano. *Photography* Own elaboration, November 2019

During the demonstrations that take place in the public space in contexts of social mobilization and revolt, the photographic record for its later dissemination plays a central role, being developed both by the protesters themselves and by communications professionals, and also by researchers and other actors in the academic world. In this sense, from the social sciences, one of the methods that appear as appropriate and relevant to the study of this type of phenomena is photoethnography, that is, the photographic record and narrative of the events that occur in the public space, trying to maintain the centrality of the image for its analysis and understanding.

In this sense, the images produced in political demonstrations, where contestatory practices appropriate and transform the public space, amplify those who become visible in several complementary dimensions. First, they are not mere records of what happens but social interpretations, which together and carefully considered allow to apprehend and narrate the political senses both of those who demonstrate and of those who produce and publish them. Following Silvia Rivera Cusicanqui (2010), when marginalized groups appropriate the public space to make themselves visible and confront those who propose to hegemonize the discursive dimension of the political, political performativity and its visual record acquire an irreplaceable value. "Images offer us interpretations and social narratives, which from pre-colonial centuries illuminate this social background and offer us perspectives of critical understanding of reality" (Rivera Cusicanqui, 2010).

Then, its irreplaceable value is based on the ability to record and make available those subjectivities and ways of existence that exceed (de la Cadena, 2015) the framework imposed by the hegemonic discourses. They make available in a rich and open way phenomena that are elusive for the word, entangling ethics and politics through a situated aesthetic. In Image 30.2 we see different visual interventions made on the wall of the Gabriela Mistral Cultural Center, and in Image 30.3, three indigenous figures (diaguita, mapuche and selk'nam) installed by the Originario collective in Plaza de la Dignidad. These images evidence modes of urban intervention collective and emerging, that visually challenge the traditional discourses that usually hegemonize the city.

Based on previous photoethnography work in political demonstrations (Fernández & Hermansen, 2009; Hermansen & Fernández, 2018, 2020), in this paper, we propose to present some fundamentals of this methodology and its application to the field of demonstrations in public space with its effects on the city and the territory. For this, we base ourselves theoretically on the idea that public space is a social, material, and symbolic construction that collective action is reconfiguring and, eventually, transforming the space of the city and its meanings (Delgado, 2007). Given the cross of knowledge that involves both the notion of public space and that of photoethnography, we position ourselves from an interdisciplinary perspective

Image 30.2 Wall of the GAM. *Photography* Own elaboration, February 2020

Image 30.3 Indigenous figures, originario collective. *Photography* Own elaboration, January 2020

that seeks to build empirical research methods based on a situated epistemological-theoretical view (Haraway, 1988), in order to advance in clearing a certain methodological fog (Hernández, 2006) that tends to obscure research through photographic images.

30.2 Fundamentals and Procedures of Photoethnography

In general terms, we understand photoethnography (Fernández & Hermansen, 2009; Hermansen & Fernández, 2018, 2020) as a method of qualitative research that is inscribed in the tradition of ethnography but that resorts to photography as the main data of fieldwork and the presentation of results, even when it always must be accompanied by ethnographic writing that allows the contextualization, description, and analysis of the photographs used in order to provide a dense description (Geertz, 1987) that favors the understanding of the phenomenon investigated.

> It is a way of knowing a certain reality and talking about it through a still image, which is associated with the text, but which at the same time maintains autonomy. There are several studies that, if carried out only with images, are insufficient. And, at the same time, there are others that, if done only with text, become insufficient or boring. What photoethnography proposes is to use text and image in collaboration. The result is a more interesting, eloquent,

beautiful, and enjoyable work. And even more accessible to the public who are not experts on the subject (Alves et al., 2021, p. 445).

Then, when photoethnography has as its preferred data the photographs that the researchers produced in correspondence with those who demonstrated while the phenomenon was happening, a particular development of the way of narrating what was observed and performing it in a particular story is obtained. When reviewing the photographs produced by non-neutral researchers—that is, we inquire in correspondence (Ingold, 2017) or alliance with those who demonstrate—the images as a whole "become the language of a lived life" (Berger & Mohr, 2016, p. 289). Images 30.4a–c, were produced from an active and situated positioning of the photoethnographer, in order to obtain photographs from different planes and positions, taking care to act in correspondence with the performativity of the bodies that in alliance demonstrate.

While there is extensive literature on the fundamentals of the photographic image, it is necessary to address some of the current problematizations on the subject, particularly to approach photography not only in its dimension of representation of reality, but also in its performative dimension (Jiménez, 2019), that is, working not only with the meanings that are inherent to it as an image, but also with its resonances, consequences and effects on the social reality studied. The performative dimension also implies assuming the conditions of production of the photography and the physical, ethical and political positioning of who produces it. In this sense, a photoethnographic narrative is not a mere representation of reality nor a neutral and objective approximation but a situated and partial view that offers a perspective that requires to make explicit its conditions of production, in which both the theoretical and methodological options of the researcher and their value and ideological positioning and their degree of closeness or distance with respect to the phenomenon investigated

Image 30.4 **a–c** Photoethnographic documentation produced in alignment with Haraway's notion of situated knowledges and in correspondence with the political performativity of the demonstrators. *Photography* Own work, November 2019

and the political collectives involved participate. As Jiménez (2019, p. 183) points out, "the visual force of photographic images facilitates these affective and aesthetic communities to promote imaginaries and lifestyles that have an impact on the material world".

Donna Haraway (1988, 1991) develops the idea of "situated knowledge" as a response to the idea of objectivity, which organizes the matrix of knowledge production in modern science and, by extension, the role of photographic data in research. For Haraway (1988), modern objectivity functions as a fiction that assumes it is possible to eliminate the personal or cultural particularities of the researcher, thus achieving access to universal truths through a vision from nowhere or everywhere. It is no coincidence that since the nineteenth century photography has been attempted to be used in research as a machine for taking samples of reality and that the part that prisms the light to fix it on the photosensitive surface is called objective. Now, the problem is not that modern objectivity is a fiction, but that its followers understand it as a mandatory step to reach universal truths—and that they have the power to impose their fiction on a planetary scale. The assumption of universality is problematic because it presupposes the effective possibility that a person can take a radical distance from the world they inhabit, like the one that those who operate microscopes, telescopes, identity recognition portraits or surveillance cameras believe they possess. On the contrary, from the perspective of situated knowledge, all perspectives are specific to a particular situation; they are situated, like those who practice photoethnography participating in the manifestations they investigate and making their own political and affective position transparent. The "situated objectivity", a more honest and responsible knowledge, is achieved by recognizing and narrating the position, the gaze and the correspondence between those who compose the phenomenon.

In terms of fieldwork, the most important issue is how to make the photographic record, both in its technical and ethical–political aspects. In technical terms, given the unpredictability of the manifestations in the public space, one must try to obtain the greatest amount and variety of possible photographs, in order to capture the different situations that arise in the field, to enable an exhaustive description of its protagonists, the actions and relationships they develop, and the contexts in which these occur. This implies an active positioning of the person making the record, in order to obtain photographic sequences with different planes and from different positions, in order to compose a global and complex vision of what is happening in the field recognizing the various positions of the bodies that manifest. In this way, a greater variety of photographs allows an exhaustive analysis that considers the different subjectivities and practices that constitute the phenomenon in question.

However, despite the public nature of the manifestations in the public space presupposes the possibility of photographic recording, one must attend to the specific conditions of appearance of the subjects who manifest, since in certain cases one would have to avoid certain types of photographs or simply not make a record of particular situations. In our experience, the transgressive and often illegal nature of the occupation of public space implies safeguarding the identity of the subjects, who even many times request, in good or bad terms, that no photographic record be made.

Given that the photographs used in photoethnographic research are produced and published by people committed to what is being made visible or claimed, the images are created and distributed in "correspondence" (Ingold, 2017) with political action, developed by subjects fully aware that they could be being recorded. Their modes of presentation, the structure of their messages, and the pace at which they unfold can be seen as a performance to be photographed. We can affirm that in developing this mode of appropriation and digitally enhanced transformation, protesters, independent journalists, and researchers in correspondence manage to communicate, but in the sense that Ingold (2017) interpreting Dewey understands communication, as the situated co-creation of the common, of that which emerges when we meet and, like photographs, entangles us beyond the place and moment in which we find ourselves. The author contrasts this form of communication with the modern understanding of what it is to communicate, namely, the linear transmission of a message from a sender to a receiver.

Finally, the presentation of the results implies significant challenges to maintaining the epistemological and methodological congruence of photoethnography. In this sense, the oral or written presentation of results should aim to present the photographs articulating analytical impartiality with the interpretative position of the researcher—to make explicit his positioning—in a manner consistent with the narrative qualities of the photoethnographic account. Following Geertz (1987), to achieve a dense description of social phenomena it is appropriate to show them in context, accounting for the meanings they have for their protagonists—which includes researchers—and seeking to propose theoretical interpretations that allow us to go beyond mere description to inscribe the results both within the body of investigative knowledge and to make them available to subjects and social collectives.

30.3 Conclusions

Given the centrality of the photographic image, not only in contemporary social protests and revolts, but in general in everyday life, we believe that the social sciences, from an effective interdisciplinary approach, should delve into the epistemological, theoretical, methodological, and ethical–political dimensions related to the production and circulation of photographs. In the case of social revolts and other phenomena of political manifestation in the public space, this also implies questioning in the first place the current conditions of the forms of occupation and appropriation of this space and how different strategies of struggle for visibility are put into play. In this sense, the struggles for visibility and the records of the appropriations and transformations of public space exceed the specific moments in which they unfold. By expanding urban and political imaginaries, these images cease to represent exceptional situations of the present to evidence an extensive time that dilutes the past/present/future temporality to prefigure that the possibility of a new society is not a utopia, but a reality in which some of its elements become present, tangible, and capable of being sustained over time.

Once the images are installed in the public domain, they acquire a new type of power, since "if those bodies on the line are not recorded elsewhere, there is no global response, and also no global form of recognition and ethical connection, and therefore, some of the reality of the event is lost" (Butler, 2012, p. 138). Butler (2012) suggests the importance of social recognition of the ability of images to transport matters that exceed discursive spectrums when she states that "receptivity (to the unexpected that images contain) is a constitutive feature of action (ethical and political)" (p. 136).

In this sense, what is manifested transcends the immediacy of being at the event through photographic recording and circulation. These records provide a narration of what occurred and, in a certain sense, make it permanent on at least two levels. Once published, the images are available to be seen and shared, to be, as Berger and Mohr (2016, p. 288) maintain, "restored to a living context: not, of course, to the original temporal context from which they were taken, that is impossible, but to a context of experience. (…) It allows what they show to be appropriated by reflection. The world they reveal, frozen, becomes manageable". The visual record of an appropriation and transformation of public space introduces the certainty that what occurred is also a possible future. By demonstrating that the hegemonic order was effectively subverted in the city, the photographs broaden the spectrum of what is politically feasible. Ways of life that are usually excluded dominate the urban landscape for a moment in public space and forever in the photographs. As Campos Medina and Bernasconi Ramírez (2021) point out regarding the social revolt in Chile, the graphic dispute in public space involved systematic strategies of erasing the graphic inscriptions of the protest, in order to restore the normality of the established order and make the memories of the revolt disappear. In this sense, the photoethnographic record not only allows us to record and analyze the phenomena it investigates but also to preserve a certain memory of them and to testify that they occurred. In other words, what is normally made invisible is described as real in the photographs and, therefore, as an effective possibility of reorganizing our ways of coexistence. In the words of Dominique Lestel (2016, p. 96):

> We must profoundly alter what it means to participate in politics. The time for reforms has failed and we need to be more radical. Doing so requires a break in our usual ways of thinking. To conceptualize the forms that politics can take in the future, we need to mobilize an excessive form of thinking, we need to adopt an a priori extravagant stance and test its plausibility.

Then, we must also question how to safeguard and highlight in our research the visual dimension of the phenomena of political occupation of public space. Using photoethnography as a mere illustration or exemplification of verbal arguments omits crucial aspects to understanding contemporary manifestations in public space. Denying the epistemological and analytical centrality to these visual data implies blocking constitutive aspects of these social phenomena and blurring the political meaning of the images.

References

Alves, F. L., Abreu, C. B. de M., Schroeder, T. M. R., & Alvarez Estrada, A. (2021). Celebration of 25 years of photoethnography: Interview with Luiz Eduardo Robinson Achutti. *Anthropological Horizons, 27*, 437–452. https://doi.org/10.1590/s0104-71832021000300015

Berger, J., & Mohr, J. (2016). *Another way of telling: A possible theory of photography*. Bloomsbury Publishing.

Butler, J. (2012). Precarious life, vulnerability, and the ethics of cohabitation. *The Journal of Speculative Philosophy, 26*, 134–151.

Butler, J. (2016). Uprising. In G. Didi-Huberman (Ed.), *Uprisings*. Gallimard/Jeu de Paume

Butler, J. (2017). Allied bodies and political struggle. In *Towards a performative theory of the assembly*. Paidós.

Campos Medina, L., & Bernasconi Ramírez, O. (2021). City, social outbreak and graphic dispute. *Atenea (Concepción), 27*, 111–128. https://doi.org/10.29393/At524-7LCCE20007

Castells, M. (2015). *Networks of indignation and hope*. Alianza Editorial.

de la Cadena, M. (2015). Earth beings. Duke University Press.

Delgado, M. (2007). Moving societies. In *Steps towards an anthropology of the streets*. Anagrama.

Didi-Huberman, G. (2016). For the desire that raises us (fragment on what raises us). In Didi-Huberman, G. (Ed.), *Uprisings*. Gallimard/Jeu de Paume.

Fernández, R., & Hermansen, P. (2009). Methodological approaches for a visual sociology from the study of practices of collective memory in the public space of Santiago de Chile. *Open Space, Venezuelan Sociology Notebook, 18*, 445–460.

Geertz, C. (1987). *The interpretation of cultures*. Gedisa.

Haraway, D. (1988). Situated knowledges: The science question in feminism and the privilege of partial perspective. *Feminist Studies, 14*, 575–599.

Haraway, D. (1991). *Simians, cyborgs, and women: The reinvention of nature*. Routledge.

Hermansen, P., & Chilet, M. (2010). Virtual city, real city: Conflict and emergence of a new urban environment. *Diseña*, 40–49.

Hermansen, P., & Fernández, R. (2018). Photo-ethnography as a research methodology for the study of commemorative protest manifestations in public space. *Universitas Humanística*, 167–196.

Hermansen, P., & Fernández, R. (2020). Photo-ethnography and political engagement: Studying performative subversions of public space. *Dearq*, 100–109. https://doi.org/10.18389/dearq26.2020.11

Hernández, R. (2006). *Arguments for an epistemology of the visual data*. Cinta de Moebio.

Ingold, T. (2017). On human correspondence. *Journal of the Royal Anthropological Institute, 23*, 9–27. https://doi.org/10.1111/1467-9655.12541

Jiménez, G. (2019). Contemporary photographic practices in digital culture: Towards a performative turn. *CIC Cuadernos de Información y Comunicación*, 175–186.

Lestel, D. (2016). *Eat this book*. Columbia University Press.

Martínez, S. (2019). *Visual culture. The question about the image*. Sans Soleil Ediciones.

Rivera Cusicanqui, S. (2010). *Ch'ixinakax utxiwa. A reflection on decolonizing practices and discourses*. Tinta Limón.

Voirol, O. (2005). The struggles for visibility: Sketch of a problematic. *Reseaux*, 89–121.

Chapter 31
Playability, Democracy and Design: Design Experience of Popular Educational Materials on the Chilean Constitution Proposal

Magdalena Manriquez Hörmann, Christian Sebastián Balmaceda, Osvaldo Almarza, Francisco Álvarez Martín, María Cristina Avilés, Francisca Balbontín, Rosa Saavedra Díaz, Pilar Grant, Laura Mena, and Pilar Saavedra

Abstract In the current context of Chile, marked by unprecedented democratic processes and two failed attempts to draft a new constitution, active reflection on the participation of all sectors becomes fundamental. The persistent inequality in Chile significantly impacts civic participation, leaving behind adults from popular sectors. Given the crucial importance of their participation for democratic strengthening, the question arises about the role of design in creating spaces that enable shared power and democratic reflection. This article delves into the design and implementation

"No one educates anyone- no one educates themself, men educate each other with the mediation of the world" (Freire, 1971/1998, p. 50).

M. M. Hörmann (✉) · F. Balbontín · P. Grant · L. Mena · P. Saavedra
Design School, Pontificia Universidad Católica de Chile, Santiago, Chile
e-mail: manemanriquez@gmail.com

F. Balbontín
e-mail: fbalbontin@uc.cl

P. Grant
e-mail: pfgrant@uc.cl

L. Mena
e-mail: lmena@uc.cl

P. Saavedra
e-mail: Pilarfundamental@uc.cl

C. S. Balmaceda
School of Psychology, Pontificia Universidad Católica de Chile, Santiago, Chile
e-mail: csebasti@uc.cl

O. Almarza · F. Á. Martín · M. C. Avilés · R. S. Díaz
Collective of popular educators, Santiago, Chile

M. Vicuña and E. Giorgi (eds.), *Intersections*, Springer Geography,
https://doi.org/10.1007/978-3-031-76402-8_31

experience of educational materials intended for reflection on the Chilean constitutional proposal in 2022. A set of four games, developed iteratively by an interdisciplinary team composed of designers, psychologists and popular educators, aimed to bring adults from popular sectors closer to dialogue and citizen participation to promote associativity and social transformation. The analysis focuses on three key aspects. First, it addresses the definition of the design principles necessary to face a challenge of this magnitude. Second, it explores the design of the mechanics and the contents of the games, highlighting the importance of collaborative workshops and careful methodological and formal approaches. Finally, it concludes with reflections on design as a discipline of democratic reach and its function in learning processes, especially in reaching the most excluded sectors, from a perspective of liberation. This work contributes to the understanding of how design can be an effective tool for the civic involvement of marginalized sectors in democratic processes, underlining the importance of addressing exclusion from an integral and participatory perspective.

Keywords Playability · Democracy · Design · Citizenship · Design against the status quo · Education · Popular education

31.1 Introduction: Constituent Proposal and Popular Sectors

The constituent process that began in Chile at the end of 2019, and whose first phase was developed for more than a year, marked an unprecedented milestone. As a result, an initial constitution proposal for Chile was generated, consisting of 388 articles, condensed into 178 printed pages. Despite the fact that the text was distributed for free in digital and printed format, and was sold throughout the country, its reading presented a challenge for anyone. The presence of technicalities, the length of the document, and uncertainties about the implications of the proposal constituted significant obstacles to citizen understanding and, therefore, to the democratic deliberation process.

This complexity facilitated political parties, various sectors, media, and campaigns to have the ability to inform and misinform about the contents, manipulate ambiguous aspects and even distort the domains of effect of a constitution (Snow et al., 1986). The difficulty in accessing the text not only limited general understanding, but also contributed to the risk of information manipulation by different actors, thus undermining the integrity of the democratic deliberation process.

This situation is especially concerning given that citizen participation is fundamental for the development of a democratic and just society. It is imperative that all sectors of society have the opportunity to actively participate in decision-making and in the construction of public policies (Altavaz Ávila, 2012). However, high levels of inequality in Chile (UNDP, 2017) perpetuate participation gaps (Corvalán & Cox,

2015), marginalizing popular sectors. Veramendi (2012) postulates that aspects such as socioeconomic level and area of residence can influence the ability of individuals to exercise, or not, and in what way, democratic citizenship. In this sense, Flanagan and Levine (2015) analyze how adults from these sectors share a lack of cumulative experiences of civic participation, which influences their psychosocial characteristics and affects their participation (COES, 2022; Lucio-Villegas & Valderrama, 2015). In fact, it has been observed in recent elections that the poorest communes in Chile have a lower percentage of voting attendance (Peralta et al., 2018; Pimentel, 2021; Sueldo et al., 2021).

Adult popular education in Latin America has a centennial history addressing these types of challenges (La Belle, 2000). Based on the postulates of Paulo Freire, various organizations have promoted the participation of people living in poverty to overcome everyday problems and transform the social structure. In this context, the Center for Research and Development of Education (CIDE) carried out popular education practices for the promotion of democracy in the adult population in Chile during the 1980s and 1990s. This team designed and implemented more than twenty programs in Chile and Latin America, addressing various community themes, including democratic strengthening at the end of the 1980s and even the design of educational materials for the constituent process of Paraguay in 1992. Faced with the needs of the 2022 constituent process, professionals from this team joined educational psychologists and designers to generate educational experiences that would help adults from popular sectors to make decisions based on reflection and dialogue about the text of the constitutional proposal.

As a result, a set of four games was created, called "Participatory Dialogues: Educational Materials and Games for Democracy" (Sebastián et al., 2022), whose design process and implementation experience we analyze below.

31.2 Playability and Democracy

The tradition of popular education has a rich history of experiences with adults from popular sectors, offering valuable epistemological and methodological perspectives to address involvement. This approach, based on the principles of Paulo Freire, emerges as a critique and alternative to the dominant educational model, seeking to "revalue culture, promote the organization and participation of vulnerable sectors for the solution of their daily problems, aiming at social transformation" (Undurraga, 2003, p. 28).

A distinctive feature of popular education in Latin America is its creative and experimental approach in non-formal educational spaces (La Belle, 2000). Through accessible materiality, in terms of cost and high reproducibility, and an active-participatory approach, interaction spaces are created where adults experience participation, understand their agency, and develop a sense of self-efficacy. Popular education integrates concepts such as game-based learning, proposing that adults engage in an experience in a more horizontal way, achieving greater involvement and avoiding identity threats (Bourgeois, 2009).

The game is understood as a free action that develops within specific temporal and spatial limits, according to mandatory but freely accepted rules. This action has its end in itself and is accompanied by a feeling of tension and joy, as well as by the awareness of being "otherwise" than in everyday life (Huzinga, 2007). In line with Bateson's concept of meta-communication (1977 in Brougère, 2012) and its development in the concept of modalization, Brougère (2002) conceptualizes this "other way" that characterizes the game as a "second degree" activity. The participant operates in a space different from everyday life, but which nevertheless refers to it in an "as if" relationship (something it has in common with, for example, narrative fiction). It is a framework delimited by a specific (and often explicit) set of rules against which the participant makes successive decisions, from deciding to enter or leave the game to each of the actions they perform during the game. The articulation between this fictional character of the game (being a second-degree experience) and decision-making is what characterizes it as a specific social and cultural practice and differentiates it from other practices (Brougère, 2012).

The explanation of why participation in games would promote learning derives from this same articulation of characteristics. By voluntarily operating following a system of rules, the participant experiences a way of doing things that is different from their everyday life, but that could be carried out in it (Vygotski, 1979). The consequences of this new way of operating are attenuated and do not pose a major risk to the participant. Getting involved in them is safer in the game (second-degree activity) than in everyday life (first-degree activity) (Brougère, 2002). In this sense, in adult learning, both in formal and non-formal contexts, the game provides a protected space, promoting learning by avoiding identity threats for participants (Bourgeois & Nizet, 1997/2007). The knowledge structure that is transformed when learning is articulated to the identity of adults. When questioning this structure, the adult feels that they are questioning themselves, as citizen learning involves the transformation of the status quo (Veramendi, 2012). Experiencing this questioning, this cognitive conflict, in a second-degree space, an "as if", allows the adult to protect their identity by leaving it in the "real" space. The adult who dares to act differently and "makes a mistake" can always say that "I was just playing".

According to Ferrara (2012), games have a motivating and engaging character for subjects in learning spaces, as long as they are designed from the understanding of why people would want to play. Among the highlighted characteristics about the game are the generation of immersion, development of autonomy and self-management skills, increase in the sense of achievement, promotion of social image, appeal to creative thinking and strong drive to social interaction and dialogue. Boghian et al. (2019) add that game-based learning also encourages critical thinking, practical/organizational skills, problem management, concentration ability, collaborative work and improvement of communication skills. Treher (2011) argues that board games are tools capable of enhancing the development of practical and direct skills and knowledge in people of all ages and in all topics.

This connection between learning and play also relates to learning aimed at promoting democratic citizenship, which is carried out through the exercise of civic competences (Milana & Sorensen, 2009). Romero-Rodríguez et al. (2016) support this connection by presenting evidence that links the development of citizen competences with gamification, arguing that it acts as a catalyst for social change by granting an interactive and transversal character to the experiences that constitute the field of citizen education.

Faced with the design challenge, the choice to create games was evident when adopting the perspective of popular education. However, Gastil and Broghammer (2021) warn that "not all efforts to gamify politics and civic engagement work well" (p. 8). There is the risk of an "excessive simplification of the topics, the limitation of user autonomy and demographic differences in participation rates" (p. 8). Along similar lines, Bourgeois (2009) argues that the tools used are not impartial in terms of their cognitive influence on people. In other words, the tools used to learn in a specific environment largely reflect the mentality of the society that created them, and the world view they promote. In addition, these same tools shape the thinking and perception of the world of those who use them (Vygotsky, 1979).

The opportunities and inherent risks of games in this context demand careful reflection on the design of each game device. Their mechanics, components, dynamics and aesthetics (Hunicke et al., 2004) can both strengthen and weaken democratic processes, especially when failed participation experiences can be more harmful than the absence of participation (Flanagan & Levine, 2015). In this sense, the definition of design principles becomes a fundamental aspect for the task at hand (Gastil, 2022).

31.3 Designing "Participatory Dialogues"

31.3.1 Design Principles for the Development of Citizenship in Popular Sectors

To define the decisions in the development of the game, we incorporate elements translated from popular education, as well as knowledge and theoretical foundations of educational psychology and adult learning. Through a contextualized analysis of this proposal, these elements have been configured as design principles at different qualitative levels, as shown in Fig. 31.1. This diagram is presented as an initial model for the creation of citizen learning experiences aimed at adults from popular sectors.

1. Construction of second-degree experiences: As previously mentioned, generating experiences that protect the identity of the adult in training is one of the keys to their involvement. This is achieved through the protected learning experience, where "the subject who learns can afford to experience worldviews, new ways of

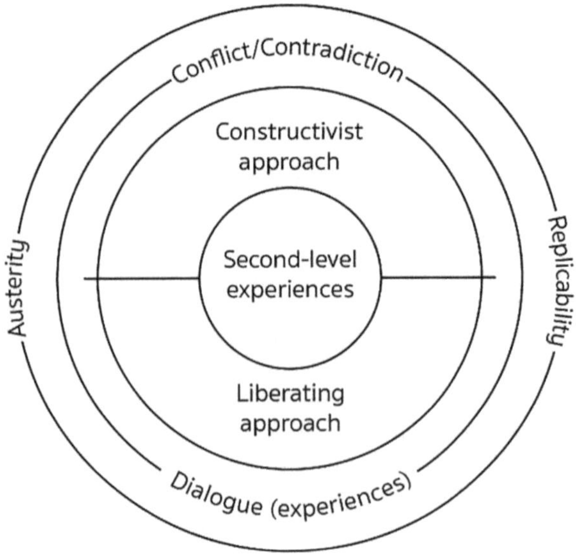

Fig. 31.1 Design principles for the development of citizenship in popular sectors. Authors' own elaboration

thinking and actions, without (too many) risks to their identity and their trajectory" (Bourgeois & Nizet, 1997/2007, p. 140). The tradition of popular education shows that the creation of these learning spaces is possible to the extent that popular education games provide second-degree experiences (Brougère, 2012).

2. Constructivist approach: Knowledge is understood as psychological processes that materialize in the interaction between human beings and their social environment. New knowledge structures (knowledge, skills) are built by transforming those with which people previously operated (Sebastián & Lissi, 2016). For this transformation to occur, a cognitive conflict based on their structures, knowledge and previous experiences must be provoked, and a resolution of this cognitive conflict must be sought in the interpsychic plane formed in the social interaction of those who learn. This knowledge structure in which individuals participate in the interpsychic plane will be progressively internalized by them to build new knowledge (Sebastián et al., 2021). Therefore, games should build notions in a staggered way, cementing knowledge and concepts for the construction of more complex ones.
3. Liberating approach: Knowledge is built in everyone's participation in a process in which distance is taken from social reality to analyze it critically. Conversation is sought starting from objects in which participants can see their experience of the place they occupy in the power relations that make up society reflected. Through the dialogued analysis of these materials, participants develop a critical awareness of these relationships that make up society. An understanding of the historical character of the way society is organized at this moment is developed, which opens the possibility of imagining other ways of organizing society that do not yet exist but could exist if some aspect of current society were transformed (Freire,

1971/1998). For this, the design approach against the 'status quo' (Khovanskaya et al., 2018) is taken, which invites to design for desired futures, changing the way we conceive the situated experience of the proposal.

4. Promotion of conflict/contradiction: the device must promote this transformation of previous knowledge structures by providing the experience of a conflict/contradiction between different perspectives and/or between the previous structure and elements of reality. Overcoming these contradictions allows the development of integrative thinking, key in wisdom (Kallio, 2011), and necessary to participate critically in democracy. The development of citizenship requires overcoming degrees of ambiguity in the relationship between knowledge and experience, that is, between theory and practice (Milana & Sorensen, 2009). Likewise, games must encourage artificial conflicts, defined by rules, to achieve measurable results (Lerner, 2014).
5. Dialogue, from their trajectories: the involvement of the participants is possible only from their life stories (Bourgeois, 2009). The device must offer a protected space (Bourgeois & Nizet, 1997/2007) so that these biographical, identity and learning trajectories can enter a constructive dialogue. This also allows the development of social capital in the participants, to face the growing social challenges (McNair, 2009).
6. Limited resources: Due to the contexts of use of this type of materials, in the face of the ephemeral nature of political processes and the need to multiply these proposals in the territories, it is necessary to design thinking about opportunities for replicability, simplicity and reuse. This simplicity and austerity also have to respond in order to be an entry door to the communities of popular sectors and connect with them from the emotional, behavioral and reflective (Norman, 2002).

31.3.2 Formalization of Games for the Constituent Proposal

These principles guided a process that started by delimiting the knowledge necessary to understand the constitutional proposal. As we indicated earlier, the "step-by-step" construction of knowledge was taken as a guide, from the most basic to the most complex. Each unit of knowledge had to provide the key conceptual construction for the next one, building independent games.[1] Thus, in an interdisciplinary discussion, the contents of each game were delimited: the first one should enable the discussion and understanding of what a constitution is; the second would promote the construction of the sense of importance of citizen participation; the third game sought the development of conceptions about fundamental rights and how they are present in the constitutions (present v/s proposal); the fourth game sought the recognition of emotions about the proposal, helping the democratic deliberation of approval/rejection.

[1] They are independent, in the sense that they can be played modularly, but dependent in the constructivist sense. The previous elaboration facilitates the next game.

31.3.3 Development and Testing of the Materials

The process of elaboration of this material was carried out in a very tight period. From the presentation of the constituent proposal to the plebiscite, we had just two months. In virtue of this temporal limitation, we chose to start with two games that did not rely on text as the central element of material and work on the mechanics of the last ones to then populate them with the content of the proposal.

In a first stage we tested quickly, within the team and in close circles. Thus, while the mechanics of the first game were being iterated, the second was being designed and so on with each of the four. In parallel, a base of local leaders from popular sectors was consolidated with whom we materialized game testing workshops. We conducted two workshops at the facilities of “Canelo de Nos” (San Bernardo) and “El árbol de la Dignidad” (Ñuñoa), with the participation of 30 local leaders from various places in Santiago and the Valparaíso region.

These workshops, added to the isolated testing experiences with some communities throughout Chile (Colectivo Cultural El brasero (Valparaíso), Taller Isla Negra Personal development and cultural change, Chileufú (Limache), Caja de Herramientas, Colectivo Paulo Freire, Vecines de Antofagasta, ABOFEM), were consolidated, from a design participatory logic, as spaces for co-construction of the games and also had a disseminating logic. These same local leaders brought copies of the materials to their communities.

When designing for adults from popular sectors, the boundaries are delicate. Just as there is a concern about the threat to identity, the graphic and formal decisions of the materials also play a key role in involvement. In this sense, we took special care with the visceral, behavioral and reflective connection of the material (Norman, 2002).

We made formal decisions about the material, including making a graphic reinterpretation of the original Popular Education material of the CIDE from the 1980s and 1990s (see Figs. 31.1 and 31.2). This work involved recognizing the basal and transversal characteristics of the references we had, a series of folders printed with cut-out materials, boards, cards. We preserved the illustration style, we reinterpreted the typographic fonts, we preserved textures and precuts while maintaining the characteristic color palette of the time. Similarly, we updated its cheap and effective way of reproduction on a small scale, through a “ *print and play*” strategy that allowed designing so that it could be easily printable in a black and white printer and letter format. All of this was also tested in the workshops, thematizing the materiality, its *affordance* (Norman, 2002) and visual communication (Fig. 31.3).

In the workshops, each group of 4–5 participants played one of the games for an hour and then explored the material of the other three games for twenty minutes (see Fig. 31.4).

In plenary, each group presented the game they experienced, indicating strengths and weaknesses. In a cross-cutting manner between the groups, questions were formulated and answered. Each workshop ended with a space for participants to

Fig. 31.2 Images of the games designed. Authors' own elaboration. Review at https://drive.google.com/file/d/1pJRPaCfdS_OPLZpLJyiqiUb3sNGTRn9D/view?usp=sharing

Fig. 31.3 Materials made by the team for the constituent process of Paraguay

freely share their experience during the workshop. The participants also provided written feedback and committed to disseminating the games.

The dissemination of the final games was done digitally, through networks of contacts, between August 4 and September 1, 2022. Each of the games included a brief evaluation questionnaire of the experience and instructions to send that feedback to the creator team through three different ways: photograph of the questionnaire sent by email; transcription of the answers also in email; and QR code linking to an online version of the same questionnaire.

31.4 Analysis of the Implementation from the MDA Framework

Given that the objective of this chapter is to analyze the process of designing materials with specific social characteristics, at a particular historical moment, we adopted the MDA framework, postulated by Hunicke et al. (2004) to carry out the task. This

Fig. 31.4 Photos of workshop held in El Canelo de Nos. Authors' own production

formalizes the design of games by deconstructing them into their various parts and allowing understanding of how certain changes impact the entirety of the game experience. This framework is traversed by the three levels proposed by Norman (2002) and allows thinking about an integration of analysis matrices. That is why it works as an optimal framework for the analysis of the participatory game system by being able to dismantle it into (1) mechanics that describe the particular components of the game (resources and rules), the (2) dynamics that describe the behaviors and interactions of the players during the execution of the mechanics, and finally the (3) aesthetics that correspond to the desirable emotional responses that are evoked in the player when interacting with the game system.

31.4.1 Mechanics

"Participatory Dialogues" materializes in a folder that contains 4 games for groups of 4–8 participants.

- "Our Political Constitution" includes 58 cards with concepts. Through dialogue, they agree on the five concepts that should be included to describe what a constitution is. This game generates basic notions about what a constitution is.

- "Let's talk about citizen participation" contains stories of everyday life situations, which are analyzed by answering questions from different roles.
- "Our right to good living" includes 30 cards with articles (paraphrased) from the 1980 constitution and the 2022 constitutional proposal. The participants must agree on which of the constitutional texts they think each card belongs to, placing them on a board. At the end, they contrast their classification with a guide that indicates the origin of each article.
- "Everyone's house" contains 36 cards with excerpts from the Constitutional Proposal. Each player draws a card and explains what the emotion they feel when confronted with that text is, placing it on the board in one of the following boxes: "Joy", "Relief", "Optimism", "Confusion", "Sadness", "Fear", "Anger", and "Another emotion".

31.4.2 Dynamics

Following the characteristic principles of popular education and its reinterpretation in design principles, the chosen mechanics and aesthetics sought to enhance the dynamic of community dialogue. In it, participants share their experiences—a key issue for adult education—in order to build a collective knowledge that recognizes and overcomes these subjectivities at the same time. The game provides that each participant "says their piece", for example, when they choose and present a card with a concept that should be in the Constitution. At the same time, the rule of agreeing on which of the presented cards will be kept requires the group to solve the problem in a democratic way. The argued and explicit deliberation as a form of conflict/controversy resolution is the dynamic that is repeatedly experienced in "Participatory Dialogues". Adults surrender to the immersion of the game, "without fear of error".

31.4.3 Aesthetics

Among the eight probable taxonomies of experiences that can be lived by players (Hunicke et al., 2004), those that appear in the implementation are *Community*, as the game provides a community structure that transcends the individual. *Discovery*, since the game constantly addresses knowledge that the player does not necessarily have a handle on; under the constructivist principles of popular education the game does not try to impose knowledge but to promote its construction. *Expression*, because to discover and discover oneself requires the exploration of experiences, knowledge and personal emotions in an area that is not a typical subject of conversation, as is the case with the Constitution.

31.5 Discussion

The democratic society still faces the challenge of establishing citizenship as a consolidated practice (Sojo, 2002). The persistent social deficiencies that negatively affect the full exercise of citizenship and, therefore, the consolidation of democracy, cannot be ignored. It is essential to ensure that all citizens are fully aware of their rights and responsibilities. Pancer (2014), in this line, raises the need to particularly investigate the civic involvement of groups that are at greater social risk, as he highlights the lack of response to questions about participation and the differences of this sector. In this framework, "Participatory Dialogues" is a materialized advance, both in initial solutions and in questions for this challenge, proposing design principles that can become a useful and careful guide towards the construction of new participatory spaces for this group of our society.

Among the initial responses, we highlight that, although it does not correspond to a formal evaluation of the effect of the games, the feedback from participants in the different iterations referred to a high valuation of these as a learning device: the games would allow training *for democracy* through an active-participative experience *of democracy*, a need deeply felt by adults and young people from popular sectors.

However, the experience also showed us important challenges and questions for this area of design. On the one hand, even when we designed an easy-to-distribute and reproduce material that included feedback mechanisms to the creative team, we are not clear on the true reach of the games. How to effectively reach a maximum of population that needs devices like this? How can we design the arrival, the weaving of networks in a context of multiple initiatives that compete for the attention and energy of citizens? An imaginable alternative is to collaborate with institutions that already have these networks, but how can we do it without "institutionalizing" (given the possible loss of credibility that this tends to bring for certain (groups of) participants)?

On the other hand, despite having included accessible mechanisms of feedback, we had very little response beyond the testing experiences. How to design means of access to popular education material that promote critical evaluation by users at the same time? Also, we have indications from the feedback from the tests that the text (explanations, instructions) of the material still requires significant adjustments to make it accessible to the vast majority of contemporary popular adults. How can we make these adjustments without losing the critical sense of the material? Without falling into the oversimplification that quickly falls into unreflective propaganda?

A possible general approach to all these questions is the multiplication of testing and redesign cycles. Beyond the discussion about the resources that this approach requires, it opens the question about the strongly situated character of a material like this. Unlike other areas of design where you can aim for the permanence of the product, the process we lived required us to design materials whose relevance was strictly linked to the ongoing social process: the exit plebiscite on the 2022 constitutional proposal. Particularly because this proposal was not approved by the citizens, but also because of the specific psychological processes it sought to promote, this set of four materials played its real relevance in the short time (two months) between the publication of the proposal and the plebiscite. How do we design under these conditions? What does it imply for the concepts and practices of design to create products of high complexity and yet destined for redesign in a very short term? How do we design so that the actors who are going to use these materials can appropriate them, recreate them according to their (sub-)cultural particularities without losing the transformative force of conceptions and practices in a field as delicate as that of citizen training? Is it possible to design the material so that a community that needs to be trained in democratic competencies can transform it without making it lose that (trans-)forming potential? What are the limits of the *affordance* of a game to promote democracy in a community that has previously experienced little to nothing of that democracy, while still maintaining relevance with that group?

We believe that when promoting democracy, design faces questions whose answers seem to be found in participatory dialogue with other disciplines, but above all with those most excluded from our societies. In the current context, we have a role not only in simplifying information, but, above all, in turning it into devices where it can be experienced, related, and reflected upon. Developing devices that materialize new dialogues and dynamics has the potential to restructure our ways of civically engaging and enable new futures that break with pre-established schemes, strengthen democracy, integrate all members of our society, and move towards social transformation. However, for that potential to be realized, it seems to not be enough to design *for* the vulnerable. We still have to create a way to design *with* them.

Acknowledgements We thank all the neighborhood organizations that believed in this project: Fundación Apachita, Territorio socioambiental, Colectivo Cultural El Brasero, Valparaíso, Taller Isla Negra Personal Development and Cultural Change, El Canelo de Nos, ABOFEM, Toolbox, Neighbors of Antofagasta, Organization the Tree of Dignity, Women of Maule, Paulo Freire Collective and Fridays for future.

References

Altavaz Ávila, A. C. (2012). Theoretical-methodological conception for pedagogical work aimed at citizen training in the education of young people and adults. Thesis for the Scientific Degree of Doctor in Pedagogical Sciences. University of Pedagogical Sciences "Enrique José Varona", Havana.

Boghian, I., Cojocariu, V.-M., Popescu, C., & Mâţ˘a, L. (2019). Game-based learning. Using board games in adult education. *Journal of Educational Sciences and Psychology, IX*(LXXI)(1), 51–57.

Bourgeois, E. (2009). Learning and transformation of the subject in training. In J.-M. Barbier, E. Bourgeois, G. Chapelle & J.-C. Ruano-Borbalan (Eds.), *(Under the direction of) encyclopedia of training*. PUF.

Bourgeois, E. & Nizet, J. (1997/2007). Learning and adult education. PUF.

Brougère, G. (2002). Game and leisure as spaces for informal learning. *Education and Societies, 10*(2), 5–20. https://doi.org/10.3917/es.010.0005

Brougère, G. (2012). Can the game be serious? Revisiting play/learn in times of serious game. *Australian Journal of French Studies, 49*(2), 117–129. https://doi.org/10.3828/AJFS.2012.10

COES. (2022). Radiography of social change: Analysis of Longitudinal ELSOC results 2016–2021. Presentation of COES Results.

Corvalán, A., & Cox, P. (2015). Participation and electoral inequality in Chile. In C. Cox & J. C. Castillo (Eds.), *Citizenship learning: Contexts, experiences and results* (pp. 175–200). UC Editions.

Ferrara, J. (2012). Playful design: Creating game experiences in everyday interfaces. Rosenfeld media.

Flanagan, C., & Levine, P. (2015). Civic engagement and the transition to adulthood. In C. Cox & J. C. Castillo (Eds.), *Citizenship learning: Contexts, experiences and results* (pp. 77–109). UC Editions.

Freire, P. (1971/1998). *Pedagogy of the oppressed*, 21st Century.

Gastil, J. (2022). To play is the thing: how game design principles can make online deliberation compelling. *American Behavioral Scientist, 67*(8), 1042–1060. https://doi.org.pucdechile.idm.oclc.org/10.1177/00027642221093592

Gastil, J., & Broghammer, M. (2021). Linking theories of motivation, game mechanics, and public deliberation to design an online system for participatory budgeting. *Political Studies, 69*(1), 7–25. https://doi.org.pucdechile.idm.oclc.org/10.1177/0032321719890815

Hunicke, R., LeBlanc, M., & Zubek, R. (2004). MDA: A formal approach to game design and game research. In *Proceedings of the AAAI Workshop on Challenges in Game AI* (Vol. 4, No. 1).

Huzinga, J. (2007). *Homoludens* (6th ed.). Alianza Editorial/Emecé Editores.

Kallio, E. (2011). Integrative thinking is the key: An evaluation of current research into the development of adult thinking. *Theory & Psychology: SAGE Journals, 21*(6), 785–801. https://doi.org/10.1177/0959354310388344

Khovanskaya, V., Dombrowski, L., Harmon, E., Korn, M., Light, A., Stewart, M., & Voida, A. (2018). Designing against the status quo. *Interactions, 25*(2), 64–67 (ACM). https://doi.org/10.1145/3178560

La Belle, T. J. (2000). The changing nature of non-formal education in Latin America. *Comparative Education, 36*(1), 21–36. http://www.jstor.org/stable/3099849

Lerner, J. (2014). *Making democracy fun: How game design can empower citizens and transform politics.* MIT Press books.

Lucio-Villegas, E., & Valderrama, R. (2015). Participatory citizenship as a practice of adult education. *Journal of Adult Education and Formative Processes, 1.* https://www.educaciondeadultosprocesosformativos.cl/index.php/revistas/revista-n-1/6-la-ciudadania-participativa-como-practica-de-educacion-de-adultos

McNair, S. (2009). Demography and lifelong learning (IFLL Thematic Paper 1). National Institute of Adult Continuing Education.

Milana, M., & Sorensen, T. (2009). Promoting democratic citizenship through non-formal adult education: The case of denmark. *Scandinavian Journal of Educational Research, 53*(4), 347–362. https://doi.org/10.1080/00313830903043125

Norman, D. A. (2002). *The design of everyday things.* Basic Books.

Pancer, M. (2014). *The psychology of citizenship and civic engagement.* Oxford University Press.

Peralta, M. I., Cuella, S., Becerra, N., Sueldo, J., Aquín, N., Del Águila, N., Giovana, E. & Bosio, M. T. (2018). Politicization of popular sectors. National University of Córdoba.

Pimentel, F. (2021). Trends in electoral behavior in post-authoritarian Chile: A bibliographic discussion. *American Crossroads, 13*(1), 7–38.

Romero-Rodríguez, L.-M., Torres-Toukoumidis, A., & Aguaded, I. (2016). Gamification and education for citizenship. Review of the significant experiences. *Educate, 56*(1), 109–128.

Snow, D., Rochford, J., Burke, E., Worden, S. K., & Benford, R. (1986). Frame alignment processes, micromobilization, and movement participation. *American Sociological Review, 51*(4), 464–481.

Sebastián, C., & Lissi, M. R. (2016). Learning as a higher psychological process. Towards a historical-cultural understanding of the development of the learning process. In P. Freire, R. Moretti & F. Burrows (Eds.), *Learning with others: Psychosocial approaches to learning in educational contexts* (pp. 19–47). Alberto Hurtado University Editions.

Sebastián, C., Vergara, M., & Lissi, M. R. (2021). The Vygotskian contribution to the construction of a general theory of human learning. In P. Fossa (Ed.), *Latin American advances in subjectivity and development. Through the Vygotsky Route* (pp. 227–248). Springer.

Sebastián, C., Almarza, O., Avilés, M. C., Álvarez, F., Saavedra, R., & Manríquez, M. (2022). "Participatory Dialogues: Materials and Educational Games for Democracy" A proposal from popular education for citizen training. UC Social Notebooks (pp. 3–7). https://drive.google.com/file/d/1PSoB9jR7O66wRCfZicqSWVsDCskSoU5L/view

Sojo. (2002). The notion of citizenship in the Latin American debate. *CEPAL Review, 76*, 25–8.

Sueldo, J., Merlo, J., & Torres, M. (2021). Popular politicization in times of crisis. *SocialConscience, 4*(8), 32–47.

Treher, E. N. (2011). *Learning with board games*. The Learning Key Inc.

Undurraga, C. (2003). How do adults learn? A psychoeducational look. UC Editions.

UNDP. (2017). Unequal. Origins, changes and challenges of the social gap in Chile. United Nations Development Program.

Veramendi, L. (2012). Education and citizenship: Analysis of the relationship between educational trajectory, literacy, ethos and political praxis in a sample of adults from a district of the Peruvian Andes. Research Advances, 6. GRADE.

Vygotski, L. S. (1979). The development of psychological processes superiors. Critique.

Chapter 32
Urban Events as Strategies for Creation and Re-Creation of the City: The Pan American Games of Santiago 2023 and Barranquilla 2027

Erick Abdel Figueroa Pereira

Abstract How are changes in urban economy and urban planning, and the urban transformations induced, directly or indirectly, by the realization of urban events such as the Pan American Games? The cases of two Latin American cities chosen as venues for the event are compared: Santiago 2023 (Chile) and Barranquilla 2027 (Colombia). The proposed study methodology is divided into two parts. One of them addresses the actions of public policy and public-private partnerships, in their interaction with the urban strategies of the event and the current urban planning. The other part of the work focuses on the articulation of the urban event with the socio-spatial organization and the mobility infrastructure. It is concluded that the event's work plan tends to move away from the urban agenda and has little relation with urban planning, so it is essential to determine what are the interests of the agents involved in the process and how their actions are articulated with the type of event chosen.

Keywords Major events · Urban transformations · Pan American Games · Santiago · Barranquilla

32.1 Introduction: Minor Urban Events in Search of Their Own Theory

Since the second half of the twentieth century, the development of urban events, some of international stature, has become one of the most commonly cited resources both to accelerate urban development processes and to promote the construction of urban facilities (Monclús Fraga, 2007). Depending on their scale and scope, urban events can induce significant transformations, reconfiguring the relationships

E. A. F. Pereira (✉)
Universidad del Atlántico, Barranquilla, Colombia
e-mail: erickfigueroa@mail.uniatlantico.edu.co

M. Vicuña and E. Giorgi (eds.), *Intersections*, Springer Geography,
https://doi.org/10.1007/978-3-031-76402-8_32

between public and private (Fessler Vaz & Berenstein Jacques, 2007; Gold & Gold, 2007; Roche, 2000). In the case of sports events, their urban effects usually go beyond the immediate surroundings of the competition venues: they modify public space, transport and telecommunications infrastructures, accommodations; they occasionally respond to cultural and environmental demands (Chalkley & Essex, 1999; Kassens-Noor, 2012; Preuss, 2004; Smith, 2012). Recent critical literature has transcended from the study of first-order events such as the Olympic Games and the Universal and International Expositions, becoming increasingly common the study of smaller scale events and international significance, but more attractive in terms of their potential impact on urban agendas, given their greater capacity to affect local economies and infrastructures. Works that question their relevance, pertinence and impact on local, regional and national finances in the context of emerging economies predominate (Grix et al., 2019; Di Vita & Wilson, 2021; Tinaz & Knott, 2021), highlighting how their choice *from above* exerts great pressure on the social fabric of the host cities, in a demonstration of political authoritarianism not exempt from strong tensions (Gusmão de Oliveira, 2021).

This work studies the urban strategies proposed for the Pan American Games. The cases of two Latin American cities chosen as venues for the Pan American Games, Santiago (Chile), which organized them in 2023, and Barranquilla (Colombia), which is still unclear whether it will hold this event in 2027. Each one of them has held a lower-ranking sports event: the South American Games Santiago 2014 and the Central American and Caribbean Games Barranquilla 2018 (see Fig. 32.1). Apart from the updating and construction of the various infrastructures required for the celebration of the competitions, both events were used as their presentation card to differentiate themselves from other cities in a growing scenario of *city branding*, with the purpose of attracting foreign investment and higher category events with potential for urban transformation. In this sense, the aim was to achieve greater international visibility of the cities, partially justified by the intention to use the event to positively impact their urban quality of life.

In order to determine the degree of articulation between politics, the economy, and urban planning for each of the cases listed, this work uses two strategies. On the one hand, the study of the impact of the event on the local and national economy, through the review of available documents on the budget allocated for the event in different areas. On the other hand, the research focuses on the analysis of urban plans, scrutinizing the degree of concentration or dispersion of the infrastructures required to organize such events, their relationship with central areas and the peripheries, and with mobility infrastructures (Figueroa-Pereira, 2013). The main sources of consultation refer to the information produced by the organizers, press news, reports and documents that define local, regional, and national budgets related to the work plan of each event, and the review of secondary literature.

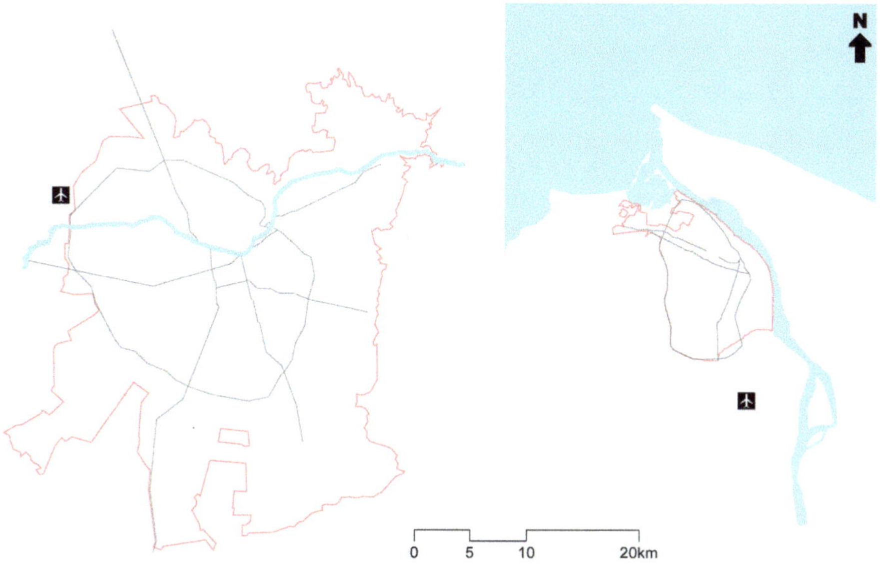

Fig. 32.1 Santiago in 2014 and Barranquilla in 2018. Comparison of the size of the cities, at the same scale. In red, urban perimeter; in gray, main roads network. Author's elaboration, 2023

32.2 Santiago 2014–2027: Events with Little Urban Impact?

Santiago,[3] capital of Chile, has about 6,000,000 inhabitants in a country inhabited by approximately 20,000,000 people (2023). There had already been two failed attempts to organize the Pan American Games in the city, in 1975 and in 1987, which could not be carried out due to serious internal circumstances accentuated under the military dictatorship (1973–1990).

After the return of democracy in 1990, in 2006 Santiago won the right to host the X South American Games, scheduled for 2014. With them, the articulation between the national, regional and local political system was tested, crucial for the logistical organization of the Pan American Games (Bravo et al., 2018). The actions related to the event (see Fig. 32.2) show two sports concentrations (National Stadium and Peñalolén Park), and a dispersion of the sports venues throughout the city, which was resolved by connecting them through the main avenues network (Américo Vespucio, Central Highway, Vicuña Mackenna, Matta, Greece, Providencia, President Kennedy). A Pan American Village was not built, so the athletes were accommodated in hotels.

The XIX Pan American Games were awarded to Santiago in 2017 and took place from October 20 to November 5, 2023. Compared with the 2014 games, few sports facilities were built, relying mainly on existing infrastructures (see Fig. 32.3). Apart from the venues built near the National Stadium, near the old Cerrillos airport, the Pan American Village was erected (Library of the National Congress of Chile, August

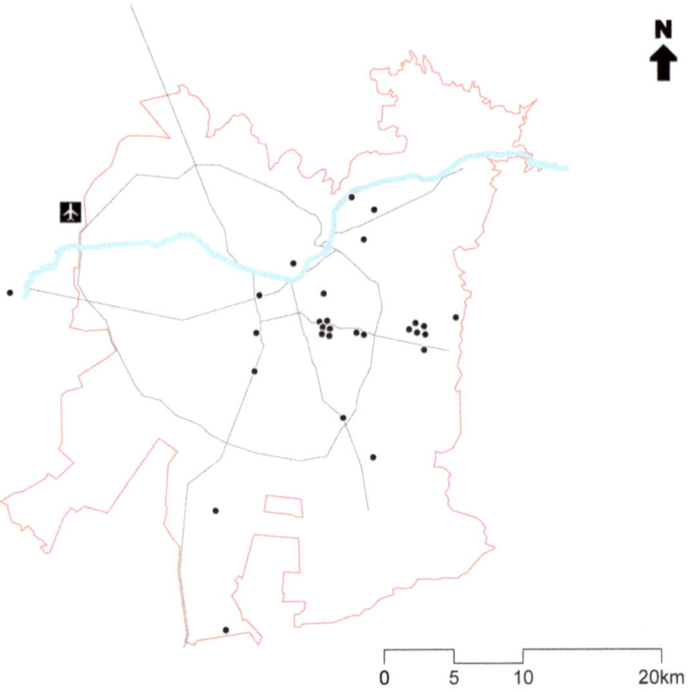

Fig. 32.2 South American Games Santiago 2014. The urban strategy. Main road network (in gray) and location of sports venues (black circles). Author's elaboration, 2023

31, 2020), product of a public-private partnership (PPP) for the development of housing projects (Curador News, Febrero 1–8, 2021).[4] The PPP assumed the design, financing, and execution of the work (La Cuarta Constructor, 2023; Muñoz H., 2023); once the event was over, the Village was incorporated into the real estate market as social housing.[5] The initiative culminated an old desire of the real estate sector: to add the lands of the old airport to the land market to increase the building density of the commune, a strategy driven by the flagship project of Parque Cerrillos.[6] In terms of connecting the housing complex with the mass land transport infrastructures, line 6 of the Santiago Metro fulfills the function of intraurban connection with the Village. The international airport, the main gateway to the event, required a radical intervention due to its inefficiency in the face of the great growth of national and international air traffic. To respond to this, the infrastructure of the air terminal was significantly expanded in the years immediately preceding the Games.

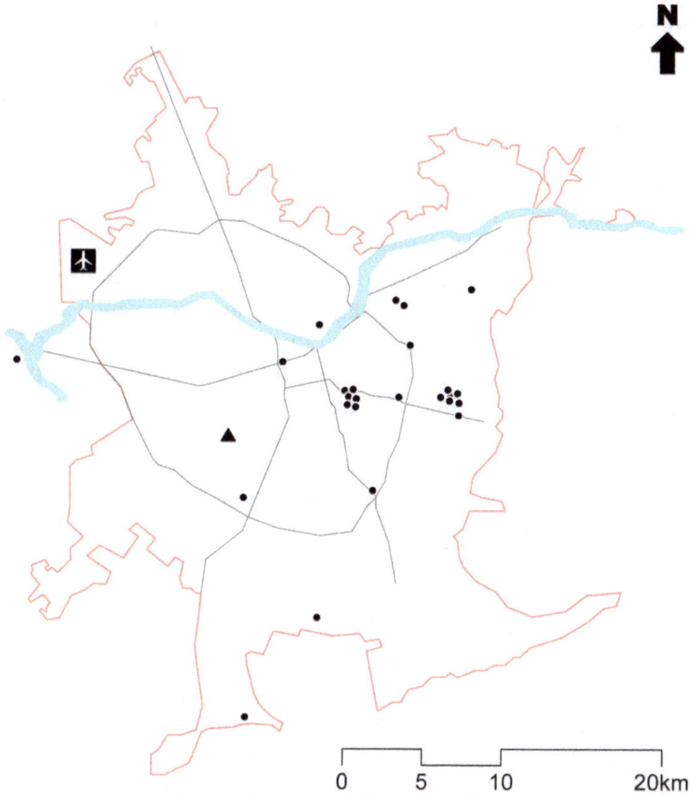

Fig. 32.3 Pan American Games Santiago 2023. The urban strategy. Main road network (in gray) and location of sports venues (black circles) and Pan American Village (black triangle). Author's elaboration, 2023

32.3 Barranquilla: Induced Transformation, Direct, Indirect, or Diffuse?

Barranquilla[7] has about 1,200,000 inhabitants and is the main port city of Colombia, a country whose population currently exceeds 50,000,000[8] (2023). The city hosted the Bolivarian Games in 1961 and the Central American and Caribbean Games in 1946 and 2018, with the latter being the basis for its management model and the construction and use of necessary equipment to bid for the XX Pan American Games 2027.

The strategy of urban insertion of sports infrastructures for the 2018 event combined two models. On the one hand, two centralities were configured (Metropolitan Stadium and northwestern complex); on the other hand, the dispersion of the venues in the city was chosen, taking advantage of some lands owned by the District of Barranquilla that met the requirements for the construction of sports infrastructures. Some of the venues were very isolated and disconnected, not only from each other but even from the Central American Village (see Fig. 32.4). This was built to the west of the city—in a new area of real estate development (now known as Alameda del Río) whose public space solution is insufficient for the large number of inhabitants who have arrived in the sector after the execution of new housing complexes in the area. The connection of this area with the intraurban road network was poorly projected, since only one entrance and exit were resolved for the entire area of future real estate development, which did not consider the growing amount of automotive traffic. Parallel to the works for the event urban monuments and citizen meeting spaces emerged on the edge of the river, which sought to stimulate the land market in the area and turn the city towards the waterfront. Since 2017, the remodeling of the international airport began, still unfinished and the subject of controversies for crimes of patrimonial detriment that are the direct responsibility of the concessionaire. No sports venues were built in popular sectors, concentrating the majority in sectors of higher purchasing power. Relevant works in mobility, such as the Circunvalar Avenue to the west, and the Via 40 to the east, have been unfinished or in repairs that never end for several years.

Unlike Santiago, where the continental event has already taken place, the location plan of the proposed venues for the 2027 Pan American Games in Barranquilla is a fairly incomplete preliminary scheme. What is known so far consists of the reuse of all the 2018 venues, but it is unknown if there will be a Pan American Village (see Fig. 32.5). It is uncertain if the event will take place in the city, due to difficulties in the articulation between the local and national governments, which to date (January 2024) have caused Panam Sports, the entity that holds the rights to the Pan American Games, to withdraw the city's right to host the next edition.

32.4 The Projected, the Expected and the Realized: Some Considerations

When the current urban morphology of Santiago and Barranquilla is compared, it is clear that both cities are being limited in their growth by the presence of significant geographical accidents,[9] raising several issues in terms of the possibilities of urban transformation (see Fig. 32.6). This should be considered in relation to the event's infrastructure plan. When the host city has enough buildable land for facilities, it is unlikely to generate urban landmarks. The large size of Santiago makes it difficult to read significant changes, so the urban impact of the 2023 Pan American Games tends to be low, as happened with the 2014 South American Games. The predominant note is a great distance between the venues, alongside their trend of moving from the center to the northeast. This fact forced the reinforcement of urban transport infrastructures. In Barranquilla, on the other hand, a large number of dispersed sports infrastructures were already made for the 2018 Central American Games, whose location responded to the promotion of expansion directions (north and northwest),

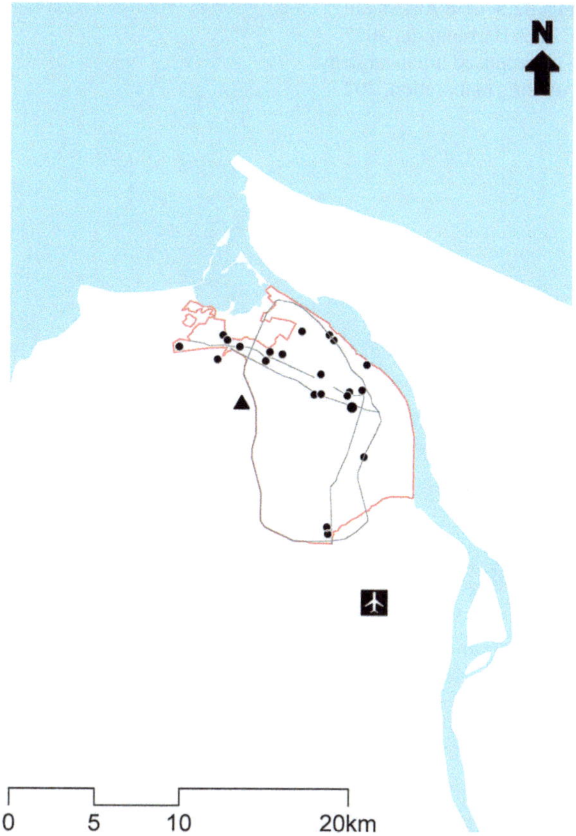

Fig. 32.4 Central American Games Barranquilla 2018. The urban strategy. Author's elaboration, 2023

some of them occupying public land. In addition, Unlike Santiago, Barranquilla has neglected the role of its airport as the main gateway to the city.

In both cases, there is little relationship between the traditional central area of the city and the location of the sports infrastructures. They tend to be located in areas far from the historic centers, wasting the possibility of configuring urban projects or urban regeneration plans.

Despite being a much smaller city than Santiago, the design of Barranquilla's road network does not favor the expedited circulation of traffic. The distances between the sports venues are significantly smaller in Barranquilla than in Santiago, and there tends to be a concentration of venues near a few main roads. However, in the Barranquilla case, the deficiencies of the road network due to its lack of continuity have a negative effect on the displacement of athletes, a fact that will possibly worsen for 2027 if the necessary corrections are not implemented.

Fig. 32.5 Pan American Games Barranquilla 2027. The proposed urban strategy. Author's elaboration, 2023

32.5 Financing an Urban Event: Lights and Shadows of Administrative Management

Although the general budgets related to sport at the national and local level do not allow a detailed financial management of the items destined for the realization of infrastructures associated with the sports events studied here, they do provide certain clues about the economic impact of the events at the local and national levels. Next, we present the findings of the study of national and local budgets by departments associated with sport, temporally discriminated between 2003 and 2024 for Chile, 2011 and 2023 for Colombia and 2011 and 2022 for Barranquilla.[10]

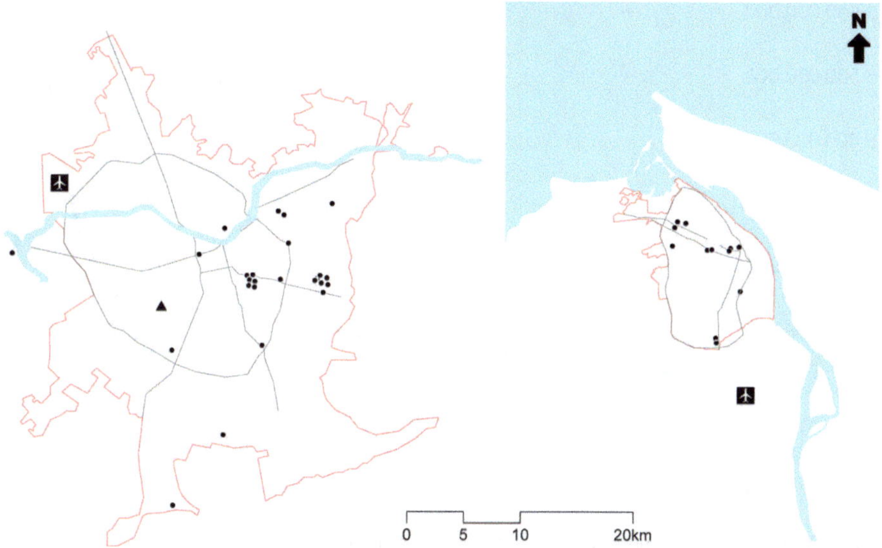

Fig. 32.6 Pan American Games Santiago 2023 and Barranquilla 2027. Comparison of urban strategies, at the same scale. Author's elaboration, 2023

32.5.1 Chile

The growth of the budget for the sports portfolio was relatively slow until 2021, the year from which there was an almost vertical increase that reached its highest point in 2023, and which will have a significant decrease towards 2024, corresponding to the accounting closure of the Organizing Committee of the 2023 Pan American Games (see Chart 32.1). The global items for Chilean sport between 2006 and 2014 had an increase of 309.97%. In relation to the Pan American Games, between the designation of the venue (2017) and the realization of the event (2023), the global budget for the sports sector increased by 369.68%.[11] In summary, what was allocated to the South American Games (2011–2014) was of the order of 3.58%, while for the Pan American Games the figure reached 35.36% of the budget for the corresponding period (2019–2024).

32.5.2 Colombia

Between 2011 and 2023, the budget for national sport increased by 663.14% (Gobierno de Colombia. Ministerio del Deporte (2010–2023). However, there were decreases in 2014, 2016 and 2017. Although between 2014 and 2018 there was an increase of 155% in the items for sports, no specific budgetary items were identified for the Central American Games, and during the analyzed period only the

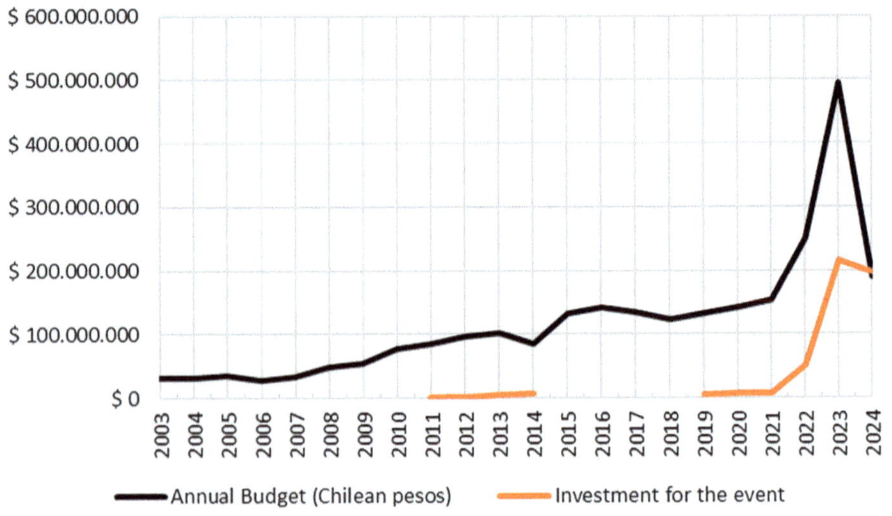

Chart 32.1 Budget for sport in Chile, 2003–2024. Author's elaboration, 2023

mention of items for "Construction, adaptation and equipment of sports venues for national and international games," equivalent to 13.50% of the 2015–2018 budget (see Chart 32.2), and for "Support for the Organization, participation and realization of international events in Colombia," corresponding to 1.22% of the budget for the period 2013–2018 (see Chart 32.3).

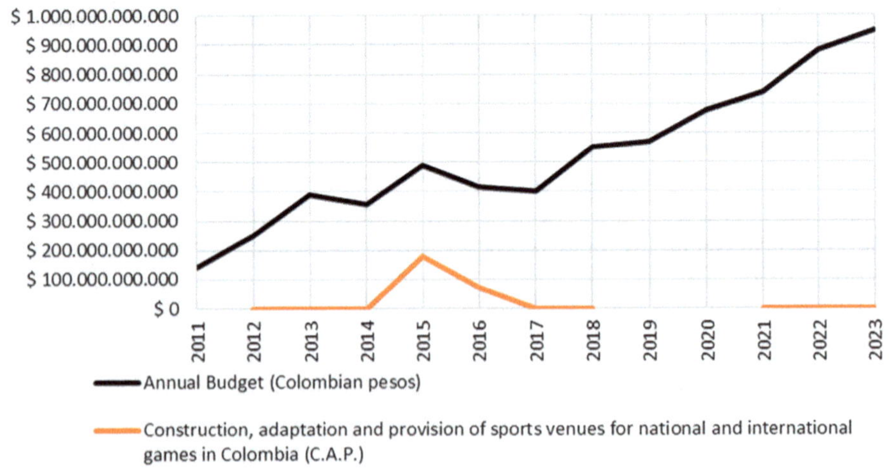

Chart 32.2 Budget for construction and provision of sports venues in Colombia, 2011–2023. Author's elaboration, 2023

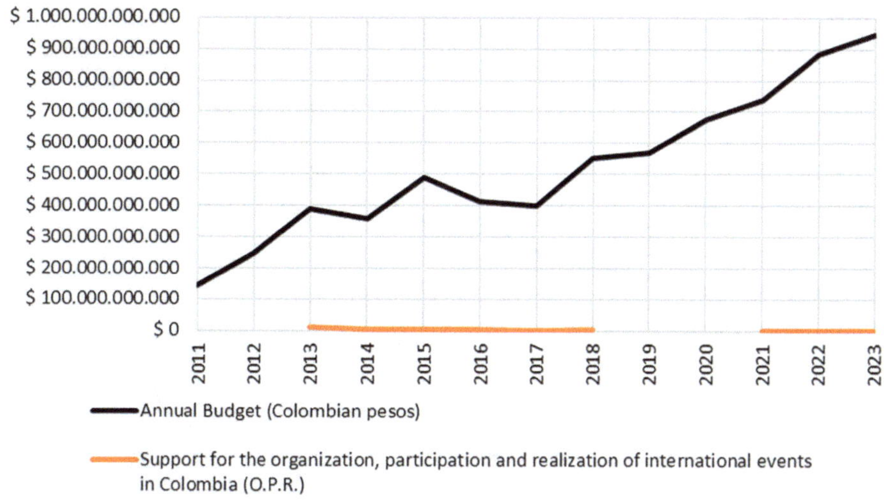

Chart 32.3 Budget for support to sports events in Colombia, 2011–2023. Author's elaboration, 2023

32.5.3 Barranquilla

Between 2011 and 2022, the general budget of the District of Barranquilla saw an increase of 261.44% in the items allocated to sports; except between the years 2014 and 2015, from 2013 to 2019 there was sustained growth. That year there was a not very pronounced decrease in the income of the Sports Secretariat, to then increase between 2020 and 2021. For 2022, the last recorded year, there was a slight decrease in the budget (see Chart 32.4).

Between 2016 and 2019, there was a 146% increase in the budget allocated to investment in sports infrastructure. Only from 2016 was the budget appropriation for the Central American Games defined, with variations ranging between 0.11% (2019) and 7.34% (2018). However, there are no records of the investments made for the Central American Games. It is striking that in the 2019 budget the construction of sports infrastructure "related to the Central American Games" is recorded. Although the work was carried out and inaugurated in 2019, it is not clear why a past event was used as justification. Normally in these cases, a portion of the budget is allocated to close the commitments of the Organizing Committee, and not to execute a work appealing to a past event. The situation of the 2027 Pan American Games, which for the moment (January 2024) have been taken away from Barranquilla, will be explained in the following section.

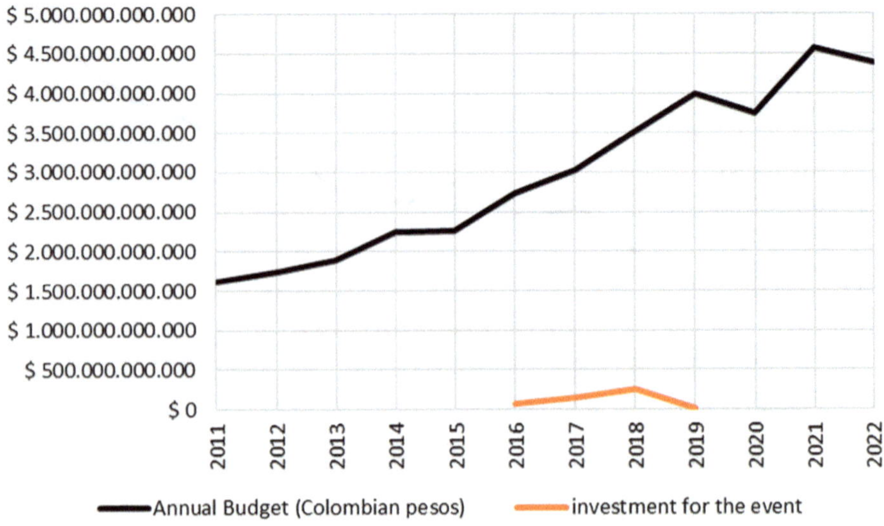

Chart 32.4 Budget for sport in Barranquilla, 2011–2022. Author's elaboration, 2023

32.6 Events and Leaders: Between Actors and Spectators

One of the crucial factors in the consolidation of public policies and their articulation with major events has to do with the solidity of state institutions, and especially the continuity of actions related to events of the scale of those discussed here. In Chile, between 2006 and 2023, just over four presidential terms, each of four years,[12] have elapsed. The stability of the executive power in the central government has not been reflected in the continuity in the sports portfolio in Chile; since its creation in 2013, the Ministry of Sports has had seven different ministers, with periods lasting between four months and just over two and a half years, with an average duration of 1.4 years.

In Colombia, where practically the same period as for Chile was covered, the situation of presidents and senior directors of national sports has been more or less similar, although the actual duration in the position of the highest authority in the country's sports leadership ranges from just over six months to close to two years. A common point in both cases is that there are usually two Sports Ministers for each four-year presidential term.[13] So far in the existence of the Ministry of Sports in Colombia (2019–2023), the average duration in the position of minister of the sector has been one year.

Compared to the Chilean case, the high rotation of the sports leadership in Colombia reveals a situation of instability of public policies in relation to sports, a fact that is reflected in the difficulty of coordinating in an articulated way the different levels of the executive power. The clearest example of this is the current situation of the tension existing between the district administration of Barranquilla and the national government, in relation to the development of an event of the magnitude of the XX Pan American Games 2027.

The recent decision of Panam Sports to withdraw from Barranquilla the venue for the Pan American Games (January 3, 2024) due to non-compliance with the payments agreed with the city, has been interpreted as a settling of scores with the Barranquilla leadership by the Presidency of the Republic, which is accused of not fulfilling the commitments signed by the previous government with Panam Sports.[14] Various voices have come together to demand that the national government unreservedly support the event (Canal RCN, October 18, 2023, October 19, 2023; Huérfano Sánchez, 2023).

It should be clarified that the current political leadership of the city of Barranquilla, which began to gain local and regional strength back in the 2000s, is of right-wing political affiliation and is questioned for its proven co-optation of political, economic, and judicial powers (Ardila Arrieta, 2023). The current national government (2022–2026), of left-wing, does not have parliamentary majorities and has been harshly questioned recently due to repeated clientelistic practices and abuses of power by parliamentarians and other members of its political leadership. The tense atmosphere of increasing reciprocal accusations between the local and national governments without assuming responsibilities for the irreversible loss of the Pan American Games 2027 venue causes controversies to go beyond the strictly sports plane and permeate other areas. This makes decisions about the non-realization of the 2027 Pan American Games enter the battlefield of political chess.

The study of similar experiences that have been successful in Colombia, such as the case of the VI Pan American Games of 1971 held in Santiago de Cali, shows that cooperation between political parties is fundamental, but in that case the political environment was very different and the actors who fundamentally influenced the outcome had links between them that went back to their school years (Figueroa-Pereira, 2013). The main difference between the cases of Cali and Barranquilla is that in the latter the actors are at both ends of the political spectrum, and neither of the parties has the slightest intention of yielding in their positions. In that sense, the evidence shows a certain air of pessimism, since it is very likely that Barranquilla will not hold the Games as a result of the intensification of the political controversy. A necessary concord between the parties and a non-aggression pact would be the main solution, so that Barranquilla's strategy to position itself as one of the main cities in the Caribbean events market does not sink.

Acknowledgments *Documentation and drawings*: Yeiser Mieles-Calderón, Daniela Salazar-Rojas, Jhonson Vidal-Jhavlosky, Robert Vides-Mercado.

Statistical research: Estefanny Paola Pulido-Escorcia, Sebastián Pérez-De Moya, Jhonson Vidal-Jhavlosky.

Notes

1. Founded in 1541 during the expansion of Spanish colonialism in the American continent, the city is located about 70 km to the east of the Pacific Ocean, on the foothills of the Andes Mountains, and has an average height of 570 m above sea level.

2. The PPP was celebrated between several companies in the construction sector and the Ministry of Housing and Urban PlanningUrban planning, represented by the Projects Unit of the Urban Development Division and by the Housing and Urbanization ServiceServices (SERVIU), owner of the lands where the Village was built.
   Design: Francisco Izquierdo Architects and MOBIL Architects.
   Developer Entity Lot B: GDI real estate management+ DLP Constructor
   Developer Entity Lot C: Mena &Ovalle real estate constructor
   Landscaping: PAUR Architects.
   Project Leader: Licancabur Investments SpA.
   Financial support: Ameris Fund.
3. According to the 2017 census data, the commune had a housing deficit of 1847 homes; the Pan American Village project consists of 1355 homes (Government of Chile, Ministry of Housing and Urban Planning, 2020).
4. As the project was developed without going through a public architecture competition for the design of the housing complex, it was strongly criticized by the Association of Architectural Offices of Chile (AOA), who considered this fact as a missed opportunity (Searle, 2020).
5. Its origin as a settlement is not entirely clear, as happens with many of the populations of the present-day Colombia. It is located practically at sea level, in proximity to the mouth of the Magdalena River to the Caribbean Sea.
6. Barranquilla became a regional capital between 1905 and 1910 and is the fourth city in the country in terms of population, after Bogotá, Medellín and Cali. Recently it has been displacing Cali in terms of its regional and national economic projection.
7. The eastern edge of Santiago is the Andes Mountain Range; that of Barranquilla, the Magdalena River.
8. In order to trace a timeline of the growth of the budget for sports in the cases mentioned, the three years prior to the designation of the venue for the events in question were taken as the starting point.
9. According to the execution of the accumulated capital expenditure to April 2023, there was an increase of 178.6% in the amount executed for Sports, with a view to the development of the Pan American and Parapan American Games. The investment showed a real expansion of 7.8%, product of the investments in projects of the Road Directorate of the Ministry of Public Works, the Ministry of Health and the Ministry of Sports (Government of Chile. General Directorate of Budgets, 2023).
10. Administrations have been divided among three rulers as re-election is not immediate, alternating between the left and the right (Michelle Bachelet and Sebastián Piñera (R.I.P., 2024) have repeated their term, while Gabriel Boric is in his first year and half of government).
11. The immediate re-election of presidents (a figure that existed between 2010 and 2018) did not alter these indicators. Although two rulers were re-elected consecutively in the periods 2002–2010 and 2010–2018, in the first case only the single director of Coldeportes who served during the second term of President Álvaro Uribe (2006–2010) was counted; during the government of Juan Manuel Santos (2010–2018) three directors of Coldeportes were appointed in a span of 8 years, one of them held the position for more than 4 years. In the presidency of Iván Duque (2018–2022), a director of Coldeportes and two ministers succeeded in a span of four years.
12. Barranquilla has only contributed $1,250,000 USD of the $8,000,000 USD for sports rights of the event, which had to be delivered in 2022. The entity warned the city's leadership that, if the situation is not resolved promptly and the total missing money is sent as soon as possible, the right to celebrate the Games would be lost. According to Huérfanó Sánchez,

    > So far, the organization of the Games in Barranquilla has only made small payments, such as 250,000 dollars for the granting of the venue (July 30, 2021), one million for sponsorship concessions (July 30, 2021) and another million for the same concept, paid on November 30, 2021.

13. Completing the picture is what was said by Canal RCN:

> It is worth noting that the organization and execution of the Games are estimated to cost 400 million dollars, as confirmed by MinDeportes. According to the communication last Wednesday to Panam Sports, this figure is divided into 70 million dollars for infrastructure works and 330 million for the expenses of organizing the Games.

14. As a point of comparison, the Lima 2019 Pan American Games cost 1234 million dollars (Lima 2019 Editorial, 2019).

References

Alcaldía Distrital de Barranquilla. Secretaría de Hacienda. (2012–2023). Ejecución Presupuestal de Rentas, 2012–2023. Retrieved from https://www.barranquilla.gov.co/hacienda/ejecuciones-presupuestales

Alcaldía Distrital de Barranquilla. Secretaría de Hacienda (2012–2023). Ejecución Presupuestal de Gastos, Vigencia 2012–2023. Retrieved from https://www.barranquilla.gov.co/hacienda/ejecuciones-presupuestales

Ardila Arrieta, L. (2023). *La Costa Nostra*. Rey Naranjo Editores.

Biblioteca del Congreso Nacional de Chile. (31 agosto 2020). Resolución 1058 Exenta. Llama a postulación en condiciones especiales para la presentación de proyectos habitacionales del Programa de Integración Social y Territorial, regulado por el DS N°19 (V. Y U.), de 2016 en terreno de propiedad de SERVIU de la Región Metropolitana, ubicado en la Ciudad Parque Bicentenario de la comuna de Cerrillos. Retrieved from https://www.bcn.cl/leychile/navegar?idNorma=1148845

Biblioteca del Congreso Nacional de Chile. (10 diciembre 2020). Resolución 1550 Exenta. Modifica Resolución N°1.059 Exenta, de fecha de 25 de agosto de 2020, del SEREMI Metropolitano de Vivienda y Urbanismo. Retrieved from https://www.bcn.cl/leychile/navegar?idNorma=1153093&idParte=10183310&idVersion=2020-12-10

Biblioteca del Congreso Nacional de Chile (12 octubre de 2021). Resolución 1220 Exenta. Modifícase Resolución N°64 exenta de la SEREMI Metropolitana de Vivienda y Urbanismo de fecha 26 de enero 2021. Retrieved from https://www.bcn.cl/leychile/navegar?idNorma=1166285

Bravo, G. A., Schonk, D. J., Silva-Bórquez, J., & González -Mesina, S. (2018). *Sport mega-events in emerging economies. The South American Games of Santiago 2014*. Palgrave Macmillan.

Canal RCN. (18 de octubre de 2023). ¡Día clave! Incertidumbre por la realización de los Panamericanos de Barranquilla 2027. Retrieved from https://www.canalrcn.com/deportesrcn/colombianos-en-el-exterior/juegos-panamericanos-de-barranquilla-2027-177316

Canal RCN. (19 de octubre de 2023). ¿Cuáles son las condiciones para que Barranquilla tenga los Panamericanos en 2027? Retrieved from https://www.canalrcn.com/deportesrcn/colombianos-en-el-exterior/barranquilla-condiciones-para-juegos-panamericanos-2027-177349

Chalkley, B., & Essex, S. (1999). Urban development through hosting international events: A history of the Olympic Games. *Planning Perspectives, 14*, 369–394.

Curador News. (1 de febrero de 2021). Adjudicada la Villa Pana y Parapanamericana - Curador News.

Curador News. (8 de febrero de 2021). Villa Panamericana: trenzar lo arquitectónico, deportivo e inmobiliario en un solo proyecto. - Curador News.

Di Vita, S., & Wilson, M. (Eds.). (2021). *Planning and managing smaller events. Downsizing the urban spectacle*. Routledge.

Fessler Vaz, L., & Berenstein Jacques, P. (2007). Contemporary urban spectacularisation. In F. J. Monclús & M. Guàrdia (Eds.), *Culture, urbanism and planning* (pp. 241–253). Ashgate.

Figueroa-Pereira, E. A. (2013). Grandes eventos como oportunidades de transformación urbana: los VI Juegos Panamericanos de 1971 en Santiago de Cali. [unpublished PhD thesis]. Pontificia Universidad Católica de Chile.

Francisco Izquierdo Arquitectos [@fia.arquitectos]. (5 de septiembre de 2023). Villa Panamericana. https://www.instagram.com/p/Cw0Y7npuhlC/?utm_source=ig_web_copy_link

Gobierno de Chile. Dirección General de Presupuestos. (2023, mayo). Informe de Ejecución Presupuestaria Mensual. Abril 2023. Retrieved from https://www.dipres.gob.cl/598/articles-311938_doc_pdf2.pdf

Gobierno de Chile. Dirección de Presupuestos (DIPRES). (2003–2024) Ley de presupuestos del Sector público, años 2006–2024. Retrieved from https://www.dipres.gob.cl/598/w3-channel.html

Gobierno de Colombia. Ministerio del Deporte. (2010–2023). Decretos Presupuesto General de la Nación, 2010–2023. Retrieved from https://www.mindeporte.gov.co/control-rendicion-cuentas/informacion-contable-financiera/presupuesto-1/presupuesto-inicial/

Gold, J., & Gold, M. (Eds.). (2007). *Olympic cities: City Agendas, planning, and the World's Games, 1896–2012*. Routledge.

Grix, J, Brannagan, P. M., & Lee, D. (2019). *Entering the global Arena. Emerging states, soft power strategies and sport mega-events*. Palgrave Macmillan.

Gusmão de Oliveira, N. (2021). *Mega-events, City and Power*. Routledge.

Huérfano Sánchez, J. S. (19 de octubre de 2023). Juegos Panamericanos: plazo y condiciones que le pusieron a Barranquilla para ser sede. https://www.antena2.com/juegos-panamericanos/juegos-panamericanos-2027-plazo-y-condiciones-que-le-pusieron-a-barranquilla

Kassens-Noor, E. (2012). *Planning Olympic legacies: Transport dreams and urban realities*. Routledge.

La Cuarta Constructor. (30 de junio de 2023). Santiago 2023. ¿Cómo será vivir en la Villa Panamericana? | La Cuarta Constructor.

Monclús Fraga, F. J. (2007). *Exposiciones Internacionales y Urbanismo. El proyecto Expo Zaragoza 2008*. UPC Edicions. Universitat Politecnica de Catalunya.

Muñoz H.,C. (22 de mayo de 2023). El fondo que levantó Ameris para construir la Villa Panamericana de Santiago 2023. | Diario Financiero. Retrieved from https://www.df.cl/mercados/fondos-de-inversion/el-fondo-que-levanto-ameris-para-construir-la-villa-panamericana-de

Preuss, H. (2004). *The economics of staging the Olympics: A comparison of the Games 1972–2008*. Edward Elgar.

Redacción Lima 2019. (2019). LIMA 2019 PRESENTÓ BALANCE DE LOS JUEGOS PANAMERICANOS Y PARAPANAMERICANOS EN EL CONGRESO. Retrieved from https://www.lima2019.pe/noticias/lima-2019-presento-balance-de-los-juegos-panamericanos-y-parapanamericanos-en-el-congreso#:~:text=Neuhaus%20destac%C3%B3%20que%20el%20presupuesto,los%204%2C200%20millones%20de%20soles

Roche, M. (2000). *Mega-events and Modernity. Olympics and Expos in the growth of global culture*. Routledge.

Santiago 2023. (2022). Gran impacto económico por Juegos Panamericanos y Parapanamericanos de Santiago 2023. Retrieved from https://www.santiago2023.org/es/noticias/portada-sitio-web/detalle/546/gran-impacto-economico-por-juegos-panamericanos-y-parapanamericanos-de-santiago-2023

Searle, R. (2020, 3 de agosto). CONCURSOS PÚBLICOS DE ARQUITECTURA. Retrieved from https://www.aoa.cl/concursos-publicos-de-arquitectura/

Smith, A. (2012). *Events and urban regeneration: The strategic use of events to revitalise cities*. Routledge.

Tinaz, C., & Knott, B. (Eds.). (2021). *Sport and development in emerging nations*. Routledge.

Part VIII
Perceptions, Emotions and Imaginaries in Architecture, Design, the City and the Territory

Chapter 33
Narratives and Public Discourses of the Right to the City: A Review from the Press to the Bicentennial Bridge Conflict—Concepción, Chile 2010–2020

Aaron Napadensky Pastene, Javier León Aravena, and Barbara Ojeda

Abstract We focus on the conflict of the Bicentennial Bridge in the city of Concepción, Chile, whose layout confronted informal settlers and the public-state sector, as an event through which to investigate, from its public arena and the discourses that structure it, the multiple narratives that entered into controversy here. Our hypothesis suggested that the narrative of the Right to the City, significantly established by supranational institutions and by local academic and professional elites, fails to constitute a narrative that substantiates or refutes the contingent processes of city-making. In the methodological strategy, we made the framing theory and content analysis converge, focusing on discovering, in an abductive and iterative way, the different communicational frames that each group of actors struggled to impose, focusing on the discursive categories defined as substantial frames, that is, (i) how they define the origin of the controversy, (II) the aspirations that the conflict brings into play, and (iii) the desired outcomes. For this, we relied on the records of the local press and the use of the software ATLAS TI, coding and categorizing discursive contents and identifying the different frames in dispute. Our results show frames, which, taking elements from the Right to the City, fail to structure themselves as a significant narrative, being quickly excluded from the public arena. Thus, the contingent discourses, of one and the other, ended up circumscribed to a hegemonic and structuring capitalist narrative, reducing the conflict to a controversy over the scale and scope of the desired modernization, thereby marking the material future of the place object of the controversy.

A. N. Pastene (✉) · B. Ojeda
Departament of Urban Planning and Design, Universidad del Bío Bío, Concepción, Chile
e-mail: anapaden@ubiobio.cl

B. Ojeda
e-mail: ojedabar@gmail.com

J. L. Aravena · B. Ojeda
Department of Social Sciences, Universidad del Bío Bío, Concepción, Chile
e-mail: jleon@ubiobio.cl

M. Vicuña and E. Giorgi (eds.), *Intersections*, Springer Geography,
https://doi.org/10.1007/978-3-031-76402-8_33

Keywords Communicational frames · Conflicts · Public arenas · Right to the city

33.1 Introduction

Urban conflicts and struggles, beyond descriptive casuistry, are tending to group under the theoretical and analytical discourse of the Right to the City. This is especially true in academic spheres,[1] supranational public and civil society.[2] Its arrival to local scale actors is less clear, as some argue (Cuberos-Gallardo, 2020; Pérez, 2019) and others refute (Benitez, 2019).

Given the context, we set out to see how much of this institutionalized narrative of the Right to the City[3] is used as a discursive argument to position demands and defend positions. For this, we conducted a case study—Chacabuco Bridge conflict, Concepción, Chile—where we used the press and its notes (2010–2020), to access its public arena and the discourses that structured it.

The research had two key questions: (i) What were the substantial and public ways of framing the conflict by the different groups of actors in the controversy? And (ii) how relevant was the narrative of the Right to the City to position demands and defend positions? The hypothesis suggested that the supra-institutionalized narrative of the Right to the City has not managed to precipitate the contingent and public discussions of city-making.

To test the hypothesis, we made the framing theory and content analysis converge, focusing on discovering, in an abductive and iterative way, the different communicational frames that each group of actors tried to impose, focusing on the category of substantial frames.

The results expose the evolution of the public arena of conflict, its metrics and communicational frames, revealing the main discourses and narratives in dispute, with special attention to the argumentative use of the Right to the City.

We conclude that the results gave credibility to the hypothesis. Although present, the Right to the City is not structured as a substantial narrative, and perhaps for the same reason, it does not constitute an alternative frame to the hegemonic narrative of capitalist modernization, which emerged strongly in our analysis.

Finally, we consider that results and conclusions account for not always evident discursive alliances that struggle to keep at the center of the public arena, the neoliberal capitalist narrative hegemony, inhibiting other narratives, limiting and performing with it the material becoming of the space in conflict.

33.2 Theoretical–Methodological Foundations

The sociological theory of "*framing*," or framing, has been widely used in media studies and analysis (Zhonghua & Xiaoling, 2021). The frame or framework, in addition to being a social construction of reality, is a device used for cognitive and communicative analysis, recognizing that confronted actors, with multiple interests and objectives, will often offer different descriptions of the same situation, which can be attributed to unequal understandings or perceptions of the situation itself, cognitive framing (Scheufele, 1999), but also to the fact that they use different interpretations to enhance their own message and position their demands (Buijs et al., 2011; Ferranti et al, 2014), communicative framing.

Here, we use the latter, differentiating it according to magnitude and complexity, into (i) the discursive ones, which can be precise and limited, contingent and mobilizing. Dedicated to identifying how actors present, conceptualize or construct the problem with the aim of mobilizing collective action and expanding their support base (Benford & Snow, 2000; Chong & Druckman, 2007; Shmueli, 2008; Snow & Benford, 1992); (ii) the narrative ones, which convey a central idea that transcends, organizes, gives meaning, enables and defines the interpretation of the problem, issue or demand (Buijs et al., 2011; Gamson et al, 1992; Lewicki et al., 2003), highlighting certain facts, arguments and experiences for its understanding (Gray, 2003).

The narrative, introduced in the studies and analysis of deliberative policies (Hajer & Versteeg, 2005), is a concept used in social movement studies to demonstrate how the change and adaptation of the positions and demands of the protest actors, are embedded in the construction of stories about the movement, its past events and future perspectives (Bruner, 1991; De Moor & Wahlström, 2019; Polletta & Chen, 2017). Allowing actors to resort to discursive categories that not only give meaning to specific physical or social phenomena (Hajer, 2005; Moulaert et al., 2016; Verloo, 2018), but also account for the stabilities and changes in the general systems of meaning and institutionalized thought (Polletta & Chen, 2017; Wittmayer et al, 2019).

The narrative is more than a discourse, constituting itself as a story with a chronological order and particular structure, beginning, middle and end (Adger et al, 2001), but that usually does not leave the interest in increasing the support bases and exercising dominance over a central theme (Buijs et al., 2011; Carragee & Roefs, 2004; Gamson, 2001, 2005; Lewicki et al., 2003), often being part of the struggle for to conquer the center of the discursive arena (Buijs et al., 2011; Idrissou et al., 2011; Lewicki et al., 2003; Pecurul-Botines et al., 2014; Von Essen, 2017).

Research on environmental conflicts has shown that the ways in which stakeholders frame conflicts explains a good part of their success or failure (Gray, 2003), possibly inducing a rethinking of the underlying assumptions about how controversy is defined and addressed (Luoma & Martela, 2021), and thereby creating new alternative understandings of the problem, challenging current assumptions and rethinking new future possibilities (Mukherjee et al, 2020; Asah et al., 2012).

33.3 Methodology

Within the theory of *framing,* there are the so-called substantial communication frames, which circumscribe those discursive struggles to impose their definition on (i) the origin of the discord, (ii) the underlying aspirations that the conflict brings into play and (iii) the desired outcomes (Shmueli, 2008). These three spheres are the ones we investigate here.

In this way, we put the framing theory into the search engine of the content analysis. Giving a foundation for this search to this methodological strategy of hermeneutic orientation and interpretive work, in which we unfold the discourses in a way different from the sum of preferences of a survey. That is, a phrase or statement can contain multiple meanings. For the same reason, we use procedures of analytical deconstruction, such as, coding[4] and categorization[5], which were iteratively discussed in continuous analytical-theoretical reductions, structuring a thematic grid under the logic of abduction (Bar, 2001).

It is pertinent to say that this method, discourse analysis, does not focus directly on the specific phenomenon itself, but rather on the statements relative to this phenomenon, on the creators of these statements and on the processes of consolidation or deconstruction of their statements (Adger et al, 2001).

Our source of information was the press, since, like all discourse, the written one, is a fact of communication framed in a social, political, historical and cultural context, which allows to approach the characteristics of the speaker, his group or organizational membership and his ways of conceiving and projecting the world (Van Dijk, 1990, 1993). However, the press, like all media, hardly abstracts completely from its interest groups, political trends and power structures, making visible and invisible certain discourses (Krippendorff, 1997), revealing an always incomplete image of power relations (Tourangeau, 2018), leaving recorded the limits of the methodology.

Regarding the press, we selected two traditional local media outlets with center-left and center-right editorial lines, which have covered the conflict the most in the ten years we defined to observe, 2010–2020 (Radio Biobío + Diario de Concepción) (Image 33.1). In addition, we set aside the writings of journalists, focusing only on press quotes issued by the actors in conflict, searching in the textual bodies, significant grammatical units, phrases and/or paragraphs, which, according to the linguistic pattern, allowed to describe through explicit or interpretable contents, a position in one of the substantial frames. In a second moment, we performed multiple theoretical reductions, identifying the narratives in controversy behind these (Table 33.1).

The analysis and coding were processes of going back and forth, marked by the emerging contingent frames and their multiple theoretical reductions (categorizations), unifying them into structural categories, defined from the central theoretical categories of entry (substantial frames) and emerging subcategories of outputs (contingent frames) (Table 33.1).

Table 33.1 Methodological structure diagram

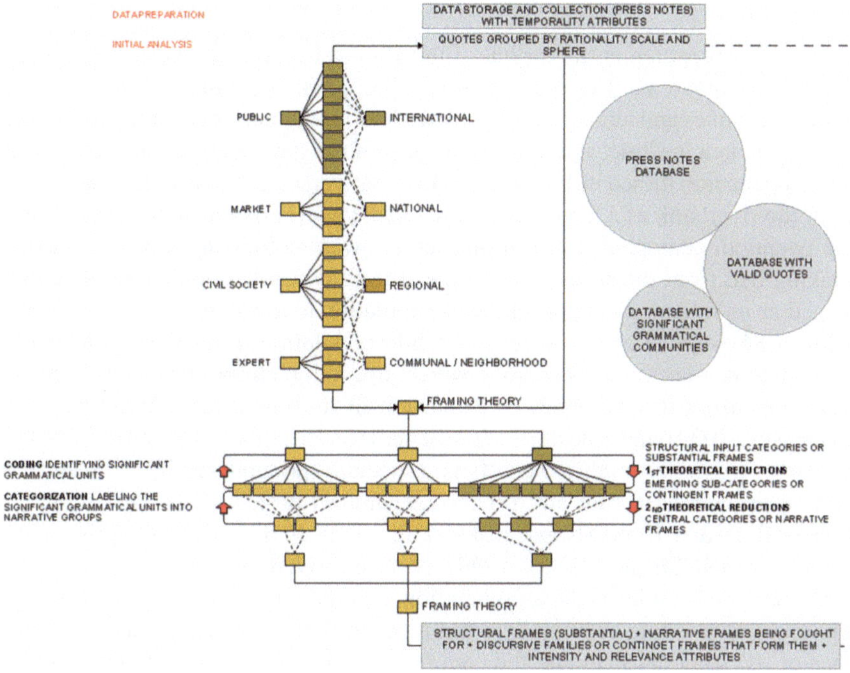

Source Own elaboration

Ciudad

Gobierno confirma entrega de puente Bicentenario para mayo de 2018

SEMINARIO: DESAFÍOS DE POLÍTICA PÚBLICA EN INNOVACIÓN Y EMPRENDIMIENTO

28 familias

Ciudad

3.000 millones de

Soterramiento de la vía férrea en Concepción no puede abandonarse

FOTO: RAPHAEL SIERRA P.

Tribunal acoge de forma parcial reclamación sobre el Bicentenario

Ciudad

En tres años Aurora de Chile tendrá un nuevo rostro

Image 33.1 Press of the conflict. *Source* Arch. Diario de Concepción

Once the above was done, we analyzed the regularities in expressions that allowed us to identify the discourses in controversy and which of them were embedded in a certain narrative.

33.4 Case Study

The city of Concepción is the capital of the Biobío Region, around it is organized the Metropolitan Area of Concepción (Image 33.2), which within its limits contains thirteen communes and almost a million and a half inhabitants, being the third largest metropolitan area in Chile, and the first in the south of the country. Since its consolidation, Concepción turned its back on the Biobío River, making the train line a border that left the riverbank as a large behind the city. However, it was in the second third of the twentieth century that precarious settlers were colonizing this space, filling and taking bed from the river, creating the Aurora de Chile population, which has remained as an informal but consolidated population to this day.

Almost 85 years after the Aurora de Chile population, it has been affected by the urban dynamics of Concepción, which like all metropolitan areas, has not been exempt from the processes of expansion of its limits, functional reordering (Napadensky, 2020) and transformation of its interior soils (Napadensky & Orellana, 2019), triggering multiple conflicts, especially precipitated by the execution of equipment, infrastructure and urban regeneration projects that have sought to reconquer the water fronts (Napadensky et al, 2022). In this sense, the urban piece that convenes us (Image 33.2), not only is on the riverbank, in the center of the metropolitan area and hosts the aforementioned population, but also since 1990 it is being object of multiple attempts at urban regeneration, reflected in various modifications of its Sectional Plan and the communal and metropolitan regulatory plans. But it is not until after the 2010 earthquake that it jumps to the public arena, with the so-called Bicentennial Bridge conflict, controversy triggered by a road layout, whose design did not contemplate the existence of the informal, but consolidated population (Aurora of Chile (Images 33.3 and 33.4)) thus initiating a media urban controversy between the State and inhabitants (Image 33.1).

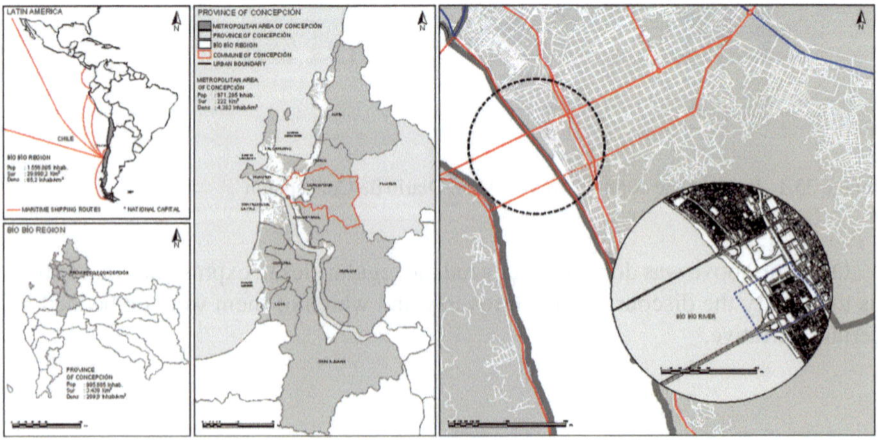

Image 33.2 Location and placement area of the case study. *Source* Own elaboration

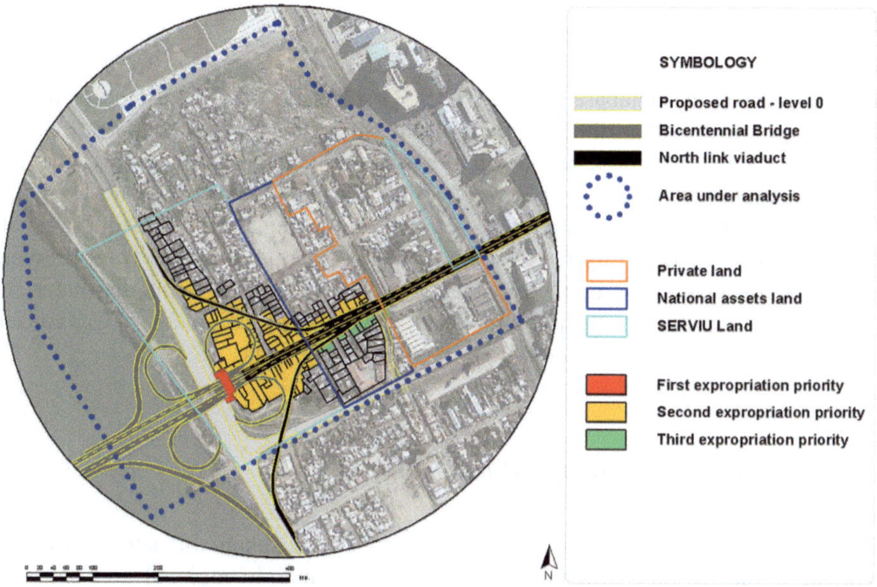

Image 33.3 Traced Chacabuco bridge and population Aurora of Chile. *Source* Own elaboration based on Arch. MOP, MINVU and SERVIU

Image 33.4 Chacabuco bridge stopped by the conflict and population Aurora of Chile. 2020. *Source* Own elaboration based on Arch. MOP, MINVU and SERVIU; and photography Arch. personal

33.5 Results

Before entering the database into the Atlas TI software, the 787 discursive quotes found were rated with attributes of temporality, scale and sphere of actor rationality, identifying a first discursive primacy in the actors of the regional scale and sphere of public institutionality, closely followed by the communal scale and civil society.

With the software, the database was refined, identifying 573 significant grammatical units, of which 225 referred to some type of substantial framing. Thus, drawing the discursive disputes to define the hegemony in terms of the origin and nature of the conflict, what this put at stake and the expected results. This analysis was carried out on these 225, whose results were organized in two parts.

33.5.1 The Public Arena of the Conflict

The quotes with attributes of temporality, scale and sphere of actor rationality evidenced the media dominance of public actors, mainly regional and national, setting, in one way or another, the hegemonic scale of the interests behind the new bridge. Also relevant is that the conflict does not publicly emerge with the demands of civil society, but with those of market actors and experts, who as we will see later, had unequal scalar approaches in terms of sought modernization (Table 33.2).

The civil society of neighborhood scale enters forcefully in year 2 of the conflict, to then be internally highlighted by discourses of communal scale and finally national. Expressing certain institutionalization of the group and its discourses (Table 33.2).

33.5.2 Narratives in Conflict and the Right to the City

We did not find explicit references to the Right to the City, but rather a minority narrative of collective rights, focused on its institutionalized denial (Table 33.3).

In contrast, a media-strong and cross-cutting capitalist narrative of modernization emerged, which, centralized in its decisions and inequitable in its distributions, presents itself as perfectible, adjustable and materially desirable. This narrative, initially driven by market actors, was intensely supported by the public and experts, constituting a hegemonic, but not monolithic, common substrate to many of the discourses analyzed, including civil society (Table 33.3).

Present at the beginning and in the frames of the origin of the conflict, the narrative of unjust subordination of rights was gradually removed from the public arena. As the conflict discursively transited towards the underlying aspirations and desired outcomes, the issue of collective rights moved to a discussion of individual rights, especially driven by the public sphere and its housing discourse.

Table 33.2 Public arena of the conflict; press quotes grouped by period, scale and sphere of discursive actor rationality

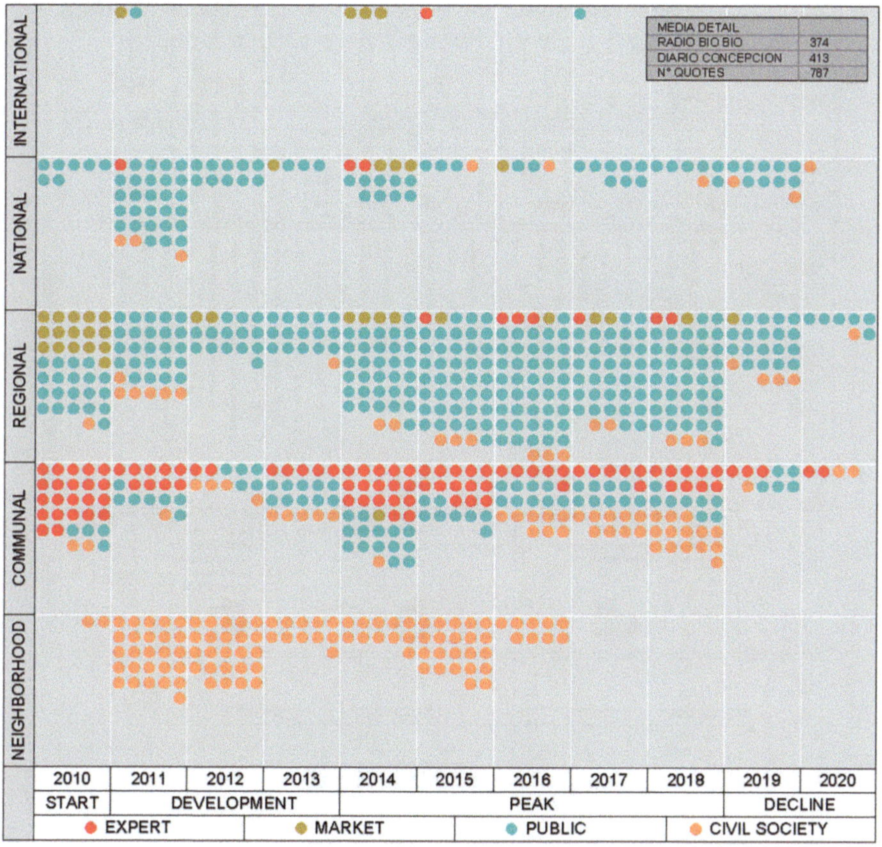

Source Own elaboration based on Atlas TI reports, and Arch. Radio Biobío and Diario de Concepción, period 2010–2020

33.6 Conclusion

The discourses, which in one way or another, alluded to the Right to the City, did not manage to constitute themselves as an alternative narrative to the hegemonic neoliberal narrative that underpins capitalist modernization. Appreciating with this, an accelerated reduction of the limits of the public discussion of the conflict, which ended up circumscribed to a controversy about the scale and scope of the desired modernization; regional, metropolitan, communal, housing, urban or social, affecting the trajectory and material outcome of the conflict.

Thus, and in line with other studies, the public arena of the conflict tended to be produced, reproduced and structured based on the discourses and narratives sustained and projected by the dominant actors (Svarstad & Benjaminsen, 2020), denoting

Table 33.3 Significant grammatical units according to intensity, framing, and type of actor

ORIGIN	PERIOD	MEDIA OUTLETS TOTAL	QUOTATIONS TOTAL	FRAMING CODES TOTAL	SUBSTANTIAL FRAMING CODE TOTAL
	2010–2020	2	660	573	225

INPUT FRAMES OR STRUCTURAL CATEGORIES

SPHERE OF RATIONALITY	01.A	01.B	01.C	01.D	02.A	02.B	03.A	03.B	03.C	03.D	04.A	04.B	04.C	04.D	05.A	05.B	05.C	05.D	06.A	06.B	06.C	07.A	07.B	08.A	08.B	09.A	10.A	10.B	TOTALS
	ORIGIN OF THE PROBLEM 01				02		03				BASE ASPIRATION 04				05				PROJECTED RESULTS 06			07		08		09	10		
EXPERTS	2	8	2	2	6	0	1	1	6	0	3	3	1	9	1	0	3	0	1	1	2	0	1	0	0	0	0	0	52
MARKET	0	4	0	0	1	0	0	0	0	0	1	0	4	1	0	0	0	0	1	0	0	1	0	1	0	3	0	0	17
PUBLIC INST.	4	7	1	0	2	2	1	2	1	0	1	0	4	12	0	2	1	2	1	0	0	1	7	10	2	2	12	19	96
CIVIL SOCIETY	11	15	3	0	1	3	0	6	5	0	0	0	1	1	1	1	4	2	0	0	0	1	0	0	0	0	5	0	60
TOTALS	17	34	06	02	10	06	02	09	12	0	06	03	10	23	02	03	08	04	03	01	02	03	08	11	02	05	17	19	225

ACTORS SCALE	01.A	01.B	01.C	01.D	02.A	02.B	03.A	03.B	03.C	03.D	04.A	04.B	04.C	04.D	05.A	05.B	05.C	05.D	06.A	06.B	06.C	07.A	07.B	08.A	08.B	09.A	10.A	10.B	TOTALS
	ORIGIN OF THE PROBLEM 01				02		03				BASE ASPIRATION 04				05				PROJECTED RESULTS 06			07		08		09	10		
INTERNATIONAL	1	0	0	0	0	0	0	0	0	0	0	0	0	0	0	0	1	0	0	0	0	0	0	0	0	0	0	0	2
NATIONAL	4	1	0	0	0	1	0	0	1	0	1	0	2	4	1	0	0	1	0	0	0	0	0	5	0	1	0	0	23
REGIONAL	6	5	1	0	3	0	1	1	0	0	1	0	5	5	0	1	1	1	2	0	1	1	6	5	1	3	18	13	80
COMMUNAL - NBHD	31	3	5	2	7	4	1	8	11	0	3	3	3	14	1	1	6	2	1	1	1	2	2	1	1	1	1	4	120
TOTALS	42	03	06	02	10	05	02	09	12	0	06	03	10	23	02	02	08	04	03	01	02	03	08	11	02	05	19	17	225

EMERGING SUBCATEGORIES OR CONTINGENT FRAME

INTENSITY OF THE FRAME IN THE ACTORS DISCOURSE (+ / –)

INTENSITY OF THE FRAME IN THE PUBLIC ARENA (+ / –)

[01] EXCLUDING AND WEAK GOVERNANCE [01.A]; CENTRALIST PLANNING [01.B]; SECTORIAL PLANNING [01.C]; SHORT-TERM PLANNING [01.D]

[02] UNFAIR URBAN COMPENSATION [02.A]; UNFAIR INDIVIDUAL COMPENSATION [02.B]

[03] NATURE [03.A]; HOUSING [03.B]; LAND USE AGREEMENT [03.C]; CITY [03.D]

[04] SOCIAL [04.A]; ENVIRONMENTAL [04.B]; ECONOMIC [04.C]; URBAN AND TERRITORIAL [04.D]

[05] CITY BUILDING [05.A]; PRESERVING HISTORY [05.B]; POLITICAL [05.C]; HOUSING [05.D]

[06] RECOVERY OF HISTORIC PROPERTY [06.A]; PROJECTING NEIGHBORHOOD ACTIVITIES [06.B]; THE RIVERBANK FOR THE NEIGHBORS [06.C]

[07] CITY CENTER EXPANSION TO THE RIVER [07.A]; FORMALIZING LAND FOR PROPERTY DEVELOPMENT [07.B]

[08] CONNECTING INTER-COMMUNAL FLOW [08.A]; CONSOLIDATING METROPOLITAN WATERFRONT [08.B]

[09] CONSOLIDATING LOAD FLOW INFRASTRUCTURE [09.A]

[10] FORMALIZING RADICALIZATION [10.A]; FORMALIZING ERRADICATION [10.B]

DISPUTE NARRATIVE

ORIGIN OF THE PROBLEM		BASE ASPIRATION		PROJECTED RESULTS	
[01] LACK OF PARTICIPATION	69	[04] CAPITALIST MODERNIZATION	41	MODERNIZATION [06] NEIGHBORHOOD	06
[02] UNEQUAL COST-BENEFIT DISTRIBUTION	15	[05] COLLECTIVE RIGHTS	16	[07] COMMUNAL	11
[03] NO RECOGNITION OF COLLECTIVE RIGHTS	23			[08] METROPOLITAN	13
				[09] REGIONAL	05
				[10] HABITATIONAL	36
	97		57		71

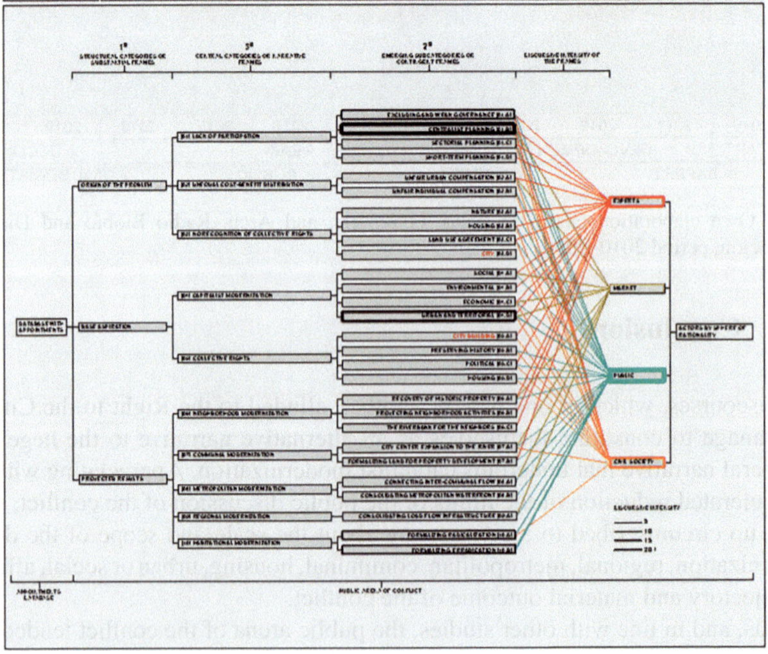

Source Own elaboration based on Atlas TI reports, and Arch. Radio Biobío and Diario de Concepción

the strength of the neoliberal narrative as the main system of truth (Adger et al., 2001), and modernization as its path to an eternal promise of individual prosperity, where the collective Right to the City and other counter-hegemonic views, do not constitute relevant counter-narratives. In this sense, the public arena, more than a space for deliberation[6], was exercised as an instrument of power and dominion, which strengthened the capacities of the actors and hegemonic narratives, to carry out their will despite the resistance of others (Weber, 1968), ending in a negotiation that was made without questioning the model or abdicating the agenda and results hegemonically desired (Gingembre, 2015; Holmes, 2007), affecting with this the course of the conflict (Aukes et al., 2020), the material transformation of the space in controversy and the possibilities of a structural change in the ways of planning and making decisions about the built environment and its developments.

In addition, we note, although it was not the focus of our research, that the public arena can help to enable or restrict the decisions and practices of the actors, who immersed in a structure of dominant knowledge, are previously circumscribed to what is expected of them (Clare et al., 2013; Rossi, 2004). In this sense and following Harvey (2007), it must be clear that neoliberalism as a hegemonic discourse is part of the common sense in which we interpret, live and understand the world, enabling some actors, arguments and ideas, while excluding others (Tourangeau, 2018).

We were able to notice a two-headed discursive sphere of public institutionality. That, on the one hand, has at the national and supranational scale, a discourse that appeals to the Right to the City as a new paradigm to concur. And on the other, the contingent discourses of regional and local actors, whose foundations are still deeply anchored to the neoliberal capitalist narrative and its tale of competitiveness and economic modernization. In the other sphere, the discourses of civil society and local communities tended to show how the collective and initial demands, which could approach the collective discourse of the Right to the City, ended up placated, deconstructed and marginalized by an instrumental, disaggregating and individual discourse of the right to housing.

Finally, we must note that despite the limitations of the method and sources, this article revealed how the press and media culture, impose biases and limitations on the news coverage, reinforcing certain discourses and frames, in this case, mostly embedded in a neoliberal state governmentality, raising a series of underlying normative assumptions, which privilege the subjectivation of certain values and norms, parts of a hegemonic regime of truth and common sense, making public discourse and its supposedly deliberative arenas, a means of control over subordinate individuals and maintenance of control over the production of urban space.

Acknowledgements The author conveys thanks to the research assistants Barbara Ojeda, Ricardo Aguilera, and Williams Jara and to the University of Biobío, its Urban Studies Laboratory, LEU, the Regular Research Project 2021, Cod. 2110239 IF/R and the Regular Research Project 2022 Cod. 2250321 IF/R.

Notes

1. See, Aguilera-Rodriguez (2022), Najman (2022), Belda-Miquel et al. (2022), Pamplona and Piedrahita (2021), Vergara-Perucich and Boano (2020), Duque-Díez (2021), among others.

2. https://hic-al.org and https://onuhabitat.org.mx/index.php/componentes-del-derecho-a-la-ciudad, to name a few.
3. According to the UN, the Right to the City is "… the right of all inhabitants to inhabit, use, occupy, produce, transform, to govern and enjoy fair, inclusive, safe, sustainable, and democratic cities, towns, and urban settlements, defined as common goods for a dignified life".
4. Operation by which each unit of analysis is assigned a code specific to a category, and the code is a combination of elements that has a certain value in a certain category system, and the association of codes that are related to each other, whether by theme, process, time, relationship, causes, or others, are called family or structural code.
5. Classification and conceptual reduction of substantive codes groupable to the same theme, thus generating the categories. And those that group theoretical codes that can better explain the variability of the data, giving meaning to these and their relationships, explaining a pattern, are called central categories.
6. Deliberation presupposes that anyone can engage in free and reasoned dialogue to identify common solutions to conflicts and disagreements over collective problems, therefore, and in principle, be open to changes (Habermas, 2005 [1996]; Olson, 2011; Parkins & Mitchell, 2005).

References

Adger, W. N., Benjaminsen, T. A., Brown, K., & Svarstad, H. (2001). Advancing a political ecology of global environmental discourses. *Development and Change, 32*(4), 681–715. https://doi.org/10.1111/1467-7660.00222

Aguilera-Rodríguez, A. R. (2022). On the right to the city and its pillars as an emerging human right. *Legal Opinion, 20*(43), 261–282. https://doi.org/10.22395/ojum.v20n43a10

Asah, S. T., Bengston, D. N., Wendt, K., & Nelson, K. C. (2012). Diagnostic reframing of intractable environmental problems: Case of a contested multiparty public land-use conflict. *Journal of Environmental Management, 108*, 108–119. https://doi.org/10.1016/j.jenvman.2012.04.041

Aukes, E., Lulofs, K., & Bressers, H. (2020). (Mis-)matching framing foci: Understanding policy consensus among. *Ocean and Coastal Management, 197*. https://doi.org/10.1016/j.ocecoaman.2020.105286

Bar, A. (2001). Abduction. The inference of discovery. *Cinta Moebio: Journal of Epistemology of Social Sciences, 12.*

Belda-Miquel, S., Pellicer-Sifres, V., & Boni, A. (2022). Building commons for the right to the city through collective social innovation in distribution and consumption: Exploring a conceptual framework and the case of Valencia. *EURE (Santiago), 48*(143) https://doi.org/10.7764/eure.48.143.07

Benford, R. D., & Snow, D. E. (2000). Framing processes and social movements: An overview and assessment. *Annual Review of Sociology, 26*, 611–639. https://doi.org/10.1146/annurev.soc.26.1.611

Benitez, J. (2019). The limits of the right to the city as a prism for understanding urban conflicts: Collective action and language of rights in the Autonomous City of Buenos Aires. *Poblac. Soc, 26*(1), 6–31. ISSN 1852-8562. https://doi.org/10.19137/pys-2019-260102

Bruner, J. (1991). The narrative construction of reality. *Critical Inquiry, 18*(1), 1–21. https://doi.org/10.1017/CBO9781107415324.004

Buijs, A. E., Arts, B. J., Elands, B. H., & Lengkeek, J. (2011). Beyond environmental frames: The social representation and cultural resonance of nature in conflicts over a Dutch woodland. *Geoforum, 42*(3). https://doi.org/10.1016/j.geoforum.2010.12.008

Carragee, K. M., & Roefs, W. (2004). The neglect of power in recent framing research. *The Journal of Communication, 54*(2), 214–233.

Chong, D., & Druckman, J. N. (2007). Framing theory. In *Annual review of political science. Annual reviews* (pp. 103–126). Palo Alto.

Clare, S., Krogman, N., & Caine, K. (2013). The “balance discourse”: A case study of power and wetland management. *Geoforum, 49*, 40–49. https://doi.org/10.1016/j.geoforum.2013.05.007

Cuberos-Gallardo, F. (2020). Citizenship, neighborhood and the right to the city. Cova da Moura, Lisbon. *Bitácora Urbano Territorial, 30*(3). https://doi.org/10.15446/bitacora.v30n3.82480

De Moor, J., & Wahlström, M., (2019). Narrating political opportunities: Explaining strategic adaptation in the climate movement. *Theory and Society, 48*(3), 419–451. https://doi.org/10.1007/s11186-019-09347-3

Duque-Díez, M. (2021). Urban violence and the right to the city: Analysis of the Medellín case. *Cities, States and Politics Journal, 8*(2), 89–103.

Ferranti, F., Turnhout, E., Beunen, R., & Behagel, J. H. (2014). Shifting nature conservation approaches in Natura 2000 and the implications for the roles of stakeholders. *Journal of Environment Planning and Management, 57*(11), 1642–1657.

Gamson, W. A., Croteau, D., Hoynes, W., & Sasson, T. (1992). Media images and the social construction of reality. *Annual Review of Sociology, 18*. http://www.jstor.org/stable/2083459

Gamson, W. (2001). Media and social movements. In J. S. Neil & P. B. Baltes (Eds.), *International encyclopedia of the social & behavioral sciences*. https://doi.org/10.1016/B0-08-043076-7/04349-7

Gamson, W. (2005). Projections of power: Framing news, US public opinion and foreign policy. *Public Opinion Quarterly, 69*(2). https://doi.org/10.1093/poq/nfi017

Gingembre, M. (2015). Resistance or participation? Fighting against corporate land access amid political uncertainty in Madagascar. *Journal of Peasant Studies, 42*(3–4), 561–584. https://doi.org/10.1080/03066150.2015.1022867

Gray, B. (2003). Framing of environmental disputes. In R. J. Lewicki, B. Gray & M. Elliot (Eds.), *Making sense of intractable environmental conflicts: Frames and cases*. Island Press.

Habermas, J. (2005) [1996]. *Between facts and norms*. Polity Press, Cambridge. https://doi.org/10.1080/1350176042000311943

Hajer, M. (2005). *The politics of environmental discourse: Ecological modernization and the policy process*. Clarendon Press.

Hajer, M., & Versteeg, W. (2005). A decade of discourse analysis of environmental politics: Achievements, challenges, perspectives. *Journal of Environmental Policy & Planning,* (7)

Harvey, D. (2007). Neoliberalism as creative destruction. *The Annals of the American Academy of Political Social Science, 610*(1), 21–44. https://doi.org/10.1177/0002716206296780

Holmes, G. (2007). Protection, politics, and protest: Understanding resistance to conservation. *Conservation and Society, 5*(2), 184–201.

Idrissou, L., Aarts, N., van Paassen, A., & Leeuwis, C. (2011). The discursive construction of conflict in participatory forest management: The case of the Agoua forest restoration in Benin. *Conservation and Society, 9*(2), 119.

Krippendorff, K. (1997). Main metaphors of communication and some constructivist reflections about their use. In M. Pakman (Ed.), *Constructions of human experience II* (pp. 107–146). Gedisa Editorial. Retrieved from http://repository.upenn.edu/asc_papers/333

Lewicki, R., Gray, B., & Elliott, M. (Eds.), (2003). *Making sense of intractable environmental conflicts: Concepts and cases*. Island press.

Luoma, J., & Martela, F. (2021). A dual-processing view of three cognitive strategies in strategic decision making: Intuition. *Analytic Reasoning, and Reframing, Long Range Planning*. https://doi.org/10.1016/j.lrp.2020.102065

Moulaert, F., Jessop, B., & Mehmood, A. (2016). Agency, structure, institutions, discourse (ASID) in urban and regional development. *International Journal of Urban Sciences, 20*(2). https://doi.org/10.1080/12265934.2016.1182054

Mukherjee, M., Ramirez, R., & Cuthbertson, R. (2020). Strategic reframing as a multi-level process enabled with scenario research. *Long Range Planning, 53*(5). https://doi.org/10.1016/j.lrp.2019.101933

Najman, M. (2022). About the effects of the location of the new social housing on the struggles for access to the city of the popular sectors. *EURE, 48* (Santiago). https://doi.org/10.7764/eure.48.145.13

Napadensky, A., & Orellana, A. (2019). Metropolization and functional organization of intermediate urban systems. Gran La Serena, Concepción and Puerto Montt. *Bitácora Urbano Territorial, 29*(1), 65–78. https://doi.org/10.15446/bitacora.v29n1.67325

Napadensky, A. (2020). An old center, for new uses? Spatial distribution of Knowledge Intensive Services [KIS] in the Metropolitan Area of Concepción [MAC], Chile. *ACE: Architecture, City and Environment, 15*(44), 9210. https://doi.org/10.5821/ace.15.44.9210

Napadensky, A., Zazo-Moratalla, A., León-Aravena, J., & Cisterna-Aravena, G. (2022). Public discourses, territorial conceptions and territorialities: Views on the conflict of the Industrial Bridge, Chile. 2008–2018. *Revista De Geografía Norte Grande,* (83). https://doi.org/10.4067/S0718340220222000300395

Olson, K. (2011). Deliberative democracy. In B. Fultner (Ed.), *Jürgen Habermas: Key concepts acumen* (pp. 140–155). https://doi.org/10.4324/9781315711461

Pamplona Sierra, M. J., & Piedrahita Bustamante, P. (2021). The right to the city in agglomerations: Views from Latin America. *Bitácora Urbano Territorial, 31*(II), 73–85. https://doi.org/10.15446/bitacora.v31n2.89561

Parkins, J., & Mitchell, R. (2005). Public participation as public debate: A deliberative turn in natural resource management. *Society and Natural Resources, 18*, 529–540. https://doi.org/10.1080/08941920590947977

Pecurul-Botines, M., Di Gregorio, M., & Paavola, J. (2014). Discourses of conflict and collaboration and institutional context in the implementation of forest conservation policies in Soria, Spain. *Biodiversity Conservation, 23*(14), 3483–3499.

Pérez, M. (2019). "One must have a house where one was born". Citizenship and the right to the city in Santiago. *EURE, 45*(135) (Santiago). https://doi.org/10.4067/S0250-71612019000200071

Polletta, F., & Chen, P. C. B. (2017). Narrative and social movements. In J. C. Alexander, R. N. Jacbos & J. Smith (Eds.), *The Oxford handbook of cultural sociology.* Oxford University Press. https://doi.org/10.1093/oxfordhb/9780195377767.013.18

Rossi, B. (2004). Revisiting Foucauldian approaches: Power dynamics in development projects. *Journal of Development Studies, 40*, 1–29. https://doi.org/10.1080/0022038042000233786

Scheufele, D. A. (1999). Framing as a theory of media effects. *Journal of Communication, 49*(1), 103–122. https://doi.org/10.1111/j.14602466.1999.tb02784.x

Shmueli, F. (2008). Framing in geographical analysis of environmental conflicts: Theory, methodology and three case studies. *Geoforum, 39*(6). https://doi.org/10.1016/j.geoforum.2008.08.006

Snow, D. A., & Benford, R. D. (1992). Master frames and cycles of protest. In A. D. Morris & C. M. Mueller (Eds.), *Frontiers in social movement theory* (pp. 133–155). Yale University Press.

Svarstad, H., & Benjaminsen, T. A. (2020). Reading radical environmental justice through a political ecology lens. *Geoforum, 108*, 1–11. https://doi.org/10.1016/j.geoforum.2019.11.007

Tourangeau, W. (2018). Power, discourse, and news media: Examining Canada's GM alfalfa protests. *Geoforum, 91*, 117–126. https://doi.org/10.1016/j.geoforum.2018.02.024

Van Dijk, T. (1990). *News as discourse. Understanding, structure and production of information.* Paidós.

Van Dijk, T. (1993). The interdisciplinary study of news and discourse. In K. B. Jensen & N.W. Jankowski (Eds.), *Qualitative methodologies for mass communication research* (pp. 135–148). Bosch Publishing House.

Vergara-Perucich, J. F., & Boano, C. (2020). The right to the financialized city. *Urban Territorial Logbook, 30*(III), 123–135. https://doi.org/10.15446/bitacora.v30n3.82437

Verloo, N. (2018). Social-spatial narrative: A framework to analyze the democratic opportunity of conflict. *Political Geography, 62*, 137–148. https://doi.org/10.1016/j.polgeo.2017.11.001

Von Essen, E. (2017). Whose discourse is it anyway? Understanding resistance through the rise of "Barstool Biology" in nature conservation. *Environmental Communication, 11*(4), 470–489. https://doi.org/10.1080/17524032.2015.1042986

Weber, M. (1968). *Economy and society: An outline of interpretive sociology*. Bedminster Press.

Wittmayer, J. M., Backhaus, J., Avelino, F., Pel, B., Strasser, T., Kunze, I., & Zuijderwijk, L. (2019). Narratives of change: How social innovation initiatives construct societal transformation. *Futures, 112*. https://doi.org/10.1016/j.futures.2019.06.005

Zhonghua, G., & Xiaoling, Z. (2021). Framing social sustainability and justice claims in urban regeneration: A comparative analysis of two cases in Guangzhou. *Land Use Policy, 102*. https://doi.org/10.1016/j.landusepol.2020.105224

Chapter 34
The Projected City: Urban Analysis from Historical Cartography—Temuco 1888–1919

Tirza Barría Catalán and Leonel Pérez Bustamante

Abstract During the so-called "pacification of Araucanía" (1862–1883), a Mapuche territory located in the south of Chile, the Operations Army proceeded to establish forts in strategic areas of the territory. These forts were the site for the development of a population associated with the colonization carried out by the Chilean state. After the creation of the provinces of Malleco and Cautín in 1887, these settlements, now turned into populations, were established as departmental head cities, with Temuco experiencing the greatest urban growth and economic boom. This chapter explores the construction of the city in light of the first urban cartographies made between 1888 and 1892. These plans account for an urban model based on the orthogonal grid of checkers and the changes that the layout underwent due to the implementation of the railway and the channeling of one of the streams. Other aspects to be evaluated will be the development of squares and avenues as possible city limits, as well as the dispersion of public and civil programs. As ongoing research, the preponderance of the foundational plan of 1888 in the subsequent development of the urban grid in later decades is determined, anticipating on paper the logics of space occupation, either in the division of lots or in the programmatic use of these. In the successive plans, the drastic intervention of the diagonal and the incorporation of avenues, squares and the railway infrastructure, all of them as pieces that mediate between the foundational area and the residential expansions after 1920. It is concluded that Temuco was born to be a city of great regional importance. In its first plans, this condition is projected, which is why the transformations carried out to adapt to the natural environment, to national transport policies and to the creation of new urban spaces associated with education, health and leisure are incorporated into the successive cartographic representations.

FONDECYT Project 1210592 "City and architecture in the Frontier: the consolidation of the Nation State in Araucanía, 1883–1974".

T. B. Catalán (✉)
Institute of Architecture and Urbanism, Universidad Austral de Chile, Valdivia, Chile
e-mail: tirzabarria@uach.cl

L. P. Bustamante
Department of Urbanism, Universidad de Concepción, Concepción, Chile
e-mail: leperez@udec.c

M. Vicuña and E. Giorgi (eds.), *Intersections*, Springer Geography,
https://doi.org/10.1007/978-3-031-76402-8_34

Keywords Araucanía · Temuco · Historical cartography · Urban layout

34.1 Introduction

In the second half of the nineteenth century, the Chilean state initiated a progressive colonization campaign in the lands of Araucanía, which for three centuries (sixteenth to nineteenth centuries) constituted a frontier between two areas: north of the Biobío River and south of the Cruces River. To achieve this, the Operations Army displaced and then annulled the border, strengthening in parallel the state institutions and the mechanisms that allowed the settlement of settlers in this area.

In this context, cities played a crucial role by serving as key elements in the state's territorialization policies, contributing to establishing an order and a territorial organization. In accordance with the above, these new cities in the Araucanía represent the reproduction of a civilizing model (Terán, 1989) and embody what was considered superior, exercising control and planning of space, in contrast to the image of the wild and uncontrolled in rural areas or naturally irrational.

The foundation of cities and towns carried out in the last decades of the nineteenth century in Araucanía corresponded to strategic places secured by a military fort. From 1880, the first plans can be identified, drawn up by the engineers of the Office of Lands and Surveys, an agency dependent on the General Inspection of Lands and Colonization. These initial plans, which are estimated to have been made between 1878 and 1885, show the basic operations that establish the urban layout known as checkerboard grid and the location of the city, with the barracks situated next to the river and the main square, as can be observed in the plans of Traiguén and Lautaro (Figs. 34.1 and 34.2).

The checkerboard grid corresponds to an orthogonal mesh or checkerboard grid consists of the alignment of streets that intersect at right angles, defining a regular size of blocks that allows for easy subdivision into smaller lots. Basically, this grid represents a regularized division between streets and blocks, as well as between public space and private space (Aguilera, 1994: 71).

The checkerboard grid, whose origin dates back to classical Greece, was widely adopted by the Spanish from the sixteenth century in the foundation of villas and Hispanic-colonial cities. This was due to its rapid execution and the advantages it offered in terms of order and distribution of sites, ensuring exposure to the sun and adequate ventilation (Guarda, 1978; Lorenzo, 1986). Hardoy (1983) considers it a "classic model" due to its ability to organize and distribute urban spaces. This organization of the plot allows equality in the size of the lots and establishes a social hierarchy based on the proximity of the lots to the main square.

During the seventeenth and eighteenth centuries, in the foundational campaigns of villas and Indian towns, they moved away from the checkerboard models used in earlier times (Durston, 1994). However, at the end of the nineteenth century, in the context of Araucanía, the checkerboard grid was readopted in the design of the

Fig. 34.1 'Traiguén', n/d. *Source* National Library, Sala Medina

first cities, with some variations mainly related to the incorporation of the railway (Barría, 2021).

In the case of Temuco, the current regional capital and subject of study of this chapter, it is recognized that the foundational model adopted was the checkerboard grid, as in the previous cases mentioned. When observing the plans identified by their dates: 1888, 1890, 1892 and 1919, changes associated with the installation of the railway and the layout of a diagonal that crosses the city, which would soon become an avenue to channel the floods of a stream, are identified in the first three plans. This foundational model will remain in force until the 1920s, when new urban areas are incorporated on the edges of the foundational core, breaking the original checkerboard by new layouts associated with the installation of working-class populations on one side and on the other, to the sale and urbanization of lots of parcels towards the western sector of the city of Temuco.

So far, no evidence has been found of the existence of a regulation that guided the design of this and other cities. Only in the decrees of 1885 and 1889 are some 'provisions for the foundation of populations and the approval of their plans' mentioned. These provisions are associated with the definition of two categories for the types of sites and the subdivision of the block lots, based on the distance to the main square.

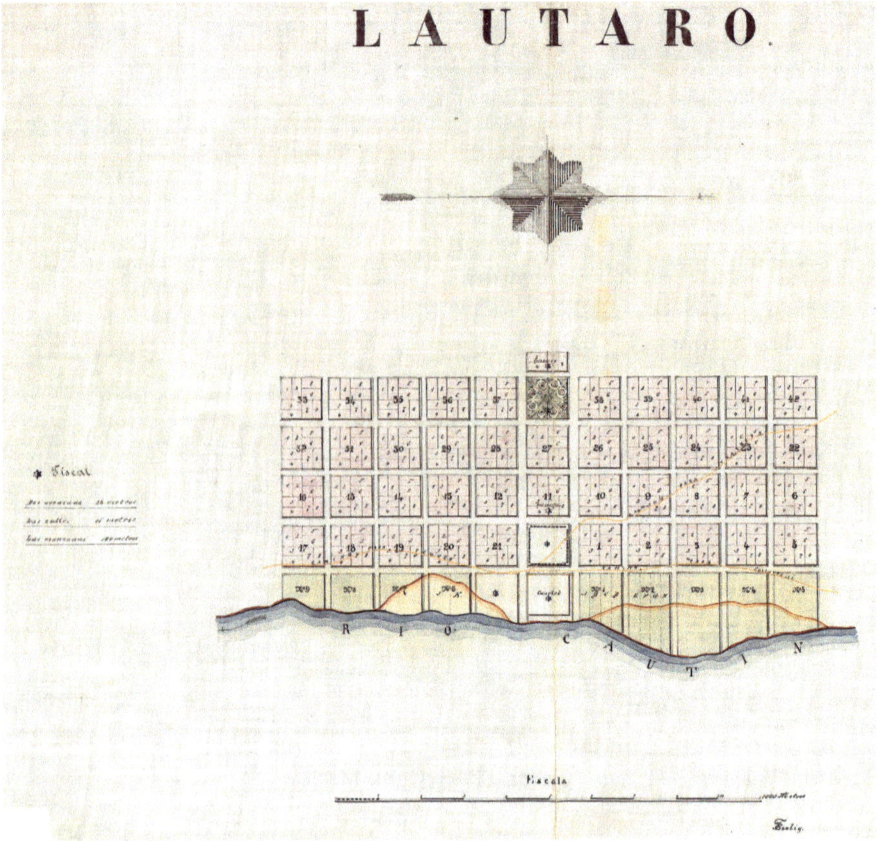

Fig. 34.2 'Lautaro', n/d. *Source* Ministry of National Assets

Therefore, the hypothesis arises that states that the first urban plan of the cities of Araucanía corresponds to an orthogonal grid model, the checkerboard, initially developed from a first layout on the ground that is fixed based on the installation of preexistences, in this case of the Fort. However, much of the rest of the city is projected on the condition of a blank slate where geographical alterations are unknown and there is a lack of background on the habitability of the places, so it is not possible to identify in the first plan, but yes in later plans the versions adjusted to the geographical reality and environmental of the sites and the incorporation of new measures in terms of urban design.

In recent times, the study of representations has moved from empirical fidelity in its representation of the real world, to the analyses that seek to understand these images as complex constructions of the world. This change has allowed a second look at cartographies, focused on the meanings of their signs. Therefore, to address this hypothesis, three historical plans of the city of Temuco will be analyzed descriptively with the aim of identifying the foundational matrix and the changes and integrations to the urban layout, deepening the contents of the texts and the graphic representation, to answer the following questions:

What are the most significant transformations in the urban morphology of Temuco that can be identified through the analysis of historical plans and cartographies of the specific period mentioned? And how can the lessons learned from the past, from the study of the mentioned sources, guide decisions and strategies to address contemporary challenges in the development and habitability of urban areas in the studied territory?

Understanding the processes of growth and change in cities from the late nineteenth century allows us to address contemporary issues that affect the habitability of certain areas, taking into account the lessons of the past to face present and future challenges in the urban sphere.

34.2 General Context About the Occupation of Araucanía

In the Frontier, a territory located south of Chile and whose area was between the Biobío River to the north and Cruces to the south, the foundation of cities and towns was executed as part of the colonization strategies of the Chilean State. The milestone of the gradual displacement of the frontier line was the refoundation of the Fort of Negrete in 1861 and the foundation of the city of Angol in 1862, the latter as an urban and military center from where men and supplies would be deployed for the success of the so-called "pacification of Araucanía" (Leiva, 1984).

On the ground, the military actions of pacification focused between 1861 and 1883, and were led first by Cornelio Saavedra and then by Gregorio Urrutia (Guevara, 1902; Lara, 1889). These consisted of advancing the frontier line from north to south by founding forts, from west to east taking the coast and from the south up the Toltén River (Antivil, 2017, 2018). The culmination of this enterprise will be the refoundation of Villarrica, a city founded in 1552, which was one of the seven cities destroyed by the Mapuches after the attack in Curalaba in 1598, and the installation of forts in the border area with Argentina. However, to ensure the colonization process, the Army remained present and in 'state of assembly' until 1887 (Lara, 1889:465).

Parallel to these actions, the Chilean State implemented occupation and colonization through mechanisms for effective settlement and forms of land profitability. Some of these were the foundation of populations, the granting of mercy titles to Mapuches and the auction and delivery of fiscal lands (Law of 1866); colonization through concessions (Law of 1874); the generation of colonies to group wandering indigenous people (Law of 1880). Other subsequent measures were Law 380 of 1896, which allowed the delivery of lands to repatriates coming from Argentina, Law 399 of 1898 for the delivery of lands to national settlers and Law 2.087 of 1908 for the delivery of property titles to national occupants (Aylwin, 1995; García, 1936).

As for the foundation of populations, the presence of a military fort was taken as a base. The larger ones were mainly associated with the north–south axis of the railway and immediate to large navigable rivers such as the Cautín, Imperial and Malleco, and the small ones are located around the railway branches built in the first decades of the twentieth century. In the decades from 1910 to 1940, these cities and populated centers experienced notable changes in the increase of their population and urban expansion, as is the case of Carahue, Nueva Imperial and Traiguén due to the implementation of roads, bridges, railway layouts. Likewise, the construction of buildings for health, education, administration, transport and tourism reflects the presence of the State to strengthen its presence in the region, materializing the ideal of progress and development of the nation as well as the standardization and homogenization of the symbols and values of Chilean society.

In accordance with the above, the city is understood as a space where the subject witnessed civilization, a representational construct, a model of territorial ordering, a form of rationality that became aware of itself, acquiring identity and a representational and imaginary structure for the community (Núñez, 2010). Therefore, the city was formed as a piece that ordered the territory and established a layout of not only productive but also social and political relationships, a situation that reduced the so-called "empty and wild" spaces of the territory.

34.3 Urban Analysis Through the Study of Maps

Maps, plans, and charts are cartographic representations, that is, spatial mental schemes that graphically depict an image of the world (Schlögel, 2007). They establish a relationship between the word and the thing represented. Therefore, the map precedes the reality it represents, which means that its creation occurs before understanding and dominating the territory. In this sense, the value of the map does not lie solely in describing reality, but in its ability to define it (Harley, 2005).

According to Harley (2005), these representations are tools that allow visualizing and communicating ideas, strategies, and objectives of a political project. They contribute to the shaping and transformation of the urban and territorial environment. Through their elaboration and use, the bases for the organization and management of spaces are established, as well as for decision-making that affects society as a whole.

Following Farinelli's perspective (2007), the map is characterized by being active and operative, exerting influence over space and acting as an ideological framework that organizes reality. For example, the urban plan, as a graphic and abstract representation of the framework of a city, is not limited to showing the physical form of the territory or the components of the urban structure, such as streets, blocks, plots, and the arrangement of buildings. It also instills a cultural model that organizes and positions the powers that govern in the city and the spaces where and how its inhabitants relate.

Depending on the era, maps were not always essentially accurate, on the contrary, their value lay in representative aspects. In this regard, Lois (2017) argues that maps are objects that visually communicate a real or imaginary sense of the geographical environment, that is, a representational expression of the space understood or of the spatial perception of it. However, according to Harley (2005), in other situations the map acquired a certain degree of reliability to reality which gives it a historical value, and this makes its information concrete for that historical moment.

34.3.1 Description of the Urban Plans of Temuco

From the selection of 4 urban plans, an exploratory analysis of a morphological and urban scale character is carried out. This analysis considers three stages of observation both in the representation and in the written background on the plan. The first stage is associated with three formal elements: the block, the street, and the lot. In a second stage, alterations in the layout related to major modifications are identified, such as the incorporation of new streets and infrastructures. Finally, the third stage is related to the distribution of services and public buildings.

Plan of 1888

A detailed analysis of the first urban plan of the city of Temuco reveals certain initial guidelines. According to the description of the plan of 1888 (Fig. 34.3), the urban layout consists of 150 blocks with sides of 100 m, of which 77 are colored and numbered. Of these, only 4 present subdivisions in lots intended for specific purposes, such as the 'intendancy', the 'temple', the 'school', and the 'German school'. In addition, space is reserved for two squares: the main one, located in the center of the grid, and a secondary one in front of the military barracks and the jail.

A second plan of this city offers more precise details, especially regarding the buildings that will make up the city. In addition to the drawing of the layout, the texts present in the perimeter of the plan explain aspects related to the dimensions of the layouts, which is not observed in the previous plan. This suggests a link with the official promulgation of the population or city decree in 1887. This demonstrates the need for a written description that establishes clear and precise guidelines on the construction of the urban layout.

1890 Plan

The second plan (Fig. 34.4), titled 'Temuco' and at a scale of 1:5,000, was approved by supreme decree No. 2,126 on May 13, 1890. In this plan, a checkerboard grid made up of 139 blocks, mostly regular, with sides of 100 m is represented. In addition, the location of the railway station and its respective track stands out as an urban piece, which anticipates the implementation of a national territorial connection as a state policy.

The names of the first occupants of properties along the banks of the Pichi Cautín stream, on the projected railway axis and on the blocks projected on the "avenue and drainage canal of the Coilaco stream, which floods the population in winter" are also indicated. The other blocks are subdivided into 8 sites, except for those located southeast of the 'station', whose blocks are divided into 4 sites. These subdivisions correspond to the future sites that will be successively delivered during the following decade.

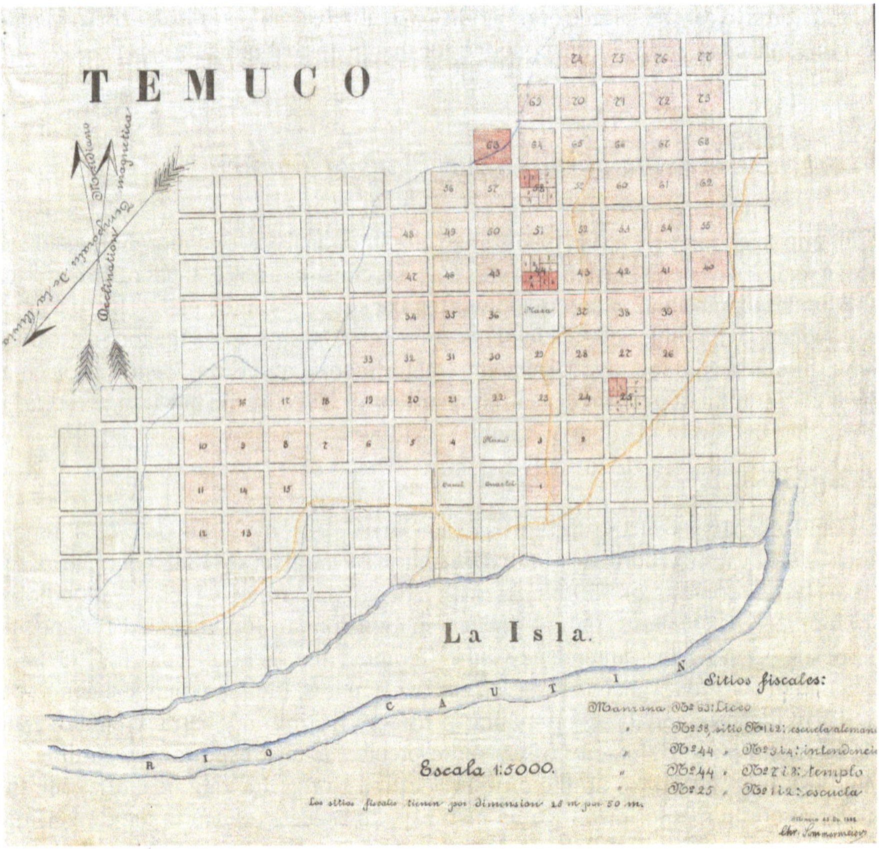

Fig. 34.3 Map of Temuco, 1888, drawn by Cr. Sommermier. Scale 1:5,000. *Source* National Library, Sala Medina

In the explanations found in the lower left corner of the plan, it is specified that the streets will be 15 m wide, while the avenues will be 20 m. In addition, in the drawing, three avenues 50 m wide can be identified, located to the west, north (with connection to the road to Cholchol) and east of the city. These avenues to some extent delimit the foundational layout and represent a border of the city that creates a new urban space, contributing to the structure and connectivity of Temuco by providing main arteries for traffic and mobility.

In the mentioned plan, several establishments and important locations in Temuco are identified. Some of them are: the high school, the superior school and the police station, located on Rodríguez Street between Mackenna and General Aldunate streets; the Franciscan Fathers' convent and the hospital, located on Portales Street; the jail and the barracks on O'Higgins Avenue, and the slaughterhouse, which is near General Lagos Street, about 100 m from the Pichi Cautín stream.

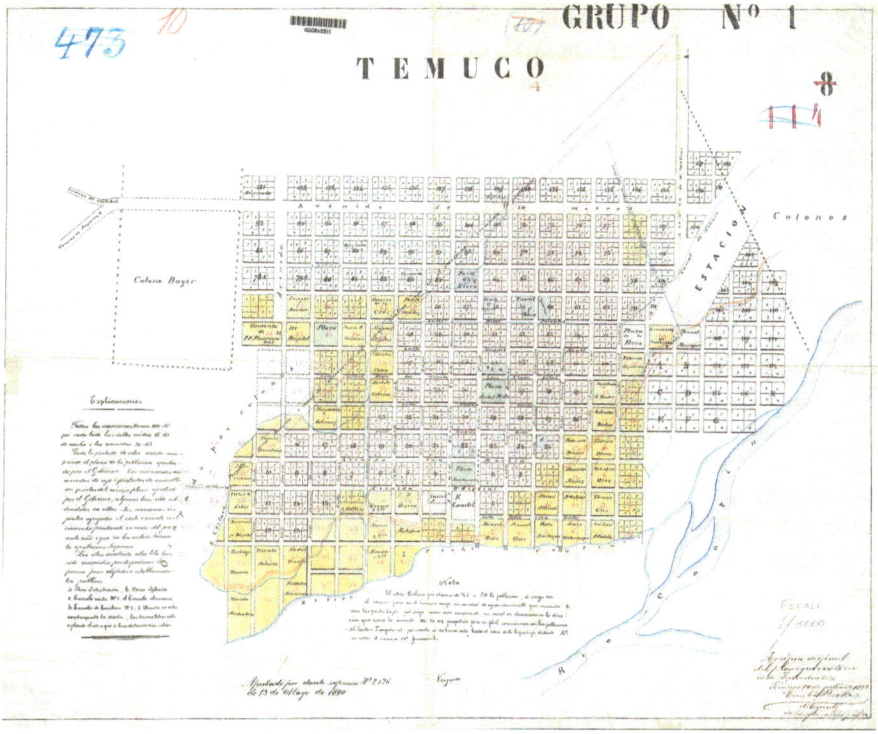

Fig. 34.4 Map of the city of Temuco, 1890. *Source* Ministry of National Assets

In addition, other places are mentioned with designated letters: the Intendancy and the Church, in front of Aníbal Pinto square; the Men's School, in front of Recabarren square; the German school, at the intersection of General Aldunate and Andrés Bello streets, and the 'Recoba' and the 'Mixed School' (without identification on the plan). Four squares are marked from east to west: 'La Feria Square', 'Aníbal Pinto Square', 'Emilio Recabarren Square', and the 'Square' in front of the Hospital. In general, the distribution of these sites ensures the location of the municipal administration, public order and other important precincts.

Regarding these seasonal floods and the projection of this canal, the legend of the plan indicates "the Coilaco stream that crosses the population from northeast to southwest, is cut off in the summer, but in the winter it collects a considerable flow of water that floods the lower parts, for which reason a drainage canal would be convenient in the direction indicated by the projected avenue to provide easy communication with the populations of Lautaro, Traiguén, etc., as it goes in a straight line to the La Esperanza stream distant kilometers without cutting the railway road".

The projection of Caupolicán Avenue in the 1892 plan is a key element to understand the urban structure of Temuco. This avenue crosses the existing checkerboard layout, giving rise to the formation of triangular blocks. This feature is maintained in all plans after 1890, including the 1892 plan and those that follow. The presence of these triangular blocks becomes a distinctive feature in the urban design of Temuco and reflects the significant influence of Caupolicán Avenue on the structure of the city.

1892 Plan

The 1892 plan of Temuco (Fig. 34.5), prepared by Ch. Sommermeier and raised by the Topographic Commission, is a valuable document that preserves the shape, extent and location of the enclosures previously identified in the 1890 plan. However, it incorporates new elements that enrich the information about the city. Among these elements is a line projected and defined as 'urban limit', which delimits the urban area of Temuco. In addition, an area designated as 'lands flooded by the Cautín' is identified, which indicates areas prone to flooding in relation to the Cautín river.

Also, a cut in the blocks is observed due to the projection of a '36 m wide avenue', which connects to the north with the 'road to Lautaro' and to the south with the 'Imperial Road'. These elements added in the 1892 plan offer a more detailed view of the urban configuration and the city limits as they were projected at that time.

Similarly, a diagonal is incorporated that crosses the checkerboard layout, reflecting the problem of floods and the proposal to channel this water course presented in the 1890 plan. This explains the fragmentation of the blocks, leaving triangular and pentagonal portions, as observed in the image on the right (Fig. 34.6).

In relation to the expansion of the urban area, 15 new blocks were incorporated in the northeastern part of the city, along with a new location for the 'La Feria square'. At the northern limit of the plan, spaces were assigned for the 'Agricultural School', the 'New Imperial Station' and the area of the park and Field of Mars, all north of the so-called '50 m avenue'. On the southern limit, the 'Reserved Island' was included, which corresponds to the current Cautín island, and the subdivisions of the sites located on the southern bank of the Cautín river, in the Padre Las Casas area, along with the projection of the railway line.

1919 Plan

In the plans after 1916, an expansion of the urban area to the east and west is evident, exceeding the established urban limit, with the creation of new populations. In the plan titled 'Plan of the city of Temuco and its populations' dated 1919 and without scale (Fig. 34.7), you can observe the expansion of the urban plant defining new limits. To the north, it is delimited by the Gibbs canal and the so-called 'fiscal mountain'. To the south, it is delimited by the Pichi-Cautín stream and the Cautín river. To the west, it is delimited by the Temuco stream, and to the east, it extends to the last blocks of the 'La Fama' population.

In general terms, in the expansion towards the east, the dimension of the foundational block of 100 m on each side was maintained. This division into 8 sites was applied in the Patske, Paredes, Kolossa, Perlwitz populations and a part of the Zlehm population, where the State Railways Workshop is also located. On the other hand, in the Osthaus, Santa Rosa and La Fama populations, located south of the railway line, the blocks were only divided into 4 sites.

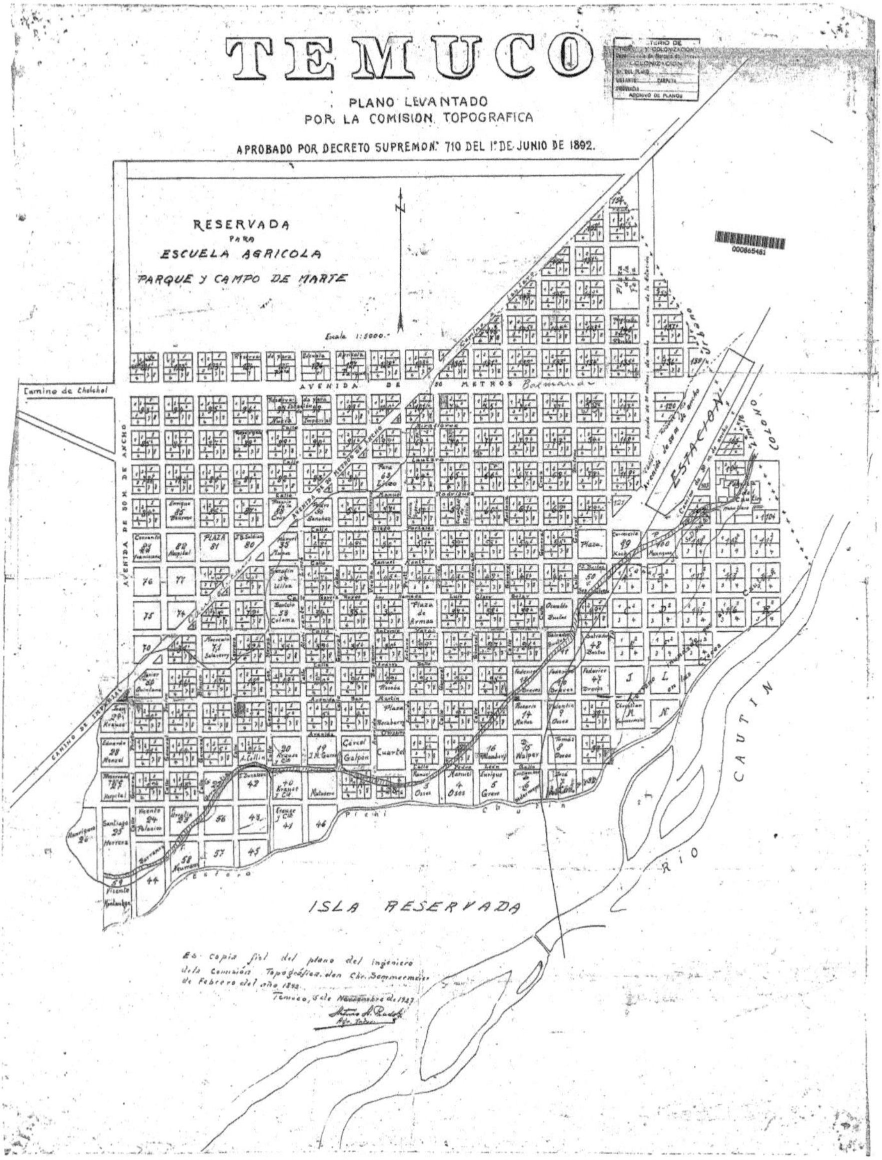

Fig. 34.5 Map of the city of Temuco, 1892. *Source* Archive of La Araucanía

South of the Cautín river, the same configuration of the Padre Las Casas population that was drawn in the 1892 plan is observed. This indicates that the distribution and subdivisions of the blocks were maintained in this area over time. In the northwest area of the urban area, the Quintas Neighborhood is incorporated, delimited by the Temuco stream and the Alemania, France and Pedro de Valdivia avenues.

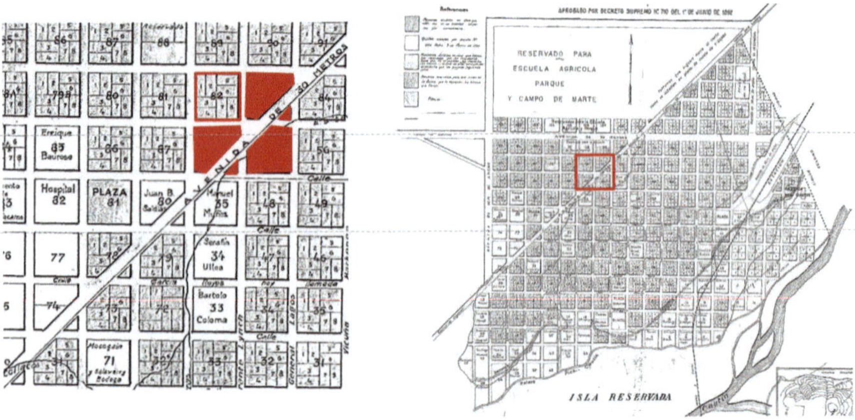

Fig. 34.6 Section of the map of the city of Temuco, 1892

Fig. 34.7 Map of the city of Temuco and its populations, 1919, Elaborated by Nicanor Boloña. *Source* National Library

This neighborhood consists of 5 large plots of approximately 300 × 200 m each, generating elongated sites of 25 × 100 m. The avenues that delimit these lands are identified, such as Avenida De La Paz, English Avenue, Italian Avenue and Spanish Avenue. To the southwest is the Dreves Population, made up of 37 blocks mostly 100 m on a side, without site divisions, and with the corresponding streets that have their respective names.

Between both sectors the settlers' parceling is drawn, where some streets are projected such as Phillipi, September, Denmark and Holland. In addition, specific locations can be found such as the School English, the German Hospital, and some estates such as Malmus, Viertel, Hochsteher, among others.

In general terms, it is observed that urban expansion in Temuco deviates from the logic of the checkerboard layout and follows a different approach, which considers factors such as topography and existing railway infrastructure. This agrees with the observations of Gutiérrez (1989), who mentions the ruptures in the postcolonial city due to the articulation with the surrounding territory, including suburban sectors and extensions. Despite these ruptures, the city continues to maintain a logic of coherent development and growth.

34.4 Conclusions

Just as the founding act in the sixteenth century represented the culmination of a phase of searching for a suitable site, guarantees of survival, and availability of necessary elements to establish a city, similarly, in Araucanía, the foundation of populations arose in the locations of the forts. An example of this was the choice of the location of the Temuco fort, situated between two mountains and a river, which allowed for strategic control of the territories of the lower and upper Mapuches. The topography of this area offered guarantees of accessibility and advantages for a good military position.

Once the minimum conditions to start a larger scale permanent settlement were established and the viability of the location confirmed, and in most cases, with the presence of a group of already established settlers, the Chilean state decreed the formation of a population or city. This act generally occurred about 10 years after the foundation of the fort and represented the maximum expression of the colonizing idea, manifesting itself in the elaboration of the first urban plan.

Regarding the contents and representations in the analyzed plans, and in relation to the first stage of observation, the distribution of blocks defined in the first plan of 1888 is preserved. However, in response to geographical conditions, modifications were made to its initial layout. An example of this was the projection and subsequent consolidation of the current Caupolicán Avenue, which is observed in the plan of 1890 and 1892, altered the homogeneity of the orthogonal grid generating residual blocks of triangular shape.

As for the second stage and in the absence of any plan that defined specific guidelines for the design of the city, the 1890 plan resolves some layout aspects by incorporating guidelines for the dimensioning of streets and avenues, and even suggests the urban solution for the floodable areas associated with the Coilaco stream that would define the location and extension of the aforementioned Caupolicán Avenue. Regarding the railway infrastructure, it is present in the 1890 plan only as a polygon and a layout line that extends over the blocks to the south, in the same way, it is represented in the 1892 plan. Comparatively, the 1890, 1892 and 1919 plans identify in the latter the displacement of the railway line to the west, avoiding the outlet channel of the Hydraulic station.

Regarding the third stage, already in the first urban plan, the organization and distribution of the sites were defined, as well as the location of civil (school and government), military (barracks) and religious (temple) institutions. In the 1890 and 1892 plans, the dispersion of programs is clearly defined, some of them being the jail, the hospital, the convent of nuns, a high school and a square as a fair, usually preceded by a square corresponding to the size of a block.

These adjustments, along with other factors related to the settlement and economic development of these cities, will contribute to consolidate the layout and determine the future expansions of the city limits.

References

Aguilera, J. (1994). *Fundación de cien ciudades hispanoamericanas*. Editorial Mapfre.

Antivil, W. (2017). Una mirada a la Araucanía: Construcciones territoriales en la colonización chilena en el siglo XIX. *Revista Urbano, 35*, 6–17. https://doi.org/10.22320/07183607.2017.20.35.01

Antivil, W. (2018). Dibujando la Araucanía. La construcción, la forma y el dominio de un territorio. [Tesis de Doctorado, Universidad Politécnica de Cataluña, España].

Aylwin, J. (1995). Estudios sobre tierras indígenas de la Araucanía: Antecedentes histórico legislativo 1850–1920. Instituto de Estudios Indígenas, Universidad de La Frontera. Serie Documentos N.° 3. http://www.memoriachilena.gob.cl/602/w3-article-9267.html

Barría, T. (2021). Araucanía, Formas y dispositivos del territorio colonizado. La Carta General de Colonización de la Provincia de Cautín, 1916. [Tesis de Doctorado, Pontificia Universidad Católica de Chile].

Durston, A. (1994). Un régimen urbanístico en la América hispana colonial: el trazado en damero durante los siglos XVI y XVII. *Historia, 28*, 59–115.

Farinelli, F. (2007). El mundo, el globo, el mapa: Los orígenes de la modernidad. En: El mundo de los mapas (pp. 41–56). Fundación Marcelino Botín.

García, V. (1936). El problema de colonización. Estudio sobre la Ley 5.604. [Tesis de Licenciatura Facultad de Leyes y Ciencias Políticas, Universidad de Chile].

Guarda, G. (1978). Historia urbana del reino de Chile. Andrés Bello.

Guevara, T. (1902). Historia de la civilización de la Araucanía. Imprenta y Litografía Barcelona. http://www.memoriachilena.gob.cl/602/w3-article-121391.html

Gutiérrez, R. (1989). *La ciudad iberoamericana en el siglo XIX. En El sueño de un orden. La ciudad hispanoamericana* (pp. 252–266). Centro de Estudios y Experimentación de Obras Públicas. MOPU.

Hardoy, J. (1983). La forma de las ciudades coloniales en la América española. En: F. de Solano (Ed.), Estudios sobre la ciudad latinoamericana.
Harley, J. B. (2005). *La nueva naturaleza de los mapas. Ensayo sobre la historia de la cartografía.* Fondo de Cultura Económica.
Lara, H. (1889). Crónica de la Araucanía. Descubrimientos y Conquista. Tomo II. Imprenta de El Progreso.
Leiva, A. (1984). El primer avance a la Araucanía. Angol 1862. Temuco: Ediciones Universidad de La Frontera.
Lois, C. (2017). Los mapas y las geometrías del espacio. Terra Brasilis (Nova Serie) (8). https://doi.org/10.4000/terrabrasilis.2029
Lorenzo, S. (1986). Itinerario de la política de poblaciones. Capítulo 1. En Origen de las ciudades chilenas: las fundaciones del siglo XVIII. 2a. ed. Andrés Bello. Recuperado de. https://www.memoriachilena.cl/602/w3-article-98902.html
Núñez, A. (2010). La ciudad como sujeto: formas y procesos de su constitución moderna en Chile, siglos XVIII y XIX. *Revista de Geografía Norte Grande, 46*, 45–66. https://revistanortegrande.uc.cl/index.php/RGNG/article/view/40891/32939
Saavedra, C. (1870). Documentos relativos a la ocupación de Arauco: que contienen los trabajos practicados desde 1861 hasta la fecha: i demás antecedentes que pueden contribuir a ilustrar el juicio de los señores diputados en la próxima discusión sobre el último proyecto del ejecutivo. Santiago. Imprenta de La Libertad. http://www.memoriachilena.gob.cl/602/w3-article-9597.html
Schlögel, K. (2007). *En el espacio leemos el tiempo. Sobre Historia de la civilización y Geopolítica.* Editorial Siruela.
Terán, F. (1989). Introducción En: El sueño de un orden. La ciudad hispanoamericana (pp. 13–21). Centro de Estudios y Experimentación de Obras Públicas MOPU.

Chapter 35
Affective Frameworks or the Construction of a Place in the Housing Complexes in Providencia Carried Out by Luciano Kulczewski

Ronald Harris Diez and Ignacio Veraguas Caripan

Abstract This chapter addresses the architectural aesthetic of Luciano Kulczewski, based on the study of two residential complexes built between 1929 and 1935 in the commune of Providencia, Santiago de Chile, which shows a concern for the emotional bond between residents and the inhabited space. Such sets, unpublished for current historiography, add to the first modern urban proposals inspired by the notion of neighborhood-garden. The affective consideration on the part of the architectural project includes what is called the creation of a "place." In fact, a place differs from a space because of a sentimental projection that, in these cases, results in a conscious search by the architect. Luciano Kulczewski's residential complexes employ recognized architectural strategies that include arranging space without sacrificing connection to the city and constantly ensuring that residents feel connected to the built environment. These approaches, in turn, correspond to the political role attributed to the ornamental management and beautification of the ensembles.

Keywords Luciano Kulczewski · Residential complex · Place · Aesthetics of architecture · History of architecture

35.1 Introduction

This article proposes a study of the architectural aesthetics in Luciano Kulczewski[1] based on the composition of the *place* in two of the residential complexes of Providencia commune, Santiago de Chile (Fig. 35.1). The notion of place and its link with

R. H. Diez (✉)
Instituto de Estética, Facultad de Filosofía, Pontificia Universidad Católica de Chile, Santiago, Chile
e-mail: roneharris10@gmail.com

I. V. Caripan
Modern Languages and Literatures, Johns Hopkins University, Baltimore, United States
e-mail: iveragu1@jh.edu

M. Vicuña and E. Giorgi (eds.), *Intersections*, Springer Geography,
https://doi.org/10.1007/978-3-031-76402-8_35

architecture is a concept that has been approached from epistemological, psychological or sociological perspectives, with emphasis on the affective dimension of the act of inhabiting a space (Espósito, 2012; Gallardo, 2013). We propose an aesthetic approach through the conception of place as an affective projection of space (Maderuelo, 2008). That is, as an intentional or *co-constructed* manifestation in the experience of inhabiting. In this article, along with the historiographic work for the understanding of the studied elements, we propose to identify the architectural decisions made by Luciano Kulczewski for the creating a place, based on the analysis of two residential complexes by the architect. In this sense, we delve into the aesthetic dimension in two of his buildings as exemplary cases of the composition of a place, noticeable in material aspects and in their own principles and guidelines that are reflected in these constructions.

The complexes we will analyze, built between 1929 and 1935, correspond to the Arturo Prat Cooperative Population of Emilio Delporte Employees (1929) (Fig. 35.2) and the Caja Nacional de Ahorros Population (ca. 1930–1935) (Fig. 35.3). The mentioned realizations are works identified thanks to previous research (Harris, 2014). The common feature that brings these creations together is the public they are

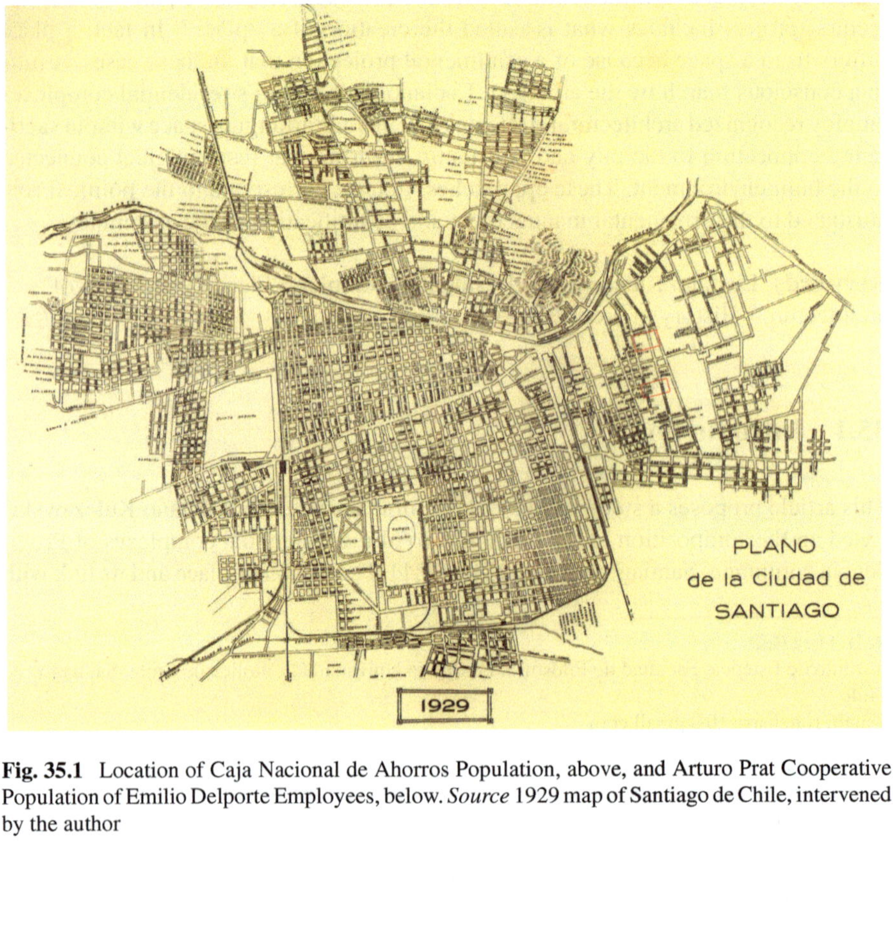

Fig. 35.1 Location of Caja Nacional de Ahorros Population, above, and Arturo Prat Cooperative Population of Emilio Delporte Employees, below. *Source* 1929 map of Santiago de Chile, intervened by the author

Fig. 35.2 Arturo Prat Cooperative Population of Emilio Delporte Employees. *Source* Author's photograph

aimed at, middle-class people, in these cases, private or public employees. Likewise, these complexes are configured as garden-neighborhood populations, a model of "appropriation" (Subercaseaux, 2004) of the garden-city proposals that were being felt from Europe and the United States (Palmer, 1984). This urban concept is introduced in those communes that, because of population growth, began to form on the old agricultural lands adjacent to the capital, as is the case with Providencia.[2]

Based on the above, we propose a study that not only focuses on heritage rescue, but also addresses the sense of space, according to the architect's own intentions.

In the garden city, the architect manages the total of a neighborhood for the first time; it is the beginning of modern urbanism. Then all the data are at stake and influence each other: the shapes of the roads and sidewalks, the size and shape of the properties, the possibility of having squares, the pairing or isolation of the houses: on a surface of soil, which is considered nature, and without mediating the abstraction of "real estate," there is the possibility of building a place, a garden that, as such must differentiate itself from the rest of the landscape (Palmer, 1984, p. 9. The emphasis is ours).

The house or home is a construction prone to understanding the affective dimensions of architecture, given the sentimental load it acquires with the time of use; likewise, it is unavoidable to attend to the proposals and principles of the architect himself that contribute to the inclusion of these aspects. We believe that these ideals, in the case of Luciano Kulczewski,[3] fostered the decision to attend, along with the

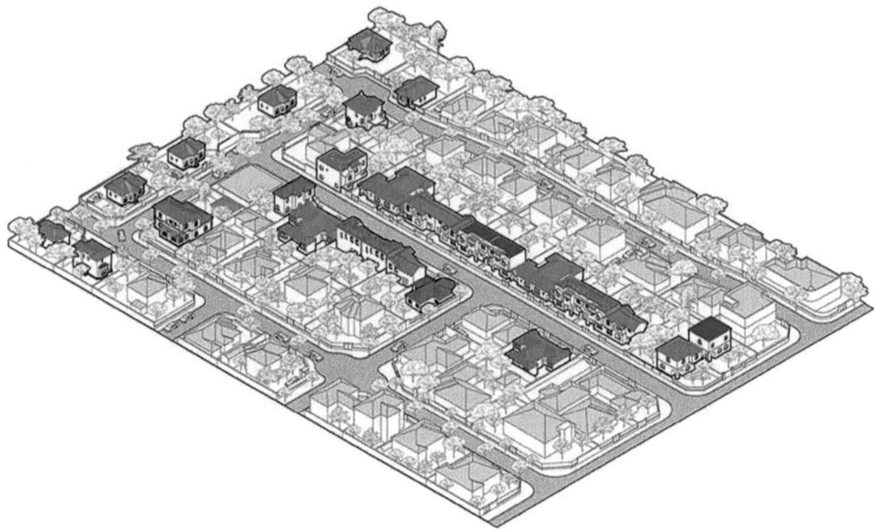

Fig. 35.3 Caja Nacional de Ahorros Population, left, and Calle Keller, right. *Source* FONDECYT, n.° 11,180,495, Responsible researcher: Ronald Harris

physical space, the affective bond between the residents and the place in which they live.

The complexes of the commune of Providencia were built based on state policies that, with the purpose of solving the housing deficit of the time, promulgated the "Cheap Housing Law" ["Decreto Ley de Habitaciones Baratas"] of 1925. In Latin America since the end of the nineteenth century and during the first half of the twentieth century, discussions and legislation on housing were a constant; in Chile these problems were assumed by the so-called social question. In 1907, Law 1838 was enacted in Chile, known as the Workers' Housing Law, legislation that created the Superior Council of Workers' Housing with a strong hygienist sense. Of course, these actions should not simply be read as indulgences of the ruling class towards the population, they also largely respond to demands, such as the urgency that denounced the existence of the tenant movement, whose first activities date from 1914 in Valparaíso (Hidalgo, 2002, p. 97). For its part, the decree of 1925, known as the Cheap Housing Law, changed the notion of "worker's housing" for "cheap housing."[4] Likewise, it fostered the creation of cooperatives, with greater influence and activity from the second half of the 1920s and the beginning of the 1930s (Hidalgo, 2002, p. 100).

The first residential complexes carried out by Kulczewski, carried out in the Yungay neighborhood between 1922 and 1924 and on Madrid Street in 1925, were inscribed in this latter type of management[5] (Harris, 2014). The Cheap Housing decree allowed the participation of savings banks and cooperatives in the management and construction of housing for their members, which in turn favored a close and real link between the clients and the architect, as it was no longer just about achievements carried out as part of private speculative operations. Working with cooperatives and savings banks usually allowed the hired architect to personally know the community that would inhabit his creations. Original planimetric vestiges, survivors in the Technical Archive of Aguas Andinas,[6] indicate to us that, for the case of the Caja Nacional de Ahorros Population, Kulczewski had early individualized who would be the resident of each of the nine houses that made up the set. This bond, as we will conclude, will be decisive when conceiving not only the solution most suitable to the needs of the client, but it is key to understanding the aesthetic dimension attended by the architect in the buildings. We can say that the architectural aesthetics of Luciano Kulczewski not only obeys the construction of a visible and stable space in physical terms, but also aims at the construction of an affective reality given within the same conditions of the work.

35.2 Housing Complexes, Dimensional Variety and Common Morphology

The Arturo Prat Cooperative Population (1929) is the first residential complex that Kulczewski carried out in the commune of Providencia. This housing savings cooperative belonged to employees of the commercial house Emilio Delporte,[7] a name that today has the main street around which the plots of the housing group were structured. Composed of 39 houses made of reinforced masonry, three architectural types were used. Each type differed in measurements and number of interior rooms with the purpose of accommodating different family configurations. Despite this variety, at a formal level there are a series of shared morphological aspects that give an identity to the set, namely, bay-windows that, located in the main room, opened to the back garden as well as underpasses that preceded access to the residences, now fenced for security reasons (Fig. 35.2). This space mediating between the exterior and the interior is worked in the Anglo-Saxon way, as it led directly to the living room of the house, moving away from the Hispano-American model of hallway, used in his previous residential realizations. The republican house in the center of the capital, organized around interior courtyards and with access from vestibules, has been left behind.

In this set, Kulczewski leaves aside, for the first time, the realization of continuous facade houses, as can be seen in the sets of the commune of Santiago, to focus on constructions of houses attached to a single party wall, in turn, preceded by a front garden. This seems to be an intermediate stage in a path of search and experimentation, which led the architect to leave the model of continuous block of houses attached to the party walls—a scheme more suitable for interventions carried out in the traditional grid of downtown Santiago, like the sets already mentioned in Yungay and Madrid street, and that would culminate in the adoption of the model of isolated house, typology apparently more suitable with the urban context of garden city, adopted by the pericentric communes.

The second residential complex carried out by Luciano Kulcewski in the Providencia commune corresponds to the Caja Nacional de Ahorros Population, made up of nine houses. It is currently located around Luis Barros Valdés Street. Although its original plan dates to 1930, its materialization seems to already account for the 1935 ordinance that will govern these complexes, in which zones of isolated buildings with eight meters of frontage and a surface area between 150 and 300 m^2 (Palmer, 1984) are established. In this way, the nine residences are configured as isolated volumes of one or two floors, located on large plots that exceed 400 m^2. He did not work with standard houses, but each house was adapted to a client who, as we have already indicated, was known in advance. However, a homogeneous morphological treatment emphasizes the sense of ensemble: volumes of multiple faces, with heavy Chilean tile roofs, give it a Mediterranean aspect.

The morphology shared by each set allows to differentiate at a first instance the inscription of Luciano Kulczewski's aesthetics. Likewise, this constructive modality highlights two aspects that, to continue, we must delve into to understand the strategies used for the configuration of a *place* in both sets. On the one hand, the buildings are versatile enough to accommodate varied families, as has already been mentioned. On the other hand, this versatility does not correspond to an indifference between each element of the set with respect to a whole, but they are brought together by means of morphological aspects alike. Each building through this familiarity is installed within the small universe that forms, in turn, each set.

35.3 Composition of a Place

At the urban level, and not only in these sets, Kulczewski worked from small interior roads that were inserted into the fabric of large blocks. In the case of Providencia, the macro blocks were the traces of the boundaries of the plots and farms that were in this rural area. Once urbanized, these delimitations served as guide to trace the avenues that today make up the commune. While Emilio Delporte was originally born as an interior road that ran between José Miguel Claro Avenue and Manuel Montt Avenue—towards the south of Francisco Bilbao Avenue—the also interior street Luis Barros Valdés, which structures the Caja Nacional de Ahorros Population, is intersected by Keller Street,[8] where the homonymous set is located (Calle Keller Population), one of Kulczewski's most known architectural achievements (Fig. 35.3). This strategy allowed the populations to enjoy a high degree of spatial autonomy with respect to the surrounding context, in the manner of what we could call a microcosm. This constructive strategy, in turn, fostered a sense of seclusion or intimacy. In addition, this compositional arrangement is not only relevant in its practical dimensions, but also takes on relevance in theoretical terms. In this case, the search for certain intimacy is not only reflected as the individual exercise of a person or family group within the shelter of a home, but that motive is also projected towards a set in its completeness, without losing sight of its inclusion within the larger system that implies the city.

Both for its contemporaneity, since it was carried out around 1930, and for its location, the Calle Keller Population[9] shares its history with the sets studied here. As has been expressed in the introduction of our study, we have chosen to develop research on two other sets located in the Providencia commune; however, to understand more clearly the urban solution proposed by Kulczewski for the Caja Nacional de Ahorros Population set, it is necessary to understand how the Calle Keller Population was generated.

The Calle Keller was originally a *cul-de-sac* that ended in some garages owned by the inhabitants of the complex (Fig. 35.4). We are unaware of the negotiations carried out—perhaps led by the architect himself—but a few years after their completion these buildings were demolished to allow access to the Caja complex, which by then was a Mediterranean plot without road continuity.[10] In this way, Keller Street, a road of intimate dimensions,[11] becomes a threshold-street that provided, in turn, intimate and secluded access to the complex. At the same time, Keller Street had at its entrance two large streetlamps (Fig. 35.4) which, unusual for the context of their time, stood like lighthouses and functioned as true "framing" devices (Saldarriaga, 2023). From these elements, a kind of transit ritual was made tangible.

In the case of the Emilio Delporte Population, the setback of the buildings located at the corners of the entrances allows the idea of thresholds to be perceived. Similarly, in the Caja Nacional de Ahorros Population, this is achieved through wide chamfers arranged on the sidewalks of the road intersections. According to Palmer, these

Fig. 35.4 Calle Keller Population. *Source* Kulczewski Archive, Chilean Architecture Archive

decisions allowed an expansion of the complex from its compact state: "thus the entrance or exit to the neighborhood and to each of its streets, is through a figure that marks a kind of decompression, like the hallway in the old houses" (1984, p. 71).

In turn, the different architectural elements used by Kulczewski in the houses: the presence of the portico, the front gardens, the scale of the houses and the tree-lined streets (Fig. 35.5), in the words of Pérez de Arce (1997), "create recognizable and measured spaces" in each one of the architect's complexes (p. 78). The combination of these factors allowed residents the emotional projection necessary for the creation of places. In this sense, through the morphological aspects mentioned in the previous section, as well as through the elements that attach the transit from the city itself to the complexes, the treatment of an emotional and affective bond to space is projected. It evidences the search to create an intimacy even in the complex itself and not only in the particularity of the interior of each home. This aspect, in turn, is what gave each residential group its own identity. We spend the past time in this last mention given that this reference today has dissolved due to new urban and road provisions: this seclusion has been interrupted by the large number of cars that occupy the interior streets as parking, saturating the interior roads.

Fig. 35.5 Section of General Córdova (above) and Luis Barros Valdés (below) streets, Caja Nacional de Ahorros Population. *Source* FONDECYT, n.° 11180495, Responsible researcher: Ronald Harris

35.4 From Inhabiting a Space to Recognizing a place

Although we do not have a considerable number of documents that attest to the thinking of Luciano Kulczewski, the compositional elements described must be understood considering the architect's ideas. In general terms, Kulczewski proposed as a relevant purpose for architecture, which purpose can be extended to the complexes, what he called the dignification of its inhabitants: "[I am] totally against the proletarianization of the common man. On the contrary [,] I want to lift him up. That's what led me […] to make all those populations with different houses from each other, but within a common resemblance" (Burmeister, 1969, s.p). Such architectural decisions previously described can be read as an objection to the "proletarianization" of the inhabitants. Kulczewski's aesthetics, in this sense, do not neglect a political positioning. However, this does not translate into the adherence to a specific ideology—indelible, in any case, to his concerns—but rather refers to a way of thinking about the way of living and its dignity, whose translation is evident in the aesthetic aspects of his buildings that do not reduce living to a necessity. In other words, it is an alternative way to the distribution of the ways in which it is imaginable and practicable the configuration of a place to live; perhaps related to what has been called with Rancière (2009) a distribution of the sensible. That is to say, the politics of architecture refers in this case to a specific way of spatial thinking, not to a partisan dimension in the execution of works. The resolution of functional problems does not suppress a formal distinction; and this distinction is not random but goes against what has been established as a general homogenization of life, in the words of the architect himself.

To the above, we must add a unique aspect in the morphological elements related to the described sets, in what has been called the eclecticism in the aesthetics of Luciano Kulczewski. The original ornamentation using elements inspired by local flora and fauna is outlined in the search for the inhabitants to recognize themselves in their territory and achieve the emotionality of the collective and individual memory necessary for this purpose.

At the level of the collective imaginary, these ornamental proposals have been linked with an art nouveau and art deco aesthetic (Boza, 1983). This is due to the high degree of geometric abstractions that range from modernist arabesques, present in the surviving metal ornaments in the false gables that served to highlight the access underpasses to the houses of the Emilio Delporte Population (Fig. 35.6), to the abstract compositions of the artistic railings that protect the openings of the houses and certain mural decorations in the Caja de Ahorros Population (Fig. 35.6). The persistence of elements like pergolas, stylized star windows and false canes present on the facades (Fig. 35.6) has been called archaic considering the construction material of the houses (reinforced masonry). This reading, influenced by the parameters proposed later by the Modern Movement, does not allow us to understand the role they fulfilled. Along with vases and grotesque columns, these elements were a romantic evocation of traditional architecture of colonial imprint, including Mudejar, which is consistent with the morphology of the set.

Fig. 35.6 Ornamental elements present in the housing complexes. *Source* Author's photograph

The favorable reception, both in its time and today, of these decorative proposals, is due in part to the magnificent execution of these realizations. Kulczewski acknowledged that he worked closely with plasterers directly on the job. This was possible thanks to the good training of these craftsmen: they were generally migrants, Spanish and Italian, who had brought with them a trade familiarly transmitted for generations (Riquelme, 1996). These aspects clearly reveal the political intention behind these practices that, linked to the socialist utopianism of the late nineteenth century, understood architecture as an art in which all trades intervened. As William Morris expressed it:

But even being so noble [architecture], and although it is the own art of civilization, it has not always existed, nor can it exist by itself in a vital and dynamic way. It must appreciate and be appreciated by all the artisan trades thanks to which men manage to make what they desire beautiful and enduring.

Such a union of the arts that help each other and subordinate themselves in a harmonious way is what I have learned to consider architecture (Morris, 2023, p. 39).

For Torrent and Atria, Kulczewski's architectural proposal is a form of "symbolic production of the new time of the big city" (2015, p. 39), understanding the value of the craftsman's work as a social act of architecture. Likewise, for Luciano Kulczewski there was a key productive relationship between art and architecture, as he expressed when asked what he understood by architecture:

> ...it is a fundamental fact in human lives: to provide housing, to give cohesion to the family, to provide a home where human lives develop; joys, sorrows, pains, all sheltered within a scope. And it is also an art. It brings together, thus, these two conditions that are sublime (quoted in Riquelme, 1996, p. 26).

Although these clarifications can be ascribed to an identity imprint, it is also possible to notice the discussion on the existential plane that this series of understandings of space produces. There are various approaches regarding an existential character in the architecture; even Peter Sloterdijk has proposed that architects do nothing more than "theory of the 'in'" (2014, 192). We are not interested, at this point, in branching our text to inquire in detail about the possible philosophical repercussions of Kulczewski's aesthetics; however, we cannot ignore the fact that such decisions are not simply solutions to situations that afflicted a certain region. The architectural strategies and political positions mentioned are also irrefutable proof of Luciano Kulczewski's singular thinking. It is, in turn, a theory about the *in*, as it implies a careful reflection on the way in which a space is inhabited and, even more, on the ways in which a space can become a place.

35.5 By Way of Conclusion

In the studied sets, the interpenetration of architectural, aesthetic and even political elements that account for the thinking of Luciano Kulczewski has been able to be dimensioned. Such recognition has been achieved by delimiting emblematic aspects of his work from the review of two residential sets in the Providencia commune. Both the Arturo Prat Cooperative Population of Emilio Delporte Employees and the Caja Nacional de Ahorros Population, works little attended by local historiography, are exemplary with respect to Kulczewski's aesthetic decisions. On the one hand, the particularity of each building is evident from the continuous dialogue with the clients and the attempt to provide heterogeneity to each dwelling. On the other hand, the circumscription of an architectural group with aesthetic affinity is also important, either as a compositional totality, and even in that sort of microcosm that we have already cited. This last aspect indicates the search for a certain intimacy not only in the space of a house, but also in the median of the transit towards the residential sets. Throughout this text, we have underscored the affective dimension of the sets since there is, both in the principles of the architect and in the constructive decisions and in specific material elements, a search for the creation of a place. That is, the aesthetic in these buildings does not refer to a beautification or a general impression, but to a reflection on dwelling in consideration of the affections involved in that continuous action.

It is not minor that such intentions are realized, in part, by the same architectural medium. It is from morphological features that the residences are treated and related to each other—with the different aspects that seal the familiarity of the dwellings and create the feeling of intimacy—as well as it is through ornamental features that Kulczewski seals the singularity of his genius. In the modern approach to works of art, it has been said that a search and delight in details or clues has prevailed over a gluttony for large totalities (Ginzburg, 2013). To some extent, our study on Kulczewski can also derive from that modern predilection that, in any case, our studied architect also shared. While we underline the sense of totality not abandoned in each of the sets, we have also mentioned the eagerness in the use of ornament in each of his buildings. It is, precisely, through such clues or details that it is possible to devote ourselves to a study of the architect. In this sense, the research presented not only falls within what is known as a heritage rescue, but also addresses the inquiry into understandings about the existential dimension of space. To the way in which one lives and dwells in cities. To the ideas materialized in the constructions that, in any case, need to be reconstructed to be brought back to light in current discussions.

Acknowledgements This chapter was carried out within the framework of a FONDECYT de Iniciación de Investigación, 2018, n.° 11180495: La vivienda colectiva en Santiago durante la primera mitad del siglo XX: pluralidad y singularidad en la obra de Luciano Kulczewski (1922–1956) [Collective housing in Santiago during the first half of the twentieth century: plurality and singularity in the work of Luciano Kulczewski (1922–1956)]. Thanks are given to the Agencia Nacional de Investigación y Desarrollo de Chile (ANID) for the funding provided.

Notes

1. Luciano Kulcewski García (1896–1972) was a renowned Chilean architect who practiced during the first half of the last century, so his work emblematically embodies the transition from the traditional academic proposal of a historicist nature, to the architecture proposed by the European avant-garde. In fact, his work from the thirties, pioneeringly embodies the premodern aesthetic that will characterize the panorama of Chilean architecture during the following decades. Along with a vast work, he had a recognized political participation throughout his life, which would culminate with the leadership of the Compulsory Worker's Insurance Fund, during the government of Pedro Aguirre Cerda.
2. Providencia is a commune with an area of 14.34 km^2, which by 1930 was home to about 42,000 inhabitants.
3. About the role and political concerns of Kulczewski, see Riquelme (1996), Harris (2014) and Harris (2020).
4. This term must be understood without the derogatory connotation it might have today and understand that it encompassed not only the problem of worker housing, but also generally that of the middle class.
5. These achievements were carried out for the client Alejandro Guttmann L., of whom no more information is possessed. In the Yungay neighborhood, on the same block, between 1922 and 1924, 4 complexes were erected aimed at buyers from different social strata, as revealed by the different residential typologies used. The Madrid population on the other hand, carried out in 1925, is a linear grouping of 15 houses, which is located on the block facing the homonymous street (contrary to common belief, the preserved planimetric documents reveal that the complexes carried out within the block, articulated around interior passages, are not a realization of Kulczewski).
6. Aguas Andinas, the current name of the old Metropolitan Company of Sanitary Works (EMOS), is a sanitary services company, responsible for the supply of drinking water in much of the city of Santiago de Chile.
7. Emile Delporte Braham was a Belgian migrant, founder of the firm Delporte y Cía, dedicated to the importation of goods and fabrics. The homonymous commercial store is located in the active commercial sector that extended between the capital's main square and the Mapocho train station.
8. Having been widely addressed by national historiography—it was early identified as a work of Kulczewski—and being a private real estate venture, it has been decided to leave it out of this study. The population was born as a result of the real estate venture carried out by Demófila Portales and Emilio Keller, mother and son, on the lands of what was their country house.
9. It is a population of 25 houses, which with front garden and backyard, are articulated in different typologies, but common formal features that clearly evoke the architecture English typical of the garden-city suburbs built around London at the end of the nineteenth century.
10. The Calle Keller and Caja Nacional de Ahorros populations are located within the macro block delimited by Manuel Montt Avenue (east), Alférez Real Street (south), Miguel Claro Avenue (west) and Alférez Real Street (north).
11. The width of Keller Street, sidewalks included, is 6.5 m, while the front gardens of the houses measure 3.5 m.

References

Boza, C. (1983). *Santiago: Styles and ornaments*. Montt and Palumbo.

Burmeister, E. (1969). Individual contributions to the development of Chilean architecture: The work of architect Luciano Kulczewski. Seminar in Art History, University of Chile.

Espósito, F. (2012). The 'affect' in architecture: The relationship between architect, place and inhabitant in the contextual experience of the project. *Architecture Review, 8*(1), 8–16.

Gallardo, L. (2013). Place and architecture. Reflection of the essence of architecture through the notion of place. *Architecture Review, 9*, 161–169.

Ginzburg, C. (2013). Clues. In *Myths, emblems and clues, Verónica Trentini*. Prometheus.

Harris, R. (2014). Luciano Kulczewski, architect. In *Eclecticism and modernizing processes in Chile in the first half of the 20th century*. Doctoral thesis. Polytechnic University of Madrid. https://oa.upm.es/32703/

Harris, R. (2016). Media characterization of the productive corpus of Luciano Kulczewski. In P. Corro and C. Robles (Eds.), *Aesthetics, mass media and subjectivities* (pp. 185–193). Institute of Aesthetics, Faculty of Philosophy of the Pontifical Catholic University of Chile.

Harris, R. (2020). The modern management of Luciano Kulczewski. The cases of the Collective Buildings for Workers of Arica and Iquique. *AUS [Architecture/Urbanism/Sustainability]*, (27), 12–17. https://doi.org/10.4206/aus.2020.n27-03

Hidalgo, R. (2002) Social housing and urban space in Santiago de Chile. A retrospective look at the action of the State in the first decades of the 20th century. *EURE, 28*(83), 83–106.

Maderuelo, J. (2008). The idea of space. *Architecture and contemporary art* (pp. 1960–1989). Akal.

Morris, W. (2023). *Architecture (collected texts)*. Pepitas ed.

Palmer, M. (1984). The commune of Providencia and the Garden City. A study of the beginnings of the current growth model of the city of Santiago. Faculty of Architecture and Fine Arts of the Pontifical Catholic University of Chile.

Pérez de Arce, R. (1997). Review of the architecture of Luciano Kulczewski. *ARQ, 35*, 77–78.

Ranciere, J. (2009). *The distribution of the sensible*. LOM.

Riquelme, F. (1996). *The architecture of Luciano Kulczewski. An essay between eclecticism and the modern movement in Chile*. ARQ Editions.

Saldarriaga, J. A. (2023). The home as an element of architecture: A history of its displacements. *Architecture Review, 28*(44), 194–207. https://doi.org/10.5354/0719-5427.2023.70030

Sloterdijk, P. (2014). City and architecture. In J. Chamorro (Trans.), *The aesthetic imperative*. Akal.

Subercaseaux, B. (2004). *History of ideas and culture in Chile. The Centenary and the Avant-garde* (Vol. III). Editorial Universitaria.

Torrent, H., & Atria, M. (2015). Luciano Kulcewski, architecture for the transforming city. *AOA Magazine, 30*, 16–39.

Chapter 36
Stereotypes and Gender Roles in Unidad Independencia Through Nacho López's Photographic Archive

Lucía Martín López and Rodrigo Durán López

Abstract With the arrival of Adolfo López Mateos to the government, a great construction plan was carried out for the Mexican Social Security Institute (IMSS) with the objective of having adequate buildings to cover the provision of social security services throughout the country, building in 1959 in the Federal District the Unidad Independencia Housing Complex, whose goal was "to give the country a new type of Mexican, with a clearer civic culture." By including a Casa de la Asegurada in its program, it was expected to continue the search for gender equality begun in 1951 by Ruiz Cortines. For the understanding and dissemination of daily life developed in the Unidad Independencia, the IMSS hired the photographer Ignacio "Nacho" López Bocanegra whose work was used as propaganda. Currently, Mexico's National Housing Program (2019–2024) proposes Unidad Independencia as a reference for affordable and quality housing. The purpose of this text is to analyze the functioning of gender roles in relation to the space of this housing complex through the lens of Nacho López in order to understand if it can be a model in this aspect. In order to obtain such results, two phases of research will be carried out: (1) The analysis of image parameters by restricting the visual codes with the stereotypes and gender roles focused on their spatial aspects; and (2) The evaluation of the archive images through a series of diagrams that translate the qualitative data and overlap them so that the analysis acquires representativeness in order to establish a comparison between the gender roles within the housing complex narrative.

Keywords Housing complex · Methodology · Photography · Photojournalism · Visual code

---

L. M. López (✉)
Universidad Rey Juan Carlos (URJC), Madrid, Spain
e-mail: lucia.martin.lopez@urjc.es

School of Architecture, Art and Design, Tecnologico de Monterrey, Monterrey, México

R. D. López
Universidad Anáhuac México, Huixquilucan, México

M. Vicuña and E. Giorgi (eds.), *Intersections*, Springer Geography,
https://doi.org/10.1007/978-3-031-76402-8_36

36.1 Introduction

The 1950s implied a great change for Mexican women. The government of Adolfo Ruiz Cortines announced the promotion of the relevant legal reforms so that women could enjoy the same political rights as men and numerous events occurred, such as the approval of the law in favor of women's suffrage in 1953, the first Feminist Congress in 1956 and the opening of the first Casa de la Asegurada (House of the Insured) of the Mexican Social Security Institute (IMSS) in 1956, which contributed towards this equality of rights (Martín López & Durán López, 2021). With the arrival of Adolfo López Mateos to the government, a major construction plan was carried out for the IMSS with the aim of having adequate buildings to cover the provision of social security services, social security throughout the country, building in 1959 in the Federal District the Unidad Independencia Housing Complex whose goal was to "give the country a new type of Mexican" (Coquet, 1961, p. 13).

This housing complex was composed of a robust program: 2,235 homes, a supermarket, a market, a set of workshops for arts and crafts, three cores of educational and basic commercial services, a sports center (with an indoor gym, bowling alley, swimming pool for adults and children, football field, frontons, basketball and volleyball courts and an archery range), a civic center, a social building (with administrative offices, post office, premises for youth clubs, library and meeting room for social gatherings), an indoor and an theaters, a movie theater, a zoo, a Casa de la Asegurada, a clinic and a large network of gardens, plazas and children's play areas. 34.5 hectares, with an average density of 450 inhabitants per hectare, where 23% of the surface was built, 67% was designated to the network of parks and gardens and the remaining 10% to streets and parking lots (Coquet, 1961, p. 6; Durán & Orensanz, 2022) (Fig. 36.1).

During the mandate of Adolfo López Mateos, the IMSS hired the photographer Ignacio López Bocanegra (Nacho López) to capture the daily life developed in the Unidad Independencia, and his work was used as propaganda, while texts were a privilege for a minority, images reached both educated and illiterate people equally (Fig. 36.2).

López's work from the 50s was characterized, according to Roberto Tejada (2007), by his "highlighted stories and picturesque essays with social criticism, a formal style of his own and a narrative approach." The ability of photographer Nacho López to work with different representative codes according to the photographic project has led to recent analyses questioning his role as a "standard" photojournalist. He is now often considered more of a photo-artist due to his frequent use of previsualizations for his shots. Previsualization "is the term that López gave to his peculiar way of combining the decisive moment captured after the click with the construction of small tableux vivant." (Limón, 2013, p. 176). That is, López adhered, according to Nieves Limón (2013, p. 179), to the "photojournalistic syntax without renouncing a particular way to capture guiding, accompanying by hand, reality."

Although the photographer, and also teacher, developed a theoretical corpus that appealed to the social responsibility of his profession (Moreno Aldan, 2007), there

Fig. 36.1 Spaces in Unidad Independencia (own elaboration, 2023)

is a great difference between the discourse constructed with the images taken prior to 1964 (when he publishes his photo report "La Ciudad de México III" in the magazine Artes de México) and his subsequent work. In the first ones, López uses previsualization and photographic technique to elaborate the imaginary of Mexico in the 50s. Nieves Limón indicates that some of his projects from this time can be considered as "photo stories about gender relations" in Mexico City. Since echoing a certain visual tradition and sheltering in a supposed and unreal naturalness, he creates an authentic visual model of the Mexican woman. So that through "the representation of these prototypes, he blurs, transforms and, ultimately, alienates these bodies" and updates the female models of the 50s (Limón, 2013, p. 184).

As Serrano et al. indicate, (2011, p. 776), "elaborating an analysis of the photographic image allows to observe the perpetuation of gender stereotypes from the relationship of visual codes, gender symbolizations and the image." The objective of the research project from which this text is extracted is to analyze Nacho López's photographic project on the Unidad Independencia to understand the functioning of gender roles transmitted through his lens in relation to the space of the housing complex. All this, to understand the cut of the visual message that the Mexican government was spreading in that period around its new projects, which in the words of Coquet (1961, p. 5), aspired "to carry out a new way of life, both individually and in its social projection." Rather than offering a model for analyzing gender through photography, the proposal here, as suggested by Serrano et al. (2011, p. 779), is to emphasize "the potential that image analysis has to make explicit habits and behaviors that are overly naturalized within society."

Before analyzing the aforementioned photographs of the Unidad Independencia under this approach, it is necessary to take a first step, the construction of the categories of analysis of roles and gender stereotypes in space. Subsequently, the

Fig. 36.2 Dissemination images of the Unidad Independencia (own assembly 2023). First row, from left to right: **a** Players training American football (MID: 77_20140827-134500:379342); **b** men heading home (MID: 77_20140827-134500:379000); and **c** family eating their meals in a dining room (MID: 77_20140827-134500:379020). Second row, from left to right: **a** Girls in an embroidery workshop (MID: 77_20140827-134500:379076); **b** couple in an outdoor restaurant (MID: 77_20140827-134500:379129); and **c** women packing food in a kitchen workshop (MID: 77_20140827-134500:379196); **d** People watching ducks in a recreational park (MID: 77_20140827-134500:379051). Third row, from left to right: **a** Worker's labor in a factory (MID: 77_20140827-134500:379487); **b** people shopping in a supermarket (MID: 77_20140827-134500:379263); and **c** men playing dominoes (MID: 77_20140827-134500:379257). *Source* Nacho López Collection—National Photographic Library. National Photographic Library INAH, Nacho López. Mexico City (1960, 1965)

images will be analyzed, and the results will be graphed so that representative information can be extracted from a gender perspective in relation to the space of the Housing Complex to understand the type of message that the Mexican government was promulgating about housing during that period.

36.2 Problem Statement

According to Papalia (2001, pp. 420–421), gender stereotypes are "exaggerated generalizations about male or female behavior." These generalizations also appear, as indicated by Serrano et al. (2011) in any photographic collection. The images construct discourses, symbols, conceptions, and ideas, and although they often create a fiction from the manipulation carried out by the photographer, they are very important for "the construction of sexual identity," due to the prototypical effect they generate through their implicit meanings (Serrano et al., 2011, p. 777).

From the moment we are born, we are exposed to numerous attitudes, behaviors, and values that we learn, according to Serrano et al., either as "modes of learning" or as "models of imitation of adult behavior." Serrano et al. (2011, p. 770) explain, "These social predeterminations have been historically structured from the binary system masculinity/femininity, whose cultural reduction promotes gender inequities that affect social coexistence and that are manifested in the domestic or massive circulation of images that construct codes and cultural representations that equivocally confront sexual identities."

Social roles and sexual identities are also reflected, and especially configured, in the household. According to Goffman (1981), the house is a "social stage," in which individuals, part of a group, assume a certain role and perform the relevant actions for its satisfaction in the form of performative activities. Roles, interactions, and symbols that not only occur in the family group but also in the social one (Ruiz Serrano et al., 2016, p. 60). In this way, and as McDowell (2000, p. 15) affirms, "spaces arise from power relations, power relations establish norms; and norms define boundaries, which are both social and spatial, because they determine who belongs to a place and who is excluded." According to Cevedio (2003, p. 91), the structures that dictate the norms on which space is constructed are the State, society, and family. And these groups are generally made up of people with common characteristics (male, white, heterosexual, middle-upper class) that strengthen their authority and perpetuate the dominance of capitalist and patriarchal power (Valero Rodríguez, 2016, p. 63). Muntañola Thornberg goes a step further indicating that space not only constructs social interaction between genders, but also controls it, so that "the effects of the power of one gender over another are perpetuated in architecture through the centuries in a stealthy and silent way, but no less efficient," (Cevedio, 2003, p. 9), just as it happens in the photographic image.

The analysis of the photographic image linked to the space represented in the scene allows us to observe the perpetuation of gender stereotypes from the relationship of visual codes, gender symbolizations, and image (Serrano et al., 2011). This is the subject of this article.

Thus, if the foundation of gender identity is the repetition of acts over time, and not a single-piece identity, as Butler indicates, (1998, p. 297), there will be possibilities to transform gender by altering the different forms of repetition, hence the interest in this research.

36.3 Method

The research is designed to obtain results in two phases: (1) Obtaining analysis parameters for Nacho López's images by restricting the visual codes (scenographic code and code of figurative elements) with stereotypes and gender roles focused on their spatial aspects; and (2) Evaluating the images from the archive through a series of diagrams that translate the qualitative data and superimpose them so that the analysis gains representativeness.

36.3.1 Parameters for the Identification of Roles and Gender Stereotypes in the Photographic Collection of the Unidad Independencia

For the compilation of the analysis categories of the photographic material, numerous texts on gender, architecture, and photography were reviewed (Table 36.1). From these, the concepts linked to gender roles and stereotypes, values and behaviors assigned to women and men by the dominant discourse, and traits, symbols, and codes attached to each identity were organized. To organize the concepts of the selected authors and relate them to each other, Table 36.1 was used. This combines, on the one hand, the ordering proposal of Macía et al. (2008) that associates roles with relationship environments linking them with different times and spaces. And, on the other hand, the organization of stereotypes ascribed to women and men that Muñoz Ruiz (2002) proposes attending to the physical and psychic configuration, the role assigned by society, the intellectual configuration, the sentimental configuration, and the religious valuation.

The images are characterized by their manifest codes and their latent codes. The manifest or visible codes in the message are those belonging to the physical gaze, while the latent or invisible codes belong to the symbolic gaze (López Díez, 2003, p. 15). In this text, only the visible codes related to gender roles will be addressed in order to use quantitative tools for their assessment. According to Millerson (2001) and López Díez (2003), the visible codes are classified into ten types: (1) Production and creation, which encompass on the one hand the dimensions, the format, and the chosen support and, on the other hand, the constitutive elements of the image. Their techniques can support in the prioritization of the subjects of the image by focusing the audience's attention on a specific person through, for example, their location in the foreground or background, their placement facing or with their back to the viewer, or their positioning standing or sitting; (2) spatial code (refers to the process of selecting the specific photographic space that is intended to be constructed through, for example, the types of shots); (3) scenographic code (Includes the chosen scene, the costumes of the people represented, the sets or natural scenarios, the figurative objects, etc.); (4) gestural code (Refers to the gestures that the people in the image represent); (5) light code; (6) chromatic code; (7) figurative elements

Table 36.1 Summary table of gender roles and stereotypes through the analyzed authors, own elaboration based on the following texts ([1] Macía et al., 2008; [2] Muñoz Ruiz, 2002, p. 141; [3] Serrano et al., 2011; [4] López Díez, 2003; [5] Amoroso, 2017; [6] Valero Rodríguez, 2016; [7] Serrano & Serrano, 2008; [8] García Cortés, 2006; [9] Wigley, 1992; [10] Cumberbatch, 1990; [11] O'Sullivan et al., 1994, pp. 132–133; [12] Martín Serrano et al., 1995; [13] Núñez, 1999; [14] Hierro, 1990; [15] Cano & Radkau, 1991; [16] Ruiz Serrano et al., 2016; [17] Serrano Barquín & Serrano Barquín, 2008; [18] Montaner & Muxí, 2011; [19] Matrix, 1984)

		FEMALE	MALE
STEREOTYPES [1]	PHYSICAL AND PSYCHIC CONFIGURATION [2]	Dependency [1] [12] [28]	Independence, autonomy [1] [3] [12] [28]
		Docile, hardworking [3]	Rebel [3]
		Selfless and reserved only to her own [12] [3]	
		Weakness [1] [3]	Greater strength and speed [1] [2] [3]
		Wariness [3]	
		Sensory skills, sensitivity [1] [2] [3]	Decision, reason [1]
		Sentimental [6] [8] [29]	Effective [6] [8]
		Impulsive yet adaptable [12]	
		Detail perception [2]	
		Delicacy [3] [28]	
		Gentleness, submission [3]	Sullen traits, sternness [3]
			Larger brain (than women) [2]
		Skills for delicate tasks [2] [28]	Skills for mechanical tasks [2]
		Memory, visual and verbal test skills [2]	Numerical, spatial and mechanical skills [2]
		Passivity, quiescence [3] [12] [16] [28]	Activity [3]
		Waiting for someone to do them "a favor" [12]	
		Calm, relaxed, less physical activity [3]	Speed, violence, combative attitude [3]
		Hidden [5] [3]	In sight and at hand [5]
		Morality [3]	
		Relationship with nature (symbolized by flowers and plants) [3] [7] [16] [28]	Related to the cultural and artificial elements of society [16]
		Icon (symbolized with children in arms) [3] [28]	
		Values: aesthetic, social, religious [2]	Values: higher religious, scientific, technical, political [2]
		Artificial [6]	Natural [6]
		Light colors [3] [7] [12] [(7]	Dark colors [3] [12]
		Light nymph [12]	Dark male [12]
		Pink color [3] [13]	Blue color [3] [13]
		Playing with dolls [3] [13]	Playing with cars [3] [13]
		Long hair [3]	Short hair [3]
		Womb, uterus [3] [4]	
		Breasts [4]	
		Buttocks [4]	
		Body [6] [12]	Mind [6] [12]
		Young (2130 years old) / woman in bloom [10] [11] [12] [28]	Mature gallant [12] [28]
		Body as sexual illustration [4]	
		Prey [12]	Hunter [12]
		Shown both to attract and to be attracted [10] [11]	
		Decorative role [10] [11]	
		Animalized or objectified women [4]	
		Beauty [6]	Work [6]
		Skirt (does not allow great activity or physical mobility, limits freedom of action) [3] [28]	Pants (allows high activity and physical mobility) [3] [28]
		Evasive look [7] [28]	Front stare [7]
		Position: seated [7] [28]	Position: vertical, upright [7] [28]
		Craftsmanship [6]	Industry [6]
		Inferiority (with respect to man) in being and knowledge [3] [26] [16] [28]	Superiority (with respect to woman) in being and knowledge [3] [26]
	ROLE ASSIGNED BY SOCIETY [2]	Motherhood [2] [4] [12] [16]	Power [2] [28]
		Family motherhood: in marriage, children, home, charity [2] [12] [13] [3] [16] [28]	Head of family (pater familias): work, family management [2] [16]
		First mother and wife, then anything else [12] [28]	
		Caring [6] [16]	Conquer [6]
	INTELLECTUAL CONFIGURATION [2]	Intuitive, quick, passionate, emotional, biased reasoning [1] [2]	Objective, reflective, total reasoning [1] [2]
	SENTIMENTAL CONFIGURATION [2]	Desire for submission and support [2] [3] [16] [28]	Desire for conquest, possession, dominance, audacity, control [2] [3] [16]
		Pudor [3]	Liberty [3] [28]
		Chaste and virginal, prenuptial purity [3] [7] [16] [28]	
		Libidinal and joyful / whorish and perverse [12] [28]	Selfrepressed and suffering [12]
	RELIGIOUS VALUE [2]	Greater mercy, more sensitive faith, charitable action [2]	Restlessness, depth, religious quest [2]
ROLES / SEXUAL DIVISION OF LABOR [1]		Associated with reproduction [1] [12] [16] [29] [30]	Associated with production [1] [12] [30]
		Unpaid labor [10] [11]	Paid labour [10] [11]
		Domestic work [10] [11] [16] [28] [29]	Domestic work represented as a special and skillful activity [10] [11]
		Educational role [3] [28]	
			Military or maritime duties [3]
TIME [1]		Circular time (no known beginning or end) [1]	Time with a known beginning and end [1]
		No timetable, but systematic [1]	Has a schedule [1]
		Has no social value [1]	Has social value [1]
		No use value (it is not integrated in official statistics) [1]	Has use value [1]
			Produces goods and services [1]
SPACE [1]		Private, house [1] [3] [5] [6] [27] [16] [28] [29] [30]	Public, city [1] [5] [6] [16] [28] [29] [30]
		Interior [3] [5] [12] [29]	Exterior [3] [5] [12] [29]
		Intimacy [3] [28]	
		Invisible, inexistent or unrealistic [1] [27] [3]	Visible, real [1] [3]
		Closed [1] [5] [3] [16]	Open [1] [5] [3]
		Isolation space [1] [6] [9] [28]	Space for relationships [1]
		Reproductive spaces [5]	Productive spaces [5]
		Private areas (kitchen, dining room, music room, garden, bedroom). Marginalized spaces (servants' quarters, laundry rooms, reading room, painting or embroidery) [16]	Domain areas (foyer, office, library, main hall, bedroom, billiard room) [16]
		Space with few social relationships [1]	Space of social power [1]
		Dynamic idea of the house [5]	Sedentary idea of the house [5]
		Country [6]	City [6]
		Conscious and responsible architecture with the surrounding world [6]	Autistic architecture, focused on what happens inside it [6]
			Hierarchical spaces [6]
			Spaces with domination and surveillance systems [6][16]
		Decorated architecture (banal, superfluous, unnecessary, immoral, impure, sensual, erotic) [6] [9]	Clean, white architecture, purity of space (pure, neutral, objective, rational, genuine) [6] [9]

Table 36.2 Parameters for the identification of gender roles and stereotypes in space according to the scenographic code and the code of figurative elements (own elaboration, 2023)

	Scenographic code	Code of figurative elements
Stereotypes	Type of action performed	Representativeness (number of people of the same gender)
		Accompaniment (alone, with women, with men, with women and men)
Roles	Sphere of daily life linked to the scene (productive, reproductive, personal, political)	
Space		Location in Unidad Independencia
	Domestic space / Public space	
	Closed space / Open space	

(interested in the objects and characters that appear in the image, it only focuses on observing the number and location of visual main characters and extras); (8) graphic and symbolic codes; (9) composition technique (information transmission strategies as visual expression of content (Correa et al., 2000, p. 97)); and (10) rhetorical code (moment in which the technique of logical argumentation gives way to iconic persuasion where symbolic codes, grouped or associated produce new meanings). This study focuses on the scenographic code and the code of figurative elements.

Based on the restriction in visual codes and their relationship with stereotypes and roles, the quantifiable parameters of Table 36.2 were selected to analyze the photographs from Nacho López's archive.

With the application of these parameters in the analysis of the photographs of the Unidad Independencia, it is intended to highlight the distorting visual codes that society assumes as "natural" to, as Serrano et al. (2011, p. 771) say, show them in order to they can be mitigated to contribute to the construction of a more equitable society.

36.3.2 Evaluation of the Images from the Photographic Collection of Unidad Independencia

In the evaluation of the images from the archive, 587 photographs of the Unidad Independencia were located within the Nacho López Collection found at the National Photographic Library of the National Institute of Anthropology and History of Mexico. From these, three types of images were discarded that were considered might alter the data: (1) Images similar to each other in composition and framing taken in "burst mode," (231 items), of which only the most representative of these was left for the study; (2) images in which no people clearly defined in the foreground so that the activities they were performing could be identified; and (3) eliminating images without people (16 items) and those in which a crowd was observed (113 items) (Fig. 36.3). In this way, 227 photographs were obtained for analysis.

To perform the analysis of the 227 selected photographs, the 1,222 people who appear in the collection were identified and classified, distinguishing them by gender and assigning them a number for their classification (Fig. 36.4).

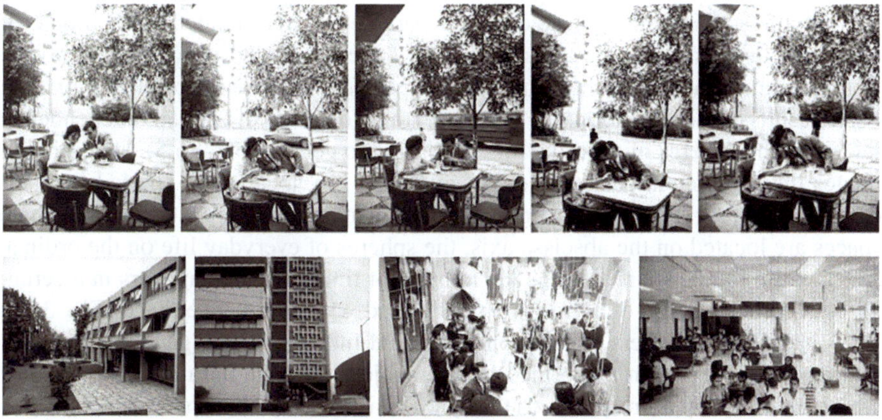

Fig. 36.3 Exclusion criteria for images (own assembly, 2023). (1) Images taken in "burst mode." First row, from left to right: **a** Couple in an outdoor restaurant (MID: 77_20140827-134500:379127); **b** couple chatting in an outdoor restaurant (MID: 77_20140827-134500:379130); **c** couple in an outdoor restaurant (MID: 77_20140827-134500:379132); and **d** couple in an outdoor restaurant (MID: 77_20140827-134500:379133). (2) Photographs where there are no people. Second row from left to right: **a** Buildings of a school, partial view (MID: 77_20140827-134500:379086) and **b** condominium of the Unidad Independencia, partial view (MID: 77_20140827-134500:379420). (3) Photographs where a crowd is shown. Second row from right to left: **a** People in the waiting room of a clinic (MID: 77_20140827-134500:379097) and **b** people during a Christmas party (MID: 77_20140827-134500:379187). *Source* Nacho López Collection—National Photographic Library. National Photographic Library INAH, Nacho López. Mexico City (1960, 1965)

Fig. 36.4 Photographs with identification number of each person according to gender (own assembly, 2023)

Once the people in each image were identified, the parameters for identifying roles and gender stereotypes in space according to the scenographic code and the code of figurative elements (Table 36.2) were analyzed for each subject using a spreadsheet. These parameters included: gender; action performed by the subject in the image; space within the Unidad Independencia where the shot is located; sphere of the person's daily life represented in the image; whether the space where the photograph was taken is domestic or public; whether the represented space is open or closed; and whether the person photographed is accompanied, including the gender of their companion.

With the acquisition of these data, the parameters were related to each other to draw conclusions about the construction of stereotypes and gender roles through four diagrams that unfold the data according to gender to facilitate the comparison of the results in each situation: (1) The diagram of locations of Unidad Independencia related to the spheres of everyday life according to gender (Fig. 36.5). Diagram that relates the parameters of the sphere of everyday life, location in Unidad Independencia, and gender representativeness. For this, a matrix diagram is used in which the spaces are located on the abscissa axis, the spheres of everyday life on the ordinate axis and the frequency represents the number of times a sphere appears in a certain space; (2) the diagram of actions by gender and where they take place (Fig. 36.6). This links the parameters of location in Unidad Independencia, gender representativeness, and the action performed by the subject, through an alluvial diagram in which the first step corresponds to the spaces and the second step to the actions; (3) the diagram of location in space and its type according to gender (Fig. 36.7). This focuses on the parameters of gender representativeness, domestic or public space and open or closed space. For this, a Sunburst type diagram is used in which the first step corresponds to gender, the second to domestic or public space, and the third to its quality of open or closed; (4) The diagram of accompaniment according to gender (Fig. 36.8). This relates the representativeness by gender, and its accompaniment with the types of spaces, the domestic versus the public, and the closed versus the open. This bar diagram is constructed from the parameters of representativeness, domestic space versus public space (or non-domestic) and accompaniment. From this diagram, for example, the degree of representativeness that single women have in public space within Nacho López's work on the Unidad Independencia complex could be observed.

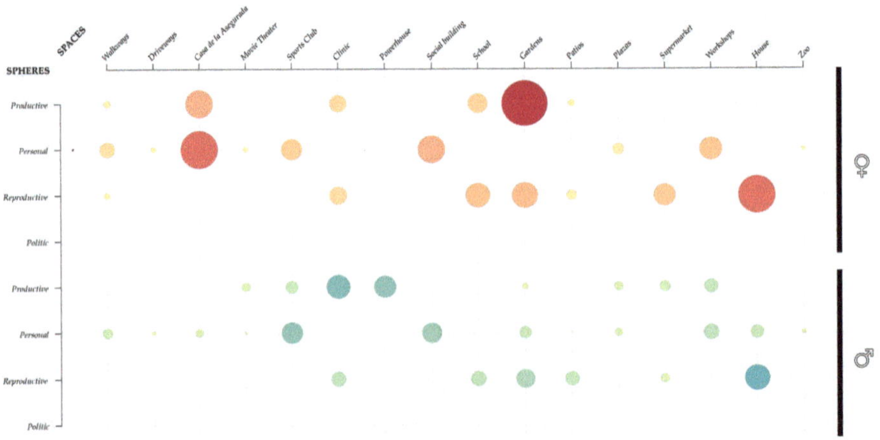

Fig. 36.5 Diagram of locations of the Unidad Independencia related to the spheres of everyday life according to gender (own elaboration, 2023)

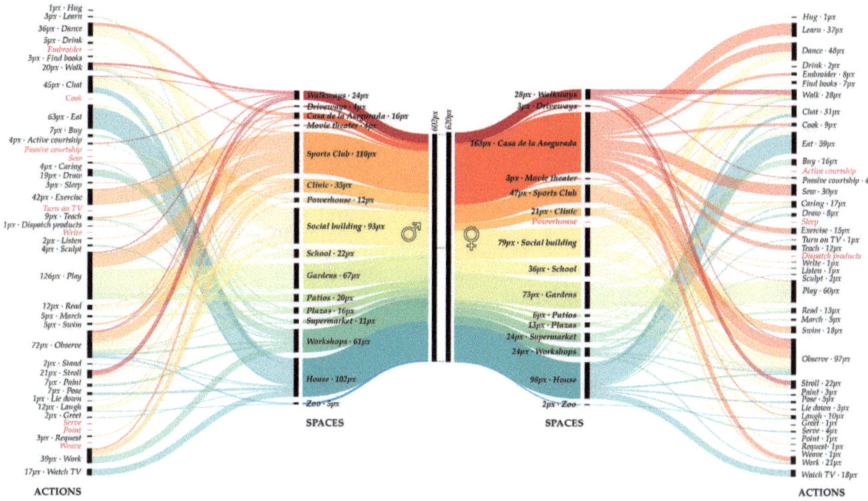

Fig. 36.6 Diagram of actions by gender and where they take place (own elaboration, 2023)

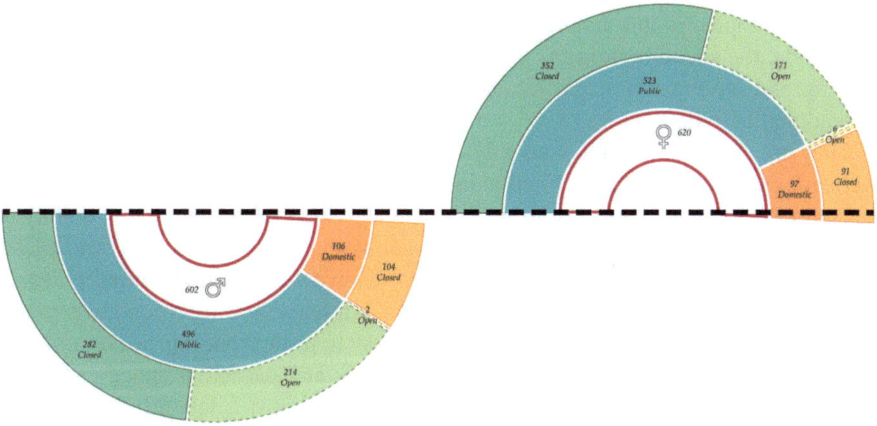

Fig. 36.7 Diagram of location in space and its type according to gender (own elaboration, 2023)

36.4 Results

Results in terms of representativeness are very balanced, as 49.26% of the 1,222 people who appear in the photographs of the housing complex are men and 50.74% are women. Women appear more frequently in the images, as shown in the four diagrams. The first thing that stands out in the diagram of locations of Unidad Independencia related to the spheres of everyday life according to gender (Fig. 36.5) is that there are three spaces that host the daily life of the female gender: the gardens, through their productive sphere in which their primary activities are directly related to

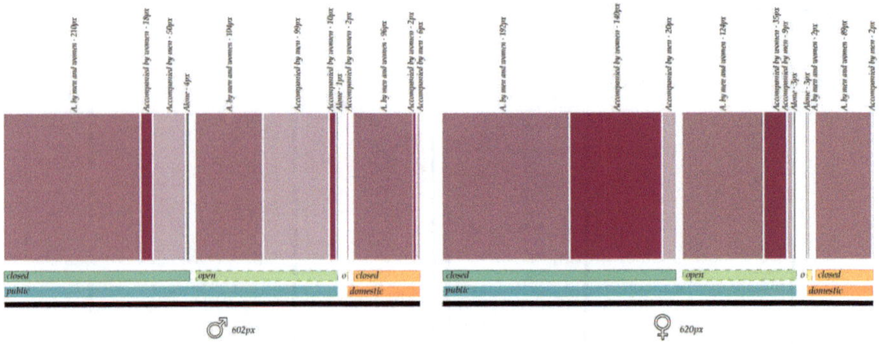

Fig. 36.8 Diagram of accompaniment according to gender (own elaboration, 2023)

working with children; the Casa de la Asegurada, through its own sphere in personal development activities such as acquiring trades or hobbies; and the house, through its reproductive sphere mainly in the care of the nuclear family. As for the productive sphere, women also appear significantly in the Casa de la Asegurada as trainers, and to a lesser extent in the school and clinic. Patios and walkways are spaces where they barely appear in this sphere. Regarding the personal sphere, after the Casa de la Asegurada, the social building is where women appear most frequently, represented dancing and having fun at parties, followed by workshops and the sports club in recreational activities.

They are also presented in walkways and squares, and barely appear in the driveways, movie theater, and zoo. Finally, in the reproductive sphere, it is in the gardens and school where they are most represented in activities related to caring, mainly of children. They are followed by the supermarket, with purchases to supply the house, and the clinic accompanying their relatives to receive care. Their minimal presence in walkways and patios taking care of minors completes their representation in this sphere. As for the male gender, it is relatively easy to notice that it is not remarkably more represented in any sphere of everyday life or any particular space in Unidad Independencia. In the productive sphere, men appear most frequently in the clinic providing medical care, followed by maintenance work in the powerhouses. Men also work with similar frequency in the workshops, the sports club, the supermarket, the movie theater, the plazas, and the gardens. In their personal sphere, their presence is well-distributed, with only four spaces where they do not appear performing activities related to leisure or personal development. Men are most often found in the sports club, with a similar number of appearances in the social building. They also appear frequently in workshops, houses, gardens, walkways, plazas, and the Casa de la Asegurada. Men are least represented in the driveways, the movie theater, the patios, and the zoo. The house is where men are most commonly found in their reproductive sphere, living with the nuclear family. They also appear frequently in gardens, patios, schools, and clinics, while they have the fewest appearances in the supermarket.

Finally, it is important to note that the photos analyzed do not show men or women living in their political sphere in any of the spaces of the Unidad Independencia Housing Complex.

The diagram of actions by gender and where they are developed (Fig. 36.6) shows the distribution of both genders in both the different buildings of the complex and the activities they perform in them. At first glance, a similar distribution can be seen in the spaces, located in the central part of the graph. The main similarities are found in the walkways, gardens, patios, driveways, movie theaters, houses, zoo, and the social building; while the most important difference is found in the strong presence of men in the sports club, and of women in the Casa de la Asegurada. Other spaces with notable differences between genders are the workshops, with a greater presence of men, the supermarket, with a greater presence of women, and the powerhouse, in which women do not appear in the photographs. The activities, located at the right and left extremes of the graph, reveal that there are considerable differences between both genders. For men, their predominant activity is playing, mainly in gardens and the sports club. For women, the most frequently seen activity in the photographs is observing, mainly in the Casa de la Asegurada, the sports club, and the social building. While there are differences in the distribution of activities performed, it is important to note the absent or unperformed activities in each gender as another way to accentuate these differences. It is observed, for example, that the male gender never appears embroidering, cooking, passively courting, sewing, serving, or weaving, among others; while the only activities that the female gender does not perform are active courtship, product dispatch, and sleeping.

The diagram of location in space and its type according to gender (Fig. 36.7) is of a simpler composition and has two semicircles, each indicating the distribution of each gender by type of space. These, in turn, are divided into public or domestic, and each of them can be open or closed. Both genders have a very similar distribution between public and domestic space, with men appearing in 82.4% of the photographs in the collection in public spaces, compared to 84.4% of women. The graph allows us to see that the primary difference in the distribution of each gender is found in the quality of the space, as it is possible to see men 43.1% of the time in open spaces, compared to only 32.7% of women. In terms of domestic space, it is also in open spaces where the differences are notable, with men appearing only in 1.9% of the photographs in these spaces (balconies or terraces of the houses), compared to 6.2% of women.

Finally, in the diagram of accompaniment according to gender (Fig. 36.8), it is possible to see once again the distribution of each gender by type of space, in this case extended horizontally and combined with vertical bars that represent the type of accompaniment. In the case of men, the graph shows that in closed spaces, photographs in which they are accompanied by both men and women predominate, while in the open public space this proportion is equal to the photos in which he appears accompanied only by men. It is also noteworthy that in open public spaces only 3.3% are accompanied by women, and the only two men who appear in the open domestic space are accompanied by women. The case of women is different, as more than a third of them (39.8%) are exclusively accompanied by other women in closed public spaces, and in those same spaces only 5.7% are accompanied by men. In the open public space, they appear predominantly accompanied by men and women, and for both genders the closed domestic space is similar, with an overwhelming majority of company of men and women (both over 90%).

36.5 Conclusions

Recently, the National Housing Program (2019–2024) of Mexico has proposed the Unidad Independencia Housing Complex as a reference for affordable and quality housing. That is why it is considered pertinent to revisit this object of study under a new approach that allows analyzing the image of the complex projected by the architectural media: the gender approach through the study of roles and stereotypes developed in space.

This article aligns with the assertion of Serrano et al. (2011, p. 779) that "the semiotic analysis of photography evidences certain modes of perpetuation of gender inequality and constitutes an educational and awareness-raising material" so that, as Valero Rodríguez (2016, p. 86) indicates, awareness of these associated connotations is crucial to visualize -and be able to eradicate- these discriminations.

In this binary context, the environment only favors the development of a few, and those who do not conform to the established canon (not focusing exclusively on gender) remain oppressed or excluded. According to García Cortés (2006, p. 128), they are "absent bodies," realities without representation, lacking their own place to realize themselves. That is why it is considered important to explore in the photographs of Unidad Independencia by Nacho López not only the discourse created around women but also the positioning on aspects such as race or social class in future works.

Finally, it is considered important that any publication or propaganda seeks balanced gender equity and reviews the manifestation and perpetuation of gender stereotypes and roles, as emphasized by DuQue Acosta (2010). Dispensing with the antagonistic binary system where the masculine and the feminine, the exterior and the interior, the natural and the artificial are opposed, it is possible that the public and private spheres will blur, allowing new realities that differ from the dominant model to emerge.

Acknowledgements We thank Carlo Emiliano Magno Estrada Romero, Diego Martín Espino García, Daniela Stephanie Muñoz Salgado, and Ana Cristina Torres Pulido for their support in the development of the classification work of the images of the Unidad Independencia belonging to the photographic archive of Nacho López.

References

Amoroso, S. (2017). De género y espacios (contenedores): hacia una deconstrucción de lo doméstico. *Asparkía, 31*, 113–130.

Butler, J. (1998). Actos performativos y constitución de género: un ensayo sobre fenomenología y teoría feminista. *Debate feminista, 18*(9), 296–314.

Cano, G., & Radkau, V. (1991). Lo privado y lo público o la mutación de los espacios (Historia de mujeres, 1920–1940). In V. Salles, & E. (Mc Phail, Textos y Pre-textos. Once estudios sobre la mujer. PIEM/The College of Mexico).

Cevedio, M. (2003). *Arquitectura y género: espacio público/ espacio privado.* Icaria.

Coquet, B. (1961). Unidad de servicios sociales y de habitación Independencia. *Arquitectura México, XVII*(73), 2–46.

Correa, R. I., Guzmán, M. D., & Aguaded, J. I. (2000). *La mujer invisible.* Grupo Comunicar Ediciones.

Cumberbatch, G. (1990). *Television advertising and sex role stereotyping: A content analysis.* Broadcasting Standards Council.

DuQue Acosta, C. (2010). Judith Butler: performatividad de gérnero y política democrática radical. *La manzana de la discordia, 5*(1), 27–34.

Durán, R., & Orensanz, F. (2022). *Ciudad Independencia Seguro Social.* Arquine.

García Cortés, J. M. (2006). *Políticas del Espacio. Arquitectura, género y control social.* IAAC & ACTAR.

Goffman, E. (1981). *La presentación de la persona en la vida cotidiana.* Amorrortu.

Hierro, G. (1990). *De la domesticación a la educación de las mexicanas.* Torres.

Limón, N. (2013). Nacho López, análisis de dos performances fotográficas. *Zer, 18*(34), 173–193.

López Díez, P. (2003). *Las mujeres en el discurso iconográfico de la Publicidad. Claves, desde la perspectiva semiótica y de género, para descodificar el lenguaje publicitario sobre las mujeres.* Emakunde.

McDowell, L. (2000). *Género, identidad y lugar.* Cátedra.

Macía, O., Mensalvas, J., & Torralba, R. (March 10, 2008). *Roles de género y estereotipos.* Retrieved August 28, 2022, from https://perspectivagenerotelecentro.wordpress.com/manual-trabajo-con-grupos-mixtos-en-el-tc/roles-de-genero-y-estereotipos/

Martín López, L., & Durán López, R. (2021). Casa de la Asegurada: A Collective Housing Facility for Women Development in Mexico. *Buildings, 11*(6), 236.

Martín Serrano, M., Martín Serrano, E., & Baca Lagos, V. (1995). Las mujeres y la publicidad. Madrid: Instituto de la Mujer

Matrix. (1984). *Making space. Women and the man made environment.* Pluto Press Limited.

Millerson, G. (2001). *Realización y producción en televisión.* IORTV.

Montaner, J. M., & Muxí, Z. (2011). *Arquitectura y política: ensayos para mundos alternativos.* Gustavo Gili.

Moreno Aldan, I. (2007). La cátedra de Nacho López. *Luna Córnea, 31*, 430–447.

Muñoz Ruiz, M. D. (2002). *Mujer mítica, mujeres reales: las revistas femeninas en España, 1955-1970.* Complutense University of Madrid (Doctoral Thesis).

Núñez, G. (1999). *Sexo entre varones, poder y resistencia en el campo sexual.* El Colegio de Sonora, PUEG/UNAM.

O'Sullivan, T., Dutton, B., & Rayner, P. (1994). *Studying the Media. An Introduction.* Edward Arnold.

Papalia, D. (2001). *Psicología del desarrollo.* McGraw Hill.

Ruiz Serrano, E., Serrano Barquín, C., Ciaramitaro, F., & Serrano Barquín, H. P. (2016). Narrativa de una casa porfiriana: sus representaciones simbólicas y de género. *Anales de investigación en Arquitectura, 6*, 47–66.

Serrano Barquín, H., & Serrano Barquín, C. (2008). La educación de género y su evidencia en la imagen fotográfica. *Espacios públicos, 11*(23), 265–278.

Serrano, H., & Serrano, C. (2008). La educación de género y su evidencia en la imagen fotográfica. *Espacios públicos, 23*(11), 265–278.
Serrano, H., Zarza, M. P., Serrano, C., Gómez, B., & Iduarte, J. (July–December 2011). Códigos visuales de género y configuraciones sexuales evidenciadas en la fotografía. *Revista Latinoamericana de Ciencias Sociales, Niñez y Juventud, 2*(9), 769–782.
Tejada, R. (2007). Nacho López: Mexican Photographer. *The Americas, 63*(4), 460–462.
Valero Rodríguez, S. (2016). *Dom[é]stica-Da. Visibilizando la cotidianeidad de la mujer en el entorno doméstico contemporáneo.* (Vol. Master's Final Project). Polytechnic University of Valencia.
Wigley, M. (1992). Untitled: The housing of gender. In B. Colomina (Ed.), *Sexuality & space* (pp. 326–389). Princeton Architectural Press.

Chapter 37
BRAT'YA: The Architecture of Dostoevsky

Jorge Sánchez Lara

Abstract Fyodor Dostoevsky told us stories about the social inequalities of the nineteenth century; he did so with tales that mix personal reflections and spatial descriptions. His work is mainly studied from literary or psychological approaches, but its representative quality shows us the evolution of other areas: "Literature… becomes a painting, in a certain sense, of course; a picture and a mirror; a mirror of passions and all intimate things: it is instruction and edification at the same time, it is criticism and it is a great human document." (Dostoyevsky, 1970) BRAT'YA is part of a critical-architectural trilogy that analyzes his narrative techniques, along with those of Marcel Proust and Juan Rulfo, to find and understand people within the environment in which they live. The authors, all of them knowledgeable about the architecture of their country, Russia, France, and Mexico, brought us closer to the defense of dignity, artistic creativity, and the link between us and nature. Dostoevsky's narrative stands out in the first part, and we will focus on it. To approach his thoughts, we read his books and the most relevant biographies, furthermore, we considered it essential, to review how the arts have interpreted the spaces, places and people to which he only referred with words: we show drawings, paintings, engravings, plans of the city of St. Petersburg, plays and films that have been made around his ideas. His writing belongs to another era and modern translations have molded it into simpler forms. To preserve temporal distances, we selected editions from different decades and countries: from the forties; the seventies, when we had a boom in his reading in Latin America; up to the most recent ones. So, along with observing the beauty of the language, we will also see the selections that we had to make between the different publications, references from his first biographers like the Russian Leonid Grossman, and interventions from scholars and critics from France, Norway, Spain, Argentina, or Mexico. The content is shown in three stages. We expose the social context, connect his stories and novels with architecture, and summarize the contributions of the novelist to our area. The second section was created in the following way: we identified his ideas and isolated phrases, organized them into themes, ordered them chronologically, that is, we followed the year of printing, and finally, we linked them through the narrative. Dostoevsky left us a fertile legacy: he described the

J. S. Lara (✉)
School of Architecture, Art and Design, Tecnologico de Monterrey, Guadalajara, México
e-mail: jsl@tec.mx

M. Vicuña and E. Giorgi (eds.), *Intersections*, Springer Geography,
https://doi.org/10.1007/978-3-031-76402-8_37

consequences of living in disparate communities, the psychic effects experienced by people who inhabit poorly maintained houses, the frustration we develop in hostile cities, the need for decent housing for our children, and the enormous desire to find compassion. Brat'ya, a Russian word, means brothers and he used it in his last novel. His stories allow us to observe and assimilate many of the current problems of our Latin American society; the parallelism between his confessions and our experiences, both reveal human fragility.

Keywords Dostoevsky · Literature · Architecture · Inequality · Art

37.1 Dostoevsky and Russia. The Writer of the Fallen

Dostoevsky (1821–1881) (Fig. 37.1) was born in Moscow, but most of his novels were written and inspired in St. Petersburg. For him, the city is a reflection of human behavior. His writing carries the marks of his life: he lost his parents early, was arrested and exiled to Siberia for ten years, suffered from epilepsy and had constant financial problems (Zweig, 1945, p. 96). In his work, also the traces of his era are found: the last period of the tsars, the abolition of serfdom, the creation of large cities and the appearance of the railway. From a young age, he identified with the disadvantaged and his stories combine humiliation, misery, and marginalization; we see how the fallen try to rise with dignity. He wrote for his people during the golden age of modern Russian literature and most of his books were translated after his death. The ambiguity of his stories, where the protagonists converse without the influence of the narrator (Bajtín, 1988, p. 15), led us to the knowledge of human depth and made his work studied everywhere (Fig. 37.2).

His life and work can be divided into three periods: the first covers his birth, the publication of his initial novel in 1846, and his arrest in 1849. The second includes his time in prison until the appearance in 1861 of *Humiliated and Offended* and

Fig. 37.1 Portrait of F. M. Dostoevsky in 1872 by V. Ya. Lauffert (FedorDostoevsky.ru, 2023)

Fig. 37.2 1880. Fontanka Ambankment, photography of Jongh Frères, published by A. Felten (https://www.rbth.com/history/331392-petersburg-photo-nineteenthcentury)

Memories of the Dead House, creations that indicate his literary maturity. Finally, the third extends until 1881, a time that contains his best books and his death.

To analyze his texts, we use the signs of Deleuze (1972) and the narrative keys of González Boixo (1984), methods of separation and analysis of contents through arguments to find the most common and important. The techniques allows us to select themes, within the entire collection, to view them from specific perspectives, the architecture for us, and thus, obtain precise observations.

Next, we will follow the proposed chronological order and weave his literary work with biographical data and thematic comments. The chosen phrases will guide us through issues related to architecture, and will be accompanied by images that different artists, mostly Russians, have made to interpret his ideas; it will be very interesting to see and discover the spaces, places, and people that were recreated from the readings of his novels.

37.1.1 Poor People (1821–1849). Disparate Communities

Dostoevsky published his first novel at the age of twenty-three, in which he described scenes from the lives of the most vulnerable. *Poor People* (1846) is an epistolary essay, the characters send each other letters, in which we learn how economic hardships affect our perception of our environment and behavior. We humans are sensitive to our surroundings, the writer will warn us from the beginning:

> When I woke up today, I felt entirely…, so, let's say, so light in body and spirit…, so cheerful and carefree. It's just that today has also been a morning! I opened the window, and the sun poured in, the birds began to sing, the air was filled with the scents of spring, and all of Nature revived… (Dostoyevski, 1970, p. 106)

This is how Makar, a state worker who will try to help a friend financially, will express himself. His situation will change rapidly and when he can't even afford to buy shoes, in his desperation, his self-love will force him to say:

> A sole is nothing, and it will never be more than a simple sole, vulgar and dirty. Boots are nothing either! The wise Greeks walked barefoot. Why should we worry about such a trivial thing? (Ibid., p. 174)

Varenka, a beautiful young woman, and companion in the story, also driven by poverty, will confess the words left to her by a visitor, and in this way, we will begin to connect with the architecture:

> Then, extending his considerations about our room; he said that it was not surprising that I was sick living in such a hovel, and he prophesied an early death for me if I continued living here. -In Petersburg all the dwellings are bad-, he said… (Ibid., p. 194)

The initial joy of the official will contrast with his final feelings upon recognizing his true situation:

> And suddenly I felt that my last energies were leaving me and that everything was lost. All dignity had disappeared: the man was annihilated in me! (Ibid., p. 186) (Fig. 37.3)

Dostoevsky's writings denounce the emotional and material consequences caused by the economic conditions of the disadvantaged; and it is likely that the architecture classes he received while studying military engineering (Grossman, 1975, p. 44) allowed him to link his characters with the rooms, corridors, and landscapes they traversed. This spirit of ethical vindication will remain throughout his work.

From this initial period, which includes about thirteen novels, we have selected excerpts from three others: *Weak Heart* (1848), *White Nights* (1848) and *Netochka Nezvanova* (1849), works that not only pose new problems, but also allow us to observe the uniformity of the arguments. The first will show us what a home can come to represent for young people; the second, the vigorous presence of streets and houses in the development of our personal relationships; and the third, the impression that neglected spaces leave in the memory of our children.

Weak Heart describes the life of a poor young bureaucrat who wants to get married. When his girlfriend accepts him, he will enthusiastically tell his best friend:

> I will take care of both of you, I will pamper you- he started again-… it is necessary to show courage and take a room for twenty-five rubles. It must be a good room brother… Good housing makes a man happy and suggests pleasant thoughts… Imagine that… -and he was so happy that his voice was already breaking-… (Dostoyevski, 1970, p. 423)

Fig. 37.3 Portrait of Margaret Catov as Fedora and Jennifer Stepanyk as Várvara in Poor Folk, 2015. Photography by Andrei Remennchuk, Fringe International Festival of New York (https://poorpeplethePlay.wordpress.com/portfolio/about-poor-people/)

The need to live with dignity will be a constant expression from now on, and the association between people's emotional condition and the spatial quality of their rooms will become increasingly strong.

St. Petersburg, the modern city, will be the setting for a failed romance, and will change colors and textures at the same pace as the feelings of the protagonist. *White Nights* is a youthful idyll; in this, a perceptive boy will see a girl on the street and try to win her over, during the process, we will learn that there is a close connection between his mental state and the places he visits and lives. When he is happy, the houses will speak to him:

> I also know the buildings. When I pass by them, it seems that each house starts to run, as soon as it sees me, taking two steps out of its facade, and looks at me through all its windows and as if it says: - Good morning, here I am! How are you? -... (Dostoyevski, 1942, pp. 10–11)

But the spaces will also have the delicacy to reveal his sadness and loneliness:

> I did not feel good on the street...and neither at home did I feel comfortable; so, I hardly knew myself. I spent two afternoons trying to figure out what was missing from me within the four walls of my house, but in vain. Why did I feel so uncomfortable in it? With an investigative gaze, I contemplated the green walls blackened by smoke, I fixed my gaze on the ceiling, where Matryona, with resounding success, protected the cobwebs; I reviewed all my furniture, especially the chairs, and mentally asked myself if that might not be the reason for my discomfort. (Ibid., p. 12)

After a while, the discouragement will dissipate, the city will capture it and anticipate a great moment:

> I took frequent and long walks through the streets, so that, according to my custom, I would forget where I was going. Once I finally found myself at the edge of the town. At that moment I was filled with great joy, and I slipped over the barrier and continued walking through the fields and meadows of crops, without feeling the least bit tired… Everyone who passed by me looked at me in a loving way… everyone seemed to be happy… And I also became happy… Such a powerful spell did nature cast on me, a sickly city dweller who feels suffocated between the walls of houses! (Ibid., pp. 14–15)

Finally, he will meet Nastenka:

> … that night was more beautiful for me than the day… Not far from me I saw a woman's figure; she was standing and leaning her elbows on the parapet of the embankment, and, apparently, she was absorbed in looking at the murky waters of the canal. She must be completely lost in her thoughts! (Ibid., pp. 16–17)

After nine nights, and despite their good friendship, they will part:

> I saw the colors of the walls pale; I saw even more cobwebs in the corners than before. I don't know why, when I looked out the window, it seemed to me that the house across the street had also aged and had become more faded and dilapidated, that the stucco on the pillars had cracked, the cornices had chapped and blackened, and the brown walls had become stained and dirty. (Ibid., p. 88)

In *Netochka Nezvanova*, we are shown the susceptibility of our childhood and the early appearance of memories. The heroine gives the novel its name: a girl who will lose her parents and must continually change homes. She will periodically remember the moments she lived before her orphanhood, her stay in a poor and dirty room:

> … from the middle of my nine years, I remember almost day by day all my past… So, I remember, for example, the little lamp that was always burning in a dim corner of our dwelling… I woke up suddenly from a nightmare and, finally, from the fear that I had since then of the silence and darkness of the night and of the mice that scurried and ran around the corners. (Dostoyevski, 1970, pp. 551–552) (Fig. 37.4)

The danger is, according to the writer, that we confuse our mood with our reality. This will be the opinion of a close friend of Netochka's father:

> … misery is almost a happiness for him, since, in his eyes, it constitutes his justification. As long as he is poor, he can assure everyone that misery is what has him sunk… he has three crimes on his conscience, for besides his own, he has ruined two other lives: those of his wife and daughter… I know he would die on the spot if he were to realize his crime. (Ibid., p. 567)

We will be able to know the hidden thoughts of our children, great observers of their environment:

> …deep down, with its gleaming reflections, the windows of the house across the street with its red curtains, its cars in front of the door, the neighing and pawing of the glossy horses, the voices and the coming and going of people on the street, the shadows of people behind the windows, standing out against the red background of the silk curtains and, in addition, a vague and remote music… -So this was that Paradise!. I thought, and therefore, where I wanted to go with my poor father… (Ibid., p. 587)

Fig. 37.4 Drawing of the mother's death, in Netochka Nezvanova, paper and oiled black chalk. 58 × 87 cm, Moscow State Art Gallery by the popular artist of the USSR ILYA GLAZUNOV (http://glazunov.ru/en/art/images-Of-russian-literature/Works/176-mother-s-death/F/2-F-Dostoevsky)

The solutions could be obvious if we could see them. In one of the many cases of moving, a woman will speak to her with love and compassion when she sees her sadness:

> Forgive me for behaving like a disobedient child today… I really don't feel well. -We are all children- she replied with a shy smile-; I am also a child… (Ibid., p. 636)

37.1.2 Dead in a House (1849–1866). Hostile Societies

The Tsarist state arrested Dostoevsky in 1849 for attending prohibited meetings (Troyat, 1961, p. 102) and did not release him until 1859 (Kietsaa, 1990, p. 136). During his imprisonment he could not publish, and some of his unfinished stories were edited later. This stage contains about eight novels and among them we will highlight two works that will show us his creative rise.

The stories of this intermediate phase resume and expand concepts that consolidate his social stance. He managed to narrate his incarcerated experience (1861); and published his first great novel, *Humiliated and Offended* (1861). In *H Notes from the House of the Dead* (1861), he denounces the spatial, physical and mental conditions of the prisoners, and establishes close links between people and places: the main character, who could be himself, will tell his arrival, stay and departure from the prison, within a process that will describe rejection, adaptation and acceptance. The beginning could not be promising:

> In the remote regions of Siberia, in the midst of steppes, mountains and impenetrable forests, there are sometimes towns of a thousand and even two thousand neighbors of wooden houses,

dilapidated and with two churches -one in the town; the other, in the cemetery-. (Dostoyevsky, 1970, p. 1163)

Now let's look at the description of his jail:

Our prison stood on the edge of the fortress, on its very bastion. If you happened to look through the cracks in the wall towards the light of God, you saw nothing…; that is, you only saw a piece of sky above the high earth bastion, shaken by the hurricanes… (Ibid., p. 1166)

The protagonist will also explain his emotional state:

The continuous spiritual unrest, the nervous excitement, the stale air of the ravine could completely destroy me… -I must go out often to get some air…I want to live, and I will live-. (Dostoievski, 2015. p. 122)

Some moments of the stay, he will tell us, were unbearable:

When we opened the door to the bath, I thought we had entered into the underworlds. Imagine a room twenty steps long and as many wide, in which perhaps up to a hundred men were gathered at a time, or at least eighty… Steam that got into our eyes, sweat, dirt, such tightness that you didn't know where to put your foot. I felt panic and wanted to turn back… (Dostoyevsky, 1970, p. 1245)

Gradually, the story will move towards a process of adaptation, and we will encounter feelings similar to those the narrator had before experiencing the ordeal:

When they closed our block, it suddenly took on a certain special aspect, the aspect of a real dwelling, a real home. Now is when I can look at my fellow prisoners, completely as if at home. (Ibid., p. 1201)

Finally, his behavior will change towards acceptance, and we will understand that the key to overcoming the experience was given to us from the beginning:

… those who manage to decipher the enigma of life, almost always stay in Siberia, and with pleasure… But the others, bewildered people who do not know how to solve the enigma of existence, soon tire of Siberia and, with sadness, they wonder why they came to it. (Ibid., p. 1163)

Dostoevsky sensed that, in one way or another, each of us has been *Humiliated and Offended* (1861) and, on this basis, he will create one of his greatest stories: he will describe St. Petersburg as a cold, gray and dark city, where the poor will struggle to maintain their dignity. The romance was written upon his return from Siberia and exposes the world of people whose souls suffer after being harassed, oppressed or aggrieved by their social or economic conditions (Cansinos in Dostoyevsky, 1970, p. 926). The publication date coincides with the emancipation of the serfs, and we will find references in the narration (Fig. 37.5).

Vanya is a young man who wants to be a writer and will try to woo Natasha, but she loves another. During the courtship, figures will appear who will take advantage of their poverty to dishonor and crush them. From the beginning, the artist's sensitivity will catch our attention, we will hear his thoughts when looking for a good place to plan his novels, and these will remind us of the vital function of the house:

Fig. 37.5 Petersburg of Dostoevski, oil on fabric, 45 × 80 cm, by Ilya Glazunov, 1956 (http://glazunov.ru/en/art/urban-cycle/Works/100-f-dostoevsky-s-petersburg)

> First of all, I needed a ventilated room with a modest rent. I also needed it to be independent and spacious, as I have been able to verify that in small rooms thoughts also suffer from narrowness. (Dostoyevski, 1973, p. 7)

His reflections will also fit into the city, and will expose us to the importance of good quality public spaces for our perception:

> When the lights are turned on, the houses seem to lose their sinister aspect, where gray and greenish tones prevail. It then seems as if our soul is illuminated, as if an invisible hand touches us… Amazing power that a flash of light has on the human soul! (Ibid., pp. 7–8).

But St. Petersburg is his rival, and pleasant accommodations will be out of his reach. A friend of his will warn him:

> I was sure that you did not live in a palace, my friend- he declared, while observing the room-, but I also did not think to find you in such a stockade, for this is a dungeon, more than a room. You will say that these are trifles, but these details can only distract you from your concerns. (Ibid., p. 191)

The actions of some characters will resemble the violent representations of the city. Vanya's executioner will tell him:

> I felt the desire to gloat by spitting all my contempt… and I wanted to do it precisely in front of you. (Ibid. p. 334)

Dostoevsky will repeatedly draw connections between mental states and urban activities to teach us the effects of poverty on those who suffer it. And we will always wonder: How did this enemy city arise? How can we be part of it without realizing it?

37.1.3 Brothers (1866–1881). Desires of Compassion

Dostoevsky wrote his most prominent works during the last stage of his life, we can consider them as a synthesis of his ongoing concerns. We will review four of his proposals, novels that transit between youthful rebellion, human disappointment, social collapse, and family disintegration.

Crime and Punishment (1866) tells us about the imbalance of a student (Kózhinov in Dostoyievski, 1977, p. 8). Raskolnikov will understand that poverty will drag his mother into misery and force his sister to marry out of necessity. He will think that stealing and killing may help them. The situation that the young man is going through will be summarized by his mother:

> What bad accommodation you have…! It looks like a tomb… I am sure that this room is half to blame for your melancholy. (Dostoyevski, 1973, p. 277)

In this story, the relationship between people's health, social conditions, and the space they inhabit is reiterated; in fact, it is one of the writings where interior descriptions are most emphasized:

> Raskolnikov headed directly towards the canal house, where Sonya (his girlfriend) lived … It was a large room, but excessively low… Sonya's room looked like a shed; it had the shape of a very irregular quadrilateral, a detail that gave it a deformed appearance. The wall, with three windows overlooking the canal, cut the room obliquely, thus forming an extremely acute angle, which was lost in depth, to the point that with the weak lighting it was impossible to distinguish anything… That huge room was almost unfurnished… Poverty was evident; even in front of the bed, the curtain was missing. (Ibid., pp. 378–379)

Raskolnikov will commit the crime and fall into an almost irreversible emotional state. He will need to leave his house and continually go out to the city to get fresh air and rest. The narration will mention dirty and dark places in Saint Petersburg; when we observe the streets and sites he will pass through during his tours, we will realize that a large part of the actions will occur very close to the house where Dostoevsky lived at the time of writing the novel.

> He was so miserably dressed that anyone else, even accustomed to poverty, would have been ashamed to show themselves in broad daylight, on the street, in such rags. But that part of the city was one that would not be surprised by anything. (Ibid., p. 7) (Figs. 37.6 and 37.7)

Three years later, Dostoevsky will ask us: What would happen if Jesus Christ returned to earth, or at least a similar character? *The Idiot* (1868) shows that we would laugh at him, at his naivety and credulity. Prince Myshkin is the principal figure and will have the qualities described, as he will feel the contempt of his friends and seek, in these moments of great sadness, refuge in nature. With discouragement he will remember:

> Once he went to the mountain; the day was clear and sunny… He had before him the bright sky, below a lake, and around the horizon, bright and vast, without end… he stretched out his arms towards that bright blue… What a feast was that what a great and permanent feast that was, an endless feast and towards which he felt attracted for a long time, always, since childhood, and in which he could never participate. Every morning the sun rises always

Fig. 37.6 Georgy Taratorkin as Raskolnikov. *Crime and Punishment scene*, 1969, by Lev Kulidzhanov (1924–2002)

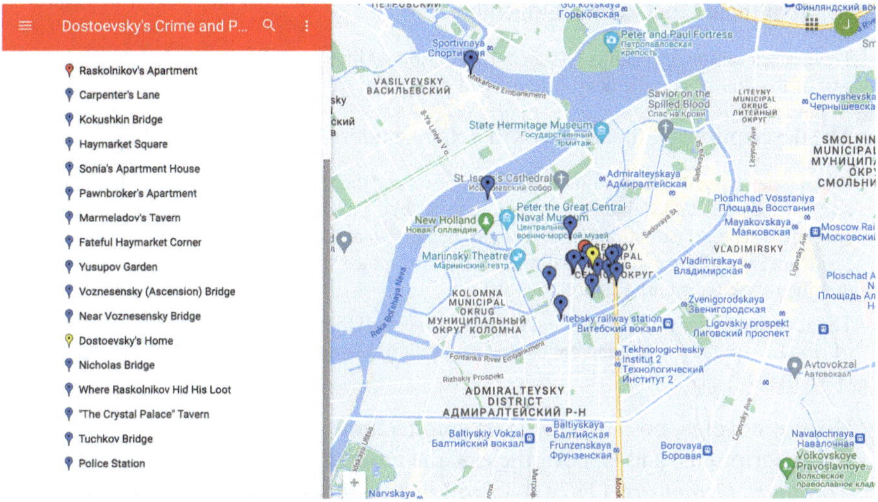

Fig. 37.7 Crime and punishment map by J. W. Connolly, 2015 (https://www.google.com/maps/d/Viewermid=1S8ttpljemdnzwzxxx9ys5gp6kyjxm&hl=en&ll=59.93520934753621%2C30.30173550000001&Z=13)

> radiant; every morning in the waterfall the rainbow forms; every evening the snowy mountain, the highest, there, far away, at the edge of the sky burns with a purple flame; every little fly that buzzes near him in the warm ray of the sun participates in all that chorus: knows its place, wants it and is happy; every little herb grows and is happy! And everything has its way, everything knows its way, it goes with a song and with a song it returns; only he knows nothing, understands nothing, neither people nor sounds, strange and outcast to everything. (Dostoievski, 1972, p. 436)

Dostoevsky places human well-being very close to nature; with a past surrounded by trees and water, many of the Prince's attitudes are understandable. Another peculiarity of *The Idiot* is the correspondence of the descriptions of people with those of their spaces, that is, houses or locations reflect the appearance and personality of the characters:

> … she was a completely empty being from whom it was impossible to get anything. (Ibid., p. 156)

Within a remote village, the poor will conspire against the rich and the rich will do so against the poor. First a house will burn and then the population. When society loses its orientation, everything decays, the novelist seems to tell us. *The Demons* of 1871 describe evil and, nevertheless, we are all receptive. We will read the following in the novel:

> Recently I saw a yellow one (leaf) with a bit of green... I opened my eyes and could not believe it, because of how beautiful it was... God and Nature are one and the same thing... The Mother of God is the great Mother Earth... Since then, every time I pray and bow down to the ground I kiss the earth. (Dostoyevski, 2019, pp. 193–309)

Dostoevsky discovered our contradictions, accepted that good and evil have the same strength and that they act together:

> The houses turned into slums and the street dissolved into a tangle of irregular alleys... this part of the city is as if the devil himself had dropped it from a basket. (Ibid., pp. 335–339)

The descriptions of the houses will help us understand the distance between them:

> When (she) arrived at the uninhabited villa, she toured the rooms accompanied by her faithful and old butler... and Fomushka, a man experienced in business and an interior decoration specialist. The consultations and projects began: what furniture to bring from the urban residence; what objects, what paintings; where to place them; where to put the flowers so that they look their best... (Ibid., p. 435)
>
> The fire, due to the strong wind and because almost all the houses in that suburb were made of wood... spread rapidly and enveloped a quarter of the neighborhood with incredible fury. (Ibid., p. 659)

His last novel is, possibly, the least architectural, as it abbreviates the interior or spatial descriptions, and is, nevertheless, the most expressive and committing. With *The Brothers Karamazov* (1879), we will approach a family conflict: a father and a son will fight over the same woman and the whole family will be affected. It was published as a book in 1880 and Dostoevsky died months later, in his final words, he identified violence and selfishness as our greatest weaknesses, in his opinion, someone among us must show empathy and compassion.

One brother asks another:

> Answer me sincerely, because I need your opinion: imagine that you are the one who raises the building of human destinies, to finally make men happy, give them peace and tranquility; but to achieve this, it is indispensable for you to torture... only a small being, that little girl who was beating her chest, and found your building on unredeemed tears. Would you agree to be the architect under these conditions? Answer me and tell me the truth! (Dostoyevski, 1978, p. 180) (Fig. 37.8)

Later, he will sentence:

> ...humans will never be able to use that freedom, because these weak insurgents will never have the gigantic forces necessary to end and finish the building of perfection... (Ibid., p. 188)

We will listen when one of the brothers reflects:

> The devil fights on this ground with God, and his battlefield is the human heart. (Ibid., p. 75)

Fig. 37.8 Karamazov scene in search of joy, theater workshop, director, Grigori Kozlov, Moscow Art Theater. In the image, Klimov as Alexey Karamazov and K. Kuznetsov as Ivan Karamazov. Photograph by P. Price, (https://ptj.spb.ru/archive/81/walk-with-classics/karamazovy-vpoiskax-radosti/)

And finally, the personal desire of the writer will be revealed to us:

> …in his heart he holds a secret of renewal for all, that power that will finally establish the truth on Earth… and there will be neither rich nor poor, nor conceited nor humiliated… (Dostoyevski, 1970, p. 42)

The dialogues of *The Brothers Karamazov* are a challenge to reflect on human absurdity, spiritual denial, exploitation of others, and our possible redemption:

> In order for man to tear his soul from the abjection and mud that envelop it, he must sign an eternal alliance with his mother, the ancient mother Earth. (Dostoyevski, 1978, p. 74)

37.2 Dostoevsky's Architecture. The Contributions of the Novelist

The novels of the Russian writer Fyodor Dostoevsky, join our material creations to our actions and thoughts. Ultimately, society creates its own environment; they are mutual reflections. His stories are a particular case in which the author uses his characters to illuminate problems, but nobody gives an explanation. They are very similar: they walk absorbed in themselves; they seem lost in their ideas; they travel through spaces, places, and cities, trying to find answers within a world that seems not to understand them.

The global reach of his work conveys a homogeneous idea of social concern. Although the narratives may seem distant in time and place, they give us an idea of the origin and development of the problems we front on today: they make visible the constant struggle to maintain dignity in the face of terrible circumstances; the beauty of being young and having great ideals; the ability of cities to exhibit our lives; the vulnerability of our children; the representation of exile; the oppression inflicted on us by barbaric populations. Dostoevsky lived in a time when many of the concepts that define the modern city were born, such as speed, technology, power, and opulence, and unlike many others who saw a promising future, he observed it with disdain. His words are, therefore, reflections on collective issues that can be seen from an architectural perspective, as they unify the rebellious desire for personal improvement, the need to preserve nature, but also the urgency to design and build shelters, as well as the prevention of social and family breakdown.

Each incident that is revealed to us occurs in a site and a place, and this forces us to understand and imagine how we have lived our spaces, how we live them, and how we could live them.

References

Bajtín, M. (1988). *Problems of Dostoeievski's Poetics* (1st ed.). Editorial Anagrama.

Deleuze, G. (1972). *Proust and signs* (1st Reprint). Fondo de Cultura Económica.

Dostoyevski, F. (1942). *White nights and Raskolnikov's Diary* (1st ed.). Cía. Editorial Espasa-Calpe Argentina, S.A.

Dostoyevski, F. (1970) *Complete works* (1st Reprint ed., Vol. I) (Vols. 1, 2 and 3). Aguilar, S.A. de Ediciones.

Dostoievski, F. (1972). *The idiot* (1st ed.). Círculo de Lectores, S.A.

Dostoyevsky, F. (1973). *Humiliated and offended* (Special). Editorial Bruguera, S.A.

Dostoyevski, F. (1973). *Crime and punishment* (Special). Editorial Bruguera.

Dostoyevski, F. (2019). *The demons* (7th Reprint). Alianza Editorial, S.A.

Dostoyievski, F. (1977). *Crime and punishment* (Vol. I). Editorial Progreso.

Dostoyevski, F. (1978). *The Brothers Karamazov* (Special). Editorial Bruguera.

Dostoievski, F. (2015). *Memories from the house of the dead* (1st ed.). National Council for Culture and the Arts.

González, J. C. (1984). *Narrative Keys of Juan Rulfo* (1st ed.). University of León.

Grossman, L. (1975). Dostoevsky. A Biography (2nd. Printing). Bobbs-Merrill Company, Inc.

Kietsaa, G. (1990). Dostoevsky. In *The life of a writer* (1st ed.). Javier Vergara Editor.

Troyat, H. (1961). *Dostoyevski* (2nd ed.). Ediciones Destino.

Zweig, S. (1945). *Three masters* (1st ed.). Editorial Hispanoamericana.

Part IX
Cultures and Aesthetics of the Public: Discussions and Alternative Paradigms in Latin America

Chapter 38
The Public and the Community in Tension in Self-built Urban Spaces: The Case of the "Operación Sitio" Settlements in Santiago de Chile

Elke Schlack and Carolina Fariña

Abstract Self-construction is a recurring theme in recent discussions about public space in Latin America. Some critical aspects of the debate are, on the one hand, the search for correspondence between the social construction of habitat and the intervention logics proposed by public policies, and on the other hand, the high level of conflict that occurs in these spaces, which is expressed in disputes in public space, causing a tension between the difficulty of building a public sphere and the withdrawal to a predominantly community life. The research presented here aims to contribute to this discussion, studying the influence of urban morphology in determining the character of these spaces which, in contexts of self-construction, are strongly defined by the agency of the inhabitants. We will focus on two dilemmas associated with this question: (a) the difficulty of building spaces from communities without losing the ability to include a wider public and (b) the complexity of articulating forms of appropriation that promote attachment to spaces and cohesion among inhabitants, which confront coercive domination phenomena of space. The evidence emerges from a socio-spatial analysis of "Operación Sitio" settlements in Santiago de Chile. However, these developments, built between the 1960–70 decades, were part of a scheme to promote progressive housing in various developing countries with the technical support of the United States and funding from the World Bank called "sites & services", which finds detractors and supporters, they are characterized by the high incidence of self-construction and their assistance by the State, which gives concrete form to the interdependence between the State's formal planning and the appropriation by the inhabitants.

Keywords Self-organization · Social production of habitat · Public spaces · Sites and services · Peripheral urbanization

E. Schlack (✉) · C. Fariña
Faculty of Architecture, Design and Urban Studies, Pontificia Universidad Catolica de Chile, Santiago, Chile
e-mail: eschlack@uc.cl

C. Fariña
e-mail: cbfarina@uc.cl

M. Vicuña and E. Giorgi (eds.), *Intersections*, Springer Geography,
https://doi.org/10.1007/978-3-031-76402-8_38

38.1 Introduction: The Problems of Public Space in Latin America and Some Conceptual Pressures

In recent decades, a debate has emerged about the so-called "peripheral developments" (Caldeira, 2017) and their value in establishing transversal logics between public policies and the agency of the inhabitants of developments with a high degree of self-construction. This issue is addressed at the Latin American level, particularly in Chile where significant reforms to neighborhood improvement policies and land densification impact the urban regeneration of old urbanizations. Regeneration practices, in tension between formal policies and the agency of its inhabitants, require a reflection on how public policies and social construction of habitat are articulated in the specific case of producing public sphere in urban spaces. There is an accusation of excessive "neighborhoodisation" of policies that seem to increase the problems of access to the city because of the lack of possibilities to solve their accurate insertion in the city from that scale of intervention (Letelier et al., 2016; Letelier et al., 2019a; Letelier et al., 2019b; Tapia Barría, 2015).

On the other hand, in peripheral sectors of Latin American cities, a high level of conflict is accused in urban spaces, which is expressed in disputes between the different uses of public space and the possibility of an experience of "otherness" free of violence or confinement to private space (Araujo, 2019; Caprón et al. 2018; Duhau & Giglia, 2008; Hiernaux & Lindón, 2016).

In the recent debate about spaces in vulnerable territories in Latin America, we have seen that the term "public space" is avoided, expressing the difficulty of embodying the public (Schlack and Araujo, 2022). Streets, squares, and other meeting places are called "open spaces", "collective spaces" and "co-places" (Giglia, 2013; Schlack & Araujo, 2022), and we find it appropriate to adhere to the term "common spaces" (Castro-Coma & Martí-Costa, 2016; Giglia, 2013; Schlack and Araujo, 2022). This allows us to account for the tension between the expectation of building public spaces—usually through state interventions—and what can be verified on the ground, the production of much less public and more community spaces, resulting from the agency of their inhabitants, which is expressed in various forms of spatial appropriation.

Regarding the social environment in which we situate this research, it was essential to consider the constitution of social networks and forms of sociability of the popular world: characterized by a high component of self-construction and community regime. We believe that these are social relationships forged from the beginning of these urbanizations and that are characterized by a "public familiarity", that is, a powerful sense of community cohesion due to a common experience in the production of space (Araos, 2016; Link et al., 2022).

The complex tension between the public and the community is addressed, understanding that, on the one hand, the public character of spaces occurs in spaces intended for an extended group of users who do not necessarily know each other and that, on the other hand, the community character of spaces when it occurs among individuals who belong to a specific community, that is, to a limited group either by family relationship, or group membership associated with a shared activity or experience.

The theoretical discussion that we consider predominant in the construction of the theoretical framework is derived from the theory of public space (Bahrdt,1979; Wehrheim, 2015) and the notion of the common (Delgado, 2008; Stavrides, 2016).

In this sense, what Wehrheim proposes is essential: "The social function of public space continues to be learning how to relate to difference through contact with otherness, that is, also learning public behavior and experiencing freedom away from the strict informal controls of local communities, understanding other models of cultural behavior through observations from a distance." (Wehrheim, 2015: 305).

This dilemma of public space, which puts tension between the public and the community, is also picked up in the theoretical field of social cohesion. In the case of neighborhoods in Chile, interesting discussions have arisen about how problematic it is to construct the figure of the "other" and aim for an integration of diversities (Link et al., 2022). To introduce a much more extensive debate exposed by these authors, we want to highlight the expectation that Massey raises about achieving a civic culture of difference and the conflict that de facto occurs in the co-residence of diversities, which do not necessarily integrate (throwtogetherness) (Massey, 2005). Similarly, Valentine argues about "otherness" that it is a wrong assumption to pretend that close encounters (neighborly proximity) promote respect for difference automatically (Valentine, 2008). Thus, as Mouratidis and Poortinga describe, in an intense neighborhood life, with the presence of pedestrians, cars and various uses, there is not necessarily a greater effective sociability (Mouratidis & Poortinga, 2020). From these points of view, it can be concluded then that a more public character of space does not ensure effective sociability and that mere physical proximity in communities does not necessarily imply integrating differences among its inhabitants.

38.2 Operación Sitio Settlements: The Layout and Self-construction as Overlapping Actions

The "Operación Sitio" settlements were part of a Chilean housing policy of the 60s, associated with the "sites & services" policy implemented in various developing countries with the technical support of the United States and the financing of the World Bank.

It consisted of providing of land to settlers and their families, who were previously organized and came from land seizures. They were relocated to formal plots built by the State (Hidalgo, 2004; Labadía, 1970; Millán, 2012; Quintana et al., 2021).

It has often been written about the "Operación Sitio" settlements and its international equivalent, valuing the impact it had on peripheral urbanizations as a response to promoting the population self-organization (Boyars & Turner, 1976; Castillo & Forray, 2014; Millán, 2012). However, its condition of progressive urbanization has also been controversial, because it required its inhabitants to complete the houses and

Fig. 38.2 Satellite image of La Faena Urbanization in Santiago, Chile. *Source* Google Earth (2023)

(b) the complexity of making different forms of appropriation coexist without being coercive for collective life.

To be able to unfold in this tension between the public and the community, the study is based on various theoretical conceptualizations linked to sociological theory and urban studies of public space.

First of all, the conception of an ideal-type of public space is known as an important reference, which evolved from urban sociology. Based on the study of social interaction in market places and the first features of an open social fabric historically traceable in European Middle Ages, a characteristic of this ideal-type of public space is unrestricted accessibility. This ideal conception is foundational in European public space theory. It constructs an idea of an ideal of social inclusion forged in the concept of tolerance, coexistence with otherness and democratic coexistence enriched by the diversity of political expressions in the city (Wehrheim, 2015).

Starting from this theoretical point of the ideal-type of inclusive public space does not exclude that public space is disputed. In this, hegemonies from the different actors involved in its ownership, production, and use become visible. While some European authors focus more on the hegemonies of the authorities as actors who exercise control over space, other authors emphasize the hegemonies of certain groups over others to make visible the dynamics of privatization of spaces. Further, we also consider the vision of those authors, interestingly from Latin American latitudes, who make visible the dynamics of resistance to dominations in public space (Foucault 2015; Habermas, 1990; Salcedo, 2002; Wehrheim, 2015).

Undoubtedly, in the territories we are studying, that is, peripheral urbanizations, the act of "taking" a front yard or putting fences so that cars do not enter a street is a phenomenon that perhaps best speaks of what alters or modifies this idea of the ideal-type public space. In urban dynamics marked by fear of the other and the difficulty of managing coexistence with it, a very recurrent phenomenon appears, that of "territorialization", which corresponds to the constant and daily effort of the inhabitants "mark the ground socially and which can express identity, but also, defense against otherness" (Araujo, 2019; Delgado, 1999). Territorialization can thus lead to expressions of freedom to use space as one decides to do it, but also to situations of domination of one group over another in space.

Paying attention to expressions of an ideal-typical public space and, at the same time, expressions of territorialization, this study is attentive to the differentiation of forms of appropriations. We consider this concept relevant to analyze a greater closeness or distance to the public character, and appropriations are also a vast wealth of vitality for collective life. However, differentiation is necessary: appropriations can promote attachment to spaces, while others express domination, forcing the withdrawal of inhabitants to individual spaces.

Bourdieu describes appropriations that operate from domination from the architectural features that symbolically express the coercion to occupy or use space; they are "places where power is reaffirmed and consumed, and it does so in the most subtle possible way: symbolic violence as non-perceptible violence (Wehrheim, 2015)".

On the other hand, other authors speak of appropriations from an idea of emancipation. For Chombart de Lauwe, urban space is a space of possibilities: "(…) the possibilities of moving freely (…), to relax, to own it, to feel things, to admire, dream, know something, do something that responds to one's desires, demands, expectations and concrete ideas and be able to realize it." (Chombart de Lauwe, 1977: 2–5; Wehrheim, 2015).

For this study, these conceptualizations permit us to discuss the conditions under which the character generates in more public or community spaces and what role do appropriations play in them so that, above all, spaces of possibilities and fewer exclusions can occur.

38.4 Methodology

The socio-spatial analysis focused on the years 2019 and 2022 and concentrated on different types of spaces in an "Operación Sitio" settlement, located in the city of Santiago de Chile. The survey consisted of a morphological study of the urban fabric, the elements that constitute the public spaces and the physical evidence of involvement of its inhabitants in the construction and management of these spaces.

The selection of eight urban spaces was based on the usual architectural typologies in these areas: main streets, passages, larger squares, small squares and facilities (sports, social, etc.). In this article, we will refer only to four of the eight spaces analyzed to illustrate in a more synthetic way the public-community gradient observable in this type of territories (Fig. 38.3). Through these chosen cases, it is possible to visualize the logics in which appropriation occurs, that is, how these different types of spaces are intervened and used.

Fig. 38.3 Axonometric of the location of the studied case. Project "Public Space and Collective Authority" within the Millennium Authority NUMAAP-Project NCS17_007 Researchers E. Schlack and C. Fariña (Drawing: Gabriela Salazar, 2004)

The methods used to identify and document the urban form and the interventions in the space are observations on the ground of the form and use of the spaces, photographic records at different times (bursts), an exhaustive planimetric record of physical objectives of the spaces, as well as semi-structured interviews and surveys of inhabitants and passers-by, leaders and urban managers who have intervened over time.

38.5 Analysis and Results

The premise of this research is that the layout, made by the State, initially expected that different forms of sociability would occur at different scales (public, community) (Lin & McLean, 2022). As we mentioned before, this assumes the existence of a correlation between a more public character in some spaces and more community in other spaces, attending to the different degrees of accessibility observable in them and their size within the urban plot. However, the evidence we will show below indicates that the agency of the inhabitants in the enablement (paving, planting, installation of furniture, courts, lighting) has been as determinant as the layout of these urbanizations and that this agency equally contributed to the production of the public and the community spaces.

The urban spaces analyzed are places that are located at different points in the neighborhood, have different degrees of accessibility and have different dimensions. The first corresponds to a square with a football field (1) located in the center of the neighborhood that has greater accessibility for being located around structured streets of the sector. The other three spaces (2, 3, 4) correspond to small squares that have restricted accessibility (only two accesses) and are in blocks surrounded by houses in different areas of the neighborhood. These small squares, despite their physical similarities in the layout, currently have a very different performance from each other. On the one hand, there is: (2) a small square equipped with a multi-court, located in an area adjacent to the center of the neighborhood. On the other hand, there is (3) the small square that has been consolidating with successive interventions managed by the inhabitants and is located on the perimeter of the neighborhood. And finally, a small square (4) devoid of equipment, except for small individual interventions, which is located on the perimeter of the neighborhood, far from the main streets. The dimensions of these spaces vary between 630 and 2,280 m^2 (Table 38.1).

Regarding the first research question about the difficulty of building spaces from the agency of communities without losing the ability to include a broader audience in those spaces, we found evidence in the cases studied that indicate that the use or

Table 38.1 Description of the four spaces analyzed in the Operación Sitio settlement

	1 Square with a football field	2 Small square with multi-sport court	3 Small square co-produced	4 Small square without equipment
Size	2,280 m^2 (field has 1,000 m^2)	630 m^2	1,320 m^2	630 m^2
Location	Center of the neighborhood, adjacent to structuring streets and passages	Center of the neighborhood, between passages	Southwest side of the neighborhood, between passages	Southeast side of the neighborhood, between passages
Infrastructure	Field: perimeter fence, synthetic grass, football goals, external bleachers, artificial lighting Square: benches, playground, trees, trash cans	Field: perimeter fence, pavement, football goals, artificial lighting Small square: exercise machines, ping pong table, benches, vegetation	Playground, artificial lighting, benches, vegetation, orchard, shrine	Vegetation, pool (temporary), pillars, artificial lighting
Activities	Field: football classes, field rental, championships, club activities Square: sitting or playing in the square, crossing, buying food at foodtruck	Football and volleyball classes, aerobox, adult and children's football, shortcut, conversation	Shortcut, conversation, play	Car parking, shortcut, conversation

the way activities are programmed in the spaces strongly affects generating a more public or collective character, respectively, modifying the initial character that was planned for these spaces.

Two of the small squares studied (2 and 3) have a similar accessibility and proximity to the surrounding dwellings. Due to these characteristics in both would be expected a similar community character. However, they are very different in terms of their lived character. In the small square containing a multi-sport court (2) (Fig. 38.4), the programs attract a wider audience beyond neighbors from its surroundings. On the contrary, the other small square (3), shows a space used only by its adjacent residents, and the feeling of privatization of the space is much greater (Fig. 38.5).

In the first small square (2), the management is in charge of a delegate from a group of neighbors who made its construction possible through the municipality assistance. These neighbors coordinate various activities both for the neighborhood's inhabitants and for outsiders. Therefore, despite this space's closed condition due to its layout, the activities inside it push to generate a public character.

On the other hand, the other small square (3) is the result of interventions to satisfy the neighbours' interests and of the active management in terms of the construction and maintenance of this square. This is expressed through appropriations such as

Fig. 38.4 Photographs of a small square with a multi-court. Project "Public Space and Collective Authority" within the Millennium Authority NUMAAP-Project NCS17_007 Researchers E. Schlack and C. Fariña. Photographies: Felipe Ibacache (2022)

Fig. 38.5 Photographs of co-produced square. Project "Public Space and Collective Authority" within the Millennium Authority NUMAAP-Project NCS17_007 Researchers E. Schlack and C. Fariña. Photographies: Felipe Ibacache (2022)

Fig. 38.6 Photographs of the square with a football field. Project "Public Space and Collective Authority" within the Millennium Authority NUMAAP-Project NCS17_007 Researchers E. Schlack and C. Fariña. Photographies: Gabriela Salazar (2004)

small family gardens, elements commemorating Christian rituals, children's games with benches, etc.

Furthermore, the assumption that a main square (1), centrally located and accessible from all four sides, should display a clear public character is not necessarily fulfilled, as its character is nuanced by the presence of a highly specialized football field used exclusively by a football club that manages its use and defines the users (Fig. 38.6). In this case, there is a hybrid situation, the section of the field that is fenced off attracts a limited group of people often excluding the inhabitants of the neighborhood. In contrast the square, which is adjacent to the field, is occupied by the inhabitants more freely and, due to its very wide accessibility, extends to people outside the neighborhood.

Regarding the second research question about appropriations and how they create conditions that promote attachment to spaces and cohesion among the inhabitants, we have discovered that different forms of appropriation determine different approaches to the public. The evidence shows that two small squares with similar conditions (2) and (4) of accessibility can suffer opposite appropriations and forms of sociability. While one exhibits cohesion among the neighbors, who have acted additively with appropriations on the space (3) (Fig. 38.5), in the other square, the domination of one group over others (4) is exhibited—in this case, a group of neighbors who exert coercion over the others and take over the square against the will of the others (Fig. 38.7). In this case, the result of this coercive relationship is the abandonment of the space by the majority of the surrounding neighbors and their retraction to the private sphere of the home.

38.6 Conclusions

The analysis of urban spaces that emerged after the urban regeneration of the "Operación Sitio" settlement in the last 50 years shows a territory with a robust community regime where the experience of a wider public space only occurs under certain conditions.

Fig. 38.7 Photographs of unconsolidated square. Project "Public Space and Collective Authority" within the Millennium Authority NUMAAP-Project NCS17_007 Researchers E. Schlack and C. Fariña. Photographies: Felipe Ibacache (2022)

Given that a diverse range of sociability forms enrich the experience of the common, it can be understood as a problematic fact that providing spaces of a more public character is an exception in territories strongly marked by self-organization. This article proposes to advance in a search for those interventions both from the inhabitants and from the urban planners responsible for their planning that can contribute to build a more complex sphere of the common.

In the study, we found that formal planning plays an important supporting role that strongly conditions the existence of a public character substrate in neighborhoods and that designates or provides, at the same time, some spaces destined for community life. At the same time, we verified that the production of the public and the community space is more complex than thought, that, contrary to what one might think, the public agency not only produces public spaces, but can produce community spaces and the agency of communities can, not only produce community spaces but also spaces of public character.

Specifically, the findings showed that spaces such as small squares, which are less accessible and open to the direct and daily intervention of the closest inhabitants, constitute a complex typology to manage due to their ambivalence. These spaces, although they contribute to qualifying the extension of the house to community spaces, are strongly exposed to dynamics of appropriation that could hypothetically constitute themselves as open spaces where neighbors from the sector and external neighbors could feel welcomed, but they can also develop in the form of spaces dominated by a single group and exclusive to others.

On the other hand, spaces with a higher level of accessibility and that convene more public programs strongly contribute to constituting public spheres within neighborhoods. Their ability to convene a wider audience is difficult to modify through appropriations. However, we realized that the way spaces are programmed from state subsidies can revert them into spaces restricted to a very community use.

Reflecting on how to remedy possible deficits of the common sphere, it is clear that the layout is a difficult condition to modify, and one of the few possible areas in which to intervene is the programming of spaces. In this aspect, programmatic decisions or the use of public spaces become especially relevant. That is why it becomes pertinent to look at the self-construction of these spaces as a potential that intervenes in this dimension, the scope of how spaces are equipped and used on different days of the week and hours of the day. Such programming and the related equipment can generate effects of abandonment and even a fragmented vision of the territory, leaving spaces invisible to the inhabitants who do not occupy them and have no notion that they exist.

To advance on this issue, it is important to understand that each place contains its own social structure, which is desirable to consider. Each intervention would require that each space be approached with a differentiated reading, understanding the diverse relationships that exist between its inhabitants and between them and the local managers and considering different models of management and maintenance of these spaces.

Finally, it is relevant to mention that neighborhood improvement policies contain many strategies that emphasize equipping public spaces. For this, they elaborate guidelines that quickly become standardized and often obsolete solutions due to the great social variability of these territories. It would be appropriate to consider the low degree of stability of the interventions, due to rethink and consider an approach that consider monitoring both physical and programmatic updates of the spaces concerning the habitat's social production demands.

An attentive look at the variability of social interaction and which actors intervene should focus on the effect on the public or community character that their acting provokes and who benefits from their agency.

Acknowledgements We appreciate the valuable contribution of neighbors, leaders of the studied settlement, and municipal officials of the studied territory. We thank ANID through its Millennium Scientific Initiative program and the Sub-Directorate of Research of the PUC School of Architecture for sponsoring this research on "Public Space and Collective Authority" within the Millennium Nucleus NUMAAP-Project NCS17_007. We appreciate the contributions of research assistants Gabriela Salgado, Felipe Ibacache, and Danitza Iturrieta in the socio-spatial survey of this research.

References

Araos, C. (2016). El Allegamiento o La Proximidad Invisible. The Invisibles. Why Poverty and Exclusion Ceased to Be Priority.] In C. Siles (Ed.), Los Invisibles. Por Qué La Pobreza y La Exclusión Dejaron de Ser Prioridad. Fundación Hanns Seidel: Instituto de Estudios de la Sociedad (1996), 36.

Araujo, K. (2019). Las Calles. Un Estudio Sobre Santiago de Chile. [The Streets. A Study About Santiago de Chile.]. LOM Editions.

Bahrdt, H. P. (1961 original text) (1979). La moderna metrópoli. Reflexiones sociológicas sobre la construcción en las ciudades. [The modern metropolis. Sociological Reflections about the buildings in cities.] Buenos Aires, Ediciones de la Universidad de Buenos Aires.

Boyars, M., & Turner, J. (1976). Housing by people: Towards autonomy in building environments. *City, 45–46*. https://doi.org/10.1017/CBO9781107415324.004

Burguess, R. (1992). *Beyond self-help housing*. Continuum International Publishing Group.

Caldeira, T. (2017). Peripheral urbanization: Self-construction, transversal logics, and politics in cities of the global south. *Environment and Planning d: Society and Space, 35*(1), 3–20. https://doi.org/10.1177/0263775816658479

Caprón, G., Monnet, J., & López, R. P. (2018). Infraestructura Peatonal: El Papel de La Banqueta (Acera). [Pedestrian Infrastructure: The Role of the Sidewalk.] *Cities,* (199), 33–40.
Castillo, M. J., & Forray, R. (2014). La Vivienda, Un Problema de Acceso al Suelo. [Housing, A Problem of Access to Land.] *ARQ (Santiago),* (86), 48–57. https://doi.org/10.4067/S0717-69962014000100007
Castro-Coma, M., & Martí-Costa, M. (2016). Comunes Urbanos: De La Gestión Colectiva al Derecho a La Ciudad. [Urban Commons: From Collective Management to the Right to the City.] *Eure, 42*(125), 131–53. https://doi.org/10.4067/S0250-71612016000100006
Chombart de Lauwe, P.-H. (1977). Aneignung, Eigentum, Enteignung. Sozialpsychologie der Raumanteignung und Prozesse gesellschaftlicher Veränderung [Appropriation, Property, Expropriation. Psychology of Spatial Appropriation and Social Transformation Processes.] *Arch+, 34,* 2–6
Davis, M. (2006). *Planet of Slums.* Verso.
Delgado, M. (1999). El animal público. [The Public Animal.] Anagrama.
Delgado, M. (2008). Lo Común y Lo Colectivo. [The Common and the Collective.] Conference held at the MediaLab Prado in Barcelona.
Duhau, E., & Giglia, Á. (2008). Las Reglas Del Desorden. [The Rules of Disorder.] Siglo Veintiuno.
Foucault, M. (2015). Vigilar y Castigar: El Nacimiento de La Prisión. [Discipline and Punish: The Birth of the Prison.] Siglo XXI.
Gianotti, E. (2014). Orígenes de Un Diseño Participativo: La Construcción de Los Barrios Populares de Santiago, 1952–1973. [Origins of a Participatory Design: The Construction of Deprived Neighborhoods in Santiago, 1952–1973.] *Revista, 180*(34), 22–29.
Gianotti, E., & Cofré, B. (2021). La Invención de La Toma, o Cómo Se Transformaron Las Ocupaciones de Terrenos En Santiago de Chile Entre 1945 y 1957. [The Invention of the Takeover, or How Land Occupations Transformed in Santiago de Chile Between 1945 and 1957.] *History, 1*(January–June 2021), 107–50.
Giglia, A. (2013). Entre El Bien Común y La Ciudad Insular: La Renovación Urbana En La Ciudad de México. [Between the Common Good and the Insular City: The Urban Renewal in Mexico City.] *Alterities, 23*(46), 27–38.
Habermas, J. (1990). Strukturwandel Der Öffentlichkeit. Untersuchungen Einer Kategorie Zur Bürgerlichen Gesellschaft. [Structural Transformation of the Public Sphere. Investigations of a Category of Bourgeois Society.]. Surkamp.
Hidalgo, R. (2004). La Vivienda Social En Santiago de Chile En La Segunda Mitad Del Siglo XX: Actores Relevantes y Tendencias Espaciales. [Social Housing in Santiago de Chile in The Second Half of the 20th Century: Relevant Actors and Spatial Trends.] (pp. 219–241). Institute of Geography, Pontifical Catholic University of Chile.
Hiernaux, D., & Lindón, A. (2016). Tratado de Geografía Humana. [Treaty of Human Geography] Daniel Hiernaux and Alicia Lindón, Directors, Anthropos and UAM. UNAM.
Labadía, A. (1970). La 'Operación Sitio'. Una Solución Habitacional de Desarrollo Progresivo. [The Operación Sitio. A Progressive Housing Solution.] *Revista Mensaje,* (192), 428–432.
Letelier, F., Patricia Boyco, Cubillos, J., Tapia, V., & Irazábal, C., (Eds.). (2019a). Lo Vecinal En Chile. [The Neighborhood in Chile]. *Serie Cuad.* Ediciones UCM.
Letelier, L. F., Micheletti, S., Boyco, P. L., & Fernández, V. (2019b). Problematización de Las Espacialidades Vecinales Como Estrategia de Intervención Comunitaria. [Problematisation of Neighborhood Spaces as a Community Intervention Strategy.] GeoGraphos. *Revista Digital Para Estudiantes de Geografía y Ciencias Sociales, 10,* 1–22. https://doi.org/10.14198/geogra2019.10.112
Letelier, L. F., Micheletti, S., & Vanhulst, J. (2016). Prácticas Instituyentes En El Espacio Vecinal: El Barrio Como Un Común. [Instituting Practices in the Neighborhood Space: The Neighborhood as a Common.] *Polis, 15*(45), 105–19. https://doi.org/10.4067/S0718-65682016000300006
Lin, T., & McLean, I. (2022). El Espacio Público Como Mediador Entre Los Pobladores y El Estado. [Public Space as a Mediator Between the Inhabitants and the State.] *INVI, 37*(106), 177–203.

Link, F., Señoret, A., & Figueroa, C. (2022). Del Barrio al Conjunto: Espacio Público y Sociabilidad En La Vivienda Social de Santiago. [From the Neighborhood to the Ensemble: Public Space and Sociability in Social Housing in Santiago.] *Revista INVI, 37*(106). https://doi.org/10.5354/0718-8358.2022.67125

Massey, D. (2005). *For space*. SAGE.

Millán, R. (2012). Operación Sitio a Cuatro Décadas de Su Constitución. Programas y Políticas de Mejoramiento Urbano En Áreas Pericentrales Del Gran Santiago. [Operación Sitio Four Decades After Its Constitution. Urban Improvement Programs and Policies in Pericentral Areas of Greater Santiago.] ISA Forum RC 43. Housing and Built Environment/Informal Settlements: Land and Housing Markets, Policies and Formalisation.

Mouratidis, K., & Poortinga, W. (2020). Built environment, urban vitality and social cohesion: Do vibrant neighborhoods foster strong communities? *Landscape and Urban Planning, 204*(September), 103951. https://doi.org/10.1016/j.landurbplan.2020.103951

Quintana, F., Salazar, B., & Bier, M. (2021). Fragmentos Urbanos de Guerra Fría En El 'Tercer Mundo.' [Cold War Urban Fragments in the 'Third World.] *Astrágalo. Cultura de La Arquitectura y La Ciudad, 29*(29), 207–232. https://doi.org/10.12795/astragalo.2021.i29.11

Salazar, G. (2004). Ferias libres: espacio residual de soberanía ciudadana. [Flea Markets: residual space of popular government.] Santiago de Chile. Ediciones Sur.

Salcedo, R. (2002). El Espacio Público En El Debate Actual: Una Reflexión Crítica Sobre El Urbanismo. [Public Space in the Current Debate: A Critical Reflection on Urbanism.] *Eure, 28*(84), 5–19.

Schlack, E. (2019). Public and collective space in 9 × 18 neighborhoods. In L. M. Vergara et al. (Eds.), *Housing and urban regeneration in deprived neighbourhoods in Santiago A North–South perspective on collaborative processes*. TU Delft OPEN.

Schlack, E., & Araujo, K. (2022). Espacio Público: Registros Alternativos Para Pensar y Construir El Espacio Público En Ciudades de Latinoamérica. [Public Space: Alternative Records to Think and Build Public Space in Latin American Cities.] *INVI Journal, 37*(106), 1–23. https://doi.org/10.5354/0718-8358.2022.68886

Stavrides, S. (2016). *Common space. The city as commons*. Zed Books.

Tapia Barría, V. (2015). ¿De Qué Hablamos Cuando Hablamos de Barrio? Trayectoria Del Concepto de Barrio y Apuntes Para Su Problematización. [What Do We Talk About When We Talk About "Neighborhood? Trajectory of the Neighborhood Concept and Notes for Its Problematisation.] *Revista Antropologías Del Sur, 3*, 121–135.

Tapia, R., Araos, C., Forray, R., Gil, D., & Muñoz, S. (2020). Hacia Un Modelo Integral de Regeneración Urbano-Habitacional Con Densificación En Barrios Tipo 9 × 18. [Towards an Integral Model of Urban-Housing Regeneration with Densification in 9 × 18 Type Neighborhoods.]. In Centro de Políticas Públicas UC (Ed.), *Propuestas para Chile. Concurso de Políticas Públicas 2019* (pp. 319–351). Pontificia Universidad Católica de Chile.

Tapia, R., Araos, C., & Salinas, C. (2012). Condominios Familiares. Una Alternativa de Solución Residencial Para Familias Allegadas En Lotes Tipo 9 × 18. [Family Condominiums. A Residential Solution Alternative for Families Living in 9 × 18 Type Lots] Propuestas Para Chile. Concurso Políticas Públicas 2012. PUC. 101.

Tapia, R., Robertson, C. (2019). Vivienda Colaborativa Informal. El Caso de Chile y La Vivienda En Lotes 9 × 18. Workshop. [Informal Collaborative Housing. The Case of Chile and Housing in 9 × 18 Lots. Workshop.]. In L. M. Vergara et al. (Eds.), *Housing and urban regeneration in deprived neighbourhoods in Santiago A North-South perspective on collaborative processes*. TU Delft OPEN.

Valentine, G. (2008). Living with difference: Reflections on geographies of encounter. *Progress in Human Geography, 32*(3), 323–337. https://doi.org/10.1177/0309133308089372

Vergara, F., & Palmer, M. (1990). El Lote 9 × 18 En La Encrucijada Habitacional de Hoy. [The 9 × 18 Lot in Today's Housing Crossroads.]. Editorial Universitaria. Facultad de Arquitectura y Bellas Artes. Pontificia Universidad Católica de Chile.

Wehrheim, J. (2015). El Carácter Público de Los Espacios de La Ciudad. [The Public Character of City Spaces]. In E. Schlack (Ed.), *POPS El uso público del espacio urbano* (pp. 284–305). ARQ UNAB Capital Books.

Chapter 39
Beginnings of Standardization in Educational Architecture: Forms of Expansion of the Republican State in La Araucanía

Pablo Fuentes Hernández and Gonzalo Rodríguez Paillape

Abstract La Araucanía, in the south of Chile, was a territory inhabited since ancient times by the Mapuche people who divided the country in two. At the end of the nineteenth century and the beginning of the twentieth century, the State undertook a series of occupation strategies aimed at consolidating modernizing processes. Among them, it addresses certain architectural experiences that show the will to consolidate the nation-state. In this context, the construction of modern school establishments as civilizing detonators is noted. The purpose of this work is to present the architectural alternatives chosen by the State to impose its official presence, among them standardization as a transformative action.[1]

Keywords Educational architecture · Standardization · State · La Araucanía

39.1 Introduction

In La Araucanía, a territory of fertile soils where Chilean sovereignty was resisted bythe Mapuche people, republican strategies were developed from the end of the nineteenth century to impose, extend and consolidate modernizing processes (Pinto, 2003). From the presidential political visions of José Manuel Balmaceda (1886–1891) to Carlos Ibáñez del Campo (1927–1931), the role of the executive power in materializing the presence of the State in La Araucanía under a republican and capitalist influence is examined.

In the process, the development of two relevant infrastructures appears: initially, a special railway development is noted, tracks, stations and bridges, particularly

P. Fuentes Hernández (✉)
Faculty of Architecture, Construction and Design, Universidad del Bío-Bío, Concepción, Chile
e-mail: pfuentes@ubiobio.cl

G. Rodríguez Paillape
Faculty of Social Sciences and Arts, Universidad Mayor, Temuco, Chile
e-mail: ruben.rodriguezp@umayor.cl

M. Vicuña and E. Giorgi (eds.), *Intersections*, Springer Geography,
https://doi.org/10.1007/978-3-031-76402-8_39

the construction of the colossal Malleco Viaduct, which extend communications, commerce and transport to the south of Chile (Balmaceda, 1888; Vargas, n/d).

Later and focus of this work—will be preferably the propagation of modern educational models that support a civilizing process of subjugation (Balmaceda, 1887; Serrano et al., 2013).

The objective of this work is to reveal the process of standardization of public infrastructure as a modernizing language used by the State in the period between the governments of Balmaceda and Ibáñez del Campo for its extensive development in fiscal works. It is about the construction of the imaginary of the nation-state in new frontier territories determined by colonization, extractivism and indigenous marginalization.

As a hypothesis, it is argued that the political, civil and military occupation of La Araucanía required spatial models and types that expressed its condition of dominance and subjugation through works with high modernizing content, for the case, railway and educational infrastructures. These are constructions whose cultural and morphological arguments materialize the ideals of a new and republican state with homogenizing purposes.

In this sense, it is worth asking, what spatial strategies did the State use to expose its civilizing language? What programs and infrastructures privileged the required impact to establish its power and dominance? What are the factors of cultural appropriation by the State of a modernity in progress, associated with consolidating nation-states?

As a methodology, presidential speeches (Montero, 2009) and governmental approaches have been examined that necessarily refer to spatial languages and responses; likewise, the use of secondary sources that examine such episodes, such as books, press and technical reports, facts that are complemented with the survey and the examination in situ of the works and their necessary valorization.

Among the results, along with a complex understanding of the territory and its interconnection possibilities, a high ideological commitment is observed. This commitment, noting the advances of modernity, both theoretical and technological, puts into action a company of significant political content as it advances, subjugates, and consolidates the presence of the State to strengthen the idea of an apparently homogeneous nation (Góngora, 1981).

In the studied period, 1886–1931, a cultural transition from an enlightened European practice to the advances of American ideologies imposing a development model is observed as a source of inspiration, knowledge and appropriate evolution.

In the studied precedents, the modern component associated with a new, standardized architecture is evident, which confronts the eclecticism of the time with repetition mechanisms inspired by economic and temporal efficiency. This representation of efficiency in terms of repetition of an already optimized type is conceivable as an example of a discourse that becomes tangible in an architectural form, in an aestheticized ideology (Van Dijk, 2003). They are products of a project management developed under a modernization project of the country, whose design transitioned from internationalized eclectic currents, to focus on project decisions that derived from architectural and constructive criteria determined by a technical and artistic vanguard.

39.2 Territorial Occupation as a Political Strategy

The military troops sent by the Chilean State to occupy La Araucanía carried a strong civilizing component, which assumed the governmental strategies of the time for the domination of the territory and subjugation of its original inhabitants (Ferrando, 2012).

> Today, the entire Araucanía is subjected, more than to material power, to the moral and civilizing power of the Republic; and at this moment, important populations are rising, destined to be commercial and industrial centers of considerable importance in the midst of virgin forests and unknown fields, which until yesterday were the impenetrable sanctuary of Araucanian pride and independence. (Santa María, 1883: 6).

In this way, the rule of law was the protection that organized the actions of old and new inhabitants, validating the exercise of occupation.[2] Likewise, the enterprise warranted a rapid intervention action, so that the advance of the troops was followed by extraction processes that ensured the exploitation of the new territories incorporated under national sovereignty. The process was developed according to the extractivist forms of fiscal and particular enrichment, in which time (understood as the speed of the process) was a decisive factor. This operation speed was determined by the extension of the railway system, a highly effective modern means that served to thread the Chilean territory from north to south. However, the stop imposed by La Frontera in its extension to the south implied serious damage in productive terms.

It was the area where an extractivist economic system of capitalist tone according to the political actions that were not only developed in Chile but were common to the modernizing evolution of the Western world, had not yet been established. Thus, the extension of the central railway network became the preferred instrument to materialize the governmental and the economic policy.

The financing of the tracks, made possible by the profits from the exploitation of nitrates and metals in the north, was considered a state policy that the rulers gave their support to year after year.

This system, organized as a central trunk line, in charge of the State, with transversal branches of particular financing, needed stops that gave meaning to the productive and urban use of the territory. In this way, the railway company gave way to the construction of some station typologies that established places of waiting, stopping, and connection (Cerda, 1983). This issue allows us to affirm that it was the first mode of standardization. Architecture on The Border.

On the other hand, the education of the child and youth population was a new line of action that captured the interests of the State at that time. It was about implementing a nation project that would foster the intellectual and moral development of the population, typical of a young country that consolidated its republican policies through the education of its citizens. Education, seen in this way, was a fundamental need for public policies on which it was necessary to develop not only its educational system in curricular terms, but also from its physical infrastructure. Although by the end of the nineteenth century some large schools had been erected, the so-called *palace schools*, due to the existing relationship in their size and architectural image associated with large institutional buildings[3]; these were concentrated mainly in Santiago, the country's capital, and in a few main cities, but in the rest of the national territory there was a significant number of school buildings that operated under precarious conditions. Many of these schools operated in rented houses in most Chilean cities and consequently their image was closer to a dwelling house—whether urban or rural—than to that of an institutional building typical of a State with ideals of progressivism.

39.3 Balmaceda School

In this context and thanks to the wealth produced by saltpeter, the liberal government of José Manuel Balmaceda, which took place between 1884 and 1890, had the possibility of having a better amount of resources, which the president partly allocated to the construction of a series of public infrastructures that improved the life of the population.

> These works of viability and the constructions of high schools, schools, prisons, administration buildings, special education buildings, roads, bridges, hospitals, temples, sanitation and so many others, correspond to a system of distribution of public wealth, which all the provinces of Chile will support and defend, because by supporting it they practice justice and by defending it they protect their most dear interests. (Balmaceda, 1890: 228).

In this area, a standard educational building was designed that had to be repeated in several Chilean cities that, under basic conditions, achieves what we can call a first attempt at architectural rationalization for serial solutions. However, this building was developed preferably in the central area, more inhabited and a fundamental part of the republic.

39.4 Architectural Standardization for Primary Schools

The Mandatory Primary Education Law of 1920 (Law 3654, 1920), decreed during the government of Juan Luis Sanfuentes, which perfected the educational programs and purposes, had an impact on the architectural design. With the aim of reaching more cities in the country, quickly and organized, it enabled the generation of some standardized models. Sanfuentes promoted the serial construction of schools in several cities of the country. This new serial system organized schools for 80, 160, 240 and 360 students. For example, the building for 160 students had some features

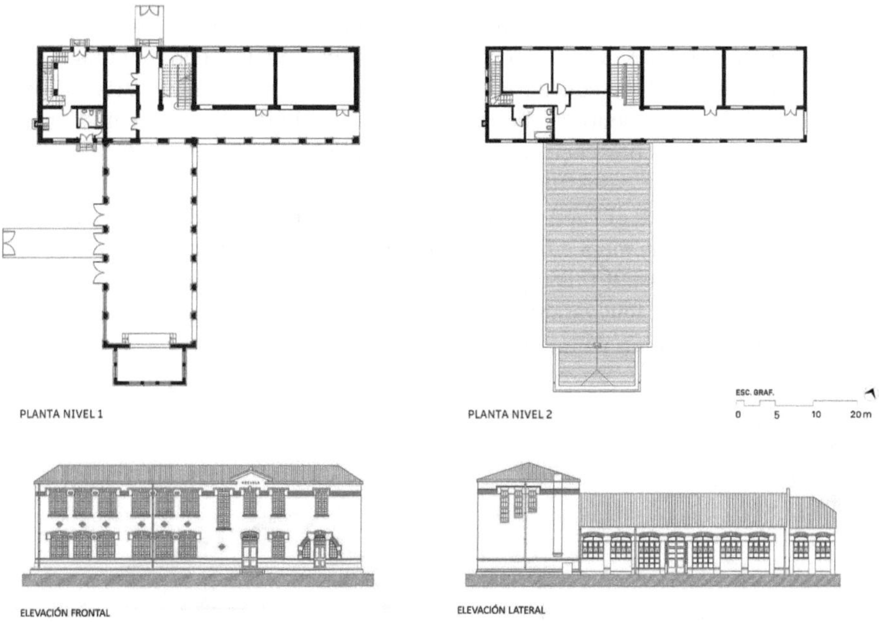

Fig. 39.1 Floor plan and typical elevations for a School of 160 students. Architect: R. del Castillo. *Source* Pablo Fuentes Hernández Archive. Drawing by Bárbara Sáez Orrego

that constituted an advance for school building. Its design, more elaborate, from the point of view of its language, managed to advance on two fronts: first, it perfected the educational institutional function through the consideration of three specific functions in the building: the organization of classrooms, the incorporation of a multifunctional space intended for support functions (auditorium), and the incorporation of a house for the director (Fig. 39.1).

Secondly, the language offered the opportunity to establish a reference image of a crucial area for the State: the education of Chilean children and youth. This reference condition was achieved through the visibility of the attributes of the new architectural language by placing these projects in places where architectural expression also became an act of exhibition.

This new language, mostly expressed in its main facade, was complemented with a sense of organization of the educational project that was developed in a building whose main body was intended for classrooms and the teacher's residence. This body reached two considerable heights. On the other hand, the multifunctional body was arranged perpendicularly at one height, aiding the formation of courtyards. Both volumes were covered by wide roofs with two and four slopes.

The language of these buildings was associated with a picturesque architecture rather than neoclassical stylistic discourses. President Sanfuentes reported that the school for 160 students was considered a "rural" school, under the 1916 law (Sanfuentes, 1919). Thus, the picturesque—visible tile roofs, wooden balconies and brackets, brick ornaments, etc.—configured a language associated with small towns. Even so, the magnitude of these new school buildings established a notable urban presence that elevated the educational function to a building rank. Its studied planning allowed its construction to be repeated in several cities in the central area of the country; to date its presence has been detected in Nancagua, Pinto, Coelemu and Los Vilos, where some cases survive.

Its presence in cities of La Araucanía constitutes a unique finding that shows not only its presence in the main cities of the new territories incorporated into the State, but also the will to install this type of works in smaller cities. In fact, this type of building we find with few variations in two small cities of La Araucanía: Renaico, with one case, and Purén with two (Figs. 39.2 and 39.3).

Fig. 39.2 Renaico School (1920). Architect: R. del Castillo. *Source* PFH Archive

Fig. 39.3 Purén Boys' School (1920). Architect: R. del Castillo. *Source* PFH Archive

These three schools installed in Purén and Renaico expose the State's will to consider these towns as future important cities by installing in them, in their hierarchical urban centers, buildings of urban prestige, a matter that has not changed over time as in all cases, they are the most outstanding buildings of the urban profile.

39.5 Modern Standardization in Educational Infrastructure

So far, most studies on educational infrastructure set in 1937 the definitive takeoff of the systematization of school buildings on the occasion of the creation of the Society for the Construction of Educational Establishments (Sociedad Constructora de Establecimientos Educacionales s.a., 1970). However, the discovery of three other typologically identical cases in the main city of La Araucanía, Temuco, gives grounds to assert that the process of architectural and modern systematization began approximately ten years earlier. Indeed, by then the presidential rise to power of Carlos Ibáñez del Campo in 1927 originated a new ideological position: the conception of the *New* Chile, which conceived development as a fact supported by the efficient action of the State, a matter that required the implementation of rigid and disciplined institutional structures. In 1928, Ibáñez implemented the practice of a technocratic modernity based on a scientific, rational and standardized education concepts he tried to translate into an architectural expression.

In particular, the ideological moment coincided with the promulgation of Law 7.500 on educational reform in 1927, which promoted state administrative efficiency under a rigid and disciplined act, but which lacked technical, financial, and teaching resources capacity. Likewise, this law stimulated school formation from the educational flexibility of the pedagogical model, entering into an operational contradiction that prompted its repeal the following year. From that moment on, Ibáñez's proposal focused on the development of an experimental educational model of primary instruction, tinged with civic and patriotic duty, promoting an improvement of school buildings.

This stage of renewal coincided with Ibáñez's interest in guiding new political directions protected by nationalist ideals whose policy he called *New* Chile. It was a form of ideology that privileged the administrative pragmatism of the state under a cloak of austerity, sobriety, and honesty, seeking to give credibility and progress to the country, thus raising an image of prestige and progress before the international public.

> The school must also instill the fulfillment of duty and love for work, and, through an environment suitable to the child's spontaneous abilities and tendencies, lead him without hindrance to the maturity of his morals and personality. (Ibáñez del Campo, 1927: 8).

The foundations of this change can be discerned when, in politics, Ibáñez's first government approached that of the United States, a matter related to strategies of collaboration, integration, and dependence. Since 1925, results of the economic influence of the United States have been recorded from the Kemmerer mission, whose criteria were materialized from 1927, in the first government of Ibáñez del Campo. Under his mandate, the economic reorganization of the Chilean State was encouraged through a series of administrative reforms that reorganized finances and public administration. The Talca earthquake of 1928 forced additional measures to be taken, which were included in the so-called "Extraordinary Plan" of public works, driven by the Minister of Finance and Public Education Pablo Ramírez. The plan dealt with the undertaking of a wide set of public works, among them, roads, bridges, drinking water and sewerage networks, agricultural irrigation, etc. (Esponda, n/d), whose purpose was analogous to that pursued by the State in Balmaceda's times in the occupation of La Araucanía back in 1890: to achieve national progress from fiscal efficiency strategies that would result in social improvement.[4]

In this framework of collaboration, the visit of Herbert Hoover in 1928, the president-elect of the United States to Chile, endorsed the use of foreign loans to finance the national modernization plan (Ortiz, 2014).[5] A few months later, the presence in the country of an American company called The Foundation Company is recorded, which in its brief presence in the country acted as a development agency. Representing this firm, the architects came to Chile in 1929. Walter Painter, John Fletcher, Gilbert Lasing, and John Medsiry, took on the task of building 601 public schools to fulfill the plans of the Ministry of Education. (Ibáñez del Campo, 1929; North American Architects…, 1929).

The Foundation Company established a new design and construction modality that broke with historicist trends in the design of schools for young people. It was a new strategy called from its beginnings *Standard* School, which established a new way of conceiving and solving a typology that concentrated some concepts of the modern rationalist ideology of architecture.

The American firm studied the essential conditions that the new buildings should meet, determining the minimum guidelines required in classrooms, workshops, laboratories, gyms, sanitary facilities, etc., spaces that contemplated lighting, ventilation, and necessary installations (DGOP, 1930). The interesting thing is that, in the majority of cases, it was the construction of the same building, with few variations, becoming a consolidated typology.

This idea of mass production seems to be related to production techniques that had been notably successful in the American economic system. It is about Fordism, a method of automotive production invented by Henry Ford from 1908 who, after the mass production of the first model, the Ford T, had an unprecedented sales success. By 1915, the system as a generalized and prevailing production model crossed over to the economic sphere to reach the political and philosophical plane, increasing the extension of the benefits of mass production to the educational plane (Mills, 1915).

The studies in Chile considered 10 types according to the capacity of the enrollment and ranged between 160 and 1,000 students. They were reviewed by the General Directorate of Public Works, who in conjunction with the intendencies of each province determined suitable lands for the construction of the models (DGOP, 1930).

Fig. 39.4 Two-story Standard School. Architect: Roberto Dávila (1929). *Source* Revista Arquitectura y Arte Decorativo, a. 1, n°. 6–7, Nov, 1929, p. 241

Given the magnitude of the works, the American firm incorporated Chilean architects, among whom Roberto Dávila Carson stands out, whose designs are recorded in the magazine *Architecture and Decorative Art* No. 1 of 1929 (Fig. 39.4).

The designs abandoned historicist decorations in exchange for the treatment of massive symmetrical bodies of reinforced concrete, two levels, with influences of art deco reliefs. Their plans reveal a clear modulation of the main rooms—an entrance flanked by side bodies that slightly protrude from the plane of the facade.

This research has been able to determine that the standardized school was also built in Iquique—currently demolished—and Linares. Jünemann (1999) reports that others were also built in Limache and Los Angeles by architect Roberto Dávila.

39.6 Situated Standardization. The Case of Temuco

Particularly, three cases are noted in the city of Temuco, main urban center of La Araucanía. This insistent repetition of the model in the city reveals a precise intention. It is about the support of the State to the construction of school buildings on a city that it has the institutional backing to become a major city, meeting the school demand for a population in full growth and urban consolidation.

The Girls' School No. 105, on Prat and Rodríguez streets in the city center—now the headquarters of the Universidad de la Frontera—consisted of classrooms, sewing and home economics rooms, a library, administration and dentist offices, a gymnasium and sanitary facilities. It is the only case that incorporates pre-Columbian ornamental motifs on its main facade. This theme was consistent in the early nineteenth century with neo-indigenous currents that promoted research into nationalist expressions of American architecture. In its geometric drawings, it is possible to detect the influences of researchers Ricardo Latcham (1928) and Abel Gutiérrez (1929) in the dissemination of aboriginal drawings and their applicability in architecture.

Schools 106 and 107[6] (Fig. 39.5), further from the center, have the particularity of being located on the same block formed by Patzke, Francisco Antonio Pinto, Francisco Bilbao, and Manuel Antonio Matta streets. It is a composition that favors the construction of the edges, leaving a central courtyard inside. Both schools are connected to the north by a gymnasium, built later in 1945, and to the south by teachers' houses.

School 106, with entrance by Matta, was also intended for girls, with a capacity for 520 students and had the same dependencies as in the previous case. School 107, with access from F. A. Pinto Avenue, is located in the mirror of school 106; it was intended for boys, with a capacity for 600 students. It had classrooms, physics and drawing rooms, rooms for manual work, library, museum, administration and dentist offices, and a gymnasium with all the advancements for physical education. It should be noted that all three cases had a central heating service.

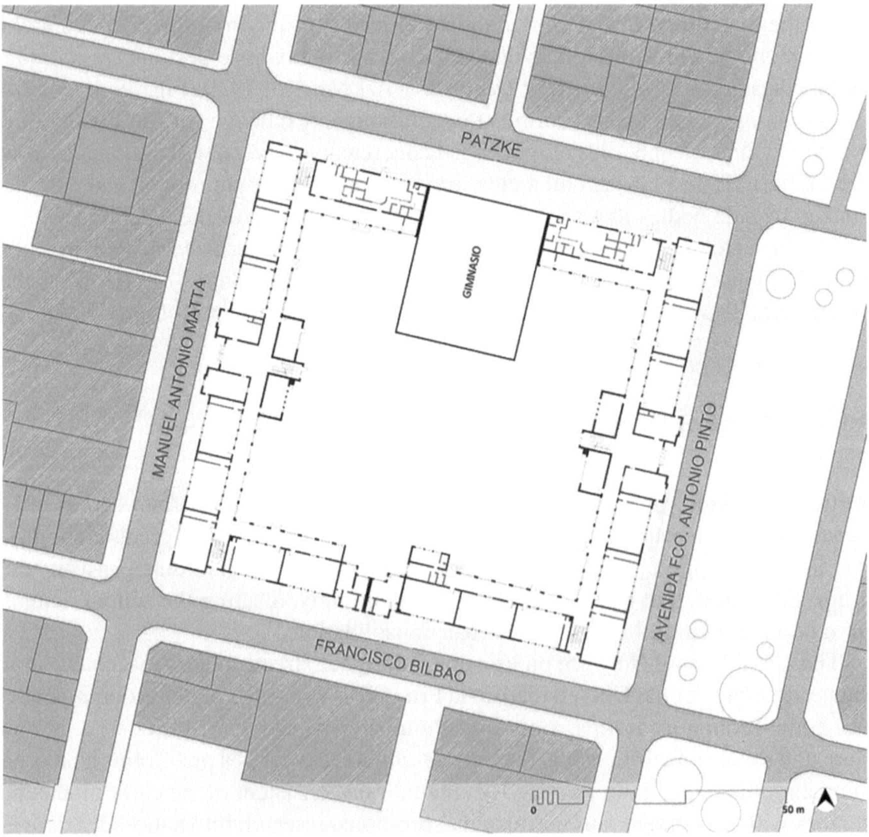

Fig. 39.5 Floor plan of standardized schools 106 and 107, Temuco. *Source* Arch. PFH. Drawing BSO

All models were built in reinforced concrete as the predominant material and structure, in response to recent seismic effects (Talca earthquake of 1928) establishing a paradigm shift in construction for public buildings. Used in the interiors in sight, it contributed a new expressive language (Figs. 39.6 and 39.7).

The buildings maintained a connection with old historicist expressions in terms of their symmetry and monumentality; although they stripped their facades of ornaments, they constituted an advance on more abstract expressions of contemporary architecture *ad portas* of its development in the twentieth century in Chile, and in particular in La Araucanía (Fig. 39.8).

Fig. 39.6 Main facade of the Standard School No. 106 of Temuco, (ca. 1931). Now Standard School D-532. *Source* PFH Archive

Fig. 39.7 Standard School No. 107 of Temuco, (ca. 1931) (later Aníbal Pinto High School). Currently not in use. *Source* PFH Archive

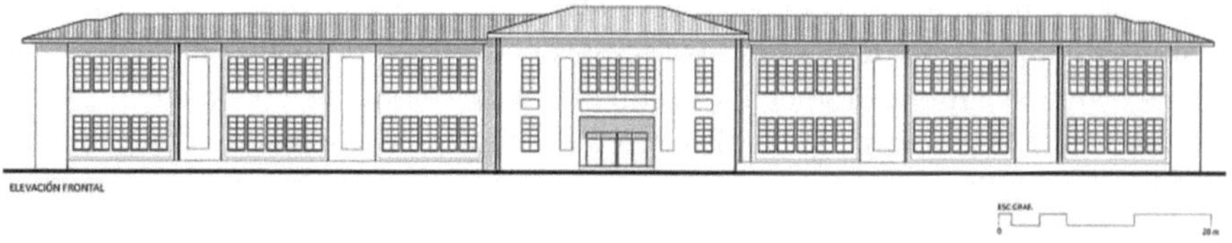

Fig. 39.8 Elevation Standard School No. 107. *Source* PFH Archive. BSO Drawing

39.7 Conclusions

In response to the initial questions, about the link between spatial strategies and civilizing language, it is established that the Chilean state undertook from the end of the nineteenth century, the interest in making the standardization of buildings a means of expression of its modern and progressive character. Among the first initiatives that qualify in this categorization, the construction of railway stations can be mentioned to manifest the process of occupation of La Araucanía. Subsequently, the incorporation of the standardization of educational infrastructure is observed, whose cases reveal the constant interest of different governments, and consequently through different currents, of the political disposition to make this action a constant civilizing.

In relation to the programs that privileged their presence, various school solutions responded quickly to the growing national educational demand. First, the appearance of the Balmaceda model is confirmed; then, that of the schools of the Department of Architecture of the Directorate of Public Works in the 1920s; and finally, that of the standardized schools of the Ibáñez del Campo government. The described examples are the prelude to a tradition that reiterated architectural types that validated the advantages of serialized architectural models.

Regarding the factors of cultural appropriation as ingredients of the development of modernity associated with the consolidation of the national state, some buildings were installed iteratively, showing their ability to propagate the benefits of repetitive systems that democratized the advantages of a rationalized architecture, capable of reaching a larger audience, in various locations. This issue, which is in the state interest to unify the republic under technically egalitarian solutions and carriers of architectural advantages, was adapted to a contemporary use that organized the massification of architectural responses.

The standardized school constitutes a turning point in the contemporary grammar of Chilean educational architecture. It can be affirmed that, having abandoned the eclectic model of palace schools—and also the picturesque one—it does not completely abandon the composition of historicist airs; that is, it remains attached to symmetry, in plan and volume, it is of stable, regular and solid body, it informs clear modular pavilions that escort its access and form its courtyard; and it manages, although almost without ornaments, to still maintain a monumental tone in its proportion and scale. These are issues that reinforce the standardization of an image in agreement with the authoritarian political moment led by Ibáñez del Campo where the totalitarian eagerness coincides with a severe and imposing aspect.

Likewise, the standard school erects an eminently urban image—and therefore, distancing itself from the rural world—thus offers a contribution to the construction of the imaginary of the modern city; an organism that represents the ideals of a developing republic where the State exercises its presence, dominion and power. It should also be added that its language is situated in a moment of transition that will soon have, in brief, a from 1937, there was a significant evolution with the creation of the Society for the Construction of Educational Establishments, which had a dominant presence in Chilean architecture for the rest of the twentieth century.

It should be noted that the gradual renewal of architectural languages in education during the twentieth century was in line with the pedagogical and ideological changes of the time, with school architecture being a paradigm of this transformation.

Regarding the relationships between architecture and political discourse as factors in the consolidation of public institutions, the architecture of education developed a transformative reaction capacity. In the Standard schools, a language is observed that represents a political ideology of fiscal austerity and civic stoicism, eager to align with values of architectural sobriety and urban ratification that initiate architectural modernity at the beginning of the thirties in Chile.

Notes

1. Fondecyt 1210592: City and architecture in La Frontera. The consolidation of the nation state in La Araucanía, 1883–1974.
2. Regarding the construction of public works in La Araucanía, it was pointed out "(…) that being of so much interest for public wealth, we must ensure, establishing population centers that strengthen there the empire of our laws and that of our industry and civilization (Pinto, 1877: 3).
3. They were monumental buildings, organized by one or more courtyards, like the convents and military barracks. They were state constructions capable of housing about 600 students; their most relevant milestones were achieved with the School Buildings Law promulgated under the government of Juan Luis Sanfuentes in 1916. Among them stands out the Federico Errázuriz School in Santiago. https://www.afda.cl/detalle_imagen.php?i=PL-000334%7C0%7C7%7C0 (Accessed on 26-12-2023).
4. Both Balmaceda and Ibáñez del Campo, in his first government, were the presidents who allocated the most resources to public works in the history of Chile until the 1930s.
5. The presence of Herbert Hoover in Chile adds to the previous advice that Edwin Kemmerer (Princeton economist) gave to the Government of Chile in 1925, laying the foundations of the economic stability of the nation, an essential requirement for obtaining financing from foreign banks.
6. Schools 105, 106, and 107, which have subsequently changed their names (105 to: Industrial School, then in 1952 headquarters of the Technical University of the State and since 1982 Headquarters of the University of the Frontier. 106 to: Girls' School, currently Standard School D-532. 107 to: Boys' High School and then to President Aníbal Pinto High School, now abandoned), were mentioned in this way in the Album-guide of the fiftieth anniversary of Temuco: 1881–1931. Letelier Printing, 1931, p. 434. In this work, for greater clarity, this original denomination has been preferred.

References

Arquitectos Norte-Americanos dirigirán la construcción de 601 Edificios Escolares. Revista Arquitectura y Arte Decorativo, Santiago de Chile, a. 1, (2), febrero de 1929, p. 66.

Balmaceda, J. M. (1887). Sesiones de la Cámara de Senadores en 1887. Biblioteca del Congreso Nacional. Diarios de Sesiones del Congreso Nacional. https://obtienearchivo.bcn.cl/obtienearchivo?id=recursoslegales/10221.3/46163/3/18870601.pdf. (Consultado el 28-05–2023).

Balmaceda, J. M. (1888). Sesiones de la Cámara de Senadores en 1888. Biblioteca del Congreso Nacional. Diarios de Sesiones del Congreso Nacional. https://obtienearchivo.bcn.cl/obtienearchivo?id=recursoslegales/10221.3/46163/10/18880601.pdf. (Consultado el 3-06-2023).

Balmaceda, J. M. (1890). La descentralización de la riqueza nacional. Discurso del Presidente de la República en la ceremonia de inauguración de la vía férrea de Collipulli a Victoria. Victoria 27 de octubre de 1890. La Nación, Santiago, 29 de octubre de 1890. En SAGREDO, Rafael y DEVÉS.; Eduardo. (1992). Discursos de José Manuel Balmaceda. Iconografía. Fuentes para la historia de la República (Vol. III, pp. 227–228). Centro de Investigaciones Diego Barros Arana. http://www.bibliotecanacionaldigital.gob.cl/visor/BND:68898 (Consultado el 30-09-2021).

Cerda, G. (1983). *Arquitectura ferroviaria de La Araucanía.* Universidad del Bío-Bío, Concepción.

DGOP, Dirección General de Obras Públicas. (1930). Revisión de proyectos y presupuestos de obras presentados por The Foundation Company, en Memoria correspondiente al año 1929. Editorial Nascimiento, Santiago de Chile.

Dirección General de Estadística. (1920). Censo de población de la República de Chile: levantado el 15 de diciembre de 1920. Disponible en Memoria Chilena, Biblioteca Nacional de Chile https://www.memoriachilena.cl/602/w3-article-82449.html. (Consultado el 30-03-2023).

"Escuelas Modernas en Chile". Revista Arquitectura y Arte Decorativo, a. 1, (6–7), noviembre de 1929, pp. 239–241.

Esponda, J. (s/f). Pablo Ramírez: figura desconocida. DIBAM. http://www.memoriachilena.cl/602/articles-123119_recurso_2.pdf (Consultado el 03-04-2023).

Ferrando, R. (2012). *Y así nació La Frontera… Conquista, Guerra, Ocupación, Pacificación 1550–1900.* Ediciones Universidad Católica de Temuco.

Góngora, M. (1981). Ensayo histórico sobre la noción de Estado en Chile en los siglos XIX y XX. Ediciones la Ciudad, Santiago de Chile.

Gutiérrez, A. (1929). Dibujos indígenas de Chile. Para estudiantes y profesores y arquitectos que quieran poner en sus trabajos el sello de las culturas indígenas de América. Imprenta Universitaria, Santiago de Chile.

Ibáñez del Campo, C. (1927). Mensaje leído por S. E. el Vicepresidente de la República en la apertura del Congreso Nacional el 21 de mayo de 1927. Biblioteca del Congreso Nacional. Diarios de Sesiones del Congreso Nacional. https://obtienearchivo.bcn.cl/obtienearchivo?id=recursoslegales/10221.3/24644/19/19270521.pdf. (Consultado el 18-12-2023).

Ibáñez del Campo, C. (1929). Sesión de las dos Cámaras reunidas el 21 de mayo de 1929. Biblioteca del Congreso Nacional. Diarios de Sesiones del Congreso Nacional. https://obtienearchivo.bcn.cl/obtienearchivo?id=recursoslegales/10221.3/24644/18/19290521.pdf. (Consultado el 18-12-2023).

Jünemann, A. (1999). *Arquitectura Del Inicio Del Modernismo: Oficina Gustavo Mönckeberg y José Aracena, Arquitectos: La Arquitectura Educacional 1920–1950.* Pontificia Universidad Católica de Chile.

"Las Escuelas standard". Álbum-guía del cincuentenario de Temuco: 1881–1931. Imprenta Letelier, 1931, p. 434.

Latcham, R. (1928). *La alfarería indígena chilena.* Universo, Santiago de Chile.

Ley 3117 de. (1916). Dispone que los bonos emitidos o que en adelante se emitan para obras de riesgo i edificación escolar servirán de caución para retirar los vales de tesorería a que se refiere la ley número 2.912 i exime a los indicados bonos del pago de todo impuesto. Biblioteca del Congreso Nacional de Chile. https://www.bcn.cl/leychile/navegar?idnorma=23751. (Consultado 11 19 del 12 de 2023).

Ley 3654 de. (1920). Ley de Instrucción Primaria Obligatoria. Biblioteca del Congreso Nacional de Chile. https://www.bcn.cl/leychile/navegar?idNorma=24146. (Consultado en 26-12-2023).
Mills, W. (1915). *American school building standards*. Franklin Educational Publishing Company.
Montero, M. (2009). Poder y palabra: Mentira implícita y accidentes en discursos presidenciales. *Discurso & Sociedad, 3*(2), 348–371.
Ortiz, E. (2014). La gran depresión, 1929, impacto en Chile. Ediciones Liberalia, Santiago de Chile.
Pinto, A. (1877). Sesiones ordinarias de la Cámara de Senadores en 1877. Biblioteca del Congreso Nacional. Diarios de Sesiones del Congreso Nacional. https://obtienearchivo.bcn.cl/obtienearchivo?id=recursoslegales/10221.3/46165/6/18770601.pdf
Pinto, J. (2003). La formación del Estado y la nación y el pueblo mapuche. De la inclusión a la exclusión. DIBAM., Centro de Estudios Diego Barros Arana, Santiago de Chile.
Sanfuentes, J. L. (1919). Mensaje leído por el Presidente de la República en la apertura de las sesiones ordinarias del Congreso Nacional. Imprenta fiscal de la Penitenciaría (pp. 4–26). https://www.bcn.cl/leychile/navegar?idNorma=23751 (Consultado el 19-12-2023).
Santa María, D. (1883). Sesiones de la Cámara de Senadores en 1883. Biblioteca del Congreso Nacional. Diarios de sesiones del Congreso Nacional (pp. 3–9). https://obtienearchivo.bcn.cl/obtienearchivo?id=recursoslegales/10221.3/46166/1/18820601.pdf (Consultado el 01-06-2021).
Serrano, S., de León, M. P., y Rengifo, F. (2013) Historia de la educación en Chile (1810–2010). La educación nacional 1880–1930. Tomo II, Aguilar Chilena de Ediciones S.A., Santiago de Chile.
Sociedad Constructora de Establecimientos Educacionales S.A. (1970). AUCA 19 (pp. 55–63). https://revistaauca.uchile.cl/index.php/AUCA/article/view/59228/62715
Van Dijk, T. (2003). Ideología y discurso. Una introducción multidisciplinaria. Ariel, Barcelona.
Vargas, E. (s/f). "Historia del Viaducto del Malleco". https://www.elferrocarrilenchile.cl/historia-viaducto-malleco (Consultado el 09-10-2022).

Part X
Pandemic and Post-pandemic: Challenges for Architecture, Design, the City and the Territory

Chapter 40
(Non) Proximate Cities, Lessons from the Pandemic: The Importance of Proximity in Urban Communities

Alfredo Hidalgo Rasmussen, Luis Natera, Edgar Egurrola, and Rossana Valdivia

Abstract The COVID-19 pandemic brought special attention to health, but also to cities through the recognition of health-related urban conditions such as walkability and diversity of uses and services in neighborhoods. This rethinking led to the development of an algorithm to measure proximity to different facilities and services considering travel time in different areas of cities from the Observatory of Cities of Tec de Monterrey. The results obtained allowed us to study the average walking times to different points of interest in a large array of cities and identify any deficiencies that may exist. This approach culminated in the construction of a tool to visualize and understand proximity in cities in an intuitive way and that has been presented in spaces with decision makers.

Keywords Proximity · 15 min city

A. Hidalgo Rasmussen (✉) · L. Natera · E. Egurrola · R. Valdivia
School of Architecture, Art and Design, Tecnologico de Monterrey, Monterrey, México
e-mail: ahidalgor@tec.mx

L. Natera
e-mail: natera@hey.com

E. Egurrola
e-mail: edgar.egurrola@tec.mx

R. Valdivia
e-mail: rvaldivia@tec.mx

A. Hidalgo Rasmussen
IAE Paris Sorbonne, Panthéon Sorbonne University, Paris, France

M. Vicuña and E. Giorgi (eds.), *Intersections*, Springer Geography,
https://doi.org/10.1007/978-3-031-76402-8_40

40.1 Introduction

The COVID-19 pandemic not only increased attention on the health and well-being of people, but it has been an accelerator of actions to rethink the way communities and people relate to their cities and environments. This process, a result of confinement strategies to reduce virus transmission, resulted in a decrease in the number of trips that each individual carried out in their day-to-day life and the way of interacting with the city to use only their closest environments. In this context, concepts such as walkability and land-use diversity in neighborhoods become relevant and lead us to rethink urbanization models (Alberti & Radicci, 2022).

These changes in people's mobility patterns depended directly on the measures taken by each city to face the pandemic (Cuadros & others, 2023). However, the main impact was due to the possibility of remotely carrying out traditional activities such as work, education, or commerce.

However, the possibility of carrying out daily activities on foot is conditioned by the distribution of land-uses in the city. The functional separation of the city according to land uses makes territorial quarantines difficult (Olivares et al., 2021), which has generated significant differences in service coverage in various areas of the city and, therefore, an inequality in the possibility of effectively carrying out a prolonged quarantine.

This condition related to the quarantine period has changed in recent years, and although at that time it was thought that it would completely change the way of living in cities, we have observed a return to normality. However, some patterns caused by the measures taken to prevent contagion, such as staying at home and the absence of permanence in workplaces or study places, have been maintained to some extent in some cities, which has caused a change in mobility patterns. According to the study on the Impact of COVID-19 on preferences for modes of transport in selected Latin American cities, through Google Mobility data, it has been observed that in Mexico City, although the time spent in workplaces has been increasing since the sharp fall it had in April 2020, it has not yet reached the same levels as in February 2020. In other cities, they have reached the same levels as before the pandemic in 2021, such as Sao Paulo, Santiago, Buenos Aires, and Bogotá.

These changes in the way activities are carried out directly impact mobility, the main loser was public transport, while individual mobility alternatives turned out to be the winners. Both in public transport in general and in the use of the Metro in particular, the drastic fall in mobility from March to April 2020 begins to recover over time. However, by mid-2022, the same levels have not yet been reached and normal levels in its use, remaining at the lowest levels of the decade (Cuadros et al., 2023).

In this sense, despite mostly resuming behaviors prior to the pandemic, the measures taken to mitigate its impact have accelerated trends such as teleworking and this change in activities left a mark on the way people move.

From this situation, where the diversity of uses of an environment influences the alternatives in mode of transport and therefore is related to the walkability of a city and the nearby environments, the research question arises: how to use the city's road network to measure the distance—and the time—to the facilities for daily activities? And with it, identify the city areas where these activities can be carried out on foot in the immediate vicinity.

Measurement of the proximity of services to housing helps identify mobility options for people based on the environment's conditions. Understanding these conditions not only provides insight into the current situation but also aids in decision-making. This research proposes using proximity measurements to develop urban mobility strategies and land use plans that local governments can promote.

This article describes how this process led to the development of an algorithm that allows the measurement of the proximity to different services and facilities from a defined point and replicating it for any intersection in the city and for any city.

For the purposes of this research, proximity is considered as the closeness between people in relation to services and their activities, unlike accessibility, understood as the possibility of reaching these services, including different means of transport or technology, regardless of the distance (Solá & Vilhelmson, 2018).

40.2 Methodology

40.2.1 Proximity Evaluation

Cities and their urban mobility infrastructure can be represented and studied as a complex network (Albert & Barabási, 2002; Newman, 2003). Specifically for their study, intersections of roads can be analyzed as nodes and streets or roads as links (Alessandretti et al., 2022; Barthelemy, 2011, 2018).

To calculate travel times, we used the pedestrian network of cities, which can be represented as a weighted graph with an adjacency matrix $= w_{\mathrm{ij}}$, where the weight w_{ij} contains the Euclidean distance between the nodes i and j, if such nodes are connected, otherwise the value is 0. The data necessary to construct the pedestrian networks was acquired from OpenStreetMaps (OSM) through OSMnx (Boeing, 2019). To calculate the shortest distance between any node to a facility or point of interest, we followed the methodology developed in (Natera et al., 2020), in such a way that we encoded the distance between the nodes i and j as l_{ij} and the travel time as $T\ t_{\mathrm{ij}} = \frac{l_{\mathrm{ij}}}{\mathrm{ps}}$, where ps is the pedestrian speed, considered as 4 km/h with a slope of 0%; however, this speed was recalculated based on the slope of the transect of the network traveled.

The developed methodology allows identifying the shortest path and the corresponding distance that must be traveled to get from any node to a given facility or point of interest.

In order to analyze cities from an integral perspective of proximity, this analysis was applied to a series of facilities and services that were grouped into different categories (Table 40.1) based on the concept of 15 min cities (Carlos Moreno, 2015),[1] which proposes that if the main daily activities are within a radius close to that time

Table 40.1 Division of categories and facilities/services evaluated to construct the 15 min index

Category	Facility/service
Education	Preschool
	Primary
	Secondary
Community services	Health
	Daycare centers
	Social assistance
Commerce	Food supply
	Personal use items
	Pharmacies
	Household items
	Complementary commerce
Entertainment	Social
	Physical activity
	Cultural

walking from homes, people significantly improve urban quality of life (Alberti & Radicci, 2022; Caselli et al., 2022; Ferrer-Ortiz et al., 2022; Weng et al., 2019).

For the selection of the 14 facilities or services used in the analysis (Table 40.1), a literature review was conducted of various academic articles that study proximity in cities around the world (Alberti & Radicci, 2022; C40, 2020; Da Silva et al., 2020; Ferrer-Ortiz et al., 2022; Graells-Garrido et al., 2021; Harper et al., 2019; Hidalgo et al., 2020; Morales et al., 2017; Moreno et al., 2021). From this foundation, we filtered out the elements essential for meeting daily needs, focusing primarily on the care and well-being of children, elderly individuals, and caregivers. This selection alsoconsiders the services identified by the Colectivo Punt 6 in their publication Spaces for Everyday Life, as of neighborhood scale for the reproductive sphere (Ciocoletto, 2014), so it is considered imperative that they are within a walkable distance to guarantee their accessibility without dependence on any other type of motorized mobility. Finally, the availability of georeferenced spatial data for all Mexican cities was evaluated, and those variables for which information was not available were eliminated.

Considering this proposal as a guideline for determination of the categories was relevant while identifying that the development and rise of the referred proposal gained notoriety in the context of the pandemic. The greatest visibility came from Paris, where Mayor Anne Hidalgo implemented these principles in her urban activation proposals during her second term; also, the C40 Cities Climate Leadership Group made it part of their program, as well as management offices such as the Agency of Ecology Urbana already mentioned the "quarter-hour cities" in their document, Mayorga (2021).

The concept of 15 min cities takes on special relevance in the post-pandemic city as a mechanism for revitalizing the city and its neighborhoods, in addition to facing new social and economic challenges post-COVID.

Added to the post-pandemic reality, the focus on 15 min cities and their analysis allows to establish special emphasis on the climate agenda and zero carbon through the reduction of private car use, applying proposals such as the Climate Action Plan and subsequently founding in 2011 the Parisian Climate Agency (APC).

The proposal of the 15-min city is related to time and distance, but also to the population and the place where they live. In the latest book by Moreno (2020[a]) Droit de cité, Richard Sennet (2020, Préface s.p.) mentions: "In the modern city, density has been subordinated to distance. Carlos Moreno proposes to reverse that relationship." The quarter-hour city allows to include the time factor in the use of the city, which has already been considered in other proposals or initiatives, such as Portland's 20-min neighborhoods (2000) or in Victoria (Plan Melbourne 2017–2050).

For the case of the Paris mayor's office, large-scale actions were first implemented referred to commitments to climate change, the specific actions in the neighborhoods have been a specific approach. Access to large volumes of urban data and its processing through mathematical models allows us to test the concept of the 15-min city in a systematic way.

40.2.2 Grouping and Analysis of Cities

Due to the volume and complexity of the data when working at the intersection level, information was aggregated to a metropolitan scale spatial resolution. With this objective, a continuous and global hexagonal grid (Uber, 2020) was used, which facilitates the analysis in a grouped way (Fig. 40.1).

From this continuous surface, a spatial aggregation of the nodes was carried out in which the average time that had to be traveled to reach each facility was calculated. Subsequently, the maximum average time to each facility was calculated and used as the time index per hexagon. In such a way that only those areas where all categories are within 15 min are considered to meet the desired vision. The methodology allows the modification of the hexagon's surface as well as the time that is considered as desired.

Fig. 40.1 Representation of the roads and intersections within a hexagon

40.3 Results and Discussion

40.3.1 Characterization of Cities and Main Findings

The results, based on the methodology developed, show that none of the Mexican cities analyzed (74 metropolitan areas recognized by INEGI) meet the assumption of the 15-min city. Figure 40.2 shows the average time of the cities analyzed; none are within the ideal range. Mexico City has the best result with an average of 35 min, while Querétaro 85 min.

These elements help make inferences about how facilities and services are distributed throughout the city. A city with a more permeable layout and with facilities evenly distributed in the territory allows shorter journeys than one with barriers and centralized facilities.

To validate and strengthen these inferences, an analysis of the relationship between population density and the average time weighted by population in the different cities was carried out. This comparative analysis has become a fundamental tool for understanding the interconnection between density population and the geographical proximity of cities, crucial elements in the study of urban dynamics.

Upon observing the resulting data (Fig. 40.3), a correlation was identified between population density and time weighted by population. Cities with higher population

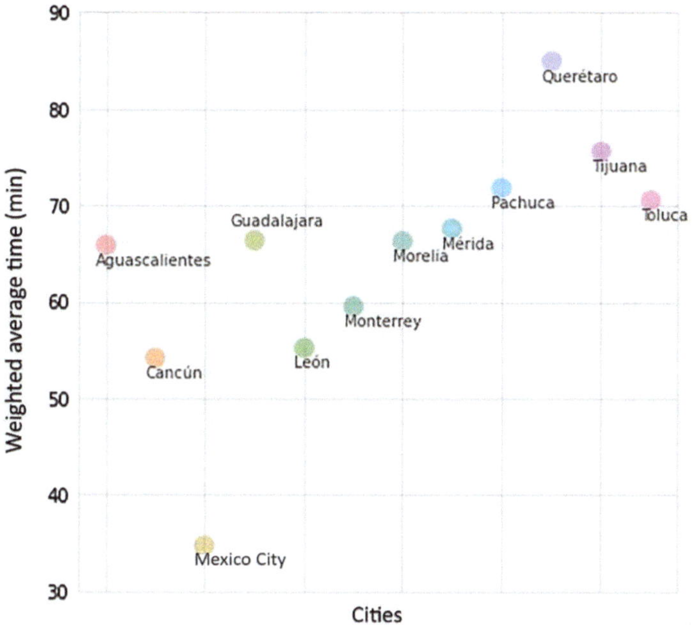

Fig. 40.2 15 min city index

densities presented lower weighted times, suggesting that the concentration of inhabitants in a given area directly impacts daily commutes and, therefore, the time people spend in transit.

The validation of this visual inspection was carried out by generating a regression line from the collected data (Fig. 40.4). This regression line yielded a negative slope, quantitatively highlighting the inverse relationship between the population density of cities and the travel time necessary to access essential services and facilities, as evaluated in the context of 15 min cities.

The inverse relationship identified can be attributed to several interrelated factors. In densely populated cities, the physical proximity of services is greater due to the concentration of inhabitants in a limited area. This implies that the evaluated services and facilities, such as health centers, educational institutions, and recreational spaces, are geographically closer to more people. Consequently, the time needed to access these services is reduced, as commutes are shortened.

In addition to the relationships of population concentrations, an analysis of the percentage of accumulated population in relation to different time intervals was carried out. This analysis allowed for a deeper understanding of the temporal distribution of the population based on their accessibility to urban services and facilities. In Fig. 40.5, a pattern of linear increase in population accumulation regarding travel time can be seen.

The pattern observed in Fig. 40.5 reveals that most of the population is not within the areas accessible in 15 or 30 min. This finding is relevant, as it suggests that

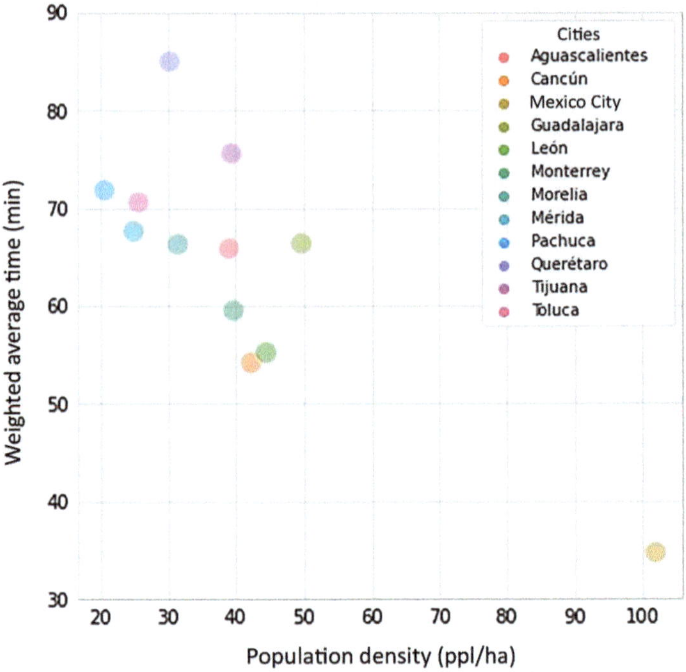

Fig. 40.3 Comparison between population density per hectare and the 15 min index

accessibility to essential services and facilities is not being optimally distributed in most of the urban areas under study. The need to improve this accessibility is evident, as ensuring that the population has adequate and efficient access to these services is fundamental for better quality of life and equitable urban development.

It is important to note that in a nearby city, we would expect a significant portion of the population to be within 15 min of essential services and facilities, with smaller increases in the percentage of people as we increase in time, following a pattern similar to that of a Pareto diagram.

On the other hand, Fig. 40.6 provides a visual perspective on the spatial distribution of different types of facilities in relation to proximity to urban areas. It is observed that commerce stands as the closest infrastructure in terms of accessibility, which can be attributed to the market logic that guides its strategic location. Shopping centers and retail stores tend to establish themselves in areas of high population density or high influx, seeking to meet demand and maximize their reach to a potentially consumer audience.

In contrast, community services and entertainment facilities are generally farther away. Specifically, entertainment options like cinemas, theaters, and cultural spaces are often scarce in many cities and, when available, are typically located more than 60 min away on average. This situation can be attributed to factors such as urban

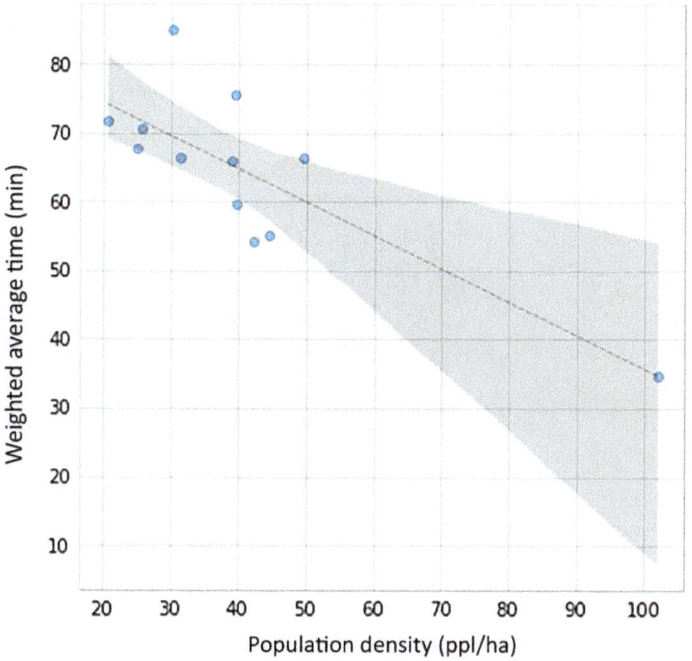

Fig. 40.4 Linear regression between population density per hectare and the 15 min index

Fig. 40.5 Percentage of population by time range in the 15 min index

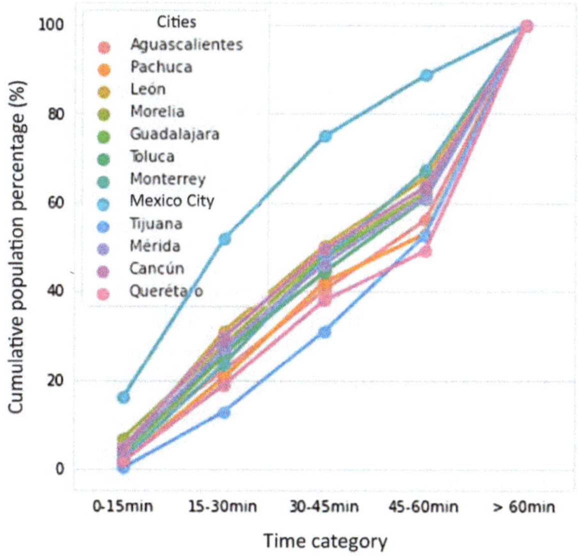

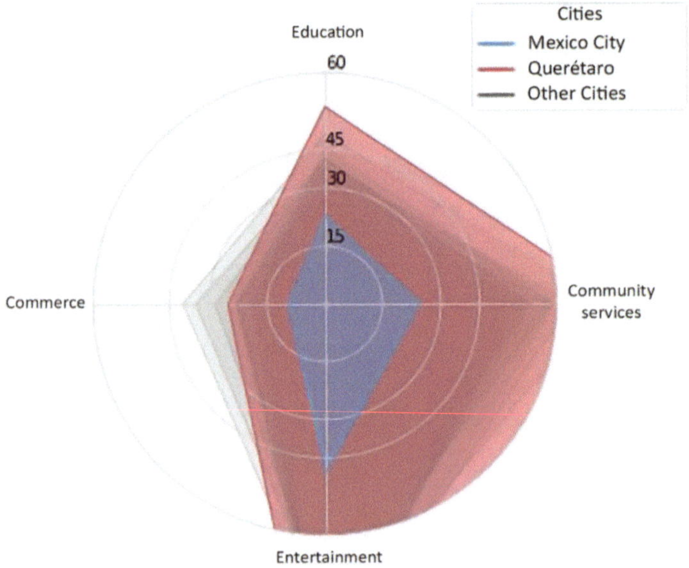

Fig. 40.6 Comparative of maximum time by category by city with emphasis on Querétaro and Mexico City

planning, local demand, and financial accessibility, which influence the availability and location of these services.

In turn, although this figure highlights the general trend in the analyzed cities, emphasis is placed on Mexico City and Querétaro, which as previously mentioned, represent the extremes with respect to the 15 min city index.

To territorialize the variables previously described, the following are the maps of the 15 min index for 5 cities. Mexico City and Cancún represent the two closest environments within the analyzed cities, while Querétaro and Tijuana are the cities with the highest indices. Finally, Monterrey is added to the analysis as an average city.

Although Mexico City and Cancún (Figs. 40.7 and 40.8) have a lower index than other cities, it is possible to observe in the maps that this is produced for two different reasons. On the one hand, Cancún is seen as a more compact city with clear urban center coverage, while Mexico City is shown as a polycentric area. However, in both cases, there is continuity of the urban area that contributes to its connectivity and its reduction in transfer centers.

On the other hand, the case of Monterrey (Fig. 40.9) also shows signs of a polycentric city, but, unlike Mexico City, it has large urban areas that are discontinuous and with low coverage of the necessary facilities and services to comply with the guidelines of the proposed 15 min city index.

This type of dispersed urbanization produces low-density environments, which exemplifies the density vs index relationships shown in Figs. 40.3 and 40.4.

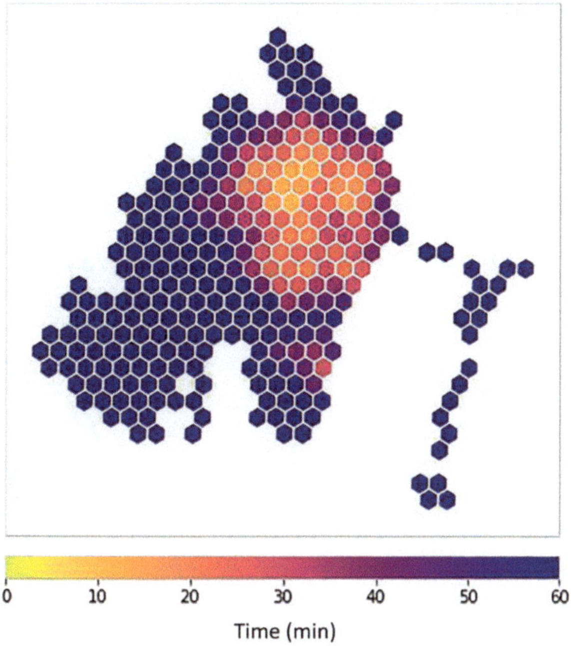

Fig. 40.7 15 min index per hexagon for Cancún

Finally, the cities of Tijuana and Querétaro, as examples of non-close cities, allow for specific comparisons against the cases of Mexico City and Cancún respectively.

With Tijuana (Fig. 40.10), continuous urban areas are observed, with some disconnected environments, but unlike Mexico City, and even Monterrey, there is a lack of evenly distributed centralities in the territory to allow pedestrian access to priority facilities and services.

Querétaro (Fig. 40.11) has a very consolidated centrality, comparable to that of Cancún. However, unlike the latter, an urban dispersion process is observed that produces environments disconnected from the rest of the city, both spatially and in terms of service and equipment coverage. These dynamics force its inhabitants to travel to urban centers to meet their daily needs.

When observing all the cities, the relevance of creating polycentric and compact environments can be highlighted, which prevents the proliferation of disconnected urban environments that encourage the use of private vehicles as the main means of transport.

40.3.2 A Tool Towards Decision Making

To share and/or compare the results, a graphic tool and an index were generated that facilitate the measurement and interpretation of urban proximity.

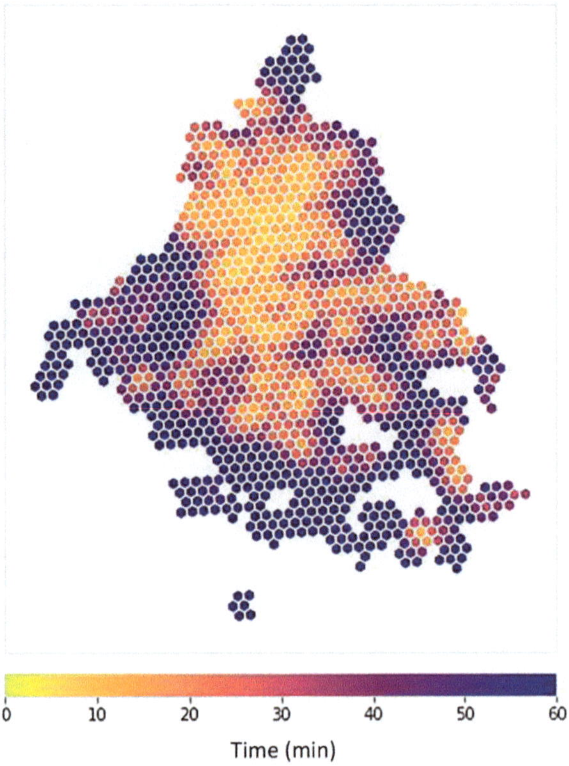

Fig. 40.8 15 min index per hexagon for Mexico City

The methodology developed for calculating proximity helps identify areas of the city with limited access to certain services or facilities, based on the maximum distance people travel for daily activities. This approach facilitates territorial decision-making at the metropolitan or municipal level by providing valuable local information.

The proximity indicator translates distance measurements into time, showing the maximum time needed to reach services on foot. This helps identify areas of the city that support various modes of individual mobility, particularly non-motorized options like walking and cycling, as well as non-combustion motorized vehicles such as scooters.

With the index and the data that make it up, an interactive tool was developed (A version of the platform is available at https://observatorio-ciudades.github.io/visor-15minutos/), which communicates proximity indicators with municipalities, local, regional, national governments and in general, decision makers. To facilitate access to information, the platform is presented from a simple and easily accessible graphic proposal.

In order not only to identify the absence or non-compliance of services, the tool also indicates the farthest equipment or service as a priority action in order to generate suitable proximity environments. The geographic viewer facilitates the consumption

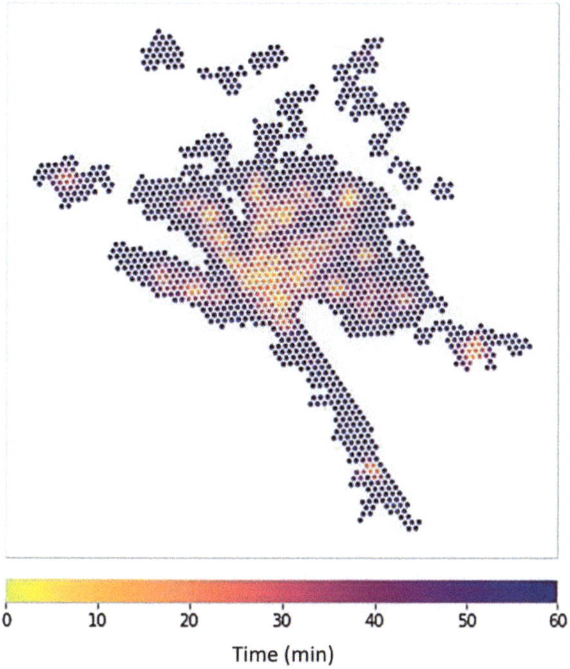

Fig. 40.9 15 min index per hexagon for Monterrey

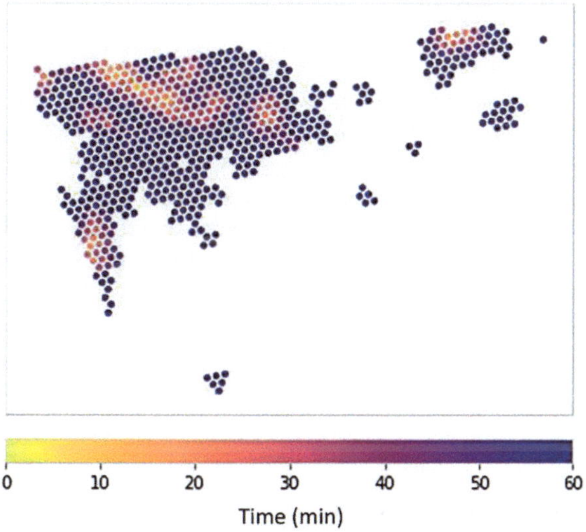

Fig. 40.10 15 min index per hexagon for Tijuana

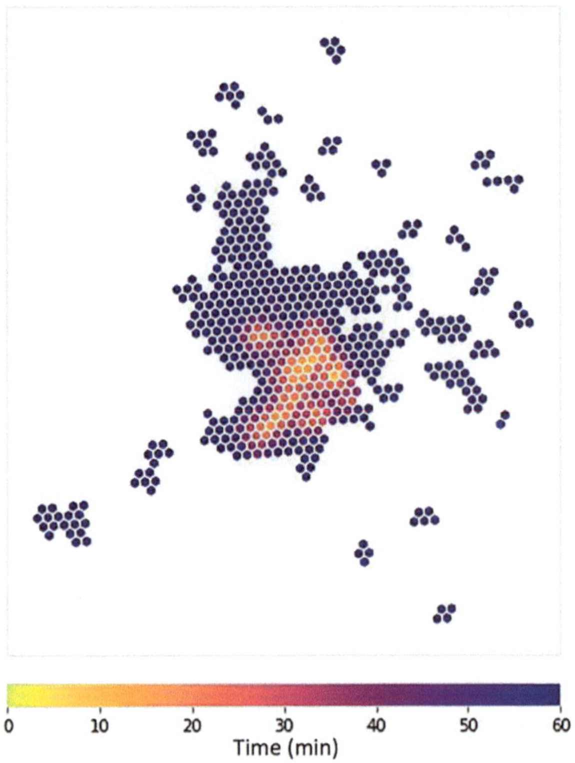

Fig. 40.11 15 min index per hexagon for Querétaro

of information by showing not only the index but also the population data. For each of the aggregated areas (hexagons), it presents the priority equipment or service, that is, the equipment or service that has less accessibility within the hexagon.

Both the development of the index and the geographic viewer allow certain adjustments to improve decision making, for example, if you wanted to perform the index to identify accessibility patterns for elderly people, it is possible to adjust the speed parameters as well as the priority equipment and services. These adjustments allow to have a specific index, which, when crossed with population and housing data gives a radiography of the territory focused on accessibility for the elderly.

40.4 Conclusions

Proximity, as a means of ensuring essential services are nearby to enhance urban quality of life, is not a new concept. However, its significance has grown since the pandemic, which emphasized staying close to home. This idea aligns with the '15 min city' concept promoted in recent projects by Paris Mayor Anne Hidalgo and the Chaire ETI at the Sorbonne, led by Carlos Moreno.

The design of the algorithm allows the measurement of the distance that must be traveled to the equipment and services, considering the topography and urban layout, and translates it into time, a measure that is related to the quality of life of people.

The designed algorithm allows the measurement of the distance to equipment and services, considering topography and urban layout, and translates this distance into time. This time-based measure is closely related to people's quality of life.

The platform simplifies the interpretation of results by detailing the proximity conditions for each hexagon and highlighting differences across various parts of the city at the metropolitan level. It enables users to understand conditions at the neighborhood scale while also providing decision-makers with insights on broader urban conditions. This helps inform public policies by establishing a basic walking range as a benchmark for designing desirable and effective urban environments.

The density is closely linked to proximity, as shown in the findings. While density alone does not ensure a high quality of life for city residents (Salcedo, 2017), a more compact and dense urban layout can enhance mobility, reduce travel times, and improve access to services.

The concept of a 'nearby city,' introduced through various models before the pandemic, is compelling due to its clear communication and relevance. It also enables the identification of key strengths and gaps in the territory by leveraging large volumes of data. This approach facilitates the identification, recognition, and formulation of actions and plans for the short, medium, and long term.

Note

1. Conceptual framework developed as leader of the Sorbonne's ETI Chair.

References

Albert, R., & Barabási, A.-L. (2002). Statistical mechanics of complex networks. *Reviews of Modern Physics,* 47–97.

Alberti, F., & Radicchi, A. (2022). The Proximity City: A comparative analysis between Paris, Barcelona and Milan. *TECHNE—Journal of Technology for Architecture and Environment,* 69–77.

Alessandretti, L., Natera, L., Saberi, M., & Szell, M. (2022). Multimodal urban mobility and multilayer transport network. *Environment and Planning B: Urban Analytics and City Science.*

Barthelemy, M. (2011). Spatial networks. *Physics Reports,* 1–101.

Barthelemy, M. (2018). *Morphogenesis of spatial networks.* Springer.

Boeing, G. (2019). Urban street network analysis in computational notebook. *The Journal of ERSA,* 39–51.

Caselli, B., Carra, M., Rossetti, S., & Zazzi, M. (2022). Exploring the 15 min neighborhoods. An evaluation based on the walkability performance to public facilities. *Transportation Research Procedia,* 346–353.

Ciocoletto, A. (2014). Spaces for everyday life. *Comanegra,* 13–18.

Cuadros, G., & others (2023). Impact of COVID 19 on transportation mode preferences in selected cities in Latin America. Project Documents (LC/TS. 2023/19) Santiago, Economic Commission for Latin America and the Caribbean (ECLAC).

C40. (2020, July). How to build back better with a 15 min city. Retrieved from C40 Knowledge. https://www.c40knowledgehub.org/s/article/How-to-build-back-better-with-a-15-minute-city?language=en_US

Da Silva, D., King, D., & Lemar, S. (2020). Accessibility in practice: 20 min city as a sustainability planning goal. *Sustainability*.

Errázuris, T., & Green, R. (Eds.). (2017). *Salcedo* (pp. 63–64). Editorial Bifurcations.

Ferrer, C., Marquet, O., Mojica, L., & Vich, G. (2022). Barcelona under the 15 min city lens: Mapping the accessibility and proximity potential based on pedestrian travel times. *Smart Cities,* 146–161.

Graells-Garrido, E., Serra-Burriel, F., Rowe, F., Cucchietti, F., & Reyes, P. (2021, March). A city of cities: Measuring how 15 min urban accessibility shapes human mobility in Barcelona. Retrieved from Cornell University. https://arxiv.org/pdf/2103.15638.pdf

Harper, L., Grodach, C., & Kamrruzaman, L. (2019, November). 20-Minute neighbourhoods: Living locally research project. Retrieved from MONASH University. https://www.monash.edu/__data/assets/pdf_file/0011/2045396/Monash-20-Min-Neighbourhood-Final-Report-29-11-19.pdf

Hidalgo, C., Castañer, E., & Sevtsuk, A. (2020). The amenity mix of urban neighborhoods. *Habitat International.*

Mayorga, M. (2021) Paris: The 15 min city. In B. Ruiz-Apilánez & E. Solís (Eds.), *On foot or by bike. Perspectives and experiences around active mobility* (pp 167–174). University of Castilla-La Mancha Editions.

Morales, J., Flacke, J., Morales, J., & Zevenbergen, J. (2017). Mapping urban accessibility in data scarce contexts using space syntax and location-based methods. *Applied Spatial Analysis and Policy,* 205–228.

Moreno, C., Allam, Z., Chabaud, D., Gall, C., & Pratlong, F. (2021). Introducing the "15-Minute City": Sustainability, resilience and place identity in future post-pandemic cities. *Smart Cities,* 93–111.

Natera, L., Deritei, D., Vancsó, A., & Vasarhelyi, O. (2020). Quantifying life quality as walkability on urban networks: The case of Budapest. In *International Conference on Complex Networks and Their Applications* (pp. 905–918).

Newman, M. (2003). The structure and function of complex networks. *SIAM Review,* 167–256.

Solá, A., & Vilhelmson, B. (2018). Negotiating proximity in sustainable urban planning: A Swedish case. *Sustainability,* 1–18.

Uber. (2020, June). H3. Hexagonal hierarchical geospatial index system. https://h3geo.org/

Weng, M., Ding, N., Jin, X., Xiao, H., He, Z., & Su, S. (2019). The 15 min walkable neighborhoods: Measurement, social inequalities and implications for building healthy communities in urban China. *Journal of Transport & Health,* 259–273.

Chapter 41
Exploring Architecture as an Exit Dispositive in Times of Crisis

Carlos Lange Valdes

Abstract Both the "Social Outbreak" declared on October 18, 2019 and the COVID-19 pandemic recognized by the State on March 18, 2020 constitute a clear expression of a crisis process associated with a high level of political, economic, social, and cultural uncertainty, which exposes the precariousness associated with the dichotomous conception of public–private spatiality sustained by neoliberal urbanism through its various production dispositives. However, within the framework of this crisis, it is also possible to observe the revaluation of a third form of spatiality, which are common spaces. Understood as those spaces that inhabitants configure from shared resources present in their ways of living, these do not necessarily arise from pre-existing communities or rationally predefined projects but emerge from the various daily practices of living and the conflicts, negotiations, and agreements generated by the inhabitants in their own territories. This article analyzes the relevance that architecture has in making visible the production of "common spaces" by the inhabitants, becoming a fundamental component of the "exit dispositives" in the context of crisis.

Keywords Crisis · Neoliberal urbanism · Dispositives · Common spaces · Living practices

41.1 Living in Crisis Contexts in Santiago de Chile

The implementation of the neoliberal development model and the subsidiary policies associated with it have made the relevance of common spaces in Chile invisible for almost 50 years. In urban terms, this has been expressed through the establishment of a logic of dichotomous spatialities of public and private character, which can be considered part of a strategic control and socio-spatial order device characteristic of

C. L. Valdes (✉)
Facultad de Arquitectura y Urbanismo, Instituto de la Vivienda, Universidad de Chile, Santiago, Chile
e-mail: clange@uchilefau.cl

M. Vicuña and E. Giorgi (eds.), *Intersections*, Springer Geography,
https://doi.org/10.1007/978-3-031-76402-8_41

neoliberal urbanism (de Simone, 2021; Janoschka & Hidalgo, 2014; Liernur, 2009; Talesnik, 2021; Vergara-Perucich & Boano, 2020).

However, both the "Social Outbreak" of October 18, 2019, the constituent process initiated on December 24, 2019, and the COVID-19 pandemic recognized by the State on March 18, 2020 have questioned this model, generating a high level of political, economic, and socio-cultural uncertainty in the country and opening a crisis process that reveals the contradictions generated by this neoliberal dichotomous conception.

On the one hand, the Social Outbreak initiated on October 18, 2019 provoked a significant revaluation of public spaces as a sphere of citizen expression and political claim associated with massive street demonstrations for demands for greater equality, social justice, and citizen rights (Manzi, 2020; Márquez, 2020).

Subsequently, the COVID-19 pandemic revalued the importance of private space from confinement and social distancing measures articulated in a quarantine strategy based on the declaration of the constitutional state of exception for public calamity (Ruiz-Hurtado, 2020).

Likewise, the debates generated around the constituent process have put into critical perspective the principles of the right to housing and the right to the city, questioning the principles of collective and individual property associated with them (Saavedra et al., 2021).

Although the events referred to above are framed in a broader historical process, it is in a brief period of approximately five years that the questioning of the prevailing modes of production of dichotomous spatialities explodes with greater force, establishing a crisis process.

For Bauman and Bordoni (2016), the concept of "crisis" reflects a moment of transition from one model of society to another, exposing its main contradictions. Applied to the case of European societies, the authors argue that the economic crisis The 2008 "subprime" crisis reveals the market's inefficiency to resolve its own limitations, as well as the State's inability to regulate them, considering that it no longer has political capacity (the ability to decide what things should be done) or power capacity (the ability to get things done). This is expressed in a significant dissociation between global and local politics, generating a high level of uncertainty about the future.

Applied to the Chilean case, the concept of "crisis" allows us to frame the questions, debates, and uncertainties, as well as the disputes, conflicts, and negotiations that emerge around the dichotomous spatialities of public and private character inherent to neoliberal urbanism.

Understanding that architecture is part of a strategic control and socio-spatial order dispositive that promotes these dichotomous spatialities—the quarantine dispositive, this essay explores its potentialities as part of an exit dispositive in the face of the crisis, which makes visible and recognizes the production of new hybrid spatialities of common character (Giglia, 2012).

41.2 Architecture as a Device

On March 18, 2020, a state of constitutional exception due to public calamity is declared in Chile, by virtue of decree Nº 104 of 2020 from the Ministry of the Interior. Due to the spread of COVID-19 contagion, this decree establishes confinement and social distancing measures that regulate and restrict the access and use of public spaces by the inhabitants.

In this context, architecture plays a political and strategic role as part of a quarantine dispositive, which articulates various control and socio-spatial order measures aimed at the production of a particular sociability and spatiality. This is expressed not only through the implementation of vaccination centers, health barriers, sanitization squads, health shelters, among others, but also in the deactivation of urban social life, making critical views and citizen demands emerged during the Social Outbreak invisible and reinforcing the subsidiary capacities of the State through the subjectivation of desocialized, depoliticized and deterritorialized individuals.

According to the definition proposed by Foucault (1991), the concept of dispositive refers to the network of knowledge/power constituted from discourses, institutions, architectural installations, regulatory decisions, laws, administrative measures, scientific statements, philosophical propositions, moral, philanthropic, among others, which has a political and strategic order function, that produces forms of subjectivation through various government technologies that define a dominant strategic position as a control-subjection mechanism in the face of an urgency.

From this perspective, it is possible to problematize the relevance acquired by architecture from the eighteenth century in function of the objectives and government techniques—the governmental rationality—of modern societies, becoming a tool to think about the order and government of territories and particularly urban spaces. For Foucault, these are faced with a set of transformations, problems, and dangers that are sought to control and regulate, such as the cholera epidemic of 1830 and the social revolts and revolutions that occurred in Paris in 1830 and 1848 (Rabinow, 2015).

While this definition coincides with the quarantine dispositive previously described due to its regulatory and coercive character, Foucault also highlights the productive and creative character of the dispositives insofar as their constitutive network articulates, complements, and enhances heterogeneous components. In the particular case of architecture, it is inscribed in a network of knowledge/power where the effective practice of freedom, the practice of social relations, and spatial distributions are articulated (Rabinow, 2015).

The distinction between the regulatory and coercive character of the dispositives contrasts with their productive and creative condition. Beyond a binary opposition, it is interesting to explore how both dimensions are articulated and intersect, forming a field of relations, as proposed by authors such as Deleuze and Agamben, among others.

The perspective proposed by Agamben regarding the concept of apparatus broadens the frameworks initially defined by Foucault, incorporating "everything that has, in one way or another, the ability to capture, direct, determine, intercept, shape, control, and secure the gestures, behaviors, opinions, and discourses of living beings (Agamben, 2012: 257), recognizing components as everyday as the pen, the cigarette, computers, mobile phones, etc.

This expansion is relevant insofar as it not only reveals the capture procedures associated with the "gigantic accumulation and proliferation of dispositives" characteristic of capitalism associated with subjectivations, subjectivations, and government strategies, but also invites us to explore the possibilities of liberation that all those components of the apparatus can acquire through their "profanation", understood this as "the counter-dispositive" that restores what the apparatus has captured, to the free and common use of men.

This approach is taken up by Boano and Astolfo (2015) for the field of architecture associating it with "free use" and the possibility of deactivating the "use value" of objects turned into commodities, freeing them from the need to be useful and leaving them inoperative and suspending their effectiveness.

This "inoperativity" in architecture, understood as a neutralization of the order forces that operate in it under the logic of neoliberal capitalism, entails an opening towards its free use. In this way, it is possible to think of an architecture that attends to the conditions of its context and is willing to reconsider its own production processes from informality, conflictuality, interdisciplinary collaborations, and collective and community design practices, among others (Boano, 2017).

This conception of inoperativity in architecture implies disarticulating work, action, and the act from its "instrumental apparatus" and reestablishing the power of beings and things "making them susceptible to being used in new ways" (Boano, 2017: 45). In this sense, the notion of "profanation" restores the common use of architecture by neutralizing the power apparatus associated with its production.

On the other hand, the notion of apparatus proposed by Deleuze (2012) highlights its ability to make visible and invisible the forms and mechanisms that produce sociabilities and spatialities, enabling ways of "seeing and making seen" framed under the model of neoliberal society, according to forms of knowledge and power instituted by it.

From this, Salazar (2004) suggests that it is possible to think of dispositives as a "machine for making seen" oriented to the "exhaustive tracking of the minimal traces, imperceptible sounds, and hidden cracks of social practices" in those fields of social experience traditionally made invisible. In the same line, Cadahia (2016) suggests that besides revealing situations, these "machines for making seen" allow creating and producing views that question the instituted conceptions of sociability and spatiality, as well as their prerogatives of objectivity.

In this line, it is interesting to rescue the perspective proposed by Stoner (2018), which allows making visible the hidden power of the spatialities produced by the inhabitants from their everyday practices of inhabiting. The "minor architecture" allows to be amazed at "the new", "the unexpected" and even "the forgotten", which is relevant to understand the emerging production of new and diverse socio-spatial orders, mainly attending to "the production of spaces within the already built" (2018: 41).

In this sense, minor architectures slide through the cracks of Euclidean convention and do not attend to the traditional idea of the formal. They also have an ephemeral and latent character, being an interesting challenge to make them appear considering that they allude to the "lowest conditions of any power structure", opening a new "order" of experiences linked to the "deterritorializing and liberating power" of the produced space. As Stoner (2018: 105) states, "a minor architecture is a becoming space more than a being forms", constituting exit dispositives at the moment of "releasing" the hidden power of objects beyond any desire for possession.

41.3 Exploring "Exit Devices" in El Pedregal Square

El Pedregal Square is located in the heart of the historic center of Santiago de Chile, being built under the auspices of the Remodelaci6n San Borja, an emblematic set of residential towers formulated from the slogans of the modern movement (Pérez de Arce, 2016). This residential project, built from 1967 and abruptly finished in 1976, contemplated within its original purposes the generation and maintenance of common spaces such as a system of terraces and walkways, under the state protection of the Urban Improvement Corporation (CORMU).

As Diaz et al. (2022) propose, although this project was dismantled by the civic-military dictatorship through the fragmentation of the land around the towers and its restrictive administration of the surrounding spaces, El Pedregal square currently stands out both for its important urban centrality and for its political and cultural relevance, which is nourished by the daily practices of the inhabitants who attend the universities, health centers, cultural centers and the formal and informal commerce that surrounds the square, in addition to its location adjacent to the main arterial road of the capital, its direct connection with the Santiago Metro Network and its location bordering San Borja Park.

Culturally, the Square is a convergence point for the deployment of various youth practices such as K-Pop dances, graffiti and mural painting, informal youth parties, etc., Likewise, its modern aesthetics currently host different expressions of urban art making it a frequent reference for the filming of commercials and music videos, as well as for artistic and advertising photography. This constitutes a particular and distinctive urban environment within the city (Arizaga, 2023).

Also, the design of its terraces that act as roofs over the ground and its proximity to public and private hospital centers, made it for a long time a shelter for homeless people, who spent the night in its surroundings through the installation of tents and other types of informal accommodations.

This diversity of uses and social practices over time caused not only a constant deterioration of the facilities and urban furniture of the square (Arze Arce & Román-Crisóstomo, 2018), but also a set of negative situations such as noise, dirt and crime that affected the quality of life of its residents, such as the murder of Daniel Zamudio carried out on the edge of the adjacent San Borja Park in 2012.

This event marks a referential milestone in the demand and constant claim of the residents of the sector for better maintenance and protection conditions of the Square. In function of this, on August 16, 2018, the Metropolitan Regional Council of Santiago (CORE) approved the sum of $606,969,000 million pesos to finance an improvement project. This investment included waterproofing works, repair of cracks and fissures that the slab where the square is located could present, finishing pavement and redesign, plus the distribution of furniture, that is, demolitions, lighting, benches, pots, trash cans, green masses, among others.[1]

However, this recovery process, initiated on August 13, 2019, was suspended due to the consequences of the Social Outbreak of October 2019. Its proximity to the main center of demonstrations and citizen protests in the country, Dignity Square, transformed El Pedregal Square into a public space constantly intervened and disputed by protesters and police forces, which is reflected in actions of high public connotation such as the burning of the supermarket adjacent to the Square or the fire of the San Francisco de Borja Church, to date used as a military chapel of Carabineros. Along with these events, many neighbors denounce that the El Pedregal Square became a recurring scene of abuses and police violence, being close to the operations center defined by the police in the context of the demonstrations.

It is precisely in this context of political and social crisis that the recovery of El Pedregal Square, through a new infrastructure provision, acquires a preponderant role for the life of the inhabitants of the sector. The remodeling and improvement process restarted on October 15, 2020, included the following provision: waterproofing of the slab; pavement change; installation of new LED lights; shade structures and new furniture (24 Lonquimay benches, 16 cube benches, 16 trash cans, 19 pagodas, 9 projectors, 30 overhead lights, 11 trees, 60 shrubs, 400 ground covers, self-watering pots, among others). Additionally, a Facade Painting project was also executed with damage from graffiti, on the first floors of Towers 5 and 6, where a total of 4,509 m^2 was intervened.[2]

To this was added the installation of a perimeter fence aimed at regulating accessibility to the square, mainly during the nights. This project was funded through the Urban Improvement Program of the Undersecretary of Regional and Administrative Development (SUBDERE) and inaugurated on April 1, 2021.

However, since its reopening to the present, the new infrastructure installed in El Pedregal Square has been strongly intervened and appropriated in different ways by various inhabitants of the sector. This process not only contradicts some of the recovery principles and actions such as cleaning and painting facades, but it has also opened new supports for expression and urban activation not contemplated in the planned recovery. In this sense, the same infrastructure installed in Pedregal Square has become an open space and support for the generation of different types of messages and graphic expressions (Fig. 41.1).

[1] https://www.coresantiago.cl/2018/08/16/consejo-regional-aprueba-proyecto-para-mejorar-historica-plaza-pedregal-del-barrio-san-borja/.

[2] https://www.munistgo.cl/municipalidad-de-santiago-inauguro-trabajos-de-mejoramiento-de-plaza-el-pedregal/.

Fig. 41.1 Urban furniture as a support for graphic expression

Likewise, it has been intervened with the aim of accommodating new artifacts not contemplated in its original improvement, which have a practical effect that allows, for example, to deposit cigarette butts and keep the floor clean of these (Fig. 41.2). Along the same lines, it is possible to find other artifacts such as a drinking fountain for animals, which becomes daily relevant considering the growing practice of walking pets in the surroundings of the square (Fig. 41.3).

It also stands out the use that some residents have made of the planters installed in the square, which have been used for the cultivation of tomatoes, in clear harmony with the use of residual spaces for urban gardens that have become common in other sectors of the city (Fig. 41.4).

This set of interventions can be analyzed from the perspectives of the "inoperative architecture" proposed by Boano and the "minor architecture" proposed by Stoner. In them, not only are the traditional canons of the discipline questioned, but a debate is also opened and encouraged aimed at releasing the potential of the intervened objects.

From the perspective of the first, Boano raises the possibility that it provides for the "deactivation" of socio-spatial regulation and control dispositives and, in turn, for the "activation" of new potentials of beings and things, opening them to new uses. This implies the opening of "new ways of looking" at spatial and urban practices, new ways of focusing and understanding them, rescuing not only their instrumental or technical relevance but also political and ethical.

Fig. 41.2 Cigarette butt collector

Fig. 41.3 Pet drinking fountain

Fig. 41.4 Tomato plantation

It is precisely in this line that Boano recognizes his crossing with Stoner's proposal, to the extent that this "... not only requires the partial deconstruction of buildings and power structures that lead to their incessant reproduction, but also the deconstruction of the architect as a subject" (2017: 55).

From the perspective of the "minor architecture" proposed by Stoner, it allows highlighting the agency capacity of the inhabitants, which goes beyond the concept of a "finished work" inherent in the disciplinary knowledge of architecture, these constitute acts of resistance and subversion against institutional urbanism.

In this sense, Stoner's proposal allows us to make visible the incomplete, unfinished, and imperfect character of these interventions, which are generated from discarded materials that constitute collective escape lines against established codes, norms, and rules. They allow us to articulate and communicate the built environment with the needs, activities, requirements, and desires of the inhabitants.

The "new ways of looking" proposed by Boano and Stoner allow us to explore and reclaim some uses of Plaza Pedregal that have not been contemplated in its recovery and remodeling project. Examples of this are the daily relevance of smoking in public spaces, walking pets, or opening and sharing spaces for community cultivation.

Likewise, "making visible" not only means revealing new socialities and spatialities but also recognizing the relevance of the use of reused materials, the self-managed character associated with the everyday knowledge and wisdom of the inhabitants, and their unfinished formal expression, permanently subjected to processes of reformulation.

Architecture understood as a "machine for making visible" allows opening spaces for dialogue and interaction between various agencies, human and non-human, as well as processes of production and transformation of the political-economic, space-territorial, and socio-cultural conditions of the territories, which makes it part of an "exit dispositive", open to permanent transformation—becoming.

It is precisely around this tension that architecture plays a fundamental role in crisis contexts. Faced with the strategic character of the quarantine dispositive and understood as a discipline oriented to understand the production of new forms of sociability and spatiality, it constitutes part of a device that allows "making visible" unforeseen possibilities to get out of the crisis, such as the production of "common spaces" (Lange and Amigo, 2020) (Elorza y Mattioli, 2020).

41.4 Final Reflections

The "exit dispositives" can be understood as networks of knowledge/power constituted from everyday practices of inhabiting that challenge the instituted socio-spatial order and promote instituting processes (Bergua, 2007) of production of sociability and spatiality.

From this perspective and understood as a discipline oriented to project and understand the production of various forms of sociability and spatiality, architecture can be considered as a fundamental component of these "exit dispositives", insofar as it allows to make visible—"make see"—unforeseen possibilities to get out of the crisis. Concepts such as "inoperative architecture" or "minor architecture" allow to support and guide the development of new looks and approaches, which attend to some of the characteristics described in the case of the artifacts and interventions produced in the El Pedregal square, recognizing in them new forms of sociability and spatiality that overcome the public/private dichotomy and that allow to explore their potentiality as "common spaces".

Common spaces arise from processes and everyday practices of collaboration that inhabitants develop through negotiations and agreements aimed at improving their neighborhood environments (Lange & Amigo, 2020). Although their production is always present in urbanization processes, they gain visibility and recognition in crisis contexts as they allow to resolve situations of deterioration and conflict (Letelier et Al, 2016) and "be quietly in disagreement" (Díaz, 2015), sustaining the production of a particular socio-spatial order.

References

Agamben, G. (2012). *What is a device?* Adriana Hidalgo Editora.

Arizaga, X. (2023). Public use spaces in modern movement housing complexes: The urban environment in three cases from Santiago de Chile. *Cities,* (26), 21–45. https://doi.org/10.24197/ciudades.26.2023.21-45

Arze Arce, G. H., & Román-Crisóstomo, S. (2018). The San Borja Remodeling in Santiago de Chile: From revolutionary promise to real estate reactivator. *Urban Territorial Logbook, 28* (1), 103–111. https://doi.org/10.15446/bitacora.v28n1.47134

Bauman, Z., & Bordoni, C. (2016). *State of crisis*. Paidós.

Bergua, J. (2007). the social instituent. In *Materials for a non-classical sociology*. University Presses of Zaragoza.

Boano, C., & Astolfo, G. (2015). A new use of architecture: The political potential of Agamben's common use. *ARQ (Santiago), 91*, 14–25. https://doi.org/10.4067/S0717-69962015000300003

Boano, C. (2017). *Any architecture*. Editions.

Cadahia, M. L. (2016). Aesthetic devices and sensitive forms of emancipation. *Ideas and Values, 65*(161), 267–285. https://doi.org/10.15446/ideasyvalores.v65n161.55200

De Simone, L. (2021). Neoliberal artifacts: construction and speculation of the city through architecture for mass consumption, 1973–1990. In D. Talesnik (Ed.), *Santiago 1977–1990: Architecture, city and politics* (pp. 18–31). ARQ.

Deleuze, G. (2012). What is a device? In G. Deleuze & Tiqquin (Eds.), *Contribution to the ongoing war*. Errata Naturae.

Díaz, F., & Carrasco, G., & Peliowski, A., (Eds.). (2022). *Let's go up. The San Borja remodeling and vertical housing in Chile*. ARQ.

Díaz, F. (2015). Defining a common space. *ARQ (Santiago), 91*, 12–13. https://doi.org/10.4067/S0717-69962015000300002

Elorza, A. L., & Mattioli, D. (2020). Territorial disputes and collective redefinition of habitat. Notes around the production of the common from the case of Parque Esperanza, Córdoba, Argentina. *Architectures of the South, 38*(58), 62–79.

Foucault, M. (1991). *Knowledge and truth*. Ediciones La Piqueta.

Giglia, Á. (2012). *The living and the culture: Theoretical perspectives and research*. Ed. Anthropos.

Janoschka, M., & Hidalgo, Rodrigo (2014). The Neoliberal City. Stimuli for critical reflection. In R. Hidalgo & M. Janoschka (Eds.), *The Neoliberal City*. Pontifical Catholic University of Chile.

Lange, C., & Amigo, M. J. (2020). Manifesto for an architecture of common spaces in Santiago, Chile. In *Collection investigations, IdPA* (Vol. 6, pp. 123–137). University of Seville.

Lange, C., & Amigo, M. J. (2021). Architecture of the common: Learnings from the inhabitants and their daily practices. *Architectures of the South, 39*(60), 48–61. https://doi.org/10.22320/07196466.2021.39.060.03

Letelier, F., Micheletti, S., & Vanhulst, J. (2016). Instituting practices in the neighborhood space: The neighborhood as a common. *Polis (Santiago), 15*(45), 105–119. https://doi.org/10.4067/S0718-65682016000300006

Liernur, J. (2009). Portals of the Labyrinth. Comments on the architecture in Chile, 1977–2007. In L. Jorge (Ed.), *Portals of the Labyrinth. Architecture and City in Chile, 1977–2009* (pp. 1–58). Andrés Bello University.

Manzi, M. G. (2020). The city of Santiago redefined as temporal communicational corporeality in times of social outburst. *Architectures of the South, 38*(57), 162–181. https://doi.org/10.22320/07196466.2020.38.057.09

Márquez, F. (2020). For an anthropology of the rubble. The social outbreak in Plaza Dignity, Santiago, Chile. *Revista, 180*(45), 1–13. https://doi.org/10.32995/rev180.Num–45.(2020).art–717

Pérez de Arce, R. (2016). The garden of the intercrossed paths: The San Borja remodeling and the schools of architecture. *ARQ (Santiago),* (92), 50–67. https://doi.org/10.4067/S0717-69962016000100007

Rabinow, P. (2015). Space, knowledge and power. Interview with Michel Foucault. *Bifurcations, 19*. https://www.bifurcaciones.cl/bifurcaciones/wp-content/uploads/2015/06/bifurcaciones_019_Reserva.pdf

Ruiz Hurtado, V. M. (2020). Sociability, cultural manifestations and conflict. *Urban Territorial Logbook, 31*(1), 125–137. https://doi.org/10.15446/bitacora.v31n1.86807

Saavedra, V., Toro, F., & Lange, C. (Eds.). (2021). *Dignified living and new constitution*. Santiago.

Salazar, C. (2004). Devices: Visibility machines. In *VVAA. Research yearbook 2003* (pp. 291–299). UAM-X.

Stoner, J. (2018). *Towards a minor architecture.* Bartlebooth.

Talesnik, D. (2021). Was postmodernism the architecture of Pinochetism? Architectures in the center of Santiago in the decade of 1980. In D. Talesnik (Ed.), *Santiago 1977–1990: Architecture, city and politics* (pp. 96–109). ARQ.

Vergara-Perucich, F., & Boano, C. (2020). Exploring the contradiction in the ethos of urban practitioners under neoliberalism: A case study of housing production in Chile. *Journal of Planning Education and Research.* https://doi.org/10.1177/0739456X20971684

Chapter 42
Urban Microbiome: Architecture and Microbiology for a Healthy Built Environment

Felipe Encinas, Katia Soto Liebe, Carmen Freed, and Juan Ugalde

Abstract While a few decades ago, it was understood that home health was related to the absence of pathogens and sterile environments, we now understand that the diversity of microorganisms plays a fundamental role in establishing healthy spaces. However, confinement within built spaces—with and without a pandemic—has strongly segregated microbial colonization, affecting immune and metabolic systems. Despite this, a direct reading from the built environment has not been established since research does not answer the relevant questions for architecture, as stated by microbiology. In this way, under the concept of "urban microbiome", this study aims to establish a first characterization of the microorganisms present inside homes in Santiago de Chile, based on the collection of microbiological samples in common areas of the house, in areas with certain architectural characteristics, to promote healthy built environments from the intersection between design and microbiology. For this, 3 sampling campaigns of bacteria and fungi (with 16S rDNA sequencing and metagenomics) were carried out on interior surfaces, plus the continuous recording of environmental variables in a social condominium in the commune of Renca, particularly in places of the houses associated with constructive pathologies. In addition, interviews were conducted with the heads of households in the houses characterized by energy poverty. About 20 genera/species of bacteria and fungi were identified.

F. Encinas (✉) · C. Freed
School of Architecture, Faculty of Architecture, Design and Urban Studies, Pontificia Universidad Católica de Chile, Santiago, Chile
e-mail: felipe.encinas@uc.cl

C. Freed
e-mail: cfreedh@uc.cl

K. S. Liebe
Faculty of Economics and Business, Universidad de Chile, Santiago, Chile
e-mail: ksotol@fen.uchile.cl

J. Ugalde
Center for Bioinformatics and Integrative Biology, Facultad de Ciencias de la Vida, Universidad Andrés Bello, Santiago, Chile
e-mail: juan@ugalde.bio

M. Vicuña and E. Giorgi (eds.), *Intersections*, Springer Geography,
https://doi.org/10.1007/978-3-031-76402-8_42

The results show a variable microbiome, highly influenced by its exterior environment and with greater diversity in better-ventilated rooms. The presence of the fungi *Aureobasidium pullulans* and *Alternaria betae-kenyensis*, which have been associated with asthma and allergies, in critical places with poor ventilation and indoor pollution. In this sense, it is evident that the houses' design, construction and expansion decisions generate impacts on the interaction between the different ecosystems and their built environment.

Keywords Urban microbiome · Ventilation · Indoor pollution · Indoor environment quality

42.1 Introduction

People living in cities spend between 80 and 90% of their time in indoor environments, which has caused a marked segregation in microbial colonization, affecting the human immune and metabolic systems. This is because microorganisms play a crucial role in modulating the immune system, so it is necessary to maintain contact with a diverse microbiome similar to the one that has been present in our environments throughout the last 300 thousand years of human evolution.

This is why confinement in built spaces, both during the pandemic and under normal circumstances, represents a concern due to adverse health effects. This is related to exposure to natural light, physical activity, and mental health, among other aspects (Diffey, 2011).

From the architectural perspective, the high use of indoor spaces poses challenges for improving their quality, where the parameters of indoor environment measurement are fundamental aspects to address. This is why various studies suggest the importance of understanding and characterizing the microbiological component in built environments to explore the fundamental relationship between human health and architecture as a relevant variable to promote health and well-being in inhabited spaces (Brown et al., 2016; Prussin & Marr, 2015; Stephens, 2016).

In this sense, the launch in recent years of global research related to the urban microbiome, such as MetaSUB, which has collected 4,800 samples from public transport systems for metagenomic sequencing in 60 cities around the world, highlights the importance of the relationship between microorganisms and built environments. This research managed to identify a set of base microorganisms in practically all the samples, which reflect important characteristics of cities and urban life and, according to the city, were driven by climatic and geographical differences. These results allow the discovery of organisms and genes of great relevance to public health (Danko et al., 2021).

In the same vein, the identification of an increase in cases of asthma, allergies, and other atopic diseases in populations residing in urban environments (Li et al., 2020; Mahdavinia et al., 2021) has highlighted the need to characterize the fraction of microorganisms present in homes and built environments in general. It has been observed that the diversity of microorganisms to which people are exposed inside homes is less than that found in natural and rural environments, which is related to the materiality of the constructions, the interior conditions, and the habits of use of the houses.

In the context of housing, the main topics of interest related to indoor health refer to physical variables, such as relative humidity and temperature, as well as chemical variables, such as volatile organic compounds released from construction materials and carbon dioxide levels (CO_2) inside the rooms. The latter has acquired special importance in the context of the COVID-19 pandemic as an indicator of the level of ventilation and, consequently, the quality of indoor air in enclosed spaces. These variables are related to the microbiological component in the house, which results from the interaction of the living beings that inhabit it, the air exchange rates, the quality of the indoor environment, and the construction materials (Fig. 42.1).

The quality of the indoor environment and its relationship with the microbiological variable are fundamental, as the inclusion of the latter in the study of housing health used to traditionally focus on the presence of pathogens, allergens, and fungi that degrade construction materials, like wood. However, we now understand that the diversity of microorganisms plays a crucial role in establishing healthy spaces.

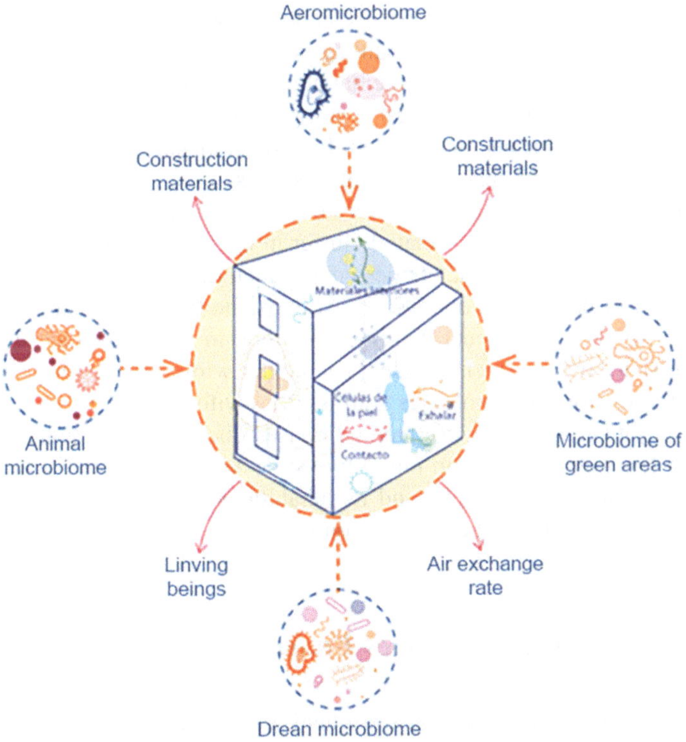

Fig. 42.1 Interactions of the house with its built environment

These microorganisms, in addition to their function as modulators of the immune system, are related to the gut microbiome, which is degraded in the face of diminished quality of life, a diet rich in processed foods, stress, and a recently established relationship with residential overcrowding (Delaroque et al., 2021). This generates a series of chronic inflammatory diseases, among which are obesity and diabetes (metabolic diseases), which have a high incidence rate in low-income sectors (Encinas et al., 2021).

In this way, from the intersection between architecture and microbiology, this study aims to first characterize the "urban microbiome" by identifying the microorganisms inside houses and measuring the quality of the indoor environment. All this is to promote built environments that foster the health of their inhabitants.

42.2 Methodology

42.2.1 Selection of the Study Area

The first stage of the study consisted of selecting the place of study, which was carried out using a spatialized sampling approach with statistical clustering tools applied to the metropolitan area of Santiago de Chile. For this, variables related to socioeconomic, urban, and environmental conditions, the constructive and thermal quality of the houses, and residential typologies were used. These variables were grouped using the K-means method with Euclidean distance, which resulted in the definition of a total of 15 clusters or conglomerates.

From this, a social condominium located in the municipality of Renca, which belongs to a predominantly associated cluster in the northeastern area of Santiago, also presents particular characteristics of the material quality of the houses and socioeconomic vulnerability conditions. At the same time, this sector exhibits one of the highest values of annual thermal amplitude in the entire metropolitan area (which translates into warmer summers and colder winters). It mainly concentrates on single-family housing typologies with families with minors.

Regarding its architecture, the studied houses obey the incremental logic. This means that families are given an initial two-story volume with the structure of the house and a minimum habitable, and its inhabitants are expected to expand them progressively, according to their own resources.

42.2.2 Selection of the Houses Under Study

To start the study, virtual meetings were held with the community board and later with a group of neighbors who expressed interest in participating in the research upon revealing scientific findings related to the role of microorganisms in preserving health.

From this and through open and voluntary recruitment, 19 houses were integrated into the study (out of a total of 42). Throughout the process, the support of the Municipality of Renca was counted on through the Community Planning Secretariat, the Community Development Directorate, and the territorial managers of the commune. All the families who participated in the research signed an informed consent, approved by the Ethical Scientific Committee of Social Sciences, Arts and Humanities of the Pontifical Catholic University of Chile.

42.2.3 Sampling and Housing Study

Each of the families, through their heads of household, was given an energy poverty characterization survey to obtain information about their energy backgrounds, usage habits, aspects related to the health of its inhabitants, and socioeconomic backgrounds.

Simultaneously, samples were taken for microbiological analysis. These samples were collected with sterile swabs in 10 × 10 cm areas on interior surfaces near the house windows. Two points were selected for the sample collection, one in the living room and another in the kitchen (Fig. 42.2), both with thermal bridges corresponding to the meeting of two different materials: the window frame, which in most cases is aluminum, and the wall, which is reinforced concrete. The samples were preserved under optimal conditions (4 °C) until their processing, which included DNA extraction in the laboratory. The molecular markers 16S rDNA and ITS were used to identify bacteria and fungi, respectively.

Additionally, indoor environmental variables (temperature, relative humidity, and CO_2 concentration) were recorded continuously for a month using the HOBO MX1102 instrument.

Finally, a planimetric survey of four houses was carried out to identify the variations that have taken place in their incremental process by the inhabitants.

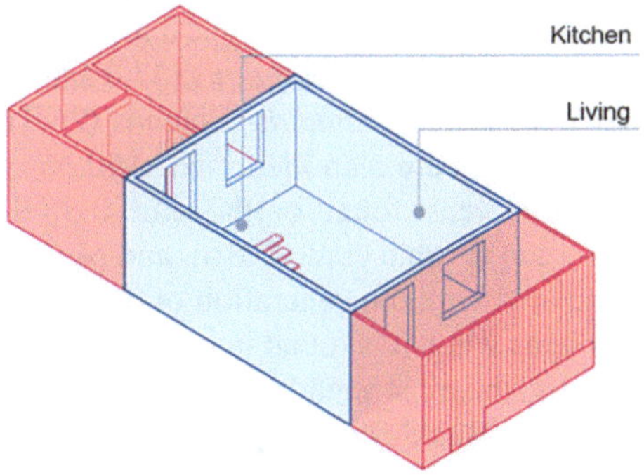

Fig. 42.2 First level volume of a typical house under study. The location of the kitchen and the living room is indicated

42.3 Results

42.3.1 Sampling Results

For the analysis of the samples, the aim was to identify the fungi and bacteria present in them. Both in the case of bacteria and in fungi, the *phylum* is a broader level of classification that groups organisms into categories based on phylogenetic similarities, while *genus* and *species* are more specific levels. At these levels, it is possible to know the functions that microorganisms fulfill and relate them to their hosts. Higher levels allow comparisons between habitats.

In each of the houses, six samples were collected from different areas. A selection of these was processed to extract genetic material (DNA) and subsequent sequencing. After processing and filtering the data, a total of 11 samples delivered quality results to be analyzed using the bacterial 16S molecular marker. Seven of them were obtained from the dining rooms of the houses and four from the kitchens (all from different homes). In the case of the dataset analyzed with the ITS molecular marker for fungi, after filtering and cleaning the data, a total of seven samples met all the criteria, five collected in the dining room and two in the kitchen of the houses.

At the level of *phylum*, the bacteriological analysis showed that the microbial communities were dominated by members of the group *Proteobacteria* (30.2% ± 11.17%), followed by *Firmicutes* (26.8% ± 9.15%), *Actinobacteriota* (24.9% ± 11.01%) and *Bacteroidota* (11.05% ± 9.29%) (Fig. 42.3). In the case of *Firmicutes* and *Fusobacteriota*, a higher relative abundance was observed in the kitchen compared to the dining rooms, while Proteobacteria had a higher relative abundance in the dining rooms compared to the kitchens. At this same taxonomic level, in the case of the ITS dataset, the analysis showed that the samples were dominated by members of *Ascomycota* (78.3% ± 11.7%) and *Basidiomycota* (21.05% ± 11.9%), where the latter group was more abundant in the samples collected in the kitchen.

In more detail, in the case of fungi at the genus level, 30 groups presented abundances of over 2%. The most frequently found groups were *Alternaria betae-kenyensis*, *Ascomycota* and *Aureobasidium* pullulans, present in 3 samples and 5 groups in at least 2 samples (Fig. 42.3). At this taxonomic level, 11 groups presented abundances over 10% in one or more samples (data not shown). The groups that presented the highest relative abundance and are related to human health were *Aureobasidium pullulans* (37.6 and 31.06%), *Alternaria betae-kenyensis* (7.76 and 18.53%) and *Candida albicans* (26.31%). Two species belonging to the genus Malassezia were also identified: *Malassezia globosa* (36.93%), characterized as a lipophilic yeast found on the skin and related to some alterations in the health of this organ (Prohic et al., 2016), and *Malassezia restricta* (10.95%), a species that is associated with the generation of dandruff (Jourdain et al., 2023) and inflammatory problems when it is found in intestinal microbiota (Limon et al., 2019). In two of the samples, the most abundant fungi were species related to plants *Dioszegia var_zsoltii* (62.49%), Buckleyzyma (11.67%) and *Pseudopithomyces rosae* (5.75%). Representative species of the groups *Clavispora* (30.81%) and *Kazachstania* (25.24%) have

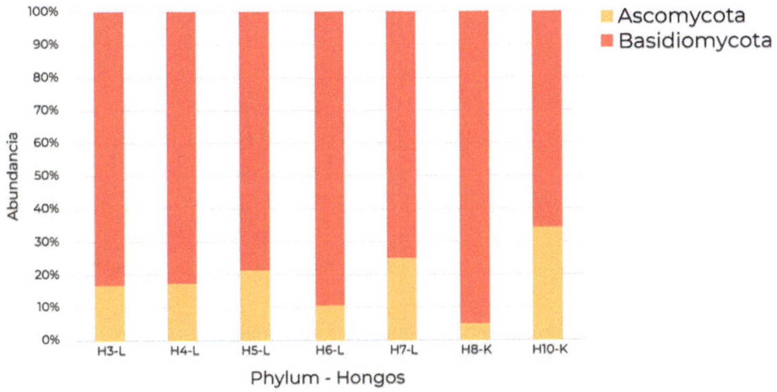

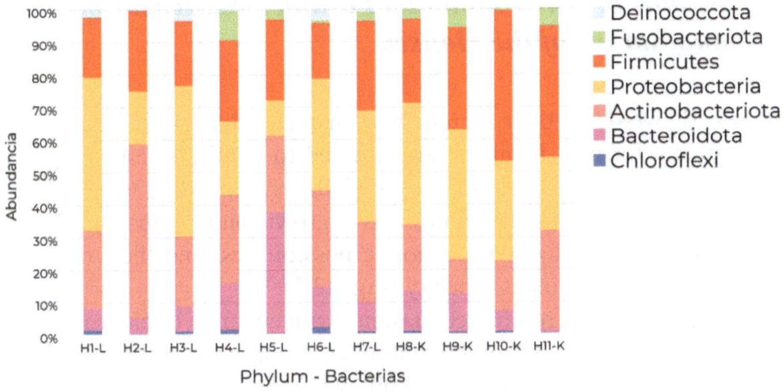

Fig. 42.3 (Top) Bacteriological Analysis, Phylum—Fungi; (Bottom) Bacteriological Analysis, Phylum—Bacteria

been characterized as pathogens but have also been isolated from plants (Drumonde-Neves et al., 2020; Kaeuffer et al., 2022; Krcmery Jr et al., 1999; Lin et al., 2022).

Regarding bacteria, at the genus level, 58 groups with abundances over 2% were identified. The most frequently found groups were *Corynebacterium* and *Sphingomonas* (7/11); *Planomicrobium* and *Streptococcus* (5/11); *Streptococcus, Streptococcus, Fusobacterium, Kocuria, Nocardioides, Paracoccus,* and *Rothia* (4/11) (Fig. 42.4). In four samples, dominant groups were observed, with abundances over 10% (Collinsella 38.74%, Prevotella 28.75%, Lacticaseibacillus 25.66%, Streptococcus 18.34%, Mycobacterium 16.50%, and Bifidobacterium 11.56%).

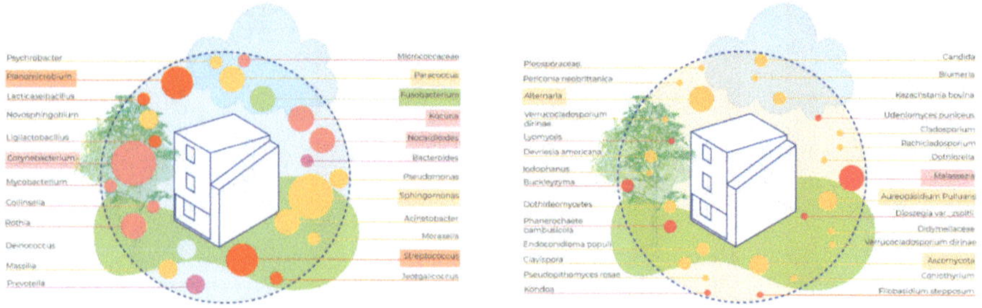

Fig. 42.4 Groups of microorganisms most frequently found among the samples analyzed at the taxonomic genus level (Left—Fungi; Right—Bacteria)

42.3.2 *Results of Environmental Parameters Measurement and Survey of the Houses*

The houses, delivered in their incremental character with 45 m^2 built, have their envelope already formed. Families are expected to build the remaining stairs and floors and integrate the missing doors and windows, expanding to a total of 69 m^2. Currently, the houses present different states of growth and expansion, which depend on the families, their individual economic possibilities, and their own technical, architectural, and management capacity (Fig. 42.5).

The houses are built of reinforced masonry with concrete bricks on the first level. In addition, due to the typology of the continuous grouping of the houses, they present firewall walls up to the second level, all built with the same materiality. On the upper floors, the exterior walls are made of partitioning with light profiles and plasterboard sheets, and the windows are of simple glass with an aluminum frame.

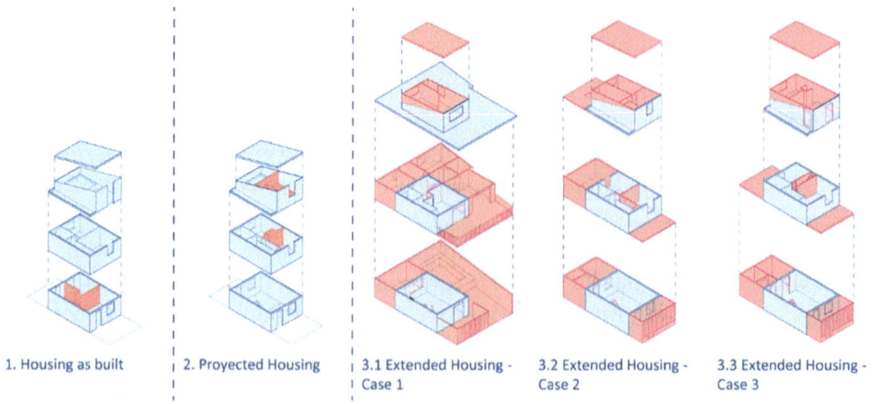

Fig. 42.5 Housing in three stages (axonometric view): 1. Housing as built, 2. Projected housing, and 3.1, 3.2, and 3.3 show how they have been extended and modified

Regarding the distribution of the houses, on the first level are the common spaces (living room and dining room) and the kitchen, in addition to a front yard that, in most of the houses, has been covered and a backyard that has been used to expand the rooms on the first floor or create new spaces, taking advantage of the available space. On the second floor are the bedrooms and the bathroom, with modifications in their distribution and use according to the needs and possibilities of each family. On the third level, there is a single bedroom that faces the main facade of the house.

Originally, a second room at this level was not considered due to the slope of the roof, which did not allow its use. However, the roof was modified in many houses by adding material to increase the available height and allow a second habitable space. These changes include rooms without ventilation or natural lighting and modifications in the envelope.

Concerning environmental parameters, continuous measurement devices were installed for temperature, relative humidity, and CO_2 concentration on each floor of the house (3 in total per house). The first was in the living-dining-kitchen area, the second in the main bedroom facing the main façade, and the third in a room on the third level, which varied according to each house. During the monitoring process, families were asked to record when they turned on the heating or ventilated the spaces under study to establish a correlation between the results of the measurements and their usage habits.

Regarding the results obtained, conditions were observed outside the comfort range, with CO_2 concentrations exceeding 5 times the 700-ppm maximum standard established by the World Health Organization (Allen et al., 2016). In addition, the relative humidity exceeded the recommended maximum of 50% (Dedesko & Sigel, 2015), which promotes the development of microorganisms that can harm health inside homes (Fig. 42.6), affecting the hygrothermal and environmental comfort of families.

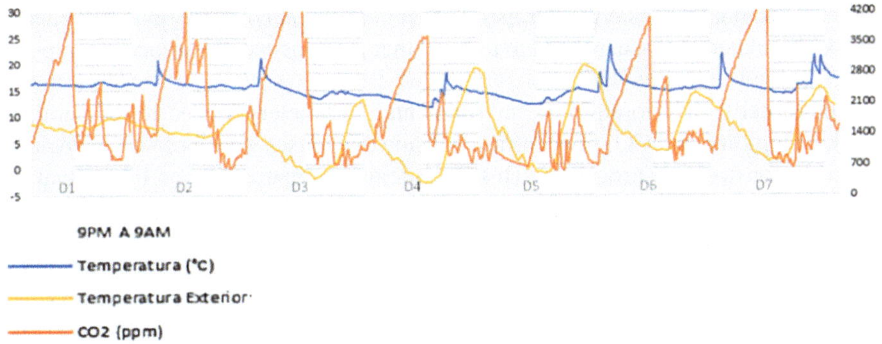

Fig. 42.6 Graph showing the temperature and CO_2 concentration inside the bedroom of a studied home

42.4 Discussion

From the sampling, a characterization of the microorganisms inside the houses has been achieved, identifying about 60 and 30 groups of bacteria and fungi at the genus level, respectively. At this taxonomic level, abundances greater than 10% were found for five groups of bacteria and 11 for fungi. The latter has been previously described in various environments, such as dust samples in air conditioning systems, animal feces, human hair, plants, and forests. The presence of fungi of the genus *Aureobasidium* and *Alternaria*, both related to asthma, as they can cause and exacerbate crises, in proportions that vary between 7 and 37% of abundance. Studies conducted with animals have shown that the microbiota can affect the inflammatory response and therefore, influence cases of asthma, as has been identified in the case of bacteria that are part of the intestinal microbiota. It is important to note that, to relate the microbiological component with the quality of the indoor environment, the study of fungi is of greater relevance, as the relative abundances of a few groups may be indicative of effects caused by poor ventilation and/or higher humidity. This is the case with the presence of *Aureobasidium* and *Alternaria; in one of the houses, their combined abundances were close to 50%, while in another house,* the groups *Dioszegia var, _ zsoltii,* and *Buckleyzyma*, both related to plants, represented 88% of the abundance of fungi. The fungi found on the skin and scalp (causing dandruff) probably account for contact with surfaces, which invites more representative sampling inside the houses, reducing the bias that this point can generate. Thus, current approaches implemented for this type of study—and our research group is currently developing—consider air filtration in conjunction with surface sampling.

In the case of bacteria, the most abundant taxonomic groups at the genus level belong to typical bacteria of the human microbiota, and it is less evident to make an association with the use of the house. The joint information is relevant when comparing the microorganisms from the kitchen and dining room samples, where it is evident that there is a greater association between species found in the kitchen (probably related to waste related to use and that is deposited on surfaces) compared to the dining room (Fig. 42.7).

Additionally, deterioration of environmental parameters related to temperature, relative humidity, and CO_2 concentration inside the houses was observed, associated with self-construction processes. This is evident in temperature conditions that are outside the comfort range, with very low temperatures during the winter inside the spaces and overheating in the summer. This is mainly in the bedrooms of the areas where the roof has been modified to incorporate new spaces. As for the concentrations of CO_2, these significantly exceed the maximum standard set by the World Health Organization. The same happens with relative humidity, which greatly exceeds the recommended maximum, promoting the development of harmful microorganisms for health inside the houses. These problems are largely related to the ventilation habits of the inhabitants, the use of heating systems based on fossil fuels, and the presence of windowless rooms resulting from the expansion processes of the houses.

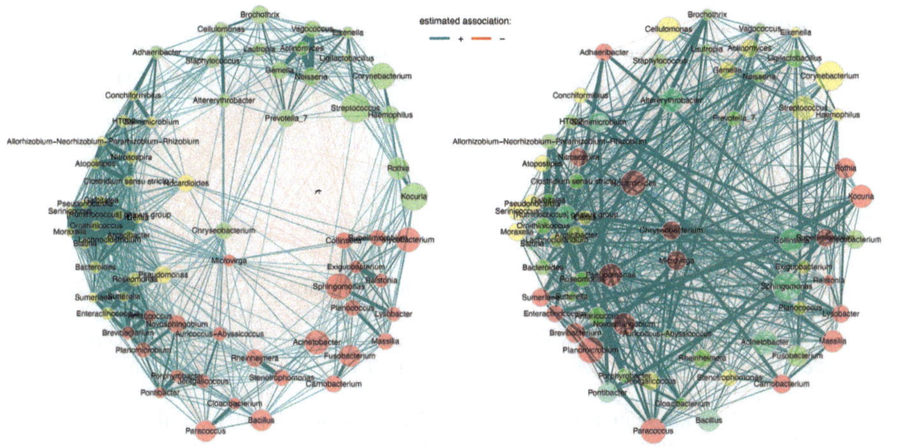

Fig. 42.7 Analysis of microbial networks for the kitchen (Left) and living room (Right) in the studied homes

In this way, these findings are presented in self-built spaces that do not contribute to improving the quality of life of households from the physical-environmental perspective, affecting the performance of the housing envelope (in its ability to model interior conditions) and resulting in indoor pollution.

42.5 Conclusions

In the development of housing policies and urban planning, the architectural project, its construction, and its evolution over time have been mostly focused on providing services and creating habitable spaces. However, the results suggest that the process of inhabiting is considerably more complex, as it involves a variety of variables that impact the built environment and its development. Among these variables are microorganisms, their relationship with the human habitat, and their link to environmental and health conditions.

These variables pose challenges to public policies, urging a multidimensional understanding that considers temporality, and the biotic and abiotic systems involved, evaluating their interactions, and rethinking their effects on the inhabitants' quality of life. In this sense, it is clear that decisions related to the design, construction, expansion, and maintenance of houses impact the interaction between the various ecosystems and their built environment. Therefore, it is imperative to provide support in the housing development processes to ensure that the living conditions meet the quality standards of the indoor environment.

Acknowledgements The authors appreciate the support for the research provided by the National Fund for Scientific and Technological Development (FONDECYT)—Project N°1201232 "Poverty-Energy-Housing Nexus"—and the Center for Sustainable Urban Development (CEDEUS).

References

Allen, J. G., MacNaughton, P., Satish, U., Santanam, S., Vallarino, J., & Spengler, J. D. (2016). Associations of cognitive function scores with carbon dioxide, ventilation, and volatile organic compound exposures in office workers: a controlled exposure study of green and conventional office environments. *Environmental Health Perspectives, 124*(6): 805–812. https://doi.org/10.1289/ehp.1510037

Brown, G. Z., Kline, J., Mhuireach, G., Northcutt, D., & Stenson, J. (2016). Making microbiology of the built environment relevant to design. In *Microbiome* (Vol. 4). BioMed Central Ltd. https://doi.org/10.1186/s40168-016-0152-7

Danko, D., Bezdan, D., Afshin, E. E., Ahsanuddin, S., Bhattacharya, C., Butler, D. J., Chng, K. R., Donnellan, D., Hecht, J., Jackson, K., Kuchin, K., Karasikov, M., Lyons, A., Mak, L., Meleshko, D., Mustafa, H., Mutai, B., Neches, R. Y., Ng, A., & Zubenko, S. (2021). A global metagenomic map of urban microbiomes and antimicrobial resistance. *Cell, 184*(13), 3376-3393.e17. https://doi.org/10.1016/j.cell.2021.05.002

Dedesko, S., & Siegel, J. (2015). Moisture parameters and fungal communities associated with gypsum drywall in buildings. *Microbiome, 8*(3): 71. https://doi.org/10.1186/s40168-015-0137-y

Delaroque, C., Chervy, M., Gewirtz, A. T., & Chassaing, B. (2021). Social overcrowding impacts gut microbiota, promoting stress, inflammation, and dysglycemia. *Gut Microbes, 13*(1).https://doi.org/10.1080/19490976.2021.2000275/SUPPL_FILE/KGMI_A_2000275_SM6531.ZIP

Diffey, B. L. (2011). An overview analysis of the time people spend outdoors. *The British Journal of Dermatology, 164*(4), 848–854. https://doi.org/10.1111/J.1365-2133.2010.10165.X

Drumonde-Neves, J., Čadež, N., Reyes-Domínguez, Y., Gallmetzer, A., Schuller, D., Lima, T., Pais, C., & Franco-Duarte, R. (2020). Clavispora santaluciae F.A., Sp. Nov., a novel ascomycetous yeast species isolated from grapes. *International Journal of Systematic and Evolutionary Microbiology, 70*(12), 6307–6312. https://doi.org/10.1099/IJSEM.0.004531/CITE/REFWORKS

Encinas, F., Soto-Liebe, K., Aguirre-Nuñez, C., González, B., Bustamante, W., Schueftan, A., Ugalde, J. A., Blondel, C. J., Truffello, R., Araya, P., & Freed, C. (n.d.). COVID-19 and city: Towards an integrated model of housing, microbiology, environment and urbanism. *ACE Architecture, City and Environment, 46*. https://doi.org/10.5821/ace.16.46.9645

Jourdain, R., Moga, A., Magiatis, P., Fontanié, M., Velegraki, A., Papadimou, C., Rahoul, V., Guéniche, A., Chopra, T., & Gaitanis, G. (2023). Malassezia restricta-mediated Lipoperoxidation: A novel trigger in dandruff. *Acta Dermato-Venereologica, 103*,. https://doi.org/10.2340/ACTADV.V103.4808

Kaeuffer, C., Baldacini, M., Ruge, T., Ruch, Y., Zhu, Y. J., De Cian, M., Philouze, G., Bachellier, P., Denis, J., Lefebvre, N., Schneider, F., Hansmann, Y., Letscher-Bru, V., Herbrecht, R., Sabou, M., & Danion, F. (2022). Fungal Infections Caused by Kazachstania spp., Strasbourg, France, 2007–2020. *Emerging Infectious Diseases, 28*(1), 29–34. https://doi.org/10.3201/EID2801.211543

Krcmery Jr, V., Mateicka, F., Grausova, S., Kunova, A., & Hanzen, J. (1999). Invasive infections due to Clavispora lusitaniae. *FEMS Immunology & Medical Microbiology, 23*(1), 75–78. https://doi.org/10.1111/J.1574-695X.1999.TB01719.X

Li, S., Yang, Z., Hu, D., Cao, L., & He, Q. (2020). Understanding building-occupant-microbiome interactions toward healthy built environments: A review. *Frontiers of Environmental Science & Engineering, 15*(4), 1–18. https://doi.org/10.1007/S11783-020-1357-3

Limon, J. J., Tang, J., Li, D., Wolf, A. J., Michelsen, K. S., Funari, V., Gargus, M., Nguyen, C., Sharma, P., Maymi, V. I., Iliev, I. D., Skalski, J. H., Brown, J., Landers, C., Borneman, J., Braun, J., Targan, S. R., McGovern, D. P. B., & Underhill, D. M. (2019). Malassezia is associated with Crohn's disease and exacerbates colitis in mouse models. *Cell Host & Microbe, 25*(3), 377-388.e6. https://doi.org/10.1016/J.CHOM.2019.01.007

Lin, M. M. H., Boss, P. K., Walker, M. E., Sumby, K. M., & Jiranek, V. (2022). Influence of Kazachstania spp. on the chemical and sensory profile of red wines. *International Journal of Food Microbiology, 362*. https://doi.org/10.1016/J.IJFOODMICRO.2021.109496

Mahdavinia, M., Greenfield, L. R., Moore, D., Botha, M., Engen, P., Gray, C., Lunjani, N., Hlela, C., Basera, W., Hobane, L., Watkins, A., Mankahla, A., Gaunt, B., Facey-Thomas, H., Landay, A., Keshavarzian, A., & Levin, M. E. (2021). Microbiota of house dust and atopic dermatitis; effect of urbanization. *Pediatric Allergy and Immunology, 32*(5), 1006–1012. https://doi.org/10.1111/PAI.13471

Prohic, A., Jovovic Sadikovic, T., Krupalija-Fazlic, M., & Kuskunovic-Vlahovljak, S. (2016). Malassezia species in healthy skin and in dermatological conditions. *International Journal of Dermatology, 55*(5), 494–504. https://doi.org/10.1111/IJD.13116

Prussin, A. J., & Marr, L. C. (2015). Sources of airborne microorganisms in the built environment. https://doi.org/10.1186/s40168-015-0144-z

Stephens, B. (2016). What have we learned about the microbiomes of indoor environments? *MSystems, 1*(4). https://doi.org/10.1128/MSYSTEMS.00083-16

Correction to: New Landscape and Emerging Dynamics of Interaction Between Users and Urban Space Produced by Digital Delivery Platforms

Rodrigo Vera

Correction to: Chapter 21 in: M. Vicuña and E. Giorgi (eds.), *Intersections*, Springer Geography, https://doi.org/10.1007/978-3-031-76402-8_21

In the original version of the book, the title of Chapter 21 has been changed from "Oliveira New Landscape and Emerging Dynamics of Interaction Between Users and Urban Space Produced by Digital Delivery Platforms" to "New Landscape and Emerging Dynamics of Interaction Between Users and Urban Space Produced by Digital Delivery Platforms". The book has been updated with the change.

The updated version of this chapter can be found at https://doi.org/10.1007/978-3-031-76402-8_21

M. Vicuña and E. Giorgi (eds.), *Intersections*, Springer Geography,
https://doi.org/10.1007/978-3-031-76402-8_43

Index

M. Vicuña and E. Giorgi (eds.), *Intersections*, Springer Geography, https://doi.org/10.1007/978-3-031-76402-8

D

E

T

U

MIX
Papier aus verantwortungsvollen Quellen
Paper from responsible sources
FSC® C105338

If you have any concerns about our products,
you can contact us on
ProductSafety@springernature.com

In case Publisher is established outside the EU,
the EU authorized representative is:
Springer Nature Customer Service Center GmbH
Europaplatz 3, 69115 Heidelberg, Germany

Printed by Libri Plureos GmbH
in Hamburg, Germany